Artificial Photosynthesis

Cambridge, UK

25-27 March 2019

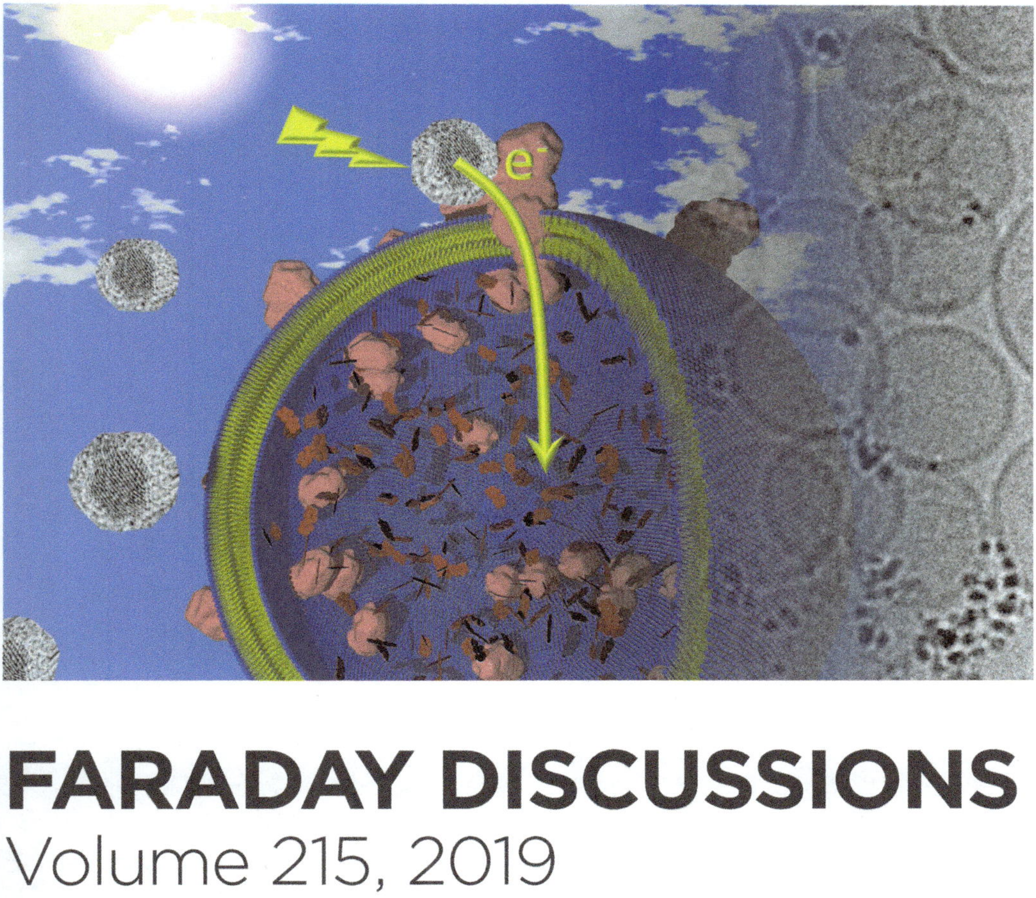

FARADAY DISCUSSIONS

Volume 215, 2019

ROYAL SOCIETY
OF CHEMISTRY

The Faraday Division of the Royal Society of Chemistry, previously the Faraday Society, was founded in 1903 to promote the study of sciences lying between chemistry, physics and biology.

Editorial Staff

Executive Editor
Philippa Ross

Deputy Editor
Heather Montgomery

Editorial Production Manager
Claire Darby

Publishing Editors
Lorna Arens, Colin King

Editorial Assistant
Aliya Anwar

Publishing Assistants
Jane Chan, Natalie Ford

Publisher
Jamie Humphrey

Faraday Discussions (Print ISSN 1359-6640, Electronic ISSN 1364-5498) is published 8 times a year by the Royal Society of Chemistry, Thomas Graham House, Science Park, Milton Road, Cambridge, UK CB4 0WF.

Volume 215 ISBN 13: 978-1-78801-671-1

2019 annual subscription price: print+electronic £1220 US $2148; electronic only £1162, US $2046. Customers in Canada will be subject to a surcharge to cover GST. Customers in the EU subscribing to the electronic version only will be charged VAT.

All orders, with cheques made payable to the Royal Society of Chemistry, should be sent to the Royal Society of Chemistry Order Department, Royal Society of Chemistry, Thomas Graham House, Science Park, Milton Road, Cambridge, CB4 0WF, UK Tel +44 (0)1223 432398; E-mail orders@rsc.org

If you take an institutional subscription to any Royal Society of Chemistry journal you are entitled to free, site-wide web access to that journal. You can arrange access via Internet Protocol (IP) address at **www.rsc.org/ip**

Customers should make payments by cheque in sterling payable on a UK clearing bank or in US dollars payable on a US clearing bank.

Faraday Discussions are unique international discussion meetings that focus on rapidly developing areas of chemistry and its interfaces with other scientific disciplines.

Artificial Photosynthesis

Faraday Discussions

www.rsc.org/faraday_d

A General Discussion on Artificial Photosynthesis was held in Cambridge, UK on the 25th, 26th and 27th of March 2019.

RSC Publishing is a not-for-profit publisher and a division of the Royal Society of Chemistry. Any surplus made is used to support charitable activities aimed at advancing the chemical sciences. Full details are available from www.rsc.org

CONTENTS

ISSN 1359-6640; ISBN 978-1-78801-671-1

Faraday
Discussions
Volume: 215

Artificial Photosynthesis

ROYAL SOCIETY
OF CHEMISTRY

Cover
See Julea N. Butt, Erwin Reisner, Lars J. C. Jeuken *et al.*, *Faraday Discuss.*, 2019, **215**, 26–38.

Liposome-encapsulated dye is photo-bleached by dye-sensitised TiO_2 after electron transfer through transmembrane multihaem electron conduits in a first step towards compartmentalised photocatalysis, shown in an electron micrograph on the right.

Image reproduced by permission of Lars J. C. Jeuken from *Faraday Discuss.*, 2019, **215**, 26.

INTRODUCTORY LECTURE

9 **Spiers Memorial Lecture**
Artificial photosynthesis: An introduction
Jan-Niclas Beller and Matthias Beller

PAPERS AND DISCUSSIONS

15 **Tuning purple bacteria salt-tolerance for photobioelectrochemical systems in saline environments**
Matteo Grattieri, Kevin Beaver, Erin M. Gaffney and Shelley D. Minteer

26 **Towards compartmentalized photocatalysis: multihaem proteins as transmembrane molecular electron conduits**
Anna Stikane, Ee Taek Hwang, Emma V. Ainsworth, Samuel E. H. Piper, Kevin Critchley, Julea N. Butt, Erwin Reisner and Lars J. C. Jeuken

39 **A kinetic model for redox-active film based biophotoelectrodes**
D. Buesen, T. Hoefer, H. Zhang and N. Plumeré

54 **Solar-driven carbon dioxide fixation using photosynthetic semiconductor bio-hybrids**
Stefano Cestellos-Blanco, Hao Zhang and Peidong Yang

66 **Biological approaches to artificial photosynthesis: general discussion**

84 **Photocatalytically active ladder polymers**
Anastasia Vogel, Mark Forster, Liam Wilbraham, Charlotte L. Smith, Alexander J. Cowan, Martijn A. Zwijnenburg, Reiner Sebastian Sprick and Andrew I. Cooper

98 **Computational high-throughput screening of polymeric photocatalysts: exploring the effect of composition, sequence isomerism and conformational degrees of freedom**
Isabelle Heath-Apostolopoulos, Liam Wilbraham and Martijn A. Zwijnenburg

111 **Visible light-driven water oxidation with a ruthenium sensitizer and a cobalt-based catalyst connected with a polymeric platform**
Zeynep Kap and Ferdi Karadas

123 **Evaluating the impacts of amino acids in the second and outer coordination spheres of Rh-bis(diphosphine) complexes for CO_2 hydrogenation**
Aaron P. Walsh, Joseph A. Laureanti, Sriram Katipamula, Geoffrey M. Chambers, Nilusha Priyadarshani, Sheri Lense, J. Timothy Bays, John C. Linehan and Wendy J. Shaw

141 **Performance of enhanced DuBois type water reduction catalysts (WRC) in artificial photosynthesis – effects of various proton relays during catalysis**
Wolfgang Viertl, Johann Pann, Richard Pehn, Helena Roithmeyer, Marvin Bendig, Alba Rodríguez-Villalón, Raphael Bereiter, Max Heiderscheid, Thomas Müller, Xia Zhao, Thomas S. Hofer, Mark E. Thompson, Shuyang Shi and Peter Brueggeller

162 **Photoinduced hole transfer from tris(bipyridine)ruthenium dye to a high-valent iron-based water oxidation catalyst**
Sergii I. Shylin, Mariia V. Pavliuk, Luca D'Amario, Igor O. Fritsky and Gustav Berggren

175 **Light induced formation of a surface heterojunction in photocharged $CuWO_4$ photoanodes**
Anirudh Venugopal and Wilson A. Smith

192 **Mechanistic insights into C2 and C3 product generation using Ni_3Al and Ni_3Ga electrocatalysts for CO_2 reduction**
Aubrey R. Paris and Andrew B. Bocarsly

205 **Iron phosphate modified calcium iron oxide as an efficient and robust catalyst in electrocatalyzing oxygen evolution from seawater**
Wei-Hsiang Huang and Chia-Yu Lin

216 **$Fe_xNi_{9-x}S_8$ ($x = 3-6$) as potential photocatalysts for solar-driven hydrogen production**
David Tetzlaff, Christopher Simon, Demetra S. Achilleos, Mathias Smialkowski, Kai junge Puring, André Bloesser, Stefan Piontek, Hatice Kasap, Daniel Siegmund, Erwin Reisner, Roland Marschall and Ulf-Peter Apfel

227 **Distinguishing the effects of altered morphology and size on the visible light-induced water oxidation activity and photoelectrochemical performance of $BaTaO_2N$ crystal structures**
Mirabbos Hojamberdiev, Kenta Kawashima, Takashi Hisatomi, Masao Katayama, Masashi Hasegawa, Kazunari Domen and Katsuya Teshima

242 **Synthetic approaches to artificial photosynthesis: general discussion**

282   **Sequential catalysis enables enhanced C–C coupling towards multi-carbon alkenes and alcohols in carbon dioxide reduction: a study on bifunctional Cu/Au electrocatalysts**
Jing Gao, Dan Ren, Xueyi Guo, Shaik Mohammed Zakeeruddin and Michael Grätzel

297   **A tandem photoelectrochemical water splitting cell consisting of $CuBi_2O_4$ and $BiVO_4$ synthesized from a single $Bi_4O_5I_2$ nanosheet template**
Yi-Hsuan Lai, Kai-Che Lin, Chen-Yang Yen and Bo-Jyun Jiang

313   **Z-scheme photocatalyst systems employing Rh- and Ir-doped metal oxide materials for water splitting under visible light irradiation**
Akihiko Kudo, Shunya Yoshino, Taichi Tsuchiya, Yuhei Udagawa, Yukihiro Takahashi, Masaharu Yamaguchi, Ikue Ogasawara, Hiroe Matsumoto and Akihide Iwase

329   **A microfluidic photoelectrochemical cell for solar-driven CO_2 conversion into liquid fuels with CuO-based photocathodes**
Evangelos Kalamaras, Meltiani Belekoukia, Jeannie Z. Y. Tan, Jin Xuan, M. Mercedes Maroto-Valer and John M. Andresen

345   **Demonstrator devices for artificial photosynthesis: general discussion**

364   **Utilising excited state organic anions for photoredox catalysis: activation of (hetero) aryl chlorides by visible light-absorbing 9-anthrolate anions**
Matthias Schmalzbauer, Indrajit Ghosh and Burkhard König

379   **Influence of carbonaceous species on aqueous photo-catalytic nitrogen fixation by titania**
Yu-Hsuan Liu, Manh Hiep Vu, JeongHoon Lim, Trong-On Do and Marta C. Hatzell

393   **p-Type dye-sensitized solar cells based on pseudorotaxane mediated charge-transfer**
Tessel Bouwens, Simon Mathew and Joost N. H. Reek

407   **Photo-generation of cyclic carbonates using hyper-branched $Ru–TiO_2$**
Stelios Gavrielides, Jeannie Z. Y. Tan, Eva Sanchez Fernandez and M. Mercedes Maroto-Valer

422   **Beyond artificial photosynthesis: general discussion**

CONCLUDING REMARKS

439   **Artificial photosynthesis – concluding remarks**
C. Bozal-Ginesta and J. R. Durrant

ADDITIONAL INFORMATION

452   **Poster titles**
458   **List of participants**

Faraday Discussions

PAPER

Spiers Memorial Lecture

Artificial photosynthesis: An introduction

Jan-Niclas Beller and Matthias Beller*

Received 29th May 2019, Accepted 29th May 2019

DOI: 10.1039/c9fd90025j

A brief introduction into artificial photosynthesis technologies is presented. Following the basic concepts of biological photosynthesis, light energy is directly or sequentially used for the synthesis of valuable chemicals with the help of man-made catalysts. Differences between artificial, hybrid and natural photosynthesis are shown and the possible advantages and disadvantages are highlighted.

Introduction

From the beginning of the 1970s until today, human energy consumption has almost doubled. If this development continues, there will be a further doubling in the next 30 years.[1] Where should all this additional energy come from? Fossil resources such as coal, oil and natural gas are not an option anymore because they generate the main greenhouse gas carbon dioxide (CO_2), which already makes a major contribution to climate change. Thus, it is clear that a shift towards alternative technologies using renewable energy is needed. In the search for possible solutions, natural photosynthesis can be used as a role model. More specifically, photosynthesis is the cornerstone of all life on this planet. Every year, more than 170 billion tons of biomass are produced, mostly by photosynthesis.[2] In this complex biological process, the energy of solar radiation is converted into a multitude of organic substances. Solar energy is available to us almost indefinitely – at least for more than 1 billion years longer. Even if only 0.1% of the energy which reaches the surface of the earth could be converted at an efficiency of only 10% it would be four times the world's total generating capacity. In other words, the total annual solar radiation falling on the earth is more than 7500 times the world's total annual primary energy demand.[3] Despite its beauty and importance, a disadvantage of natural photosynthesis is the low efficiency with which light energy is used to produce organic compounds. For most conventional plants, this is less than 1%.[4] Therefore, it makes sense to simplify this multifaceted process and improve its efficiency by artificial photosynthesis technologies.

Leibniz-Institut für Katalyse an der Universität Rostock, Albert-Einstein-Straße 29a, 18059 Rostock, Germany. E-mail: Matthias.Beller@catalysis.de

In this article, we will briefly introduce the reader to the differences between artificial, hybrid and natural photosynthesis and present their possible advantages and disadvantages. Besides, we will describe the state of the art and possible technology options in the future.

A short glimpse at biological photosynthesis

The overall process of photosynthesis takes place in several steps, which are spatially and temporally separated from each other. In the so-called light reaction, sunlight is used to split water, while at the same time electrons are provided for the dark reaction. This latter process uses the resulting reducing agents in the form of reduced nicotinamide adenine dinucleotide phosphate (NADPH) to convert carbon dioxide into biochemical platform molecules and subsequently to all kinds of organic matter. A key role in the dark reaction is played by the enzyme rubisco (ribulose-1,5-bisphosphate-carboxylase/-oxygenase), which controls carbon dioxide fixation. Although the theoretical maximum efficiency of the light reaction is about 10%, in reality it is <1% on an annual average for crops.[5] In other words, in the vast majority of plants more than 99% of light energy is lost through scattering, reflection and heat. In addition, plants do not absorb high-energy ultraviolet and infrared light. This means that large amounts of energy are lost that otherwise could be used. It is somewhat surprising that, despite their relatively low efficiency, numerous plant species are cultivated today for energy usage. These "energy-rich" crops are used directly for the production of heat and/or electrical energy, as well as indirectly for the processing of biofuels. Specific examples include both arable crops, for which varieties are cultivated that are partly optimized for energy use (*e.g.* rapeseed and maize), and cultivated crops that have not yet been used, or are hardly used at all, such as Chinese reed. Potential energy crops for the future are soya, oil palm, purgonium nut and sugar cane.

Modified biological photosynthesis

By using modified biological photosynthesis,[6] fuels and valuable substances can be produced using genetically improved single-cell organisms (bacteria, fungi, algae, *etc.*). This concept must not be confused with the well-known production of biofuels by converting biomass into biogas, biodiesel or bioethanol, where natural microorganisms are used. For precise modification of photosynthetic organisms, modern tools of genetic engineering and/or synthetic biology are used. Ideally, in this way the known biological processes should not only be optimized but breakthrough solutions be sought, for example the implementation of a completely new metabolism for the conversion of carbon dioxide. In such a way, this approach also aims to increase the energy efficiency of the light reaction.

In addition, the coupling of non-biological with biological components to hybrid systems is being actively pursued at the moment.[7] In most of these combinations, a process driven by light is combined with a biological dark reaction. For example, electricity generated by photovoltaics is typically converted into hydrogen (H_2) and carbon monoxide (CO) by the electrolysis of water and carbon dioxide. Subsequently, microorganisms can produce organic substances from these raw materials. Following this principle, researchers from Siemens and Evonik elegantly combined electrolysis and biotechnology. By 2021, an initial pilot plant is to go into operation in Marl, North Rhine-Westphalia, producing chemicals such as butanol or hexanol – both starting materials for speciality plastics or dietary supplements, for example. The companies announced that the next step could be to build a plant with a production capacity of up to 20 000 tons per year. The production of other speciality chemicals or fuels is also conceivable.[8]

Artificial photosynthesis

In a recent white paper from the German Academies of Sciences, artificial photosynthesis was defined as a technological toolbox which serves to produce chemical energy carriers and valuable products using sunlight as the sole energy source in integrated systems.[9] Related to this idea are so-called "Power to X" technologies,[10] whereby renewable energies should be transformed into fuels and recyclables. Here, an energy surplus arising from wind power or photovoltaics should be converted into better storable energy vectors such as methane, methanol or e-fuels. Basically, "Power to X" pursues similar goals as artificial photosynthesis; however, at present the former focus is on sequential processes, while artificial photosynthesis is pursuing a system approach similar to that of photosynthesis. The particular strength of this general approach lies in the provision of renewable energy to materials, which can be easily stored and transported. This is achieved by mimicking a central principle of the biological model: combining light-induced charge separation with catalytic processes for the production of energy-rich compounds. To be clear, the general goal of artificial photosynthesis is not to imitate biological processes in detail, but instead to fundamentally improve them in order to create a higher degree of efficiency. Nowadays a variety of approaches are summarized under the general term of artificial photosynthesis.[11] Most of them have it in common that reduction and/or proton donation processes take place through the use of external (light) energy. After the initial light absorption in artificial photosynthesis the following processes do not generate structurally complex organic substances as in normal photosynthesis. Instead, simple but valuable energy carriers as well as platform chemicals such as methane or methanol are made by reducing CO_2. Even the production of ammonia as an energy vector by reducing atmospheric nitrogen is being discussed.[12] In this case the only raw materials required are nitrogen, protons and electrons.

Electron transfer in artificial photosynthesis processes always consists of two half reactions: (a) the oxidation of water molecules to give electrons and protons and (b) the reduction of carbon dioxide (or nitrogen) as proton and electron acceptor. For light absorption, both naturally-occurring light-absorbing pigments and "high-chem" materials, typically applied in photovoltaic systems, can be used. Advantages of the latter materials are their stability and luminous efficacy. To develop more efficient integrated devices, the binding of the light-absorbing unit (so-called antenna systems) to the catalytically active reduction unit has been actively pursued in recent years. Due to their availability and price, 3d metal-based (molecular) catalysts are of current interest especially.[13] After suitable immobilization on stable solid-state supports, e.g. carbon, quantum dots, metal oxide surfaces or semiconductors, the resulting materials are used for electro-, photo-, and photoelectrocatalytic synthesis of fuels and bulk chemicals.[14] With respect to stability, artificial inorganic photovoltaic units (e.g. based on silicon or semiconductors) are advantageous and require little maintenance. However, the maximum efficiency of the current state of the art is only 33%, which is not realistically achievable.

In the quest to create a more efficient integrated device, several research groups are trying to develop so-called "artificial leaves". Pioneering work from the group of Nocera used cobalt and nickel as catalysts for water splitting analogous to PS II.[15] However, this system and following versions are not commercially competitive, yet. In 2015, Lewis and co-workers described another system consisting of a photoanode (for water oxidation) and a photocathode (hydrogen generation from protons) with a membrane in between.[16] Despite significant advancements and comparably good incident photoefficiencies, for the key step of the process, the splitting of water, the catalyst – often based on expensive precious metals such as platinum – efficiency, stability and thus costs still do not meet the requirements for large-scale industrial applications.

Challenges in the future

As described *vide supra*, the efficiency for converting the energy of visible sunlight into chemical energy in natural photosynthesis is comparatively low. For the vast majority of plants it is less than 1%. In contrast, technologies related to artificial photosynthesis make it possible to achieve significantly higher efficiencies already today. By using commercially available solar panels and well-established water electrolysis, "green electrons" can be used to split water into oxygen (as in PS II) and hydrogen with overall efficiencies of 7–10% (almost 10 times more than in the natural system). Nevertheless, the price of the so-produced hydrogen is still higher by a factor of 2–5 compared to that of hydrogen production *via* reforming of fossil resources. In addition, one must also take into account that hydrogen has a number of disadvantages compared to liquid fuels. Thus, the upscaling and operation costs of these sequential processes remain challenging at the moment. Obviously, a globally operating carbon dioxide tax on fossil fuels will make the present technologies economically more interesting. However, at the moment the main uses will be smaller and decentralized energy applications. In the mid- to long-term future (>2030), integrated devices, which directly convert carbon dioxide with the aid of sunlight (renewable energy) to organic products, might be advantageous. To apply integrated artificial photosynthesis technologies on a global scale, we will also have to find out how carbon dioxide can be most efficiently converted into organic products. In this respect, the amount of energy required for the direct use of CO_2 from the atmosphere (only 0.04 percent, *i.e.* 400 ppm) is also a critical factor for the success of the new technologies. Estimates assume a considerable energy input here (at least 20 kJ mol^{-1}).

Conclusions

The artificial photosynthesis technologies described here rely on basic concepts of biological photosynthesis. In contrast to natural processes, light energy is directly or sequentially used for the synthesis of valuable chemical substances with the help of man-made catalysts (*e.g.* photocatalysts). In the next decades, artificial photosynthesis could contribute to replacing large quantities of fossil fuels or raw materials. It thus has the potential to make a significant contribution to more environmentally friendly energy generation and climate protection. In recent years, considerable progress has been made by researchers all over the world. Nevertheless, a number of challenges need to be mastered before large-scale technical and economic implementation is possible. These include, in particular, the efficient use of carbon dioxide from the atmosphere and the production of cost-effective and long-term stable photocatalysts. Finally, it is worth mentioning that progress in this field will likely trigger the advancement of different research fields like photoredox synthesis, photoreforming, *etc.*[17]

Conflicts of interest

There are no conflicts to declare.

Notes and references

1 IRENA (2019), *Global energy transformation: A roadmap to 2050*, International Renewable Energy Agency, Abu Dhabi, 2019.
2 D. Peters, *Chem. Ing. Tech.*, 2006, **78**, 229–238.
3 https://www.worldenergy.org/wp-content/uploads/2013/10/WER_2013_8_Solar_revised.pdf.
4 R. Blankenship, D. Tiede, J. Barber, G. W. Brudvig, G. Fleming, M. Ghirardi, M. R. Gunner, W. Junge, D. M. Kramer, A. Melis, T. A. Moore, C. C. Moser, D. G. Nocera, A. J. Nozik, D. R. Ort, W. W. Parson, R. C. Prince and R. T. Sayre, *Science*, 2011, **332**, 805–809.
5 Here, the energy efficiency of photosynthesis is considered to be the overall efficiency of the energy conversion of the visible spectrum of light into organic products.
6 O. Kruse, J. Rupprecht, J. H. Mussgnug, G. C. Dismukesc and B. Hankamer, *Photochem. Photobiol. Sci.*, 2005, **4**, 957–970.
7 See for example: (*a*) D. Leister, *Plant Physiol.*, 2019, **179**, 778–793; (*b*) K. P. Sokol, W. E. Robinson, A. R. Oliveira, J. Warnan, M. M. Nowaczyk, A. Ruffinês, A. C. Pereira and E. Reisner, *J. Am. Chem. Soc.*, 2018, **140**, 16418–16422.
8 T. Haas, R. Krause, R. Weber, M. Demler and G. Schmid, *Nat. Catal.*, 2018, **1**, 32–39.
9 *Artificial Photosynthesis*, ed. M.-D. Weitze, acatech – National Academy of Science and Engineering, German National Academy of Sciences Leopoldina, Union of the German Academies of Sciences and Humanities, Munich, 2018, p. 74, see also: https://www.acatech.de/wp-content/uploads/2018/05/en_KPH_web_neu.pdf.
10 V. Eveloy and T. Gebreegziabher, *Energies*, 2018, **11**, 1824.
11 For recent reviews see: (*a*) Z. N. Zahran, Y. Tsubonouchi, E. A. Mohamed and M. Yagi, *ChemSusChem*, 2019, **12**, 1775–1793; (*b*) B. Zhanga and L. Sun, *Chem. Soc. Rev.*, 2018, **48**, 2216–2264; (*c*) S. Remiro-Buenamañana and H. García, *ChemCatChem*, 2019, **11**, 342–356; (*d*) J.-W. Wang, D.-C. Zhong and T.-B. Lu, *Coord. Chem. Rev.*, 2018, **377**, 225–236; (*e*) S. Fukuzumi, Y.-M. Lee and W. Nam, *Biochem. Soc. Trans.*, 2018, **46**, 1279–1288.
12 H. Hirakawa, M. Hashimoto, Y. Shiraishi and T. Hirai, *J. Am. Chem. Soc.*, 2017, **139**, 10929–10936.

13 For an excellent recent review see: (*a*) K. E. Dalle, J. Warnan, J. J. Leung, B. Reuillard, I. S. Karmel and E. Reisner, *Chem. Rev.*, 2019, **119**, 2752–2875; (*b*) C. Steinlechner, A. F. Roesel, E. Oberem, A. Päpcke, N. Rockstroh, F. Gloaguen, S. Lochbrunner, R. Ludwig, A. Spannenberg, H. Junge, R. Francke and M. Beller, *ACS Catal.*, 2019, **9**, 2091–2100.

14 See for example: (*a*) K. P. Sokol, W. E. Robinson, J. Warnan, N. Kornienko, M. M. Nowaczyk, A. Ruff, J. Z. Zhang and E. Reisner, *Nat. Energy*, 2018, **3**, 944–951; (*b*) S. Kreft, R. Schoch, J. Schneidewind, J. Rabeah, E. V. Kondratenko, V. A. Kondratenko, H. Junge, M. Bauer, S. Wohlrab and M. Beller, *Chem*, 2019, **3**, DOI: 10.1016/j.chempr.2019.04.006.

15 D. G. Nocera, *Acc. Chem. Res.*, 2012, **45**, 767–776.

16 E. Verlage, S. Hu, R. Liu, R. J. R. Jones, K. Sun, C. Xiang, N. S. Lewis and H. A. Atwater, *Energy Environ. Sci.*, 2015, **8**, 3166–3172.

17 E. Reisner, *Angew. Chem., Int. Ed.*, 2019, **58**, 3656–3657.

PAPER

Tuning purple bacteria salt-tolerance for photobioelectrochemical systems in saline environments

Matteo Grattieri, [ID] [a] Kevin Beaver,[b] Erin M. Gaffney[a] and Shelley D. Minteer [ID] *[a]

Received 31st October 2018, Accepted 7th January 2019

DOI: 10.1039/c8fd00160j

The development of photobioelectrochemical systems is an exciting field requiring a combination of electrochemical, biological and material science knowledge. One of the main advantages of applying anoxygenic photosynthetic microorganisms *versus* non-photosynthetic bacteria is the possibility to utilize sunlight as the energy source, while removing organic contaminants from a solution. Since bacterial cells utilize energy to maintain the intracellular osmolarity, bacterial species that do not rely on organic species as an energy source have an advantage over species requiring them for their sustainment. Herein, we discuss the possible use of *Rhodobacter capsulatus*, an extremely versatile photosynthetic purple bacteria, for application in environments within a range of low to moderately high salinity ($0-25$ g L^{-1} NaCl). Bacterial cells' capability to adapt to changing salinity, and effects on bioelectrochemical performance will be presented, as well as major drawbacks and research needs to drive future efforts and discussions.

Introduction

Solar power constitutes an extremely attractive renewable energy source, and impressive technological achievements, with record photoconversion efficiencies, have recently been reported.[1] Photobioelectrochemical systems (photo-BES), among different technologies to harvest solar energy, are emerging as an interesting possibility to utilize biological entities capable of converting photons into electrical current, based on their ability to grow, reproduce, and contain "self-repairing mechanisms" that are not present for other systems harvesting sunlight.[2] Specifically, these systems rely on the capability of photosynthetic microorganisms to use an electrode surface as electron acceptor (and/or donor),

[a]Departments of Chemistry and Materials Science & Engineering, University of Utah, 315 S 1400 E Rm2020, Salt Lake City, Utah 84112, USA. E-mail: minteer@chem.utah.edu

[b]Departments of Biology and Chemistry, Lebanon Valley College, 101 N College Ave, Annville, Pennsylvania 17003, USA

transferring electrons liberated after photon absorption.[3] While current outputs reported so far are relatively low (in the range of 10–100 μA cm^{-2}),[4] it should be noted that extracellular electron transfer mechanisms taking place at the microorganism–electrode interface remain poorly defined,[5,6] and their better understanding could expand photo-BES applications for solar energy harvesting. Herein, to better distinguish among the possible photo-BESs, a distinction is made between oxygenic photosynthetic organisms, which do not require an organic substrate and capture light to carry out charge separation of water,[7] and anoxygenic photosynthetic organisms, where light is used as the energy source to perform substrate oxidation. In this current work, focus will be placed on the latter category and we refer the reader interested in advancements in oxygenic photosynthetic organisms-based photo-BESs to other relevant publications.[8–10]

Photo-BES for saline environment remediation and monitoring

The main advantage of using anoxygenic photosynthetic organisms is their possible application in photo-BES for solution decontamination and monitoring. Specifically, photoheterotrophic organisms could be applied for organic substrate removal while converting sunlight into electrical energy, obtaining a biosensor (or ultimately a self-powered biosensor with no need of an external power source),[11] using the generated current as an indicator of the degradation process or hazardous compounds' presence.[12,13] This would provide an online and easy to measure signal, which would allow for an early warning in case of toxic events (drastic drop in the current response) and planning of interventions. It is critical to remark that by applying photo-BES as a biosensing tool, we are not claiming that this should be a substitute for highly sensitive and selective techniques, such as liquid chromatography or mass spectroscopy, but rather suggesting an integration of the two technologies, for better management of environmental hazardous situations.

One of the challenging aspects to applying photo-BESs as a biosensor is the effect of changing environmental factors on the bioelectrochemical performance of the devices. Changes in temperature, salinity, or light intensity could strongly influence bioelectrocatalytic properties of the utilized microorganisms, leading to false-positive warnings, and it was previously shown that salinity as low as 10 g L^{-1} could inhibit microbial bioelectrocatalysis.[14] This motivates our choice of an extremely versatile anoxygenic photosynthetic organism, *Rhodobacter capsulatus* (*R. capsulatus*). *R. capsulatus* is a non-sulfur (it does not utilize sulfur compounds as substrates), purple photosynthetic bacterium capable of growing in anaerobic photosynthetic conditions, aerobic dark conditions, and saline conditions. Herein, we are interested in the anaerobic photoheterotrophic metabolism of this organism, for sunlight-powered bioremediation of contaminated solutions with different salinities. *R. capsulatus* has been heavily studied due to an interesting ability of mediating extracellular horizontal gene transfer between cells *via* the gene transfer agent (rcGTA) represented in Fig. 1.[15] The rcGTA is a phage-like vesicle that carries ~4.5 kb of double stranded DNA to neighbouring cells to allow for horizontal gene transfer.[16] The 4.5 kb DNA segment consists of random segments of the genome with no correlation between the genetic position of the DNA segments and the frequency of being packaged into the rcGTA.[17] Maximum production of rcGTA is achieved when *R. capsulatus* growth reaches the stationary phase,[18] but its production can also be induced by common cellular signals, such

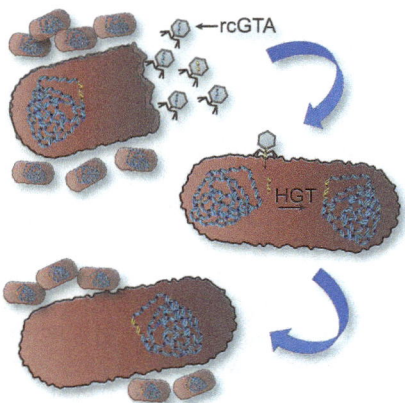

Fig. 1 Schematic of the gene transfer agent (rcGTA) in *R. Capsulatus*, where the orange DNA strand represents a random 4.5 kb segment that could account for the environmentally relevant trait leading to adaptation. Cells reaching the stationary phase, thus being adapted to the environment, can transfer DNA from one self-lysed *R. capsulatus* cell to another live cell through the rcGTA. When DNA enters the cell through the rcGTA, it is incorporated into the genome, and accumulation of the environmentally relevant traits could lead to evolution of *R. capsulatus* to adapt to an environmental stress.

as *N*-acyl homo serine lactones involved in quorum sensing.[19] The horizontal gene transfer mechanism in *R. capsulatus* would allow for an expedited evolution of a culture of *R. capsulatus*. This rapid exchange of genetic information could describe the adaptation of *R. capsulatus* to a changing environment.

A challenge for *R. capsulatus* application in a bioelectrochemical system is the location of its active redox center, which is buried below the outer and inner cellular membrane, hindering the direct extracellular electron transfer with an electrode surface. Accordingly, exogenous redox mediators are required to sustain the extracellular electron transfer (EET).[20,21] Specifically, our group has recently demonstrated that the limiting electron transfer step takes place in the lipophilic membrane of *R. capsulatus* cells, with a proton-decoupled single electron transfer process.[22] Accordingly, the redox mediator utilized to support the EET must be carefully selected considering its solubility, chemistry, and redox potential in a lipophilic environment. Based on this finding, we utilized soluble *p*-benzoquinone as the exogenous redox mediator, and we investigated bioelectrocatalytic properties and biophotocurrent generation under a salinity range from 0 to 25 g L^{-1} NaCl. *R. capsulatus* cells showed a particularly rapid adaptation to increasing salinities, and the effects of such adaptation to salinity are discussed. Furthermore, the possibility to utilize bioinformatics tools for studying how *R. capsulatus* cells adapt to salinity is presented, opening exciting new research possibilities.

Experimental

Chemicals

All chemicals were obtained from Sigma-Aldrich, except $MgCl_2$ (Fisher Scientific), K_2HPO_4 (VWR Analytical), KH_2PO_4 (Fisher Scientific), and KOH (Macron Chemicals), and used with no further purification.

Rhodobacter capsulatus growth

R. capsulatus strain ATCC 33303 was obtained from American Type Culture Collection (ATCC). *R. capsulatus* cells were grown in a liquid broth medium at different salinities, ranging between 0 and 25 g L^{-1} NaCl. Composition of the medium was as follow (for 1 L of MilliQ water): 20 mg EDTA, 1.0 g $(NH_4)_2SO_4$, 4.0 g malic acid, 200 mg $MgSO_4 \cdot 7H_2O$, 75 mg $CaCl_2 \cdot 2H_2O$, 12 mg $FeSO_4 \cdot 7H_2O$, 1 mg thiamine, 15 mg biotin, 0.9 g K_2HPO_4, 0.6 g KH_2PO_4, and 1 ml of trace element solution. The composition of the trace element solution was (per 250 ml of MilliQ water): 700 mg H_3BO_3, 398 mg $MnSO_4$, 188 mg $Na_2MoO_4 \cdot 2H_2O$, 60 mg $ZnSO_4 \cdot 7H_2O$, 10 mg $Cu(NO_3)_2 \cdot 3H_2O$, and 50 mg $CoCl_2$. The pH of the medium was adjusted to 6.8 using 10 M KOH prior to sterilization at 125 °C for 25 min. Trace elements were added after sterilization, by filtration through a 0.20 μm filter (VWR International). Bacteria cells were grown in sterile 20 ml scintillation vials, which were sealed with airtight rubber stoppers. An incandescent light bulb was used to maintain a light intensity of 3000 LUX during the growth. After 72 hours, the cells were collected by centrifugation at 5000 g for 20 min (Allegra X-15R benchtop centrifuge, Beckman Coulter). To restart a new growth, cells were resuspended in 2 ml of growth medium at the desired salinity concentration. Adaptation of *R. capsulatus* was investigated restarting a culture at a specific salinity for two times (secondary and tertiary growth). For the electrochemical characterization, cells collected after the first centrifugation step were resuspended in 1 ml of 20 mM MOPS buffer (pH 7) + 10 mM $MgCl_2$ + 50 mM malic acid, and further concentrated by centrifugation at 15 000 rpm for 10 min (Eppendorf Centrifuge 5424R). Finally, a solution with a bacterial cell concentration of 1 g ml^{-1} was prepared using 20 mM MOPS buffer (pH 7) + 10 mM $MgCl_2$ + 50 mM malic acid. Salinity of the electrolyte for electrochemical experiments was adjusted using NaCl as needed.

Electrodes and electrochemical setup

Electrodes for this study were obtained by cutting a Toray carbon paper electrode (TGP-H-060 non-wet proof, Fuel Cell Earth) with an area of 1 cm^2. The electrodes were sterilized by exposure under UV-light. 30 μL of the 1 g ml^{-1} solution of bacterial cells grown at the desired salinity were deposited on every electrode. The solution was allowed to dry for 1 hour under N_2 gas atmosphere. The dry electrodes were stored under N_2 gas until use in the electrochemical experiments. It should be noted that prolonged desiccation and storage processes could result in the dehydration of bacterial cells leading to their death. From preliminary experiments we observed very little difference in biophotocurrent generation for electrodes desiccated and stored under N_2 gas for a maximum of 4 hours. Storing of the electrodes with bacterial cells for longer than 4 hours resulted in a gradual decrease of the biophotocurrent, until no more significant production was obtained (approximately after 6–7 hours). Accordingly, the electrodes utilized for the present study were always employed within 3 hours from their preparation. All the electrochemical measurements reported in this study were performed at 20 ± 1 °C. Biophotocurrent generation of *R. capsulatus* with *p*-benzoquinone as a soluble mediator was investigated in a three-electrode electrochemical cell by cyclic voltammetry (CV) and amperometric *i*–*t* tests (CH660 potentiostat, CH

Instruments). *p*-Benzoquinone was utilized in previous studies as a monomeric redox mediator to support the electron transfer between photosynthetic organisms or apparatus and an electrode surface.[21,23,24] Our group recently reported that *p*-benzoquinone undergoes a proton-decoupled single electron transfer in the lipophilic membrane of *R. capsulatus*, and that its chemical and electrochemical properties allow maximizing the EET process compared to other soluble quinone-based mediators.[22] A concentration ranging from 30 to 200 μM was utilized for *p*-benzoquinone, by dissolving it in a stock solution using the MOPS buffer. The working electrode was Toray carbon paper with bacterial cells immobilized on the surface, or in the absence of bacterial cells (sterile electrode) for the control experiments. The counter electrode was a Pt mesh, and the reference electrode was a saturated calomel electrode (SCE, CHI 150, CH Instruments Inc.). All of the potentials in this work refer to this reference electrode. The electrolyte used for the study was 20 mM MOPS buffer (pH 7) + 10 mM $MgCl_2$ + 50 mM malic acid with differing concentrations of the soluble redox mediator. A Dolan-Jenner Fiber-Lite lamp (model 190-1 quartz-halogen illumination system with an optical light guide providing a light intensity of 76 mW cm^{-2}) was used to excite the photocurrent generation of *R. capsulatus* on the working electrodes. CVs were performed at 1 mV s^{-1} and amperometric *i–t* tests were performed by applying an anodic overpotential of 200 mV (η) based on the CV tests. After a dark step of 900 s, an illumination step of 500 s was performed, and the current was valuated at the end of the illumination step.

Biophotocurrent calculation

Biophotocurrent from the amperometric *i–t* test was calculated by subtracting the current obtained with bacterial cells immobilized on the Toray paper electrode from the current obtained with the bare Toray paper electrode (control experiment), at the end of the first illumination step (500 s). All of the experiments (with and without bacterial cells) were performed in triplicate. The average value was calculated and standard deviation was used to calculate the error.

Horizontal gene transfer study

A biochemical assay for gene transfer agent activity was adapted from Solioz *et al.*[18] After obtaining a tertiary cells growth adapted to a salinity of 15 g L^{-1} NaCl, we separated bacterial cells by filtration using a sterile 0.45 μm filter (VWR International Sterile Syringe Filter w/0.45 μm polyethersulfone membrane) and collected the filtrate in a sterile centrifuge tube. New growths at 10 g L^{-1} NaCl were started using cells obtained from a growth at 0 g L^{-1} NaCl and adding 2 ml of the filtrate. The amount of cells obtained after an incubation of 72 hours (early-stationary phase) was compared with the cells obtained when no filtrate was added. Comparison of the growths was performed by mass weight of the pellets after separating bacterial cells by centrifugation at 5000 g for 20 min and discarding the supernatant. As a control experiment, bacterial growth was started using only sterile medium and 2 ml of the filtrate, to ensure that this did not contribute to initial cell density. The control experiment showed absence of growth after an incubation of 100 h, confirming removal of cells during the filtration process.

Results and discussion

Growth adaptation to salinity

Effects of increasing salinity could first be observed by monitoring bacterial cell growth. A drastic increase of salinity content from 0 to 15 g L^{-1} NaCl resulted in a dramatic decrease in cell growth after 72 h, with a consequent decrease in bacteriochlorophyll production and loss in red colour of the growth (Fig. 2, left). Furthermore, an immediate increase of salinity up to 25 g L^{-1} resulted in a complete absence of cell growth. Conversely, if bacterial cells were allowed to adapt to salinity (secondary and tertiary growth), no significant effects on cell growth were observed up to 20 g L^{-1} NaCl. The tertiary growth allowed complete adaptation of bacterial cells (up to 20 g L^{-1} NaCl), with no significant decrease in bacteriochlorophyll as shown in Fig. 2 for the adapted cells (right).

When salinity was increased up to 22 g L^{-1} NaCl, a substantial decrease in cell growth was observed even for the tertiary growth, and further raising the salinity resulted in the complete inhibition of *R. capsulatus* growth. It is possible that a longer adaptation time, by transferring the cell growth for a higher number of times, or quorum sensing could result in cell growth even at salinity higher than 22 g L^{-1} NaCl. Future research will be focused on clarifying these aspects.

Photobioelectrocatalyis in saline conditions

Photobioelectrocatalysis at increased salinity was first investigated by performing cyclic voltammetry tests in the dark and under illumination, both with *R. capsulatus* and sterile electrodes. Specifically, the biophotocurrent obtained with 200 μM *p*-benzoquinone in solution in the presence of 0 (red) and 22 g L^{-1} NaCl (blue) is shown in Fig. 3A, where dashed lines represent the response in dark conditions and continuous lines indicate response under illumination. Fig. 3B shows CVs performed with sterile electrodes under the same conditions of illumination and salinities. From a first analysis of the results, a significant decrease in the current response can be clearly noted passing from 0 to 22 g L^{-1} NaCl. Specifically, the current under illumination at +0.35 V was −2.9 ± 0.7 and −1.6 ± 0.1 μA cm^{-2} for *R. capsulatus* in 0 and 22 g L^{-1} NaCl, respectively. Conversely, the current from CVs with sterile electrodes was −0.4 ± 0.2 and −0.9 ± 0.1 μA cm^{-2} in 0 and 22 g L^{-1} NaCl, respectively, thus confirming

| Not adapted to salinity | After adaptation |

| 0 | 10 | 15 | 0 | 20 |

Fig. 2 Salinity effects on *R. capsulatus* grown in photoheterotrophic conditions. Numbers below bottles indicate NaCl content in g L^{-1}. Adaptation was performed by recovering *R. capsulatus* cells exposed to a specific NaCl concentration and restarting the growth at the same salinity (secondary and tertiary growth).

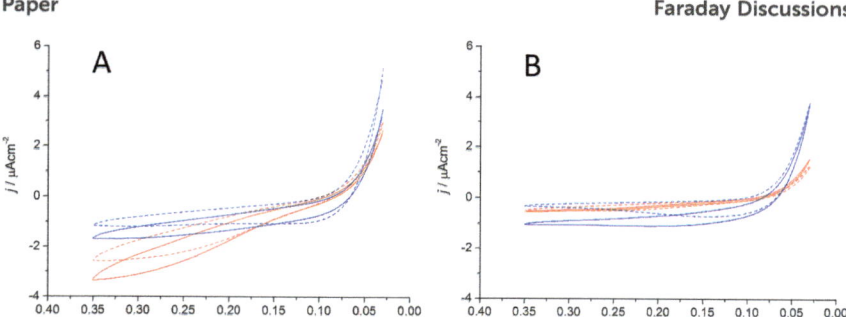

Fig. 3 Representative cyclic voltammetry test at 0 (red) and 22 g L^{-1} NaCl (blue), under dark (dashed lines) and light conditions (continuous lines) for *R. capsulatus* (A) and sterile electrodes (B). Scan rate: 1 mV s^{-1}, counter electrode: Pt, reference electrode: SCE.

R. capsulatus bioelectrocatalysis for malic acid oxidation in the salinity range investigated. It can also be noted that CVs performed at 22 g L^{-1} NaCl showed a considerably higher capacitance compared to the test in the absence of NaCl. This result is expected due to the higher ionic strength of the electrolyte.

Adaptation effects on biophotocurrent generation

It has to be noted that complete adaptation to 22 g L^{-1} NaCl was not reached, even after tertiary growth, as previously discussed. This motivated our interest in investigating adaptation effects on biophotoelectrocatalysis. Accordingly, we selected a potential of 0.36 V (200 mV anodic overpotential for the oxidation of *p*-benzoquinone) to conduct amperometric *i*–*t* test with *R. capsulatus* cells that were grown for the first time (primary growth) in the presence of 10 g L^{-1} NaCl (red) and cells adapted to 20 g L^{-1} NaCl (dark) obtained from the tertiary growth (Fig. 4). The difference in biophotocurrent between cells not adapted to 10 g L^{-1} NaCl and cells adapted to 20 g L^{-1} NaCl changed depending on the concentration of *p*-benzoquinone utilized. Specifically, a significant difference was obtained at low *p*-benzoquinone concentrations (0.2 ± 0.2 and 0.6 ± 0.1 µA cm^{-2} with 30 µM *p*-benzoquinone for not adapted and adapted cells, respectively), while high concentrations of *p*-benzoquinone led to an increased variability in the response, possibly due to toxic effects of the mediator on bacterial cells (1.5 ± 0.4 and 2.1 ± 0.3 µA cm^{-2} with 200 µM *p*-benzoquinone for not adapted and adapted cells, respectively). The observed trend indicates that adaptation plays a critical role on biophotoelectrocatalysis and biophotocurrent generation, and it is critical to note that *R. capsulatus* showed a particularly fast adaptation to increasing salinity, requiring less than two cell transfers to adapt to high salinities, further motivating our interest for its application in environments with variable salt content.

A comparison of biophotocurrent obtained at 0, 15 and 20 g L^{-1} NaCl for adapted cells (Fig. 5, black, red, and blue, respectively) revealed that increasing salinities negatively affected the bioelectrochemical performance as previously introduced. However, after the decrease in biophotocurrent obtained for salinities as low as 10–15 g L^{-1} NaCl, current generation for the cells adapted to the salinity was stable in the range of NaCl concentration investigated (NaCl up to 20 g L^{-1}). Specifically, biophotocurrents were 3.3 ± 0.5, 2.2 ± 0.1, and 2.1 ± 0.3 µA cm^{-2} at 0, 15, and

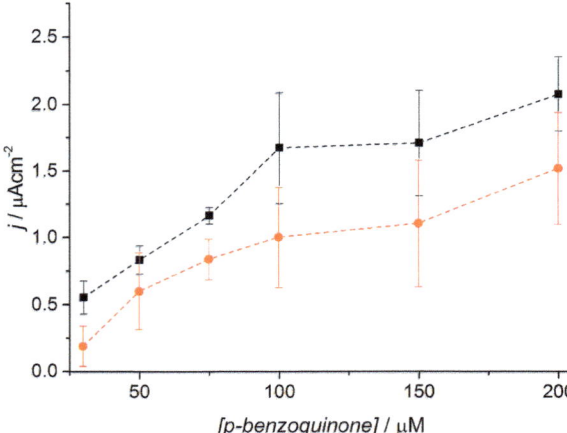

Fig. 4 Biophotocurrent at different *p*-benzoquinone concentrations for *R. capsulatus* not adapted (primary growth) to 10 g L^{-1} NaCl (red) and adapted (tertiary growth) to 20 g L^{-1} NaCl (black). Applied potential: 0.36 V *vs.* SCE, counter electrode: Pt, reference electrode: SCE. Average values are displayed. Error bars indicate the standard deviation over a set of three independent replicate experiments.

20 g L^{-1} NaCl, respectively, for 200 μM *p*-benzoquinone. This result further motivates our interest in studying the possibility to adapt *R. capsulatus* cells to higher salinities as discussed below, for potential biophotocurrent generation in hypersaline environments (total salt content higher than 35 g L^{-1}).

Representative amperometric *i–t* tests showing biophotocurrent generation at 200 μM *p*-benzoquinone for different salinities are reported in Fig. 6A.

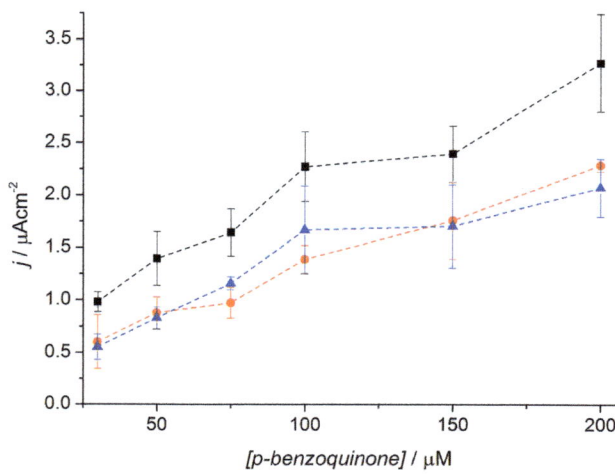

Fig. 5 Biophotocurrent at different *p*-benzoquinone concentrations for *R. capsulatus* at 0 g L^{-1} NaCl (black), and adapted (tertiary growth) to 15 (red) and 20 g L^{-1} NaCl (blue). Applied potential: 0.36 V *vs.* SCE, counter electrode: Pt, reference electrode: SCE. Average values are displayed. Error bars indicate the standard deviation over a set of three independent replicate experiments.

This journal is © The Royal Society of Chemistry 2019

Specifically, the results obtained at 0, 15, and 20 g L^{-1} NaCl are presented (black, red, and blue lines, respectively). For comparison, i–t tests performed with sterile electrodes at the same salinities are also shown (Fig. 6B). All of the amperometric i–t tests were performed with illumination steps of 500 s, which followed a 900 s step in dark conditions to allow stabilization of the system during polarization. From a first analysis of the results, it can be noticed that the highest photocurrent was obtained in the test at 0 g L^{-1} NaCl, reaching 4.8 ± 0.2 μA cm^{-2}; however, passing from 15 to 20 g L^{-1} NaCl resulted in an increasing current response, reaching 2.9 ± 0.1, and 3.2 ± 0.2 μA cm^{-2} for 15, and 20 g L^{-1} NaCl, respectively. The same trend was obtained for the control experiments in sterile conditions (with 1.5 ± 0.4, 0.7 ± 0.1, and 1.11 ± 0.08 μA cm^{-2} for 0, 15, and 20 g L^{-1} NaCl, respectively.) This result could be explained by the positive effects of higher salinities on the conductivity of the electrolyte. Furthermore, biophotocurrent generation was stable over the time-lapse investigated (approximately 9 min), similar to other studies utilizing photosynthetic organisms.[10,21,25] While optimization of the performance over long-term operation was not the objective of this work, future studies will be focused on maximizing the stability of the system for prototype development and in-field applications. Use of rationally designed redox polymers to mediate extracellular electron transfer and facilitate bacteria immobilization would represent a significant improvement for the system.

Reasons for adaptation to salinity

The bacterial phage-like gene transfer agent is small in size relative to the bacterial cells. Thus, after filtering the liquid growth through a 0.45 μm filter the gene transfer agent, if present, is separated from the bacterial cells. It was previously seen that exposure to the filtrate containing the gene transfer agent allowed the transfer of antibiotic resistance between *R. capsulatus* cells.[18] Accordingly, to unveil if the gene transfer agent of *R. capsulatus* plays a role in its rapid adaptation to salinity, we exposed cells that were never previously exposed to NaCl to the filtrate. The addition of the filtrate resulted in a 15 ± 2% increase in growth yield. It should be noted that this assay provides an initial insight into the

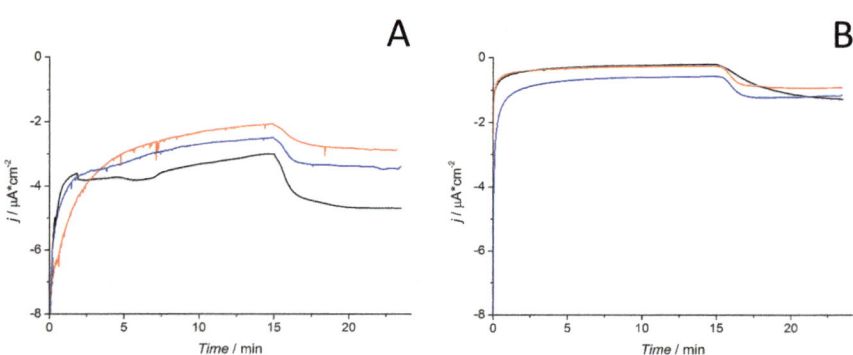

Fig. 6 Representative amperometric i–t test with *R. capsulatus* on the electrode surface (A), and with sterile electrodes (B) at 0, 15 and 20 g L^{-1} NaCl (black, red, and blue lines, respectively). Applied potential: 0.36 V *vs.* SCE, counter electrode: Pt, reference electrode: SCE.

activity of the gene transfer agent. We cannot exclude the possibility that the addition of the filtrate provided other biological components, potentially influencing the adaptation. Another possible mechanism of adaptation that could also play a role is the increased expression of proteins to allow for resistance to the osmotic pressure associated with high salinity. Proteins associated with adaptation to high salinity synthesize compatible osmotic solutes, such as trehalose, and import compatible solutes from the extracellular medium, such as glycine betaine and potassium. Specifically, these proteins are overexpressed during high salinity adaptation in the closely related bacterial strain *Rhodobacter sphaeroides*.[26] Future studies to elucidate the complete mechanisms of adaptation to salinity could include common biochemistry techniques, such as RNA sequencing, to confirm gene expression, and analysis of the extracellular medium to determine the presence of rcGTA. RNA sequencing on cultures of *R. capsulatus* growing at different salinities could determine a differential expression of any genes potentially accounting for the adaptation process. Additionally, since *R. capsulatus* is a well-studied microorganism, microchip arrays of the genetic sequence or RNA sequencing could help in determining mRNA expression.[17] Genetic analysis could also help determine if the adaptation strategy relies on evolution of the genome to become resistant to high salinity, which might be a result of the rcGTA mechanism, generating a halotolerant strain of *R. capsulatus*.

Conclusions

The possibility to develop photobioelectrochemical systems for operation in saline environments is of the utmost importance, as it would allow their application in seawater quality monitoring, as well as self-sustained decontamination of saline wastewater with online monitoring of the process. Herein, we have demonstrated the influence of adaptation to salinity on anoxygenic photosynthetic purple bacteria cells for photobioelectrocatalysis. After an initial decrease of biophotocurrent generation, possibly due to the additional energy required for maintaining the intracellular osmolarity, consistent electrochemical performance was achieved up to 20 g L^{-1} NaCl. While adaptation to higher salinity gave more complicated results, investigating biochemical adaptation strategies could extend the salinity range for biophotoelectrochemical systems development, with the opening of exciting future research possibilities.

Conflicts of interest

The authors declare no conflicts of interest.

References

1 K. Yoshikawa, H. Kawasaki, W. Yoshida, T. Irie, K. Konishi, K. Nakano, T. Uto, D. Adachi, M. Kanematsu, H. Uzu and K. Yamamoto, *Nat. Energy*, 2017, **2**, 17032.
2 M. H. Ham, J. H. Choi, A. A. Boghossian, E. S. Jeng, R. A. Graff, D. A. Heller, A. C. Chang, A. Mattis, T. H. Bayburt, Y. V. Grinkova, A. S. Zeiger, K. J. Van Vliet, E. K. Hobbie, S. G. Sligar, C. A. Wraight and M. S. Strano, *Nat. Chem.*, 2010, **2**, 929–936.

3 M. Rosenbaum, Z. He and L. T. Angenent, *Curr. Opin. Biotechnol.*, 2010, **21**, 259–264.

4 M. Rasmussen and S. D. Minteer, *J. Electrochem. Soc.*, 2014, **161**, H647–H655.

5 G. Pankratova and L. Gorton, *Curr. Opin. Electrochem.*, 2017, **5**, 193–202.

6 A. Kumar, L. H.-H. Hsu, P. Kavanagh, F. Barrière, P. N. L. Lens, L. Lapinsonnière, J. H. Lienhard V, U. Schröder, X. Jiang and D. Leech, *Nat. Rev. Chem.*, 2017, **1**, 0024.

7 A. J. McCormick, P. Bombelli, R. W. Bradley, R. Thorne, T. Wenzel and C. J. Howe, *Energy Environ. Sci.*, 2015, **8**, 1092–1109.

8 K. L. Saar, P. Bombelli, D. J. Lea-Smith, T. Call, E.-M. Aro, T. Müller, C. J. Howe and T. P. J. Knowles, *Nat. Energy*, 2018, **3**, 75–81.

9 M. Grattieri and S. D. Minteer, *Nat. Energy*, 2018, **3**, 8–9.

10 G. Saper, D. Kallmann, F. Conzuelo, F. Zhao, T. N. Toth, V. Liveanu, S. Meir, J. Szymanski, A. Aharoni, W. Schuhmann, A. Rothschild, G. Schuster and N. Adir, *Nat. Commun.*, 2018, **9**, 2168.

11 M. Grattieri and S. D. Minteer, *ACS Sens.*, 2018, **3**, 44–53.

12 M. Grattieri, K. Hasan and S. D. Minteer, *ChemElectroChem*, 2017, **4**, 834–842.

13 L. Giotta, A. Agostiano, F. Italiano, F. Milano and M. Trotta, *Chemosphere*, 2006, **62**, 1490–1499.

14 O. Lefebvre, Z. Tan, S. Kharkwal and H. Y. Ng, *Bioresour. Technol.*, 2012, **112**, 336–340.

15 B. Marrs, *Proc. Natl. Acad. Sci. U. S. A.*, 1974, **71**, 971–973.

16 F. Chen, A. Spano, B. E. Goodman, K. R. Blasier, A. Sabat, E. Jeffery, A. Norris, J. Shabanowitz, D. F. Hunt and N. Lebedev, *J. Proteome Res.*, 2009, **8**, 967–973.

17 A. P. Hynes, R. G. Mercer, D. E. Watton, C. B. Buckley and A. S. Lang, *Mol. Microbiol.*, 2012, **85**, 314–325.

18 M. Solioz, H.-C. Yen and B. Marrs, *J. Bacteriol.*, 1975, **123**, 651–657.

19 C. A. Brimacombe, A. Stevens, D. Jun, R. Mercer, A. S. Lang and J. T. Beatty, *Mol. Microbiol.*, 2013, **87**, 802–817.

20 K. Hasan, S. A. Patil, K. Gorecki, D. Leech, C. Hagerhall and L. Gorton, *Bioelectrochemistry*, 2013, **93**, 30–36.

21 K. Hasan, K. V. R. Reddy, V. Eßmann, K. Górecki, P. Ó. Conghaile, W. Schuhmann, D. Leech, C. Hägerhäll and L. Gorton, *Electroanalysis*, 2015, **27**, 118–127.

22 M. Grattieri, Z. Rhodes, D. P. Hickey, K. Beaver and S. D. Minteer, *ACS Catal.*, 2019, **9**, 867–873.

23 K. Hasan, Y. Dilgin, S. C. Emek, M. Tavahodi, H.-E. Åkerlund, P.-Å. Albertsson and L. Gorton, *ChemElectroChem*, 2014, **1**, 131–139.

24 J. Z. Zhang, P. Bombelli, K. P. Sokol, A. Fantuzzi, A. W. Rutherford, C. J. Howe and E. Reisner, *J. Am. Chem. Soc.*, 2018, **140**, 6–9.

25 G. Longatte, F. Rappaport, F.-A. Wollman, M. Guille-Collignon and F. Lemaître, *Electrochim. Acta*, 2017, **236**, 337–342.

26 M. Tsuzuki, O. V. Moskvin, M. Kuribayashi, K. Sato, S. Retamal, M. Abo, J. Zeilstra-Ryalls and M. Gomelsky, *Appl. Environ. Microbiol.*, 2011, **77**, 7551–7559.

Faraday Discussions

PAPER

Towards compartmentalized photocatalysis: multihaem proteins as transmembrane molecular electron conduits†

Anna Stikane, [iD] [ab] Ee Taek Hwang,[ab] Emma V. Ainsworth,[c] Samuel E. H. Piper,[c] Kevin Critchley,[bd] Julea N. Butt, [iD] *[c] Erwin Reisner*[e] and Lars J. C. Jeuken [iD] *[ab]

Received 4th November 2018, Accepted 12th December 2018

DOI: 10.1039/c8fd00163d

The high quantum efficiency of natural photosynthesis has inspired chemists for solar fuel synthesis. In photosynthesis, charge recombination in photosystems is minimized by efficient charge separation across the thylakoid membrane. Building on our previous bioelectrochemical studies of electron transfer between a light-harvesting nanoparticle (LHNP) and the decahaem subunit MtrC, we demonstrate photo-induced electron transfer through the full transmembrane MtrCAB complex in liposome membranes. Successful photoelectron transfer is demonstrated by the decomposition of a redox dye, Reactive Red 120 (RR120), encapsulated in MtrCAB proteoliposomes. The photoreduction rates are found to be dependent on the identity of the external LHNPs, specifically, dye-sensitized TiO_2, amorphous carbon dots (a-CD) and graphitic carbon dots with core nitrogen doping (g-N-CDs). Agglomeration or aggregation of TiO_2 NPs likely reduces the kinetics of RR120 reductive decomposition. In contrast, with the dispersed a-CD and g-N-CDs, the kinetics of the RR120 reductive decomposition are observed to be faster with the MtrCAB proteoliposomes and we propose that this is due to enhancement in the charge-separated state. Thus, we show a proof-of-concept for using MtrCAB as a lipid membrane-spanning building block for compartmentalised photocatalysis that mimics photosynthesis. Future work is focused on incorporation of fuel generating redox catalysts in the MtrCAB proteoliposome lumen.

[a]School of Biomedical Sciences, University of Leeds, Leeds, LS2 9JT, UK. E-mail: L.J.C.Jeuken@leeds.ac.uk

[b]The Astbury Centre for Structural Molecular Biology, University of Leeds, Leeds, LS2 9JT, UK

[c]Centre for Molecular and Structural Biochemistry, School of Chemistry and School of Biological Sciences, University of East Anglia, Norwich, NR4 7TJ, UK

[d]School of Physics and Astronomy, University of Leeds, Leeds, LS2 9JT, UK

[e]Department of Chemistry, University of Cambridge, Lensfield Road, Cambridge, CB2 1EW, UK

† Electronic supplementary information (ESI) available. See DOI: 10.1039/c8fd00163d

Introduction

Global research efforts are continuously advancing strategies for harnessing solar energy into sustainable electricity, solar fuels and solar chemicals.[1-3] The light harvesting stage, *i.e.*, photo-induced charge separation and electron (or hole) transfer to the electrode or catalyst, remains the principal efficiency-limiting step in these strategies.[3] In contrast, the stunning efficiency of biological light-harvesting systems results from a very precise and sophisticated arrangement of photosynthetic components: organic photosensitizers (*e.g.*, P680, P700, *etc.*), electron relay (chlorophyll, pheophytin, quinones, tyrosine, *etc.*) and biocatalytic conversions (Q_B reduction/water splitting).[4] These components are optimized in the dimensions of space (the relative location of the components), energy (excited-state and redox properties) and time (the rates of competing processes).[5,6] The composition of biological photosynthetic assemblies allows efficient photon absorption at light harvesting antennae, after which energy is passed along a series of chromophores to the reaction centres (*e.g.*, plant photosystems I and II), where it is used for excitation of the P680/P700 cofactors.[4] Electrons ejected from P680/P700 are relayed along an electron transfer chain and the light energy is ultimately stored as a transmembrane proton gradient and reduced redox-active molecules such as NADPH[7] (Fig. 1a).

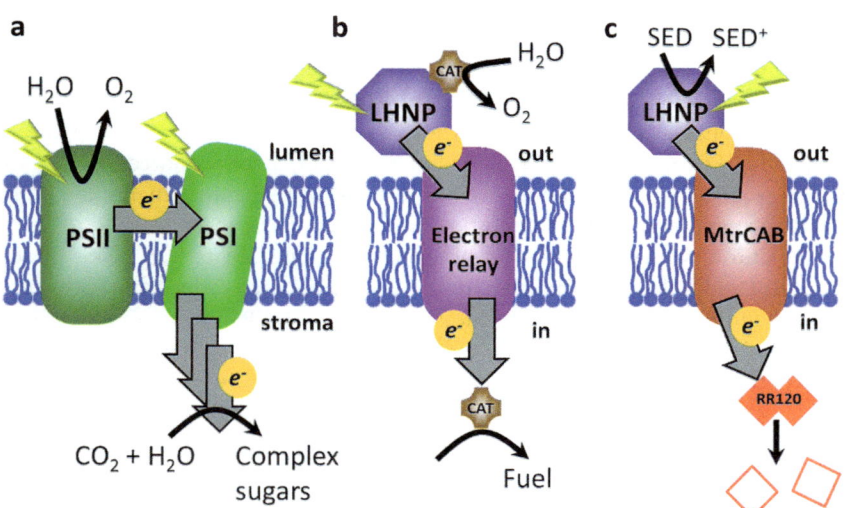

Fig. 1 A schematic of light-driven electron transfer across the lipid membrane in nature (a), in the envisioned biomimicking system (b), and as presented in this study (c). (a) Photosystems I and II (PSI and PSII) are photo-excited and electrons are transferred *via* several electron acceptors across the membrane, where they are ultimately used for CO_2 conversion into complex sugars. (b) External electrons are supplied photochemically from a light-harvesting nanoparticle (LHNP), which is regenerated by a water-oxidising catalyst (CAT). Electrons are relayed across the membrane to a catalyst leading to fuel generation within the compartment. (c) Electron transfer across the lipid bilayer is ensured *via* the transmembrane protein complex MtrCAB and monitored following reductive bleaching of an internalised red azo dye, Reactive Red 120 (RR120). SED – sacrificial electron donor.

Features such as near-unity quantum yields and environmentally friendly operation put biological light-harvesting systems above any other known system with regard to the initial steps of light harvesting for production of solar electricity and/or chemical synthesis.[7] Hence, there is a lot of interest in directly exploiting natural or genetically modified organisms[3,8–11] or their components for energy harvesting in artificial bio-hybrid systems.[12] Natural systems such as photosystem I, photosystem II and whole plant thylakoid membranes have been directly coupled to electrodes and inorganic catalysts in various photosynthetic devices to directly produce electricity, fuel (e.g., molecular hydrogen) or evolve oxygen.[12–19] However, light-induced damage and degradation limits the use of pigment–protein complexes, especially photosystem II.[12,20] Alternative approaches are being developed where synthetic light-harvesting analogues are interfaced to (bio) catalysts to biomimic the general principles of natural photosynthesis.[6,7] Reported systems include examples in which photosensitizers (PSs) (e.g., porphyrins) and light-harvesting nanoparticles (LHNPs) (e.g., quantum dots) are interfaced with various conductive materials ranging from graphene to peptide nanotubes to semi-conductor nanoparticles, fuel producing enzymes and electron mediators to regenerate cofactors for redox enzymes.[7,21] Efforts are also being made to explore the effects of photosynthetic component spatial organisation by mimicking natural systems such as stacked plant thylakoid membranes[22] and chlorosomes of green sulfur bacteria.[23]

In this work, we aimed to mimic another aspect of plant photosynthesis, i.e., the use of a lipid membrane to arrange and spatially separate the photosynthetic components between the different environments of the thylakoid lumen and stroma (Fig. 1a).[4] Specifically, the objective was to spatially separate the photo-oxidation and reduction reactions in the external and internal space of the liposome compartments, respectively (Fig. 1b). Thus, the envisioned system requires four components: (1) PSs or LHNPs to harvest light energy on the outside of liposomes, (2) transmembrane electron transfer, (3) a reduction catalyst within the liposome compartment (lumen) to chemically store the light energy and (4) an oxidation catalyst outside the liposome to regenerate the PS or LHNP.

We chose synthetic LHNPs over their natural equivalents (e.g., plant photosystems I and II) because they are simpler and cheaper to produce and because of their stability and chemical inertness.[24–26] Three LHNPs were compared: dye-sensitised TiO_2 nanoparticles and two types of carbon dots. Dye-sensitized TiO_2 nanoparticles are well-studied and among the most active photocatalyst materials.[24] We used TiO_2 nanoparticles photosensitized with a Ru(ɪɪ)(bipyridine)$_3$ dye in which one of the bipyridines is phosphonated in the 4,4′-position to enable chemisorption to TiO_2 (RuP-TiO_2, see Hwang et al.[27]). Carbon dots form another group of emerging light-absorbing nanomaterials showing remarkable photostability, water solubility, low toxicity and a sustainable and cost-effective synthesis, avoiding the use of rare metals.[25,26,28] Here we tested amorphous carbon dots (a-CD)[25] and graphitic carbon dots with core nitrogen doping (g-N-CDs).[28,29]

To transfer electrons across the lipid membrane after light harvesting, an icosa-haem transmembrane protein, MtrCAB, was employed (Fig. 1c), which provides an electron-transfer relay through the otherwise insulating lipid membrane.[30,31] MtrCAB is a heterotrimeric protein (MtrA, MtrB and MtrC) found in the bacterium *Shewanella oneidensis* MR1, where it forms a 20 haem long conductive molecular 'wire' across the bacterial outer membrane.[30,31] This enables the bacterium to use insoluble minerals such as iron and manganese oxides as external electron acceptors for its anaerobic metabolism.[32] When incorporated to span the lipid bilayer of a proteoliposome, MtrCAB exhibits fast transmembrane electron transfer, estimated to be 10^3–10^4 electrons per second by spectroscopic reduction of encapsulated methyl viologen.[33] We have previously demonstrated efficient electron exchange between the soluble decahaem subunit MtrC and LHNPs.[34-36]

In the present study, electrons are transferred *via* MtrCAB to the liposome lumen, where we envision they could generate fuel (such as hydrogen) by a fuel-generating catalyst. In this proof-of-concept study, electron transfer is optically monitored (539 nm) by destructive reduction of an encapsulated azo dye, Reactive Red 120 (RR120, Fig. S1†).[37-39] RR120 contains two azo bonds (R–N=N–R′), each of which requires a transfer of four electrons in order to be cleaved to a colourless (pale yellow) product (Fig. S1c†), *i.e.*, 8 electrons per RR120.[38] The optical signatures revealing haem redox status (Fig. S2†) are also monitored.

Materials and methods

Unless stated otherwise, all of the chemical substances were obtained from commercial suppliers and used without further purification: 3-(*N*-morpholino) propansulfonic acid (MOPS, >99.5%), sodium sulphate (Na$_2$SO$_4$, analytical reagent grade), *N,N*-dimethyldodecylamine N-oxide (LDAO, BioXtra, >99%), sodium hydrosulfite (DT, >82%) and Reactive Red 120 azo dye (RR120) were purchased from Sigma-Aldrich. Ethylenediaminetetraacetic acid disodium salt dehydrate (EDTA, >99.5%) and *n*-octyl glucoside (OG, laboratory grade) were acquired from Melford and Triton X100 detergent (electrophoresis grade) was purchased from Fisher Chemicals. A Milli-Q system was used to generate ultra-pure water (resistance 18.2 MΩ cm) which was used throughout. Ruthenium (Ru) dye sensitized TiO$_2$ anatase nanoparticles (RuP-TiO$_2$, diameter 6.8 ± 0.7 nm), g-N-CD (diameter 3.1 ± 1.1 nm) and a-CD (diameter 6.8 ± 2.3 nm) were synthesized an characterized as described previously.[25,27,29,40,41] *Shewanella oneidensis* MR1 protein MtrCAB was purified in Triton X-100 as described before.[42] The detergent exchange into 5 mM LDAO and additional purity resolution were performed using a Superdex 200 Increase SEC column (GE Healthcare) eluted with 5 mM LDAO, 20 mM HEPES pH 7.8. The purity of the purified MtrCAB was confirmed by SDS-PAGE with the protein visualized by Coomassie and haem stain.[43] *Escherichia coli* polar lipid extracts were purchased from Avanti Polar Lipids and stored in 5 mg dry aliquots under a nitrogen atmosphere at −20 °C.

Preparation of MtrCAB proteoliposomes

5 mg of *E. coli* polar lipid extract was dissolved by vigorous vortexing for up to 20 minutes in 294 µL of MOPS buffer (20 mM MOPS, 30 mM Na$_2$SO$_4$, pH 7.4) containing 6.6 mM RR120 and 85 mM OG. 50.5 µL of 10 µM MtrCAB (or 5 mM LDAO for control liposomes) was added to the lipid solution and kept on ice for a further 10 min. The sample was then rapidly diluted while mixing in 50 mL of ice-cold 20 mM RR120 in MOPS buffer. The sample was transferred to an ultracentrifuge tube (polycarbonate) and centrifuged for 100 min at 71 000g at 4 °C. The

supernatant containing most of the non-encapsulated RR120 was discarded and the pellet was re-suspended in 500 μL of MOPS buffer. The resulting sample was then centrifuged at 5000g for about 5 min to pellet any aggregates. The remaining non-encapsulated RR120 was removed by two consecutive rounds of 60 min sample incubation with 0.6 g of Bio-Beads (Bio-Rad SM-2) per 1 mL of sample at 4 °C on a rolling shaker. The experiments were performed within 2 days of the liposome preparation.

Liposome characterization

The concentration and size distribution of the liposomes was determined by nanoparticle tracking analysis (NTA) using Nanosight (NS300, Malvern Panalytical). The liposome size was also determined by dynamic light scattering (DLS) using a Zetasizer Nano Z (Malvern Panalytical). The size and volume of the liposomes were estimated by treating the liposomes as spherical particles, with the average diameter based on NTA data.

The amount of reconstituted MtrCAB was determined using a BCA assay (ThermoFisher Scientific). As the absorbance of encapsulated RR120 overlaps with the BCA reagent absorbance, the liposomes were first lysed with 0.1% v/v Triton X100 and RR120 was removed by two consecutive desalting columns (0.5 ml Zeba™ Spin, ThermoFisher) according to the manufacturer's protocol. The effectiveness of the desalting columns was confirmed using a control sample of RR120 loaded liposomes without MtrCAB.

Reduction of RR120 encapsulated in the MtrCAB proteoliposomes

Samples for the photo-reduction experiments were assembled under a nitrogen atmosphere (glovebox, O_2 < 0.1 ppm) to ensure an anaerobic environment. MtrCAB proteoliposome samples were diluted 10-fold in MOPS buffer containing 50 mM sacrificial electron donor (EDTA). An appropriate amount of 10 mg ml^{-1} photosensitiser stock (27 μmol NP per L RuP-TiO$_2$, 476 μM g-N-CD or 44 μM a-CD; mass of particles is estimated based on size determined by EM and density of material) was added to 1 μmol LHNP per L final concentration. The cuvette was then sealed airtight and removed from the glovebox for UV-vis absorbance spectroscopy (Cary 5000 UV-Vis-NIR, Agilent) fitted with an integrating sphere (Internal DRA-900, Agilent). UV-vis absorbance spectra were measured after 10 s, 50 s, 60 s, 120 s, or in some cases 300 s, of sample irradiation using a cold light source holding a 150 W (15 V) halogen lamp (OSRAM) with a fibre optic arm (Krüss KL5125). The sample was placed 10 cm from the light source and irradiated. The light intensity at the sample under these conditions was approximately 450 ± 40 mW cm^{-2} at 400 nm. Afterwards, the chemical reductant DT was added (final concentration of 27 mM) to monitor further possible reduction of the RR120. Finally, Triton X100 detergent was added (final concentration 0.045% v/v) to lyse the lipid vesicles and observe the reduction of any remaining RR120. Control experiments testing reduction by DT (*i.e.*, without LHNP) were also performed. Photo-reduction control experiments with non-encapsulated RR120 were performed as above, but with 10 μM RR120, 50 mM EDTA and 1 μmol LHNP per L PS in MOPS buffer. The recovery yield of MtrCAB was observed to vary between the proteoliposome preparations (see Results). To account for this, comparisons of the encapsulated RR120 (photo)reduction by DT and the photosensitisers were made based on proteoliposomes from the same preparation.

Treatment of UV-visible spectroscopy data

The spectroscopy data were corrected for sample dilution and for variation in the background signal (by setting the absorbance at 750 nm as the zero absorbance for each spectrum). Absorbance at 539 nm was selected to follow changes in the RR120 absorbance over time, because it is less influenced by the absorbance of reduced MtrCAB (α- and β- haem peaks at 552 and 522 nm). In order to correct for the contribution of liposome scattering, the optical density outside the RR120 absorbance peaks was measured at 440 and 610 nm, *i.e.*, either side of the RR120 absorbance, and the average value was subtracted.

Results

Characterization of MtrCAB liposomes

MtrCAB proteoliposomes loaded with the dye RR120 were prepared as described in the methods section. (Proteo)liposomes from each preparation were characterized to determine their size, concentration and the amount of reconstituted MtrCAB and encapsulated RR120, as described in the materials and methods. Although the size of the MtrCAB proteoliposomes showed some batch-to-batch variation, the proteoliposomes were consistently between 100 and 200 nm in diameter (Fig. S3†). The reconstitution protocol generated about 10^{13} liposomes per mL and thus an estimated total lumen volume in the order of 10–30 µL per mL of sample. Approximately $43 \pm 13\%$ of initial MtrCAB was present in the reconstituted proteoliposomes with an estimated ratio of 10–50 MtrCAB proteins per liposome (depending on the liposome size) assuming an even distribution across the liposomes.

Estimation of the amount of RR120 encapsulated in the MtrCAB proteoliposomes was performed spectroscopically using optical absorbance at 534 nm ($\varepsilon_{534 \text{ nm}} = 31.8$ mM^{-1} cm^{-1} was determined here using titration). It was estimated that, on average, the RR120 concentration in the liposome lumen was ~10 mM, *i.e.*, the same order of magnitude as during the liposome formation.

MtrCAB provides electron transfer across the bilayer

The ability of MtrCAB to transfer electrons across the membrane and reductively degrade RR120 was confirmed using excess chemical reductant (DT; Fig. 2). DT (E_m approximately -0.41 V *vs.* SHE at pH 7.4)[44] reduced MtrCAB (haem potential window ranging from -0.45 to 0 V *vs.* SHE)[45] within the time resolution of the experiment (<20 s), as indicated by a shift of the MtrCAB Soret peak due to haem absorbance (from 410 to 420 nm, Fig. 2a). This was followed by a slower (minutes) decrease in RR120 absorbance (450–570 nm, RR120 becomes reductively bleached at ≤ -0.4 V *vs.* SHE)[46], confirming the destructive reduction of the encapsulated RR120 (Fig. 2a). Only ~10% of RR120 was reduced in the control experiments using liposomes without MtrCAB, indicating that RR120 is protected from reductive bleaching when inside the liposomes and that reduction of encapsulated RR120 proceeds only if MtrCAB is present (Fig. 2b). As a positive control, detergent (Triton X100, TX) was added at the end of the experiment to lyse

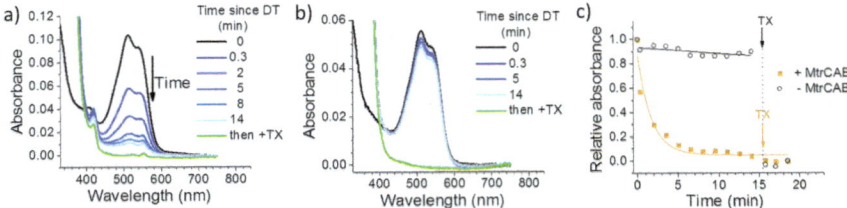

Fig. 2 Chemical reduction of encapsulated RR120 by sodium dithionite (DT) with (a) and without (b) MtrCAB. Reduction is followed optically by monitoring the absorbance of the MtrCAB haems (oxidised peak at 410 nm, reduced peak at 420 nm) and RR120 (oxidised 450–570 nm region). Black – oxidized sample; blue – intact liposomes after the addition of sodium dithionite; green – sample after disruption of the proteoliposome bilayer by detergent (Triton X100, TX). Time points indicate the time passed since the addition of DT. (c) The decrease of RR120 absorption ($\lambda = 539$ nm) over time using liposomes with and without MtrCAB. The yellow and black lines show the exponential and linear fits to the data, respectively.

the liposomes. This was followed by the immediate reductive bleaching of any remaining and now released RR120 (Fig. 2a and b, green lines). The rates of reduction of encapsulated RR120 were observed to vary between the MtrCAB proteoliposome preparations, likely due to the fact that the MtrCAB recovery yields varied (see above). For this reason, (photo)reduction of encapsulated RR120 by different reductants (*i.e.*, DT, LHNPs) was compared using proteoliposomes from the same preparation. In such studies the relative rates of RR120 reduction by the different LHNPs are as reported by the representative data shown below.

Photoreduction across the membrane

Three different LHNPs, *i.e.*, RuP dye-sensitized TiO$_2$ nanoparticles (RuP-TiO$_2$),[27] amorphous carbon dots (a-CD)[25] and graphitic carbon dots with core nitrogen doping (g-N-CDs),[28,29] were tested for photoreduction of RR120 encapsulated in liposomes with and without MtrCAB (Fig. 3a and b). All of the LHNPs have been

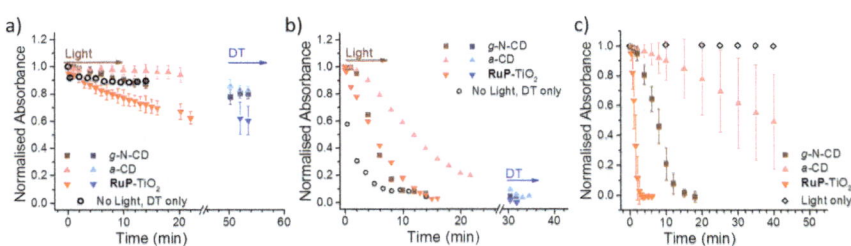

Fig. 3 Photoreduction of RR120 encapsulated in liposomes without MtrCAB (a) and in MtrCAB proteoliposomes (b) followed by a decrease in the RR120 absorbance at 539 nm. Squares – g-N-CD; upward triangles – a-CD; downward triangles – RuP-TiO$_2$; red – sample after irradiation; blue – sample after addition of DT; black circles – chemical reduction using DT added at $t = 0$ and without irradiation. Time points indicate the cumulative time of irradiation. In the case of DT, the time of DT addition is arbitrarily set to 50 and 30 min for (a) and (b), respectively, and the following time points indicate the time passed since the addition of DT. (c) Direct photo-reduction of 10 μM RR120 in solution by LHNPs. White rhombus – irradiation of RR120 without LHNPs.

previously shown to have sufficiently low reducing potential (< -0.45 V $vs.$ SHE)[29,47] to be able to reduce methyl viologen, and thus MtrCAB and RR120. Consistent with the data above, in the absence of MtrCAB, the majority (>70%) of RR120 was protected from photoreduction inside the liposome compartments (Fig. 3a). However, subsequent addition of DT to all of the samples showed that slightly more RR120 was reduced in the samples exposed to g-N-CD and RuP-TiO$_2$ compared to the 'DT only' control (compare the black open circles to the blue data points in Fig. 3a). This could suggest that small amounts of RR120 are released from the liposomes due to interactions between the RuP-TiO$_2$/g-N-CD and the liposomes. To further quantify this, well-established vesicle leakage assays were performed using a self-quenching dye, carboxyfluorescein.[48] No significant leakage was observed upon addition of any of the LHNPs, indicating that no, or very limited, damage is incurred to the vesicles by the LHNPs.

In the presence of MtrCAB, all three of the LHNPs photo-reduced the encapsulated RR120 (Fig. 3b). These experiments used 1 μM LHNPs, with an estimated ratio of 45 ± 2 LHNP per MtrCAB. RuP-TiO$_2$ and g-N-CD showed the fastest photoreduction, but with a lower rate compared to that with DT. Both the g-N-CDs and a-CD showed a short 1–2 min delay from the start of irradiation until the onset of RR120 photoreduction. This delay is further referred to as the 'lag phase' throughout this paper. Quantification of MtrCAB haem photoreduction by all three of the LHNPs was also attempted. Unfortunately, haem difference spectra could not be used due to spectral overlap with changes in RR120 and DT absorbance. Instead the first derivatives of all of the spectra were used instead as this is less sensitive to the background absorbance (Fig. S2 and S4†). This approach suggested that most MtrCAB is photoreduced by RuP-TiO$_2$ within the first minute of irradiation. In the case of the g-N-CDs and a-CD, it appeared that MtrCAB became reduced after several minutes, a time that coincides with the initial lag phase of RR120 reduction. After the lag phase, MtrCAB appeared to be fully reduced by the g-N-CDs, whereas only partial MtrCAB photo-reduction seems to be observed by the a-CDs. This suggest that with a-CD, photo-reduction of RR120 is in a large part rate limited by the photo-reduction of MtrCAB.

Finally, the photo-reduction of RR120 in the MtrCAB proteoliposomes was compared to the direct photo-reduction of non-encapsulated RR120 (Fig. 3c). RuP-TiO$_2$ showed faster photoreduction compared to the MtrCAB proteoliposomes, clearing >90% in less than 2 min, in line with the conclusion that reduction in the proteoliposomes is rate limited by the interaction between RR120 and MtrCAB. In contrast, the g-N-CDs and a-CDs took significantly longer to directly photo-reduce RR120 compared to the MtrCAB proteoliposomes, $i.e.$, about 20 min for the g-N-CDs and for the a-CDs it took more than 40 min to reduce even 50% of the RR120. Both of the LHNPs also showed longer and more variable kinetics, with lag phases of up to 5 min for the g-N-CDs and 10–20 min for the a-CDs. These variations in photo-reduction could reflect heterogeneity within the carbon dots, as observed before.[49,50]

Discussion

In plant photosynthesis, a lipid membrane is used as scaffolding to arrange and spatially separate photosynthetic components between the different environments of the thylakoid lumen and stroma.[4] Here we mimic such physical separation and show biomimetic photo-reduction across an insulating lipid membrane, where energy generated by external LHNPs is transferred across the lipid membrane *via* MtrCAB conduits to reduce electron acceptors located in the lumen of liposomes.

This system has several interfacial electron transfer steps: (1) LHNP to MtrCAB, (2) MtrCAB to RR120 and (3) SED to LHNP (Fig. 1c). All of the experiments used excess amounts of SED (50 mM EDTA) and we have previously shown that the SED is not rate limiting for the photo-reduction of MtrC by RuP-TiO$_2$.[27] As MtrCAB provides the electron relay across the membrane, the observed rate of RR120 reduction within the liposomes will be dependent on the amount and distribution of MtrCAB within the liposome population. Chemical reduction of MtrCAB with DT was instantaneous with respect to the time resolution of the experiments reported here. MtrCAB reduction by DT thus represents the fastest possible RR120 reduction within each liposome sample. The photoreduction by all three of the LHNPs was slower than the reduction by DT, confirming that the overall rate of RR120 reduction was at least partly limited by the electron supply from the LHNP to the MtrCAB. However, for the RuP-TiO$_2$ and g-N-CD, MtrCAB was almost fully reduced during the photo-reduction experiments, suggesting that the reductive bleaching kinetics of RR120 were also rate limited by the reduction of RR120 by MtrCAB. MtrCAB orientation in liposomes is not known and likely random, possibly further complicating the observed kinetics.

TiO$_2$ has high affinity for Glu/Asp protein residues,[51–53] and RuP-TiO$_2$ has been shown before to bind strongly to MtrC and MtrCAB.[36] In addition, RuP-TiO$_2$ showed the best direct photo-reduction of non-encapsulated RR120. Despite this, the photoreduction of RR120 in the MtrCAB liposomes with RuP-TiO$_2$ was slower compared to the chemical reduction with DT. We attribute the slower photoreduction of RR120 in the MtrCAB proteoliposomes to the self-agglomeration or aggregation of the RuP-TiO$_2$ particles, as observed with cryo-electron microscopy analysis (Fig. S5†). Hence, interaction between MtrCAB and RuP-TiO$_2$ might have been impaired. In contrast, the interaction between MtrCAB and both the g-N-CDs and a-CDs is likely to be transient as no aggregation was detected upon mixing of the particles with the MtrCAB liposomes. Nevertheless, for both of the carbon dots, relaying the electrons *via* MtrCAB improved the bleaching rate of RR120 remarkably, which is up to four times faster in the MtrCAB proteoliposomes compared to the direct photoreduction of RR120. Encapsulation of RR120 at mM concentration in the small lumen of the liposomes (compared to 10 μM RR120 in the control experiments with direct photoreduction) will enhance the reduction kinetics by MtrCAB and, indeed, reduction of RR120 by MtrCAB was not observed to be rate limiting for a-CD. The enhanced photobleaching kinetics in the proteoliposomes are thus due to the faster reduction of MtrCAB by a-CDs compared to reduction of RR120 by a-CD (at concentrations ≪10 μM). We propose that this enhancement is due to the MtrCAB conduit, which can accumulate multiple electrons on its 20 haems, improving the rate of the multi-electron reduction required to bleach each RR120 molecule. In this respect, MtrCAB is able to stabilise the charge separated intermediate for the photo-reduction of RR120, mimicking the role of the chlorophyll/pheophytin/Q_A electron relay of the natural photosystems I and II.

These results provide an insight into how control over the nano-device organization and assembly can be used in artificial photosynthesis and solar-fuel catalyst design to enhance catalytic and quantum efficiencies. This work adds to the ongoing work in which the organisation of different photosynthetic components is exploited for (bio-)nanocatalysis.[7] For example, stacked multi-layers of lipid membranes containing PSII[22] have been shown to increase production of ATP due to highly efficient exchange of substrates, while limiting the diffusion of photo- and catalytic centres. Besides lipid membranes, various other template materials such as viruses, graphene and peptide fibres have been used to gain control over precise physical distribution of porphyrin PSs and catalytic reaction centres (e.g., Pt, TiO_2 and IrO_2 clusters).[54–59] A 10-times higher yield for selective CO_2 conversion into methanol was reported using hollow graphene-doped nanofibers (G-fibers).[59] In this case, multiple enzymes required for methanol generation were confined within the nanofibers, and the photo-excited electrons were transported through the graphene fibers from photosen-sitizers located on the outside.[59] In a similar approach, photo-oxidation was separated from photo-reduction reactions by employing hierarchical cobalt oxide–silica core–shell nanotube arrays, where water oxidation and photo-reduction were confined to the inner and outer surface of the nanotubes, respectively.[60] Many other ideas for building architectures with isolated environments for separated photo-oxidation and reduction can be drawn from the field of artificial nano-compartments, which has reported the use of various materials ranging from labile biological liposomes, protein cages and virus capsids to rigid synthetic polymersomes and hybrid vesicles.[61–63]

Conclusion and future perspective

Here, we show a proof-of-concept of using the transmembrane MtrCAB conduit for compartmentalized photo-reduction. Three LHNPs demonstrated efficient photo-reduction of a liposome-encapsulated dye using MtrCAB as an electron relay. The rate with which two different carbon dots photo-reduced the encap-sulated dye was improved in the liposome system. This example demonstrated how incorporation of a scaffolding material to separate photo-oxidation and reduction reactions can be beneficial for the overall efficiency of solar energy harvest. In particular, we propose that MtrCAB can aid in the stabilisation of the charge separated state, improving the quantum yield. Such a component could be beneficial to further advance artificial photosynthesis strategies and other (bio-)nanocatalysis applications. To further explore the potential use of MtrCAB conduit and nano-compartments, MtrCAB compartments should be tested for photosynthetic production of solar fuels or solar chemicals. In this case, a catalyst can be encapsulated in the liposomes, which enables a PS/LHNP to function in a separate environment from the fuel-generating catalyst. Finally, the lipids and/or MtrCAB could be replaced by synthetic components to explore other com-partmentalised and structured molecular nano-architectures.

Conflicts of interest

The authors declare no conflicts of interest.

Acknowledgements

This work was supported by the BBSRC (DTP studentship 1827308; grants BB/K009753/1, BB/K010220/1 and BB/K009885/1) and the Engineering and Physical

Sciences Research Council (PhD studentship 1307196). We thank Dr Simone Payne for purifying MtrCAB, Dr Manuela A. Gross for synthesizing RuP, Dr Daisuke Hojo for the TiO_2 particles and Dr Benjamin C. M. Martindale for providing the carbon dot samples.

References

1 M. A. Green, *Nat. Energy*, 2016, **1**, 15015.
2 S. Ardo, D. Fernandez Rivas, M. A. Modestino, V. Schulze Greiving, F. F. Abdi, E. Alarcon Llado, V. Artero, K. Ayers, C. Battaglia, J.-P. Becker, D. Bederak, A. Berger, F. Buda, E. Chinello, B. Dam, V. Di Palma, T. Edvinsson, K. Fujii, H. Gardeniers, H. Geerlings, S. M. H. Hashemi, S. Haussener, F. Houle, J. Huskens, B. D. James, K. Konrad, A. Kudo, P. P. Kunturu, D. Lohse, B. Mei, E. L. Miller, G. F. Moore, J. Muller, K. L. Orchard, T. E. Rosser, F. H. Saadi, J.-W. Schüttauf, B. Seger, S. W. Sheehan, W. A. Smith, J. Spurgeon, M. H. Tang, R. van de Krol, P. C. K. Vesborg and P. Westerik, *Energy Environ. Sci.*, 2018, **11**, 2768–2783.
3 D. Lips, J. M. Schuurmans, F. Branco Dos Santos and K. J. Hellingwerf, *Energy Environ. Sci.*, 2018, **11**, 10–22.
4 N. Nelson and A. Ben-Shem, *Nat. Rev. Mol. Cell Biol.*, 2004, **5**, 971–982.
5 M. Şener, J. Strümpfer, J. Hsin, D. Chandler, S. Scheuring, C. N. Hunter and K. Schulten, *ChemPhysChem*, 2011, **12**, 518–531.
6 V. Balzani, A. Credi and M. Venturi, *ChemSusChem*, 2008, **1**, 26–58.
7 J. H. Kim, D. H. Nam and C. B. Park, *Curr. Opin. Biotechnol.*, 2014, **28**, 1–9.
8 W. Wei, P. Sun, Z. Li, K. Song, W. Su, B. Wang, Y. Liu and J. Zhao, *Sci. Adv.*, 2018, **4**, eaap9253.
9 M. K. Sarma, S. Kaushik and P. Goswami, *Biomass Bioenergy*, 2016, **90**, 187–201.
10 L. M. Lassen, A. Z. Nielsen, B. Ziersen, T. Gnanasekaran, B. L. Møller and P. E. Jensen, *ACS Synth. Biol.*, 2014, **3**, 1–12.
11 D. J. Vinyard, J. Gimpel, G. M. Ananyev, S. P. Mayfield and G. C. Dismukes, *J. Am. Chem. Soc.*, 2014, **136**, 4048–4055.
12 E. Musazade, R. Voloshin, N. Brady, J. Mondal, S. Atashova, S. K. Zharmukhamedov, I. Huseynova, S. Ramakrishna, M. M. Najafpour, J.-R. Shen, B. D. Bruce and S. I. Allakhverdiev, *J. Photochem. Photobiol., C*, 2018, **35**, 134–156.
13 M. Miyachi, S. Ikehira, D. Nishiori, Y. Yamanoi, M. Yamada, M. Iwai, T. Tomo, S. I. Allakhverdiev and H. Nishihara, *Langmuir*, 2017, **33**, 1351–1358.
14 F. Zhao, F. Conzuelo, V. Hartmann, H. Li, M. M. Nowaczyk, N. Plumeré, M. Rögner and W. Schuhmann, *J. Phys. Chem. B*, 2015, **119**, 13726–13731.
15 M. Miyachi, K. Okuzono, D. Nishiori, Y. Yamanoi, T. Tomo, M. Iwai, S. I. Allakhverdiev and H. Nishihara, *Chem. Lett.*, 2017, **46**, 1479–1481.
16 L. M. Utschig, N. M. Dimitrijevic, O. G. Poluektov, S. D. Chemerisov, K. L. Mulfort and D. M. Tiede, *J. Phys. Chem. Lett.*, 2011, **2**, 236–241.
17 S. C. Silver, J. Niklas, P. Du, O. G. Poluektov, D. M. Tiede and L. M. Utschig, *J. Am. Chem. Soc.*, 2013, **135**, 13246–13249.
18 T. Noji, H. Suzuki, T. Gotoh, M. Iwai, M. Ikeuchi, T. Tomo and T. Noguchi, *J. Phys. Chem. Lett.*, 2011, **2**, 2448–2452.
19 K. P. Sokol, W. E. Robinson, J. Warnan, N. Kornienko, M. M. Nowaczyk, A. Ruff, J. Z. Zhang and E. Reisner, *Nat. Energy*, 2018, **57**, 10595–10599.

20 F. Zhao, S. Hardt, V. Hartmann, H. Zhang, M. M. Nowaczyk, M. Rögner, N. Plumeré, W. Schuhmann and F. Conzuelo, *Nat. Commun.*, 2018, **9**, 1973.

21 S. H. Lee, D. S. Choi, S. K. Kuk and C. B. Park, *Angew. Chem., Int. Ed.*, 2018, **57**, 7958–7985.

22 Y. Li, J. Fei, G. Li, H. Xie, Y. Yang, J. J. Li, Y. Xu, B. Sun, J. Xia, X. Fu and J. J. Li, *ACS Nano*, 2018, **12**, 1455–1461.

23 M. Wang, J. Chen, T. Lian and W. Zhan, *Langmuir*, 2016, **32**, 7326–7338.

24 S. H. Lee, J. H. Kim and C. B. Park, *Chem.–Eur. J.*, 2013, **19**, 4392–4406.

25 B. C. M. Martindale, G. A. M. Hutton, C. A. Caputo and E. Reisner, *J. Am. Chem. Soc.*, 2015, **137**, 6018–6025.

26 Y. Wang and A. Hu, *J. Mater. Chem. C*, 2014, **2**, 6921–6939.

27 E. T. Hwang, K. Sheikh, K. L. Orchard, D. Hojo, V. Radu, C.-Y. Lee, E. Ainsworth, C. Lockwood, M. A. Gross, T. Adschiri, E. Reisner, J. N. Butt and L. J. C. Jeuken, *Adv. Funct. Mater.*, 2015, **25**, 2308–2315.

28 G. A. M. Hutton, B. C. M. Martindale and E. Reisner, *Chem. Soc. Rev.*, 2017, **46**, 6111–6123.

29 B. C. M. Martindale, G. A. M. Hutton, C. A. Caputo, S. Prantl, R. Godin, J. R. Durrant and E. Reisner, *Angew. Chem., Int. Ed.*, 2017, **56**, 6459–6463.

30 M. J. Edwards, G. F. White, C. W. Lockwood, M. C. Lawes, A. Martel, G. Harris, D. J. Scott, D. J. Richardson, J. N. Butt and T. A. Clarke, *J. Biol. Chem.*, 2018, **293**, 8103–8112.

31 M. Breuer, K. M. Rosso, J. Blumberger and J. N. Butt, *J. R. Soc., Interface*, 2015, **12**, 20141117.

32 J. K. Fredrickson, M. F. Romine, A. S. Beliaev, J. M. Auchtung, M. E. Driscoll, T. S. Gardner, K. H. Nealson, A. L. Osterman, G. Pinchuk, J. L. Reed, D. a Rodionov, J. L. M. Rodrigues, D. a Saffarini, M. H. Serres, A. M. Spormann, I. B. Zhulin and J. M. Tiedje, *Nat. Rev. Microbiol.*, 2008, **6**, 592–603.

33 G. F. White, Z. Shi, L. Shi, Z. Wang, A. C. Dohnalkova, M. J. Marshall, J. K. Fredrickson, J. M. Zachara, J. N. Butt, D. J. Richardson and T. A. Clarke, *Proc. Natl. Acad. Sci. U. S. A.*, 2013, **110**, 6346–6351.

34 C.-Y. Lee, B. Reuillard, K. Sokol, T. Laftsoglou, C. Lockwood, S. Rowe, E. T. Hwang, J.-C. Fontecilla-Camps, L. J. C. Jeuken, J. Butt and E. Reisner, *Chem. Commun.*, 2016, **52**, 7390–7393.

35 B. Reuillard, K. H. Ly, P. Hildebrandt, L. J. C. Jeuken, J. N. Butt and E. Reisner, *J. Am. Chem. Soc.*, 2017, **139**, 3324–3327.

36 E. V. Ainsworth, C. W. J. Lockwood, G. F. White, E. T. Hwang, T. Sakai, M. A. Gross, D. J. Richardson, T. A. Clarke, L. J. C. Jeuken, E. Reisner and J. N. Butt, *ChemBioChem*, 2016, **17**, 2324–2333.

37 F. Zhang, A. Yediler, X. Liang and A. Kettrup, *J. Environ. Sci. Health, Part A: Toxic/Hazard. Subst. Environ. Eng.*, 2002, **37**, 707–713.

38 A. B. dos Santos, F. J. Cervantes and J. B. van Lier, *Bioresour. Technol.*, 2007, **98**, 2369–2385.

39 M. C. Costa, F. S. B. Mota, A. B. Dos Santos, G. L. F. Mendonça and R. F. do Nascimento, *Quim. Nova*, 2012, **35**, 482–486.

40 S. A. Trammell, J. A. Moss, J. C. Yang, B. M. Nakhle, C. A. Slate, F. Odobel, M. Sykora, B. W. Erickson and T. J. Meyer, *Inorg. Chem.*, 1999, **38**, 3665–3669.

41 C. X. Guo, D. Zhao, Q. Zhao, P. Wang and X. Lu, *Chem. Commun.*, 2014, **50**, 7318.

42 G. R. Heath, M. Li, I. L. Polignano, J. L. Richens, G. Catucci, P. O'Shea, S. J. Sadeghi, G. Gilardi, J. N. Butt and L. J. C. Jeuken, *Biomacromolecules*, 2016, **17**, 324–335.

43 P. E. Thomas, D. Ryan and W. Levin, *Anal. Biochem.*, 1976, **75**, 168–176.

44 S. G. Mayhew, *Eur. J. Biochem.*, 1978, **85**, 535–547.

45 R. S. Hartshorne, C. L. Reardon, D. Ross, J. Nuester, T. a Clarke, A. J. Gates, P. C. Mills, J. K. Fredrickson, J. M. Zachara, L. Shi, A. S. Beliaev, M. J. Marshall, M. Tien, S. Brantley, J. N. Butt and D. J. Richardson, *Proc. Natl. Acad. Sci. U. S. A.*, 2009, **106**, 22169–22174.

46 C. Guarantini, A. G. Fogg and M. V. B. Zanoni, in *Proceedings of the Symposium on Chemical and Biological Sensors and Analytical Electrochemical Methods*, Pennington: Electrochemical Society Inc., 1997, vol. 97, ch. 19, pp. 467–476.

47 J. Willkomm, K. L. Orchard, A. Reynal, E. Pastor, J. R. Durrant and E. Reisner, *Chem. Soc. Rev.*, 2016, **45**, 9–23.

48 J. N. Weinstein, R. Blumenthal and R. D. Klausner, in *Methods Enzymology*, Academic Press, 1986, vol. 128, pp. 657–668.

49 Y. Zhou, P. Y. Liyanage, D. L. Geleroff, Z. Peng, K. J. Mintz, S. D. Hettiarachchi, R. R. Pandey, C. C. Chusuei, P. L. Blackwelder and R. M. Leblanc, *ChemPhysChem*, 2018, **19**, 2589–2597.

50 J. B. Essner, J. A. Kist, L. Polo-Parada and G. A. Baker, *Chem. Mater.*, 2018, **30**, 1878–1887.

51 E. Reisner, J. C. Fontecilla-Camps and F. a Armstrong, *Chem. Commun.*, 2009, 550–552.

52 E. Reisner, D. J. Powell, C. Cavazza, J. C. Fontecilla-Camps and F. A. Armstrong, *J. Am. Chem. Soc.*, 2009, **131**, 18457–18466.

53 Y. Kang, X. Li, Y. Tu, Q. Wang and H. Ågren, *J. Phys. Chem. C*, 2010, **114**, 14496–14502.

54 Y. S. Nam, T. Shin, H. Park, A. P. Magyar, K. Choi, G. Fantner, K. A. Nelson and A. M. Belcher, *J. Am. Chem. Soc.*, 2010, **132**, 1462–1463.

55 Y. S. Nam, A. P. Magyar, D. Lee, J. Kim, D. S. Yun, H. Park, T. S. Pollom, D. A. Weitz and A. M. Belcher, *Nat. Nanotechnol.*, 2010, **5**, 340–344.

56 K. Liu, R. Xing, Y. Li, Q. Zou, H. Möhwald and X. Yan, *Angew. Chem., Int. Ed.*, 2016, **55**, 12503–12507.

57 K. Liu, M. Abass, Q. Zou and X. Yan, *Green Energy Environ.*, 2017, **2**, 58–63.

58 J. H. Kim, M. Lee, J. S. Lee and C. B. Park, *Angew. Chem., Int. Ed.*, 2012, **51**, 517–520.

59 X. Ji, Y. Kang, Z. Su, P. Wang, G. Ma and S. Zhang, *ACS Sustainable Chem. Eng.*, 2018, **6**, 3060–3069.

60 W. Kim, B. A. McClure, E. Edri and H. Frei, *Chem. Soc. Rev.*, 2016, **45**, 3221–3243.

61 S. Schmidt, K. Castiglione and R. Kourist, *Chem.–Eur. J.*, 2018, **24**, 1755–1768.

62 L. Klermund and K. Castiglione, *Bioprocess Biosyst. Eng.*, 2018, **41**, 1233–1246.

63 S. Khan, M. Li, S. P. Muench, L. J. C. Jeuken and P. A. Beales, *Chem. Commun.*, 2016, **52**, 11020–11023.

Faraday Discussions

PAPER

A kinetic model for redox-active film based biophotoelectrodes†

D. Buesen, (iD) T. Hoefer, (iD) H. Zhang (iD) and N. Plumeré (iD) *

Received 5th November 2018, Accepted 18th December 2018

DOI: 10.1039/c8fd00168e

Redox-active films are advantageous matrices for the immobilization of photosynthetic proteins, due to their ability to mediate electron transfer as well as to achieve high catalyst loading on an electrode for efficient generation of electricity or solar fuels. A general challenge arises from various charge recombination pathways along the light-induced electron transfer chain from the electrode to the charge carriers for electricity production or to the final electron acceptors for solar fuel formation. Experimental methods based on current measurement or product quantification are often unable to discern between the contributions from the photocatalytic process and the detrimental effect of the short-circuiting reactions. Here we report on a general electrochemical model of the reaction–diffusion processes to identify and quantify the "bottlenecks" present in the fuel or current generation. The model is able to predict photocurrent–time curves including deconvolution of the recombination contributions, and to visualize the corresponding time dependent concentration profiles of the product. Dimensionless groups are developed for straightforward identification of the limiting processes. The importance of the model for quantitative understanding of biophotoelectrochemical processes is highlighted with an example of simulation results predicting the effect of the diffusion coefficient of the charge carrier on photocurrent generation for different charge recombination kinetics.

1 Introduction

Photosynthetic proteins have evolved toward having near perfect light harvesting and charge separation properties, which makes them potentially valuable as photoactive components in devices for conversion of sunlight into electricity or solar fuels.[1-3] A variety of biophotocathodes has been reported which typically follow the same general design. A natural or artificial electron mediator is used to shuttle electrons between the electrodes and the donor side (D/D^+) of the photosynthetic protein. An electron acceptor then recovers the electron at the acceptor site (A/A^+) of the protein which can subsequently act either (i) as a charge

Center for Electrochemical Sciences (CES), Faculty of Chemistry and Biochemistry, Ruhr University Bochum, Universitätsstr. 150, D-44780 Bochum, Germany. E-mail: nicolas.plumere@rub.de

† Electronic supplementary information (ESI) available. See DOI: 10.1039/c8fd00168e

carrier that diffuses to the collector electrode to close the electrical circuit and generate electricity or (ii) as a redox catalyst that generates chemical energy in the form of a solar fuel in a follow up coupled reaction (Fig. 1).

In both cases, the overall energy conversion efficiency is closely related to the rates of electron transport defining the photocurrent and to the redox potential of the various components defining the light-induced potential difference within the electrochemical half-cell and thus the potential energy gain. From a practical perspective, it is advantageous to immobilize the electron mediator and the photosynthetic proteins within thin redox films on the electrode surface to allow for efficient electrical wiring and for high catalyst loading and thus obtain high photocurrent generation.[4,5]

However, besides the photocatalytic process, possible competitive pathways may have a detrimental impact on the performance of such biophotocathodes. One of the general challenges in photoelectrochemical systems is related to charge recombination processes.[6] Light-induced charge separation at the photosystem produces a high energy electron that is ideally transferred to the charge carriers or to the redox catalyst with minimal energy loss. However, the large driving force imposed by the light-induced potential difference favors recombination of these reduced electron acceptors with oxidized components of the redox matrix or with the electrode surface (Fig. 1, red pathway).[7,8] These short-circuiting processes lower the photocurrents and hence the overall power output or solar fuel generation of the devices. Therefore, in-depth understanding of the processes involved in photocurrent generation, including such short circuit pathways, is an essential pre-requisite for the rational design and optimization of biophotoelectrochemical systems.

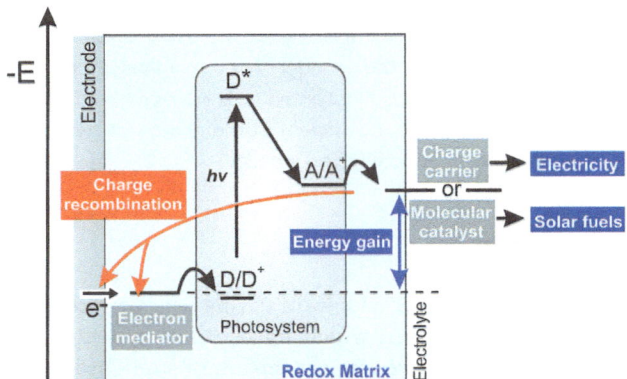

Fig. 1 Energy level diagram and schematic illustration of reactions in a biophotoelectrode based on photosynthetic proteins. The light-induced charge separation at the photo-synthetic protein triggers the electron transfer chain. The electron mediator refills the hole at the donor site (D/D+) of the photosystem and a charge carrier or a molecular catalyst recovers the electron from the acceptor site (A/A+). The electron mediator shuttles the charge from the electrode while the charge carrier can be used to generate electricity or the molecular redox catalyst catalyzes reactions for solar fuel generation. The energy gained from light leads to a potential difference between the electron mediator and the charge carrier defines the driving force for charge recombination of the charge carrier at the mediator or at the electrode.

The short-circuiting processes are often invisible to electroanalytical methods since their contribution do not provide any net photocurrent. Therefore, it is paramount to establish theoretical models for biophotoelectrochemical systems which consider both the processes generating the photocurrent and the processes competing with photocurrent generation. The ability to simulate and deconvolute the various contributions, including photocathodic processes and recombination processes, would enable the pinpointing of bottlenecks in the power generation process. Ideally, such a model would not only include simulations for the entire observed signal, but would also contain the development of dimensionless groups which are useful for summarizing the rates of the major processes in the system, and in particular, predict how a given parameter may impact photocurrents and thus how it could be modulated to achieve energy conversion enhancement.

Several models have been previously developed for biophotoelectrochemical systems, which considered photosynthetic proteins[9-12] or whole photosynthetic cells immobilized on electrodes or in solution.[13,14] In these previous reports, electronic communication between the photosystems and the electrode were modelled based on freely diffusing electron mediators. Here, we establish a model for biophotoelectrodes with both the photosystems and the electron mediators confined in redox films on the electrode surface based on a reaction scheme that is generally applicable and relevant for multiple experimental cases.[1,4,15,16] In particular, we consider both an outer-sphere electron transfer between the photosystem and the electron acceptor (which is typically relevant for photosystem 1 based biophotocathodes) as well as photoenzymatic reactions (which are typically relevant for biophotocathodes based on purple bacterial reaction centers). Moreover, we include the possibility for either electron transfer to a charge carrier that subsequently diffuses to the bulk of the solution or for electron transfer to a redox catalyst followed by subsequent catalytic reduction of a final electron acceptor generating solar fuels. The model is built upon previous models for bioelectrochemical systems[17,18] in which we integrate the effect of light induction of electron transfer and the associated charge recombination processes to predict the time dependent photocurrent generation and the associated concentration profile. Dimensionless groups are developed for understanding the limiting processes. We highlight the usefulness of this modeling tool with an example of simulation results predicting the effect of the diffusion coefficient of the charge carrier on photocurrent generation with different charge recombination kinetics.

2   System schematic and reactions

The process generating the photocurrent (Fig. 2, in black) includes the redox mediator $M_{red}|M_{ox}$ and the photosystem $P_{red}|P_{ox}$ which are immobilized in the redox film, as well as the electron acceptors Y_{ox} and Z_{ox} and their respective reduced forms Y_{red} and Z_{red} which are freely diffusing within the film and in the bulk of the solution. The charge hopping between the redox mediators is assumed to follow the rules of diffusion and is described by an apparent diffusion coefficient D_M. The partition of both substrate–product pairs $Y_{red}|Y_{ox}$ and $Z_{red}|Z_{ox}$ is neglected and their diffusion coefficients (D_Y and D_Z, respectively) remain unchanged whether they are in or out of the redox matrix.

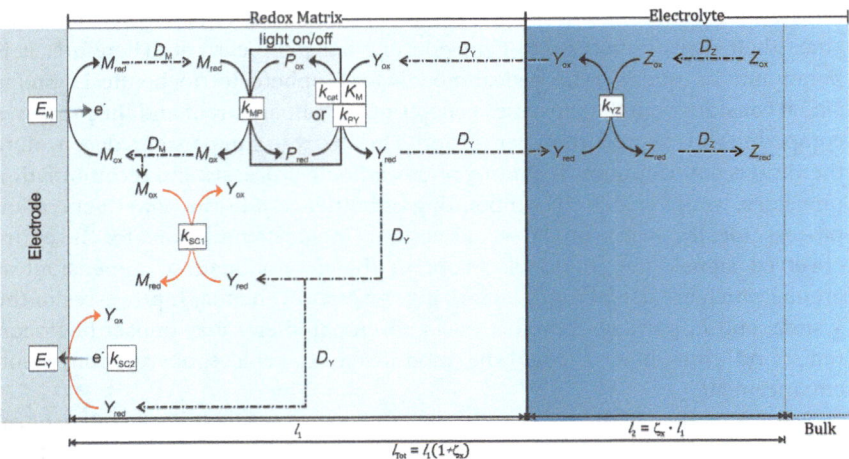

Fig. 2 Schematic illustration of electron transfer pathways for a biophotocathode based on photosynthetic proteins immobilized within a redox film containing electron mediators for shuttling electrons between the protein and the electrode. The process contributing to photocurrent generation is given in black while the short-circuiting processes are given in red.

Furthermore, we assume a steady state between the oxidized and reduced form of the photosystems while catalysis is taking place. The model considers only one electrochemical half-cell in which a constant potential is applied at the electrode. The heterogeneous electron transfer kinetics of $M_{red}|M_{ox}$ are modeled according to Butler–Volmer kinetics. M_{red} and P_{ox} react in a bimolecular reaction with a rate constant k_{MP} in single one-electron steps. This reaction is followed by a light-induced reaction between the reduced form of the photosystem with Y_{ox}, which is either modeled according to Michaelis–Menten kinetics (k_{cat} and K_M) or by a simple bimolecular reaction (k_{PY}). This reaction leads to Y_{red} which may serve as a charge carrier that diffuses in the bulk of the solution or as a redox catalyst that reduces another electron acceptor Z_{ox} in a subsequent reaction cascade leading to Z_{red}. In the latter case Y_{red} reacts in a bimolecular reaction with Z_{ox} with a kinetic constant k_{YZ}. Because both Y_{red} and Z_{ox} are free to diffuse in and out of the film this reaction can take place either in the redox matrix or in the surrounding electrolyte solution. The produced solar fuel Z_{red} can then diffuse into the bulk of the system.

In order to consider the possibility for charge recombination we consider two short-circuiting reactions involving Y_{red} (Fig. 2, in red). The redox potential of $Y_{red}|Y_{ox}$ (E_Y^0) is more negative than the redox potential of $M_{red}|M_{ox}$ (E_M^0). Therefore, the potential difference favors the reduction of M_{ox} by Y_{red}. We model this first short-circuiting pathway (SC1) as a bimolecular reaction with a kinetic constant k_{SC1}. The second possible short-circuiting pathway (SC2) takes place at the electrode and leads to the reoxidation of Y_{red}. The heterogeneous (multi-) electron transfer is also modeled according to Butler–Volmer kinetics with a heterogeneous rate constant k_{SC2}^0 and an apparent electron transfer coefficient α_Y.

The reaction stoichiometry between Y_{ox} and P_{red} is fixed as one-to-one, but the number of electrons transferred in the reaction (ν_Y) is variable. Additionally, the number of electrons that are transferred from P_{ox} to M_{red} (ν_M) is flexible depending on the properties of the mediator. The number of electrons transferred between Y_{red} and M_{ox} in the SC1 process depends on the ratio of ν_Y and ν_M.

3   Modeling equations

3.1   Space and time domains

The total space (l_{tot}) is divided into two space domains: (1) the redox film, and (2) the stationary surrounding solution. As shown in eqn (1), the length of the film domain is l_1, and the length of the second domain has a length which is set as a multiple of the film thickness (ζ_x).

$$l_{tot} = l_1 + l_1\zeta_x = l_1(1 + \zeta_x) \tag{1}$$

As shown in eqn (2), the total time (t_{tot}) is divided into three time domains: (1) when the light is initially off (t_{eq}), (2) when the film is under photoillumination (t_{exp}), and (3) when the light is off once again (t_{rec}).

$$t_{tot} = t_{eq} + t_{exp} + t_{rec} = \tau_{tot}t_{exp} \tag{2}$$

By means of the factor τ_{tot}, the total time is expressed as a multiple of the exposure time, which is the primary time related variable of interest. Information regarding the calculation of τ_{tot} is given in the ESI.†

3.2   Main equations

The main modeling equations were derived from material balances on the reduced form of the mediator (M_{red}), on the oxidized form of the first electron acceptor (Y_{ox}), and on the oxidized form of the final electron acceptor (Z_{ox}). Equations for these species in the film domain are shown in eqn (3)–(5) respectively. They constitute a system of coupled partial differential equations (PDEs) with time, space, and concentration as the primary variables. The equations account for transient behavior by the presence of first order time derivatives. They also account for 1-dimensional spatial variation related to diffusion by the presence of second order spatial derivatives. These are considered within the context of several interrelated chemical and electron transfer reactions. The equations were scaled with respect to their maximum possible values, so that the range of the major dimensionless variables (time, space, and concentrations) is between 0 and 1.

Reaction stoichiometry is explicitly included in the model, and allows for flexibility with regards to the number of electrons that can be transferred to the electrode by the mediator, and by the first electron acceptor ($Y_{red}|Y_{ox}$), which appear as z_M and as z_Y in the modeling equations, respectively. The stoichiometry between the first electron acceptor and the second electron acceptor ($Z_{red}|Z_{ox}$) is fixed as one-to-one.

These modeling equations shown for the film domain represent the most complex form of these equations, and the corresponding equations for simpler reaction schemes, or for the main equations in the solution domain can be deduced by setting the relevant kinetic terms or concentrations to zero. For example, the main equation for Y_{ox} in the solution domain can be deduced by setting κ_{CAT}^Y and κ_{SC1}^Y in eqn (4) to zero.

$$\begin{bmatrix}\omega_M\\\tau_{tot}\end{bmatrix}\frac{\partial M_{red}}{\partial t}=\begin{bmatrix}\frac{1}{(1+\zeta_x)^2}\end{bmatrix}\frac{\partial^2 M_{red}}{\partial x^2}-\frac{(\nu_M)\kappa_{cat}^M M_{red} Y_{ox}}{\left(\frac{1}{\nu_M}\right)\mu M_{red}+\left(\frac{1}{\nu_M}\right)M_{red}Y_{ox}+\theta_{MM}Y_{ox}}$$

$$-\frac{(\nu_M)\kappa_{MP}^M M_{red}Y_{ox}}{\left(\frac{1}{\nu_M}\right)M_{red}+\theta_{MP}Y_{ox}}+\left(\frac{\nu_M}{\nu_Y}\right)\kappa_{SC1}^M(1-M_{red})(1-Y_{ox}) \tag{3}$$

$$\begin{bmatrix}\omega_Y\\\tau_{tot}\end{bmatrix}\frac{\partial Y_{ox}}{\partial t}=\begin{bmatrix}\frac{1}{(1+\zeta_x)^2}\end{bmatrix}\frac{\partial^2 Y_{ox}}{\partial x^2}-\frac{\left(\frac{1}{\nu_M}\right)\kappa_{cat}^Y M_{red} Y_{ox}}{\left(\frac{1}{\nu_M}\right)\mu M_{red}+\left(\frac{1}{\nu_M}\right)M_{red}Y_{ox}+\theta_{MM}Y_{ox}}$$

$$-\frac{\left(\frac{1}{\nu_M}\right)\kappa_{MP}^Y M_{red}Y_{ox}}{\left(\frac{1}{\nu_M}\right)(\theta_{MP})^{-1}M_{red}+Y_{ox}}+\left(\frac{\nu_Y}{\nu_M}\right)\kappa_{SC1}^Y(1-M_{red})(1-Y_{ox})$$

$$+\kappa_{YZ}^Y Z_{ox}(1-Y_{ox}) \tag{4}$$

$$\begin{bmatrix}\omega_Z\\\tau_{tot}\end{bmatrix}\frac{\partial Z_{ox}}{\partial t}=\begin{bmatrix}\frac{1}{(1+\zeta_x)^2}\end{bmatrix}\frac{\partial^2 Z_{ox}}{\partial x^2}-\kappa_{YZ}^Y Z_{ox}(1-Y_{ox}) \tag{5}$$

3.3 Contributions to current

The total current is the sum of two contributions, as shown in eqn (6). The photocatalytic contribution to the current (ι_{cat}) is calculated based on the concentration gradient of the reduced form of the mediator at the electrode surface as shown in eqn (7). The SC2 contribution to the current (ι_{SC2}) is calculated based on the concentration gradient of the oxidized form of the first electron acceptor Y_{ox} at the electrode surface, as shown in eqn (8).

$$\iota_{tot}=\iota_{cat}+\iota_{SC2} \tag{6}$$

$$\iota_{cat}=\frac{1}{(1+\zeta_x)}\left(\frac{\partial M_{red}}{\partial x}\right)_{x=0}=\sigma_b^M(M_{red})_{x=0}-\sigma_f^M\left[1-(M_{red})_{x=0}\right] \tag{7}$$

$$\iota_{SC2}=\frac{1}{(1+\zeta_x)}\left(\frac{\partial Y_{ox}}{\partial x}\right)_{x=0}=\sigma_b^Y\left[1-(Y_{ox})_{x=0}\right]-\sigma_f^Y(Y_{ox})_{x=0} \tag{8}$$

In contrast to the SC2 recombination process, current loss due to the SC1 recombination process is manifested indirectly as a reduction in the main current. In order to isolate the effect of SC1 recombination, each simulation is run twice: one time with the SC1 kinetic constant set to zero, and a second time with the SC1 kinetic constant set to its nominal value, where the corresponding catalytic currents are subtracted.

3.4 Boundary conditions

The boundary conditions for M_{red} and Y_{ox} at the electrode surface are shown in eqn (7) and (8). Z_{ox} cannot undergo electron transfer at the electrode surface; therefore, it has a "no-flux" boundary condition.

The boundary condition at the film/electrolyte interface is different depending on the species involved. Since the mediator is immobilized in the film, the intermediate boundary condition for M_{red} is "no-flux", which requires that the concentration gradient be equal to zero at the film/electrolyte interface when approaching the film boundary from domain I on the left side (l_1^-), as shown in eqn (9). The boundary condition for M_{red} in the bulk is that its concentration is zero, the same value that it has in the surrounding solution domain.

$$\left(\frac{\partial M_{red}}{\partial x}\right)^{I}_{x=l_1^-} = 0 \tag{9}$$

Since $Y_{red}|Y_{ox}$ is freely diffusing throughout the system, in particular, between the film and the surrounding solution, and without mass transfer resistance, the intermediate boundary condition for Y_{ox} is "perfect-flux", where the concentration gradients just before and just after the film/solution interface are equal, as shown in eqn (10).

$$\left(\frac{\partial Y_{ox}}{\partial x}\right)^{I}_{x=l_1^-} = \left(\frac{\partial Y_{ox}}{\partial x}\right)^{II}_{x=l_1^+} \tag{10}$$

The bulk semi-infinite boundary condition holds for Y_{ox} and Z_{ox}, therefore, their concentrations at the bulk must remain undisturbed at their initial values, and with a slope of zero. At the end of each simulation, the concentration profiles at the end time are inspected, and if necessary, the simulation is repeated with a greater distance.

4   Dimensionless groups

After scaling of the major variables with respect to their maximum possible values, the dimensional parameters were formed into dimensionless groups. Two steps were deliberately taken in an effort to simplify and to unify the treatment of the dimensionless groups: (1) classification of the groups into "types", where the respective rate ratios are as transparent as possible, and (2) the use of double script notation, where it is easily seen which rates are being compared.

Inspection of the scaled main equations and the scaled electrode surface boundary conditions shows three main kinds of dimensionless groups: "κ" type groups, "ω" type groups, and "σ" type groups. As was done in a reference model,[19] various κ type groups are used related to the various reaction–diffusion processes, which occur within the volume of the film or in the surrounding solution. In the model, single script notation was used to denote the particular chemical process of interest. The use of ω groups was inspired from another reference model,[20] which focuses on transient electron transfer within redox-active films. Finally, σ type groups denote electron transfer processes at an electrode surface. One example of each of these group types is described in detail in the following sections. A summary of the dimensionless groups in the model is included in the ESI.†

4.1 κ_{SC1}^Y as an example of a κ type group

The κ_{SC1}^Y group is shown in its most simplified form in eqn (11). The similarity of this expression to λ/σ^2 in a reference model,[17] and to κ in another reference model[18] were helpful for interpreting this new dimensionless group as the ratio of the rate of SC1 to the rate of diffusion of Y.

$$\kappa_{SC1}^Y = \frac{l_1{}^2 k_{SC1} M_{tot}}{D_Y} \tag{11}$$

This ratio can be more clearly demonstrated after multiplication of the numerator and denominator by Y_{tot} and rearrangement; the result is shown in eqn (12). In this equation, the units of the numerator and the denominator are mol cm^{-3} s^{-1}; the numerator therefore represents the maximum possible rate of SC1 (when both reaction species are at their maximum possible concentrations), and the denominator represents the maximum molar diffusion rate of SC1 in a basis area of $l_1{}^2$.

$$\kappa_{SC1}^Y = \frac{k_{SC1} Y_{tot} M_{tot}}{\left[(D_Y\,Y_{Tot})/l_1{}^2\right]} \propto \frac{SC1\ \ reaction\ \ rate}{Y\ \ diffusion\ \ rate} \tag{12}$$

The inverse square root of κ_{SC1}^Y, which is shown in eqn (13), is also of interest because it allows for one to think of the same process in terms of the SC1 reaction layer.

$$\left(\kappa_{SC1}^Y\right)^{-1/2} = \frac{\sqrt{D_Y\,(k_{SC1}\,M_{tot})^{-1}}}{l_1} \propto \frac{SC1\ \ reaction\ \ layer}{film\ \ thickness} \tag{13}$$

The reaction layer concept, which was introduced and emphasized in a foundational reference model,[17] and was also used in a later reference model[18] was useful for the interpretation of this dimensionless group as the fractional distance in the film that a formed Y_{red} molecule will be able to travel within the film before undergoing SC1, ignoring all other processes in the system.

4.2 ω_Y as an example of a ω type group

The ω_Y group is shown in its simplest form in eqn (14). It can be rearranged as a ratio of two time scales, as shown in eqn (15). In this form, ω_Y can be interpreted as the ratio of the minimum time needed for a film of a basis area of $l_1{}^2$ to be fully saturated with Y by diffusion *versus* the exposure time. As such, ω_Y is an indicator of how much of the film is accessible to Y by diffusion at the given experimental time scale.

$$\omega_Y = \frac{l_1{}^2}{D_Y t_{exp}} \tag{14}$$

$$\omega_Y = \frac{\left(l_1{}^2/D_Y\right)}{t_{exp}} \propto \frac{minimum\ \ diffusion\ \ time}{experimental\ \ time\ \ scale} \tag{15}$$

The inverse square root of ω_Y, which is shown in eqn (16), is also of interest because it allows for one to think of the same process in terms of the Y diffusion layer.

$$(\omega_Y)^{-1/2} = \frac{\sqrt{D_Y(t_{exp})}}{l_1} \propto \frac{\text{Y diffusion layer}}{\text{film thickness}} \tag{16}$$

Comparison to the w dimensionless group in a reference model[20] was helpful for the interpretation of this new dimensionless group as the fractional distance in the film that a Y_{ox} molecule will be able to travel within the film in the time given, ignoring all other processes in the system.

4.3 σ_{SC2}^Y as an example of a σ type group

The σ_{SC2}^Y group arises from the non-dimensionalization of the Butler–Volmer equation, and therefore is composed of dimensionless groups related to k_Y^0 (υ_Y, shown in eqn (17)), the overpotential (ε_Y, shown in eqn (18)), and to α_Y (the apparent electron transfer coefficient).

υ_Y is the only factor in σ_{SC2}^Y that is outside of an exponent in eqn (19), therefore the units of this term will determine the overall units of σ_{SC2}^Y. Multiplication of the numerator and denominator of υ_Y by Y_{tot} and rearrangement results in units of mol cm^{-2} s^{-1} in the numerator and the denominator. This allows for the interpretation of this dimensionless group as the rate of heterogeneous electron transfer of Y at the electrode surface *versus* the rate of diffusion of a surface plane of Y, without considerations related to overpotential and the apparent electron transfer coefficient.

$$\upsilon_Y = \frac{l_1 k_Y^0}{D_Y} = \frac{k_Y^0 Y_{tot}}{(D_Y Y_{tot}/l_1)} \tag{17}$$

$$\varepsilon_Y = \frac{(E_{hold} - E_Y^0)}{(RT/z_Y F)} \tag{18}$$

In keeping with the notation from a reference textbook,[21] a reduction at the electrode surface is considered as a "forward" reaction, and oxidation is conversely regarded as a "backwards" reaction. Since SC2 occurs through oxidation of Y_{red}, the "backwards" reaction is SC2; therefore σ_b^Y can be regarded as σ_{SC2}^Y for this case as shown in eqn (19), and can be interpreted as the rate of SC2 *versus* the diffusion rate of Y, which includes the effects of the applied overpotential and of the apparent electron transfer coefficient on the electron transfer rate.

$$\sigma_b^Y = \sigma_{SC2}^Y = \upsilon_Y \exp[\varepsilon_Y(1 - \alpha_Y)] \propto \frac{\text{SC2 reaction rate}}{\text{Y diffusion rate}} \tag{19}$$

4.4 Secondary dimensionless groups: μ and θ_{MM}

A secondary group, μ, is shown in eqn (20). As the ratio of the Michaelis–Menten constant and the maximum substrate concentration, it denotes the fractional degree of enzyme saturation.

$$\mu = \frac{K_M}{Y_{tot}} \tag{20}$$

θ_{MM} is a secondary group because it can be represented as a ratio of two individual κ groups. Starting from its simplified form, shown in eqn (21), multiplication of the numerator and the denominator by P_{tot} demonstrates how θ_{MM} can be interpreted as a ratio of the catalysis and electron transfer rates, as shown in eqn (22).

$$\theta_{MM} = \frac{k_{cat}}{k_{MP} \, M_{Tot}} \tag{21}$$

$$\theta_{MM} = \frac{k_{cat} P_{tot}}{k_{MP} M_{tot} P_{tot}} \propto \frac{\text{catalytic reaction rate}}{\text{electron transfer rate}} \tag{22}$$

Since κ groups are ratios of reaction and diffusion rates, κ_{cat}^Y and κ_{MP}^Y can be defined by eqn (23) and (24).

$$\kappa_{cat}^Y = \frac{k_{cat} P_{tot}}{\left(D_Y Y_{tot}/l_1^2 \right)} \propto \frac{\text{catalytic reaction rate}}{\text{Y diffusion rate}} \tag{23}$$

$$\kappa_{MP}^Y = \frac{k_{MP} M_{tot} P_{tot}}{\left(D_Y Y_{tot}/l_1^2 \right)} \propto \frac{\text{electron transfer rate}}{\text{Y diffusion rate}} \tag{24}$$

Similarly, κ_{cat}^{MP} denotes the ratio of the catalytic reaction and electron transfer rates as shown in eqn (25).

$$\kappa_{cat}^{MP} \propto \frac{\text{catalytic reaction rate}}{\text{electron transfer rate}} \tag{25}$$

κ_{cat}^{MP} can then be expressed as the ratio of κ_{cat}^Y and κ_{MP}^Y, which is equal to θ_{MM}, as shown in eqn (26).

$$\kappa_{cat}^{MP} = \left(\frac{\kappa_{cat}^Y}{\kappa_{MP}^Y} \right) = \frac{k_{cat} P_{tot}}{k_{MP} M_{tot} P_{tot}} = \theta_{MM} \tag{26}$$

4.5   The SC1/SC2 ratio by the use of groups with double notation

Double script notation can be useful for deriving expressions for rate comparisons of interest. For example, a dimensionless expression which is indicative of the relative rates of SC1 in the film and SC2 at the electrode surface can be derived, for the case where the oxidation of Y_{red} is strongly favored. This can be calculated from the ratio of κ_{SC1}^Y (eqn (12)) and σ_{SC2}^Y (eqn (19)), in which the Y diffusion rate cancels out; the final result is shown in eqn (27).

$$\frac{\kappa_{SC1}^Y}{\sigma_{SC2}^Y} = \frac{l_1 k_{SC1} M_{tot}}{k_Y^0 \exp[\varepsilon_Y(1 - \alpha_Y)]} \propto \frac{\text{SC1 reaction rate}}{\text{SC2 reaction rate}} \tag{27}$$

The appearance of l_1 in the numerator of eqn (27) implies that for an extremely high value of l_1 (extremely thick films), recombination is much more likely to be by SC1 than by SC2; this is physically reasonable since for very thick films, most of the Y_{red} would be generated further away from the electrode. However, the likelihood of SC1 *versus* SC2 also depends on the relative kinetic parameters, as well as on the apparent electron transfer coefficient. For example, extremely high SC2 kinetics together with extremely low SC1 kinetics can therefore result in a higher likelihood for SC2 over SC1, even in a very thick film.

The derivation of expressions such as eqn (27) are useful because they are order of magnitude estimates of the individual rate ratios of interest. However, such expressions do not negate the need for full simulations, which include simultaneous considerations of all competing rates in the system, and which therefore generate exact results regarding the behavior of the system under a given set of conditions.

5   Numerical solution of the modeling equations

Because of the "no-flux" film/solution boundary condition for the reduced form of the mediator (eqn (9)), it was not possible to use built-in Matlab PDE solver functions, which ignore no-flux internal boundary conditions. Therefore, the space variable of the PDE system was discretized, resulting in a series of simultaneous first-order ordinary differential equations (ODEs) in time according to the method of lines.[22] The discretization of the space variable was performed using a vertex-centered finite volume scheme with variable x-spacing.[22-24] Custom functions for single and double exponentially expanding grids were constructed, according to a method suitable for electrochemical simulations.[25] The finite volume method was chosen due to its strengths related to spatial discontinuities (since it is an integral-based method), and to its strengths related to adherence to the conservation equations (since the governing equation is solved in "conservative" form). The finite volume method schematic and the finite difference equations are included in the ESI.†

The system of ODEs was then solved using a Matlab built-in ordinary differential equation solver (ode15s), which is designed specifically for systems in which concentration profiles increase steeply over short distances (*i.e.* numerically "stiff"[22,26]). The time discontinuities within the system (*i.e.* light on and off) were managed within ode15s by time-dependent coefficients which were changed by steep linear on/off ramps. Numerical solution of the system allowed for the calculation of a "deconvoluted" current–time curve, which shows all contributions (direct and indirect) to the observed total current, for calculated concentration profiles at specified times, and for the generation of concentration profile animations at all time points of the simulation.

After implementation of the simulation was completed, a series of calculations were performed in order to verify the correctness of the model. As much as possible, this verification was performed in a "piecewise" way, in which the model was simplified for direct and quantitative comparison to the results from relevant reference models. For example, for verification of the correct implementation of the heterogeneous electron transfer at the electrode by SC2, the kinetics for all chemical processes was set to zero, and the resulting current–time curve was compared to expected results from an analytical expression for quasi- and irreversible electron transfer in a potential step experiment.[21] For verification of the SC1 process, reduction of the model was not possible. Therefore, in this case, a general material balance which simultaneously considers the initial and final concentration profiles as well as the corresponding current–time curve was used. More details and examples related to the piecewise verification are given in the ESI.†

6 Applications of the model

Such a model is useful for its predictive power, especially for situations or conditions where the outcome is unclear with qualitative estimation. Due to the large number of parameters, as well as the uniqueness of each individual experimental system, it is useful if simulations can be run to explore how parameters affect the photocurrent generation. In order to facilitate this, a standalone app was developed (see ESI,† Standalone App for Simulations).

As a case study we performed simulations to predict the effect of the diffusion coefficient of the charge carrier (D_Y) on photocurrent generation as a function of the kinetic constant for the recombination (k_{SC1}) of the reduced charge carrier Y_{red} with the electron mediator M_{ox}. The schematic illustration of the reactions is shown in Fig. 3A. In this particular example, we model the reaction between the photosynthetic protein and the charge carrier by means of Michaelis–Menten kinetics. The recombination of Y_{red} at the electrode (SC2) is set as zero to unambiguously reveal the effect of D_Y and k_{SC1} on photocurrent generation. As highlighted in Fig. 3A, D_Y is involved in two competing processes. In the photo-catalytic portion of the scheme, D_Y defines the flux of Y_{ox} to the photocatalytic reaction layer and thus an increase in D_Y would be expected to be beneficial to the photocatalytic process. However, D_Y also defines the flux of Y_{red} to the reaction layer for recombination of Y_{red} with M_{ox}, so an increase in D_Y leads to a faster recombination rate. Since the rate of the photocatalytic process and the rate of recombination have opposite effects on photocurrent generation, the impact of D_Y on the system cannot be predicted based on a qualitative comparison of the two processes. Instead, simulations are required in order to quantitatively predict the effect of D_Y on the photocurrent generation.

The simulations were performed for a set of parameters (see the ESI†) that ensure that mass transport of Y_{ox} is limiting the photocatalytic process. 77 current–time curves were generated for 7 different values of D_Y and 11 different values of k_{SC1} while all other parameters where kept constant. Examples of deconvoluted current–time curves for increasing k_{SC1} values are shown in Fig. 3B for $D_Y = 6 \times 10^{-6}$ cm^2 s^{-1}. For the lowest value of k_{SC1} (below 10^2 M^{-1} s^{-1}), a steady state photocurrent is obtained which is mostly overlaying with the pre-dicted current for the one corresponding to k_{SC1} set to 10^1 M^{-1} s^{-1}. As k_{SC1} is increased, the photocurrent–time curves increasingly deviate from the pure photocatalytic curve. At transition values for k_{SC1} (for instance 10^5 M^{-1} s^{-1} and 10^7 M^{-1} s^{-1}) the photocathodic current is lower and decreases over time during illumination while an anodic current appears in the following dark phase. These features are characteristic for recombination processes. At the highest k_{SC1} values (above 10^9 M^{-1} s^{-1}), both the photocurrent and the dark current completely vanish. The same qualitative trend of photocurrent decrease with increasing k_{SC1} is observed for all investigated values of D_Y.

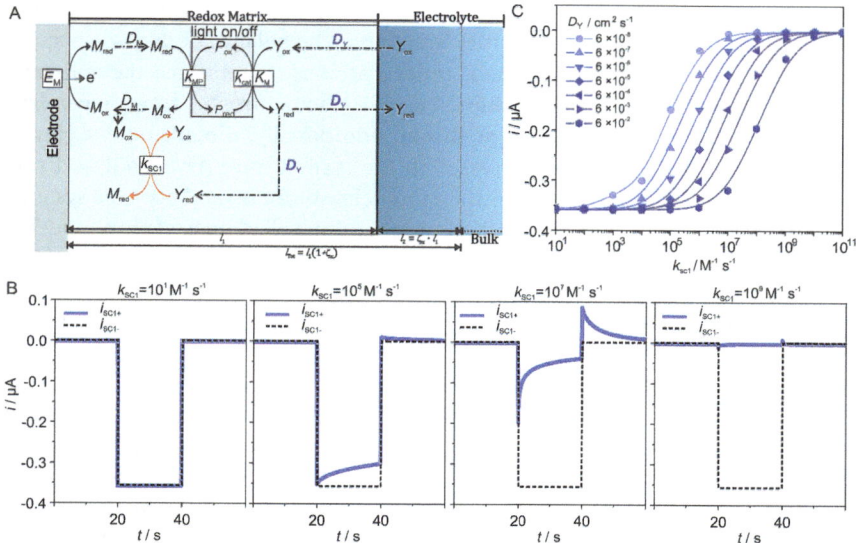

Fig. 3 (A) Schematic illustration of the reactions for the biophotocathode based on a photoenzyme generating a charge carrier that diffuses into the bulk of the electrolyte. The recombination pathway is limited to the reaction between the reduced charge carrier and the oxidized electron mediator. (B) Photocurrent predicted for the photocatalytic process alone (black dashed line) and for the recombination pathway associated to the photocatalytic process (blue solid line) for increasing values of k_{SC1} at $D_Y = 6 \times 10^{-6}$ cm^2 s^{-1}. (C) Photocurrent at 40 s (at the end of the illumination period) plot $vs.$ log(k_{SC1}) for increasing values of D_Y.

Quantitative analysis of the impact of the diffusion coefficient is performed by plotting the photocurrent values (before the dark phase) against the k_{SC1} values for each value of D_Y (Fig. 3C). The i $vs.$ k_{SC1} plots confirm the current cancelling effect of the charge recombination process irrespective of the value of D_Y. However, the most important feature is that the transition in photocurrent loss in the i $vs.$ k_{SC1} curves is shifted to higher k_{SC1} values as D_Y is increased. For a 10-fold increase in k_{SC1}, a given value for the photocurrent can be maintained if D_Y is increased by a factor of 100. These results demonstrate the ability to accommodate for increasing k_{SC1} by increasing D_Y according to the relationship shown in eqn (28).

$$ i \propto \frac{(D_Y)^2}{k_{SC1}} \tag{28} $$

7 Conclusions

A model was developed for biophotoelectrodes based on photosynthetic proteins, in which the photocurrent can be simulated for the entire experiment, together with a deconvolution of the individual contributions to the total observed current, either by the catalytic process, or its loss through recombination processes within the redox film or at the electrode surface. Because of its strengths related to discontinuities in space (*i.e.* at the film/solution interface) and the balance of flux, the finite volume method within the context of the method of lines was used to solve the system of partial differential equations. The balance of flux is especially important for this problem because under stationary conditions, the substrate can diffuse in and out of the film. Besides photocurrents, time dependent concentration profiles were also predicted. Additionally, dimensionless groups which summarize the major processes in the system were developed and presented. The model is flexible and is therefore relevant for several possible systems with respect to the reaction between catalysts and the electron acceptor (modeled either as a bimolecular reaction, which is relevant for photosystem 1, or as a Michaelis–Menten enzymatic process, which is relevant for purple bacterial reaction centers). Similarly, the model can be adjusted either for direct generation of a charge carrier relevant for biophotovoltaic cells, or for an additional redox catalyst for the generation of a solar fuel as a final product.

The simulations were compiled into a stand-alone app, which can be used to investigate the effects of different parameters on photocurrent generation. The example given here, in which the effect of increasing charge carrier diffusion coefficient on the ability of the system to withstand increasing mediator–charge carrier recombination kinetics was investigated, shows the ability of the simulation to predict the performance of the system for complex situations where it is not possible by means of qualitative reasoning. The same simulation approach can be carried out for predicting the effects of any other parameters described in the model. Therefore, the model developed in this work will be helpful for the rational design and further optimization of biophotoelectrodes for maximal energy conversion efficiency.

Author contributions

All authors contributed in conceiving the research. D. B. supervised all modeling activities and developed the model on "charge carrier" generation. T. H. developed the model on "solar fuel" generation and the stand-alone app. H. Z. performed the electrochemical simulations. All authors contributed to writing the manuscript.

Conflicts of interest

There are no conflicts to declare.

Acknowledgements

This work was supported by the European Research Council (ERC Starting Grant 715900), by the DFG within the framework of the Cluster of Excellence RESOLV (EXC 1069), by the DFG/ANR within the projects SHIELD PL746/2-1 and No. ANR-15-CE05-0020. H. Z. is grateful for the support by the China Scholarship Council (CSC) (No. 201306950049). The authors thank Dierk Gruhn for helping with the simulations.

Notes and references

1 V. M. Friebe and R. N. Frese, *Curr. Opin. Electrochem.*, 2017, 5, 126–134.
2 P. J. D. Janssen, M. D. Lambreva, N. Plumeré, C. Bartolucci, A. Antonacci, K. Buonasera, R. N. Frese, V. Scognamiglio and G. Rea, *Front. Chem.*, 2014, 2, 36.

3 K. Nguyen and B. D. Bruce, *Biochim. Biophys. Acta, Bioenerg.*, 2014, **1837**, 1553–1566.

4 T. Kothe, S. Pöller, F. Zhao, P. Fortgang, M. Rögner, W. Schuhmann and N. Plumeré, *Chem.–Eur. J.*, 2014, **20**, 11029–11034.

5 K. P. Sokol, D. Mersch, V. Hartmann, J. Z. Zhang, M. M. Nowaczyk, M. Rögner, A. Ruff, W. Schuhmann, N. Plumeré and E. Reisner, *Energy Environ. Sci.*, 2016, **9**, 3698–3709.

6 N. Plumeré and M. M. Nowaczyk, Biophotoelectrochemistry of Photosynthetic Proteins, in *Biophotoelectrochemistry: From Bioelectrochemistry to Biophotovoltaics. Advances in Biochemical Engineering/Biotechnology*, ed. L. Jeuken, Springer, Cham, 2016, 158, pp. 111–136.

7 V. Proux-Delrouyre, C. Demaille, W. Leibl, P. Sétif, H. Bottin and C. Bourdillon, *J. Am. Chem. Soc.*, 2003, **125**, 13686–13692.

8 A. F. Janzen and M. Seibert, *Nature*, 1980, **286**, 584–585.

9 M. T. Robinson, D. E. Cliffel and G. K. Jennings, *J. Phys. Chem. B*, 2018, **122**, 117–125.

10 P. N. Ciesielski, D. E. Cliffel and G. K. Jennings, *J. Phys. Chem. A*, 2011, **115**, 3326–3334.

11 R. Caterino, R. Csiki, A. Lyuleeva, J. Pfisterer, M. Wiesinger, S. D. Janssens, K. Haenen, A. Cattani-Scholz, M. Stutzmann and J. A. Garrido, *ACS Appl. Mater. Interfaces*, 2015, **7**, 8099–8107.

12 F. Milano, F. Ciriaco, M. Trotta, D. Chirizzi, V. de Leo, A. Agostiano, L. Valli, L. Giotta and M. R. Guascito, *Electrochim. Acta*, 2019, **293**, 105–115.

13 G. Longatte, M. Guille-Collignon and F. Lemaître, *ChemPhysChem*, 2017, **18**, 2643–2650.

14 G. Longatte, A. Sayegh, J. Delacotte, F. Rappaport, F. Wollman, M. Guille-Collignon and F. Lemaître, *Chem. Sci.*, 2018, **9**, 8271–8281.

15 O. Yehezkeli, R. Tel-Vered, J. Wasserman, A. Trifonov, D. Michaeli, R. Nechushtai and I. Willner, *Nat. Commun.*, 2012, **3**, 742.

16 K. R. Stieger, S. C. Feifel, H. Lokstein and F. Lisdat, *Phys. Chem. Chem. Phys.*, 2014, **16**, 15667–15674.

17 C. P. Andrieux, J. M. Dumas-Bouchiat and J. M. Savéant, *J. Electroanal. Chem. Interfacial Electrochem.*, 1982, **131**, 1–35.

18 P. N. Bartlett and K. Pratt, *J. Electroanal. Chem.*, 1995, **397**, 61–78.

19 V. Fourmond, S. Stapf, H. Li, D. Buesen, J. Birrell, O. Rüdiger, W. Lubitz, W. Schuhmann, N. Plumeré and C. Léger, *J. Am. Chem. Soc.*, 2015, **137**, 5494–5505.

20 K. Aoki, K. Tokuda and H. Matsuda, *J. Electroanal. Chem. Interfacial Electrochem.*, 1983, **146**, 417–424.

21 A. J. Bard and L. R. Faulkner, *Electrochemical Methods: Fundamentals and Applications*, Wiley, New York, 2nd edn, 2001.

22 M. E. Davis, *Numerical Methods and Modeling for Chemical Engineers*, Wiley, New York and Chichester, 1984.

23 V. Ruas, *Numerical Methods for Partial Differential Equations: An Introduction Finite Differences, Finite Elements and Finite Volumes*, John Wiley & Sons, Chichester, West Sussex, United Kingdom, 2016.

24 W. J. Lick, *Difference Equations From Differential Equations: Volume 41 of Lecture Notes in Engineering*, Springer Verlag, 1989.

25 M. Rudolph, D. P. Reddy and S. W. Feldberg, *Anal. Chem.*, 1994, **66**, 589A–600A.

26 L. F. Shampine and M. W. Reichelt, *SIAM J. Sci. Comput.*, 1997, **18**, 1–22.

Faraday Discussions

PAPER

Solar-driven carbon dioxide fixation using photosynthetic semiconductor bio-hybrids

Stefano Cestellos-Blanco,[a] Hao Zhang[b] and Peidong Yang (ID) *[abc]

Received 19th November 2018, Accepted 28th January 2019

DOI: 10.1039/c8fd00187a

Solar-driven conversion of carbon dioxide to value-added carbon products is an ambitious objective of ongoing research efforts. However, high overpotential, low selectivity and poor CO_2 mass transfer plague purely inorganic electrocatalysts. In this instance, we can consider a class of biological organisms that have evolved to achieve CO_2 fixation. We can harness and combine the streamlined CO_2 fixation pathways of these whole organisms with the exceptional ability of semiconducting nanomaterials to harvest solar energy. A novel nanomaterial–biological interface has been pioneered in which light-capturing cadmium sulfide nanoparticles reside within individual organisms essentially powering biological CO_2 fixation by solar energy. In order to further develop the photosensitized organism platform, more biocompatible photosensitizers and cytoprotective strategies are required as well as elucidation of charge transfer mechanisms. Here, we discuss the ability of gold nanoclusters to photosensitize a model acetogen effectively and biocompatibly. Additionally, we present innovative materials including two-dimensional metal organic framework sheets and alginate hydrogels to shield photosensitized cells. Finally, we delve into original work using transient absorption spectroscopy to inform on charge transfer mechanisms.

Fossil fuel-derived energy has powered industrial development over the past two centuries. However, the exploitation of these energy reserves presents an indisputable complication, as these reserves are finite. Semiconductor-based devices have advanced to aptly capture inexhaustible solar energy in the form of electricity.[1] Although solar cells have been tested industrially at scale, they suffer from a persistent lack of adequate energy storage solutions.[2] Nature evolved to solve this issue in photosynthesis as organisms adapted metabolic pathways to reduce CO_2 and store solar energy in ensuing chemical bonds.[3] The longevity and specificity of carbon fixation in biology is unrivalled by inorganic catalysts.[4] Reaction specificity is established by enzymatic conformations that stabilize CO_2 reduction intermediates while steric hindrances assist in specific product

[a]Department of Materials Science and Engineering, University of California, Berkeley, CA 94720, USA. E-mail: p_yang@berkeley.edu

[b]Department of Chemistry, University of California, Berkeley, CA 94720, USA

[c]Kavli Energy NanoSciences Institute at the University of California, Berkeley, CA 94720, USA

formation.[5] Furthermore, the genetic code ensures that this metabolism is exactly replicated. At the same time, however, engineered semiconducting materials outpace the light-harvesting ability of natural photosynthesis.[6] For these reasons, combining light-absorbing materials with CO_2-reducing organisms offers an appealing solution for solar-based CO_2 fixation.[4-6]

Progress has been made in coupling light harvesting devices with whole cell organisms to fix CO_2 into value-added chemicals.[7,8] This work relies on the ability of electrotrophic bacteria to take up reducing equivalents from inorganic sources, including nanostructured electrodes.[9] These organisms are commonly paired with a cathode in photoelectrochemical systems where they consume electrons to power the conversion of CO_2 into upgradeable carbon products. Exemplarily, acetogens receive reducing equivalents from solar harvesting electrodes and secrete acetate as a by-product of CO_2 fixation.[7] While these efforts confirm the applicability of an integrated cell-semiconductor system, further improvement may be limited by the extracellular nature of the interface between the cell membrane and the electrode. For instance, it is important to increase the CO_2-reducing current while at the same time maintaining a stable interface between the cells and the cathode. However, increasing the CO_2-reducing current is hampered by the fact that high currents create a local basic environment around the proton-consuming cathode.[7] The local basic environment irreparably damages the cell-semiconductor interface. Furthermore, extracellular electron uptake is not fully understood and therefore difficult to optimize. Although membrane-bound proteins take up electrons, a substantial part of these electrons may be lost due to the sluggish kinetics of charge transfer across the membrane.[10] Collectively these limitations indicate that a paradigm shift is needed to redesign the interface between light-absorbing semiconductors and microorganisms.

In a landmark study, Sakimoto et al. confirm the ability to enhance acetogen *Moorella thermoacetica* (*M. thermoacetica*) with cadmium sulfide (CdS) nanoparticles.[11] The nanoparticles, which lie on the surface of the cell membrane, capture solar energy and supply reducing equivalents directly into the bacterium. The consumed reducing equivalents jump-start the cell metabolism, which is reliant on a source of electrons, and enable CO_2-to-acetate conversion via the Wood–Ljungdahl pathway (WLP).[3] As this study marked the first report on a photosensitized microorganism for CO_2 reduction, further thorough elaboration and improvement is essential. Here, we discuss our studies that build on the first proof of concept. Briefly, we employ gold nanoclusters instead of cadmium sulfide as less cytotoxic and truly intracellular light harvesters. Secondly, we explore cytoprotective materials to shield bacteria from photooxidation due to harsh light and reactive oxygen species (ROS). Finally, we make inroads in elucidating the charge transfer mechanism at the biologic–inorganic interface using transient absorption spectroscopy.

Sakimoto and coworkers report on the first successful study to incorporate light-absorbing nanoparticles within acetogenic bacteria to enable CO_2-to-acetate conversion.[11] *M. thermoacetica* induces the precipitation of CdS nanoparticles with the addition of Cd^{2+} and a source of sulfur such as cysteine. Through an enzymatic process, the sulfur is reduced to sulfide and readily reacts with Cd^{2+} to form high quality CdS nanoparticles [Fig. 1(a)]. These nanoparticles are primarily anchored on the membrane of *M. thermoacetica* [Fig. 1(b)]. Upon illumination, the CdS nanoparticles deliver photogenerated electrons to metabolic pathways that

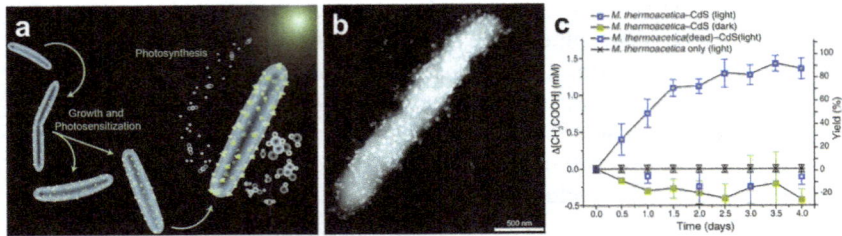

Fig. 1 (a) Model *M. thermoacetica*–CdS construct illustrating early growth stage, CdS nanoparticle ripening (yellow) and CO_2 reduction into acetate. (b) STEM image of *M. thermoacetica*–CdS hybrid. (c) Acetate obtained from *M. thermoacetica*–CdS during photosynthesis with deletional controls. Reproduced with permission.[11] Copyright 2016 The American Association for the Advancement of Science.

realize CO_2 reduction and secrete exogenous acetate [Fig. 1(c)]. Further cysteine is added to serve as a hole scavenger to improve charge separation at the nanoparticle interface. The discovery of photosensitized microorganisms for CO_2 fixation effectively sets up a novel line of investigation. It conceptualizes a new interface between whole microorganisms and semiconductors while confirming light-induced intracellular charge transfer in non-photosynthetic microorganisms. While photosensitized microorganisms enable the photoreduction of CO_2, they also present opportunities for investigating the interface between nanomaterials and whole-cell organisms. Electron transfer without added or produced molecular mediators, such as H_2, viologens or membrane-bound proteins, spurs questions on whether we can inject electrons at various points in biologic charge transport chains. For example, if light-activated nanomaterials jump-start CO_2 reducing pathways, could we potentially target more specific parts of the cell metabolism by material design? Additionally, could we use the light-controlled nanomaterials to set up studies to explore how organisms take up electrons? Applications beyond CO_2 fixation of photosensitized whole-cells merit more in depth discussion.

Furthermore, CdS nanoparticles, while effective light absorbers, pose a known environmental hazard and are acutely cytotoxic to bacteria as they induce oxidative stress.[12] Moreover, a more effective system could be realized with intracellular particles that permeate the whole cell instead of mostly just the membrane.

Gold nanoclusters (AuNCs) are sub-nanometric particles that consist of several gold atoms bound together into a network by ligands such as glutathione. At this level of confined atomicity, AuNCs harbor chromophore-like discrete energy states.[13] Therefore, the clusters act as viable light harvesting centers. Additionally, through synthetic manipulation, their core size may be tuned to optimize for passage through the cell membrane thus enabling intracellular localization of the clusters.[14] Interchangeable surface ligands also enable exquisite control over the biochemical properties of the cluster. $Au_{22}(SG)_{18}$ (SG is glutathione), a type of AuNC, offers water solubility and high fluorescence,[15] a marker for light harvesting, establishing it a prime candidate for microorganism photosensitization[16] [Fig. 2(a)].

Studies have previously confirmed that cells take up AuNCs with high efficiency and low cytotoxicity, as gold and glutathione are individually

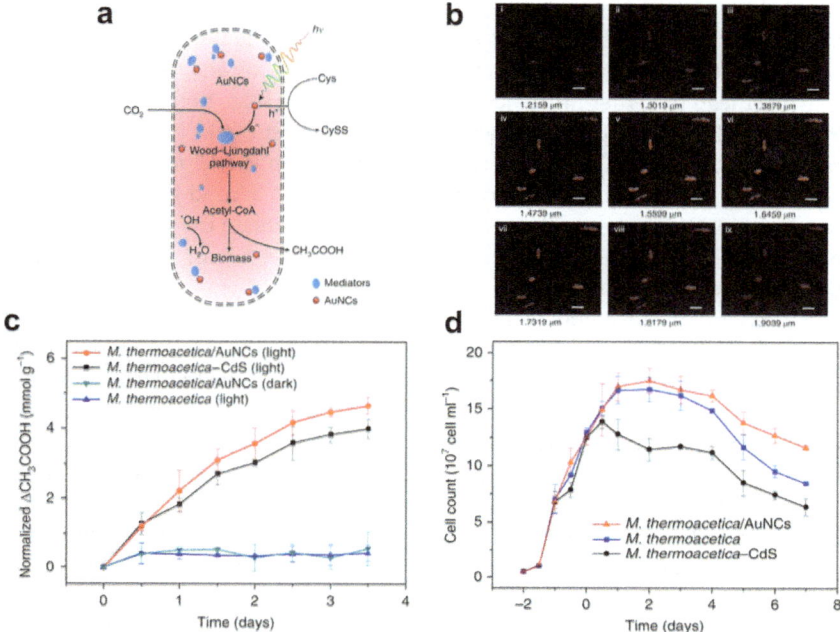

Fig. 2 (a) Schematic of *M. thermoacetica*–$Au_{22}(SG)_{18}$ highlighting pathways to obtain acetate from solar energy. (b) Structure illumination microscopy (SIM) images at different focal planes with emission at 540 nm from $Au_{22}(SG)_{18}$ in *M. thermoacetica*. (c) Normalized acetate production from *M. thermoacetica*–nanoparticle constructs in continuous photosynthesis. (d) Cell counts from *M. thermoacetica*–nanoparticle cultures. Reproduced with permission.[16] Copyright 2018 Nature Publishing Group.

biocompatible.[17] $Au_{22}(SG)_{18}$ specifically, when added to a pre-exponential culture of *M. thermoacetica*, is taken up with over 90% efficiency and remains stable over a period of at least seven days.[16] We have found ~4×10^{-7} mg of $Au_{22}(SG)_{18}$ per cell to be an optimal concentration for our experiments. Furthermore, $Au_{22}(SG)_{18}$ does not inhibit growth or other primary cell functions of *M. thermoacetica* at optimal concentrations. In fact, cell numbers and colony-forming units are higher in cultures of *M. thermoacetica*–$Au_{22}(SG)_{18}$ than in *M. thermoacetica*–CdS [Fig. 2(d)]. Advantageously, it has been determined that certain conformations of AuNCs quench radicals, including ROS.[18] In photosensitized bacteria, ROS is created as a toxic by-product upon photo-activation of the nanoparticles.[19] This effect is particularly destructive with CdS. Results show that during photosynthesis the level of intracellular ROS is much lower in *M. thermoacetica*–$Au_{22}(SG)_{18}$ as compared with *M. thermoacetica*–CdS. Altogether, these metrics of improved cell culture viability indicate that CO_2-to-acetate yield may be higher in *M. thermoacetica*–$Au_{22}(SG)_{18}$ constructs.

As hypothesized, $Au_{22}(SG)_{18}$ nanoclusters permeate the entire cellular structure and are found localized beyond the cell membrane.[16] Structure illumination microscopy is employed to resolve the placement of AuNCs in the cells by detecting fluorescence from the AuNCs under excitation at 540 nm. Fig. 2(b) shows fluorescence emission emanating quasi-uniformly in the cells. Images of

AuNCs in *M. thermoacetica* at a series of focal planes demonstrate that AuNCs can indubitably be found throughout the cells. Critically, the improved cell viability and intracellular penetration through the application of $Au_{22}(SG)_{18}$ as a photosensitizer lead to higher rate of acetic acid production [Fig. 2(c)]. After a period of four days of constant photosynthesis, the yield of acetic acid in the *M. thermoacetica*–$Au_{22}(SG)_{18}$ culture is appreciably higher than in the *M. thermoacetica*–CdS construct.[16] Ultimately, AuNCs function as a powerful second-generation photosensitizer. Additional consideration is needed to elucidate the transport mechanism of $Au_{22}(SG)_{18}$ into the cytoplasm, as well as the role of individual charge uptake pathways within the cell.

AuNCs are effective photosensitizers in bacteria for solar-driven CO_2 fixation. While the rate of photooxidation is diminished with the use of AuNCs, it still persists in the system as a whole. Non-photosynthetic bacteria such as *M. thermoacetica* are not evolved to handle the high photon flux required for successful photosynthesis. Furthermore, cysteine, which acts as the hole scavenger in both CdS and AuNC systems, becomes depleted and currently limits the total output of acetic acid.[11,16] A strategy has been developed in which cystine is reduced back to cysteine by photoactive TiO_2 nanocatalysts.[20] Although the photoreduction of cystine increases the availability of cysteine, which allows for a higher acetate yield, it comes at the expense of much higher photooxidation in the system. The TiO_2 nanocatalysts are responsible for photoanodic water oxidation that produces O_2 and toxic amounts of ROS.[19,21] For these reasons, it is necessary to identify cytoprotective materials that will shield the photosensitized bacteria from photooxidative damage.

Advantageously, the cysteine/cystine redox shuttle allows for physical segregation of the photosensitized bacteria and the TiO_2 nanocatalysts. A selective membrane that quenches ROS and restricts O_2 passage while maintaining the diffusion of CO_2 and the redox shuttle would be beneficial. Photosynthetic microorganisms such as cyanobacteria emit extracellular polymers that contain UV-blocking molecules.[22,23] Additionally, a plethora of robust microbes resist changes in temperature, pH, and salinity by forming spores.[24] Diatoms, for instance, synthesize siliceous exoskeletons to protect themselves from environmental stressors. Taking a cue from nature, encapsulation of the photosensitized bacteria would improve their ability to withstand harsh circumstances.[25]

Hydrogels, including alginate, are commonly employed to encapsulate a unit of cells as they allow for unencumbered proliferation. Bacteria grow without restriction as they form microvoids in the soft alginate hydrogel.[26] This offers the advantage that bacteria would not lose their protection as they replicate, and dormant states are not induced by otherwise tight armor.[27] Fig. 3(a) demonstrates the growth of *M. thermoacetica* inside of alginate microspheres produced by a microfluidic injection process (see Experimental section below). *M. thermoacetica* creates microvoids inside of the alginate microspheres as determined by environmental scanning electron microscopy. Moreover, as alginate primarily consists of water, its CO_2 diffusive properties are the same as those of the aqueous culture media. The viability of *M. thermoacetica* decreases over the duration of whole photosynthesis with CO_2 reduction coupled with O_2 evolution. This is due to the strict anaerobic nature of *M. thermoacetica*, as it does not have a protective mechanism against O_2 or ROS. The alginate scavenges and attenuates the concentration of superoxides, hypochlorites and peroxides. This leads to an

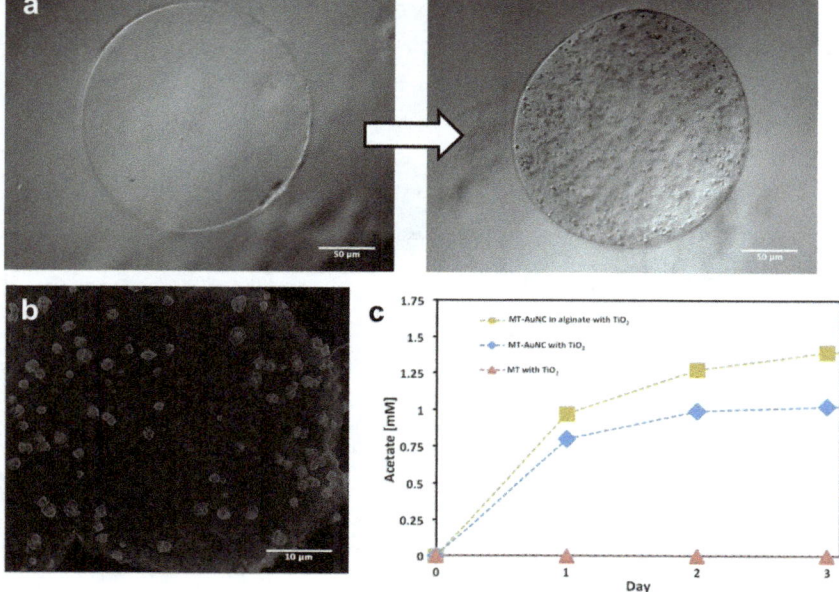

Fig. 3 (a) Differential interference contrast microscopy illustrates *M. thermoacetica* growth from day 0 to day 4 encapsulated in alginate microspheres. (b) Environmental scanning electron microscopy image shows detailed microvoids of an alginate sphere populated with *M. thermoacetica*. (c) CO_2-to-acetate production by encapsulated photosensitized *M. thermoacetica* with light-activated TiO_2 nanoparticles to reduce cystine to cysteine molecular redox shuttle.

increase in acetate during photosynthesis by *M. thermoacetica* [Fig. 3(c)]. A further protective layer may be synthesized on the surface of the bacteria-filled hydrogel.[28] For example, polydopamine and silica have been patterned directly on the surface of alginate to shield cells from stresses.[29]

Metal organic framework (MOF) materials have received consideration as cytoprotective materials that are directly crystallized on or wrap around biological structures.[30,31] Our group has developed a strategy to uniformly wrap photosensitized *M. thermoacetica* with a nanometric thick MOF monolayer for cytoprotection.[32] The MOF monolayer maintains a dynamic wrapping during cell elongation, separation and even of newly grown cell surfaces [Fig. 4(a)]. Importantly, *M. thermoacetica* covered with the MOF monolayer sustains a smaller drop in viability under increasing concentrations of a model ROS, H_2O_2, as compared to the bare *M. thermoacetica* [Fig. 4(c)–(e)]. The viability is enhanced due to the catalytic activity of the MOF enclosure toward decomposition of the ROS. The MOF monolayer contains a zirconium cluster, which like zirconia particles rapidly decomposes H_2O_2 [Fig. 4(b)]. The MOF-wrapped *M. thermoacetica*–$Au_{22}(SG)_{18}$ can continuously produce acetate from CO_2 reduction under oxidative stress. MOF-wrapped photosensitized *M. thermoacetica* was shown to produce up to 200% more acetate than its bare counterpart under whole photosynthesis conditions. Therefore, the MOF cytoprotective method addresses the inherent vulnerability of anaerobes to oxidative stress, and provides a suitable platform for implementing

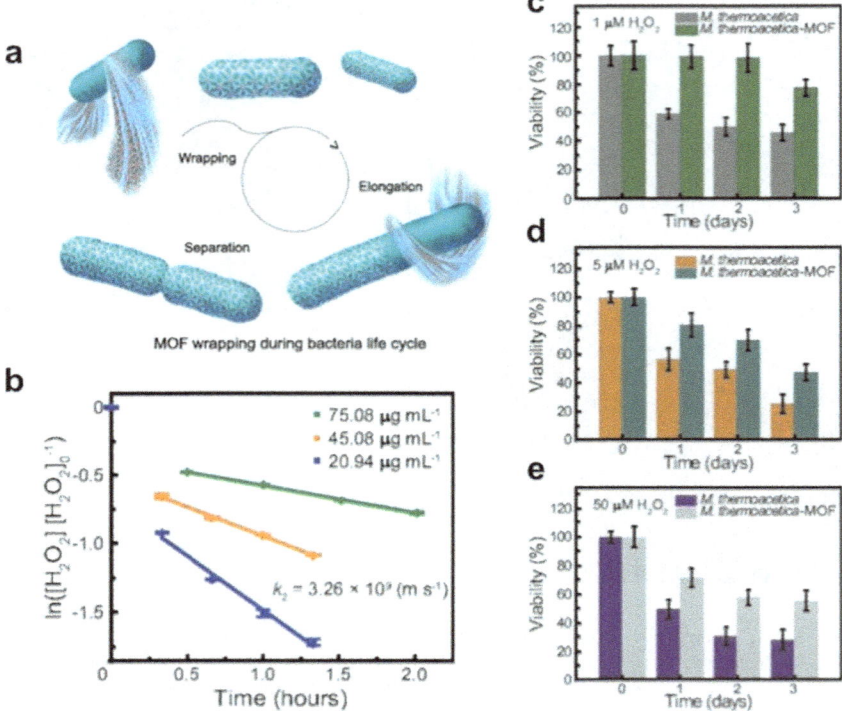

Fig. 4 (a) Depiction of spontaneous zirconium-based MOF-wrapping of *M. thermoacetica* by bonding to membrane-linked phosphate units. (b) Hydrogen peroxide decomposition by zirconium-based MOF. (c)–(e) Enhancement of viability of MOF-wrapped *M. thermoacetica* compared to bare *M. thermoacetica* at various hydrogen peroxide concentrations. This figure reproduced with permission.[32] Copyright 2018 the National Academy of Sciences of the United States of America.

a whole artificial photosynthesis with both carbon dioxide reduction and oxygen evolution reaction.

A final aspect of the photosensitized microorganism system that is worth exploring is the mechanism of electron transfer between the inorganic light harvesting nanoparticles and the cell. While electron transfer processes between a poised electrode and electrotrophic bacteria have been investigated, those studies are limited by the inability to replicate and observe the processes *in situ* by spectroscopic means.[33] Photosensitized bacteria lend themselves as a platform to undertake transmittance-based transient absorption (TA) spectroscopy to study charge carrier lifetimes, as these colloidal suspensions are translucent and modular.[34] By applying controls the molecular basis of charge uptake can be inferred. The biochemical activity of proteins confirmed to take up reducing equivalents can be correlated with the results obtained with TA. In the event of hydrogenase involvement, the photoexcited electrons would help generate H_2, which is then used in the WLP to convert CO_2 into acetate whereas the direct uptake of those electrons by membrane-bound proteins including cytochromes, ferredoxins and flavoproteins would facilitate ATP synthesis[35] [Fig. 5(a)]. In the first of its kind study, Kornienko and colleagues subjected *M. thermoacetica*–CdS constructs to hydrogen incubation pre-photosynthesis at varying lengths of time in order to ramp up hydrogenase activity.[34] The rate of acetate production is highest in the first three hours for those samples with no hydrogen incubation but the average rate of acetate activity over 48 hours is highest in the samples with the longest hydrogen incubation. While seemingly paradoxical, the results suggest that two electron uptake processes take place. In samples with limited hydrogenase activity, the electrons feed directly into membrane bound proteins accelerating acetate production but ultimately cannot sustain metabolic activity due to a lack of high energy reducing equivalents [H_2, NAD(P)H, Fd]. Whereas samples with sufficient hydrogenase activity, a CdS-to-hydrogenase electron transfer pathway is established to produce high energy reducing equivalents. Accordingly, TA kinetic results demonstrate faster decay kinetics in *M. thermoacetica*–CdS hybrids after hydrogen incubation than in *M. thermoacetica*–CdS grown with glucose and bare CdS [Fig. 5(b)]. These observations support the notion of electron transfer to an acceptor site on hydrogenase.

Moreover, due to the direct intracellular interface formed in *M. thermoacetica*–$Au_{22}(SG)_{18}$, this hybrid system could provide more insight into electron transfer processes. Strong quantum confinement effects and single electron transitions in AuNCs are manifested in their physicochemical properties such as discrete energy levels, multiple absorption bands and enhanced photoluminescence.[36] However, insufficient progress has been made in uncovering photoelectron transfer mechanisms. More efforts are needed to fully understand AuNCs excited state interactions, especially for energy conversions. We have established a model system as our starting point to elucidate the electron-donating ability of $Au_{22}(SG)_{18}$ to methyl viologen (MV^{2+}). The stability of its radical cation ($MV^{+\bullet}$) and the ease of its spectroscopic detection makes MV^{2+} favorable for fundamental studies.[37] TA allows for the spectroscopic determination of electron-transfer yield and associated kinetics of charge transfer in reduced methyl viologen with a characteristic absorption at 600 nm and 390 nm. More controlled experiments could provide new insights in understanding the intracellular photoexcited electron transfer process.

Altogether, photosensitized microorganisms offer a pioneering approach to convert CO_2 to upgradeable hydrocarbons using solar energy. This biohybrid

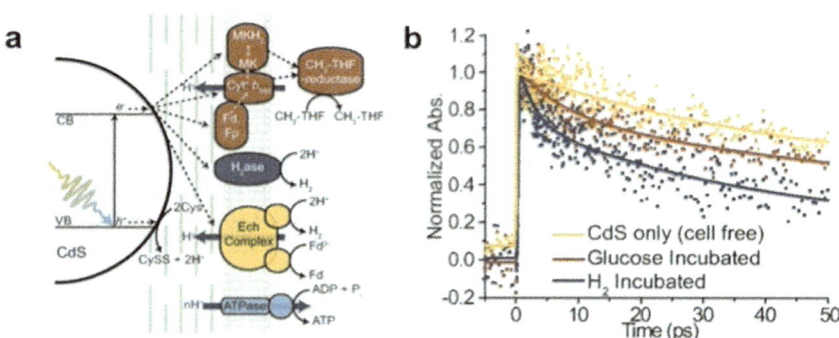

Fig. 5 (a) Schematic featuring possible active photoelectron uptake mechanisms. (b) Transient absorption plot of bare CdS, glucose and H_2 incubated *M. thermoacetica*–CdS. Fig. 5(a) and (b) reproduced with permission.[34] Copyright 2016 the National Academy of Sciences of the United States of America.

approach exploits the replication, self-healing and specificity of CO_2-fixing whole bacteria and the extraordinary solar capture of semiconducting nanoparticles. The intracellular interface, particularly in the AuNC constructs, circumvents the sluggish kinetics associated with extracellular charge uptake. Progress in cytoprotective materials will enable the application of this technology. Additionally, understanding gained through spectroscopic studies will shape future iterations of photosensitizers as well as optimize the pairing process. Finally, we hope engaged readership and research will be drawn to our highlighted areas of active investigation.

Experimental

AuNCs synthesis, *Moorella thermoacetica* growth and photosensitization

Red-emitting $Au_{22}(SG)_{18}$ synthesis is outlined by Zhang and coworkers.[16] Solutions of $HAuCl_4$ (12.5 mL, 20 mM) and glutathione (7.5 mL, 50 mM) were combined into 180 mL of ultrapure water and stirred vigorously. After 120 seconds, the pH of the reaction solution was adjusted to 12.0 with 1 M NaOH and subsequently 0.24 mg $NaBH_4$ was added. The pH of the solution was lowered to 2.5 with 0.3 M HCl after 30 minutes and allowed to react while stirring for 6 hours. Isopropyl alcohol was added at 1 : 1 (v/v) in order to precipitate the AuNCs. The solution was centrifuged at 14 000 rpm in order to collect the AuNCs. Finally, the AuNCs were further washed with methanol.

Extensive *M. thermoacetica* culturing protocols are described in detail in Sakimoto *et al.*[11] Briefly, an initial inoculum of *M. thermoacetica* (ATCC 39073) was cultured in 10 mL anaerobic heterotrophic media at 52 °C after slow thawing from -80 °C. The headspace of the 15 mL volume tube was filled with 80 : 20 mixture of $N_2 : CO_2$ at 150 kPa. After two subsequent rounds of culturing 0.5 mL of 4 mg mL^{-1} $Au_{22}(SG)_{18}$ was added to each culture tube when $OD_{600} = 0.28$. After 24 hours, each culture was washed at 2500 rpm for 10 min, and the cell pellet was resuspended in 3.5 mL of autotrophic medium supplemented with 0.1 wt% cysteine and 1% (w/v) sodium alginate.

Encapsulation of photosensitized *Moorella thermoacetica*

As previously described, 1% (w/v) sodium alginate was directly added to the autotrophic media which is used to resuspend the AuNCs-containing bacteria. The microbeads were formed by injecting the alginate solution (5 mL per hour) through a 25G needle parallel to a flowing stream of N_2 gas into 6.5 mL of stirred autotrophic media supplemented with 2% $CaCl_2$. We then distributed each 10 mL total solution (alginate spheres and liquid autotrophic media) into 25 mL anaerobic tubes, each with 5 mg of TiO_2 anatase nanopowder. In order to promote enzymatic activity, each tube was pressurized with 150 kPa of 80 : 20 $H_2 : CO_2$ and incubated for 12 hours at 52 °C.

Photosynthesis, measurements and sample characterization

Before the light experiments, the headspace of each sample tube was exchanged with 80 : 20 $N_2 : CO_2$ and pressurized to 150 kPa. The light intensity of a 75 W xenon lamp (Newport) with an AM 1.5 G filter was calibrated to 0.2% sun. Acetate concentrations were measured by quantitative ^{1}H-NMR with a sodium 3-

(trimethylsilyl)-2,2′,3,3′-tetradeuteropropionate internal standard. All spectra were processed with MestreNova software. Furthermore, alginate microbeads were imaged with differential interference contrast microscopy using a Zeiss Axioimager upright microscope. Sample preparation for environmental SEM consisted of microtoming individual alginate microspheres with a cryostat (Leica CM3050S). E-SEM was undertaken with an FEI Quanta microscope (Thermo Scientific).

Conflicts of interest

There are no conflicts to declare.

Acknowledgements

This work was supported by NASA, Center for the Utilization of Biological Engineering in Space, under Award NNX17AJ31G. H. Z. acknowledges the Suzhou Industry Park Fellowship.

References

1 D. Kim, K. K. Sakimoto, D. Hong and P. Yang, Artificial Photosynthesis for Sustainable Fuel and Chemical Production, *Angew. Chem., Int. Ed.*, 2015, **54**, 3259–3266.

2 T. R. Cook, D. K. Dogutan, S. Y. Reece, Y. Surendranath, T. S. Teets and D. G. Nocera, Solar energy supply and storage for the legacy and nonlegacy worlds, *Chem. Rev.*, 2010, **110**, 6474–6502.

3 A. Das and L. G. Ljungdahl, Electron-Transport System in Acetogens, in *Biochemistry and Physiology of Anaerobic Bacteria*, Springer-Verlag, New York, 2003, pp. 191–204.

4 B. Tian, S. Xu, J. A. Rogers, S. Cestellos-Blanco, P. Yang, J. L. Carvalho-De-Souza, F. Bezanilla, J. Liu, Z. Bao, M. Hjort, Y. Cao, N. Melosh, G. Lanzani, F. Benfenati, G. Galli, F. Gygi, R. Kautz, A. A. Gorodetsky, S. S. Kim, T. K. Lu, P. Anikeeva, M. Cifra, O. Krivosudský, D. Havelka and Y. Jiang, Roadmap on semiconductor-cell biointerfaces, *Phys. Biol.*, 2018, **15**, 031002.

5 K. K. Sakimoto, N. Kornienko and P. Yang, Cyborgian Material Design for Solar Fuel Production: The Emerging Photosynthetic Biohybrid Systems, *Acc. Chem. Res.*, 2017, **50**, 476–481.

6 K. K. Sakimoto, N. Kornienko, S. Cestellos-Blanco, J. Lim, C. Liu and P. Yang, Physical Biology of the Materials-Microorganism Interface, *J. Am. Chem. Soc.*, 2018, **140**, 1978–1985.

7 C. Liu, J. J. Gallagher, K. K. Sakimoto, E. M. Nichols, C. J. Chang, M. C. Y. Chang and P. Yang, Nanowire-bacteria hybrids for unassisted solar carbon dioxide fixation to value-added chemicals, *Nano Lett.*, 2015, **15**, 3634–3639.

8 C. Liu, B. C. Colón, M. Ziesack, P. A. Silver and D. G. Nocera, "Water splitting–biosynthetic system with CO_2 reduction efficiencies exceeding photosynthesis, *Science*, 2016, **352**, 1210–1213.

9 D. R. Lovley, Powering microbes with electricity: Direct electron transfer from electrodes to microbes, *Environ. Microbiol. Rep.*, 2011, **3**, 27–35.

10 T. J. Silhavy, D. Kahne and S. Walker, The bacterial cell envelope, *Cold Spring Harbor Perspect. Biol.*, 2010, **2**, a000414.

11 K. K. Sakimoto, A. B. Wong and P. Yang, Self-photosensitization of nonphotosynthetic bacteria for solar-to-chemical production, *Science*, 2016, **351**, 74–77.

12 Z. X. Lu, L. Zhou, Z. L. Zhang, W. L. Shi, Z. X. Xie, H. Y. Xie, D. W. Pang and P. Shen, Cell Damage Induced by Photocatalysis of TiO_2 Thin Films, *Langmuir*, 2003, **19**, 8765–8768.

13 M. A. Abbas, P. V. Kamat and J. H. Bang, Thiolated Gold Nanoclusters for Light Energy Conversion, *ACS Energy Lett.*, 2018, **3**, 840–854.

14 X. D. Zhang, Z. Luo, J. Chen, S. Song, X. Yuan, X. Shen, H. Wang, Y. Sun, K. Gao, L. Zhang, S. Fan, D. T. Leong, M. Guo and J. Xie, Ultrasmall glutathione-protected gold nanoclusters as next generation radiotherapy sensitizers with high tumor uptake and high renal clearance, *Sci. Rep.*, 2015, **5**, 8669.

15 Y. Yu, Z. Luo, D. M. Chevrier, D. T. Leong, P. Zhang, D. E. Jiang and J. Xie, Identification of a highly luminescent $Au_{22}(SG)_{18}$ nanocluster, *J. Am. Chem. Soc.*, 2014, **136**, 1246–1249.

16 H. Zhang, H. Liu, Z. Tian, D. Lu, Y. Yu, S. Cestellos-Blanco, K. K. Sakimoto and P. Yang, Bacteria photosensitized by intracellular gold nanoclusters for solar fuel production, *Nat. Nanotechnol.*, 2018, **13**, 900–905.

17 J. Wang, G. Zhang, Q. Li, H. Jiang, C. Liu, C. Amatore and X. Wang, *In vivo* self-bio-imaging of tumors through *in situ* biosynthesized fluorescent gold nanoclusters, *Sci. Rep.*, 2013, **3**, 1157.

18 B. Santiago-Gonzalez, A. Monguzzi, J. M. Azpiroz, M. Prato, S. Erratico, M. Campione, R. Lorenzi, J. Pedrini, C. Santambrogio, Y. Torrente, F. De Angelis, F. Meinardi and S. Brovelli, Permanent excimer superstructures by supramolecular networking of metal quantum clusters, *Science*, 2016, **353**, 571–575.

19 D. Meissner, R. Memming, B. Kastening and D. Bahnemann, Fundamental problems of water splitting at cadmium sulfide, *Chem. Phys. Lett.*, 1986, **127**, 419–423.

20 K. K. Sakimoto, S. J. Zhang and P. Yang, Cysteine–Cystine Photoregeneration for Oxygenic Photosynthesis of Acetic Acid from CO_2 by a Tandem Inorganic–Biological Hybrid System, *Nano Lett.*, 2016, **16**, 5883–5887.

21 Z. X. Lu, L. Zhou, Z. L. Zhang, W. L. Shi, Z. X. Xie, H. Y. Xie, D. W. Pang and P. Shen, Cell Damage Induced by Photocatalysis of TiO_2 Thin Films, *Langmuir*, 2003, **19**, 8765–8768.

22 M. Ehling-Schulz and S. Scherer, UV protection in cyanobacteria, *Eur. J. Phycol.*, 1999, **34**, 329–338.

23 S. P. Adhikary and J. K. Sahu, UV protecting pigment of the terrestrial cyanobacterium Tolypothrix byssoidea, *J. Plant Physiol.*, 1998, **153**, 770–773.

24 S. H. Yang, D. Hong, J. Lee, E. H. Ko and I. S. Choi, Artificial spores: Cytocompatible encapsulation of individual living cells within thin, tough Artificial shells, *Small*, 2013, **9**, 178–186.

25 J. C. Rooke, A. Léonard, C. F. Meunier and B. L. Su, Designing photobioreactors based on living cells immobilized in silica gel for carbon dioxide mitigation, *ChemSusChem*, 2011, **4**, 1249–1257.

26 P. Allan-Wojtas, L. Truelstrup Hansen and A. T. Paulson, Microstructural studies of probiotic bacteria-loaded alginate microcapsules using standard electron microscopy techniques and anhydrous fixation, *LWT–Food Sci. Technol.*, 2008, **41**, 101–108.

27 J. H. Park, S. H. Yang, J. Lee, E. H. Ko, D. Hong and I. S. Choi, Nanocoating of single cells: From maintenance of cell viability to manipulation of cellular activities, *Adv. Mater.*, 2014, **26**, 2001–2010.

28 J. Niu, D. J. Lunn, A. Pusuluri, J. I. Yoo, M. A. O'Malley, S. Mitragotri, H. T. Soh and C. J. Hawker, Engineering live cell surfaces with functional polymers *via* cytocompatible controlled radical polymerization, *Nat. Chem.*, 2017, **9**, 537–545.

29 B. J. Kim, T. Park, S. Y. Park, S. W. Han, H. S. Lee, Y. G. Kim and I. S. Choi, Control of Microbial Growth in Alginate/Polydopamine Core/Shell Microbeads, *Chem.–Asian J.*, 2015, **10**, 2130–2133.

30 X. Lian, Y. Fang, E. Joseph, Q. Wang, J. Li, S. Banerjee, C. Lollar, X. Wang and H. C. Zhou, Enzyme-MOF (metal–organic framework) composites, *Chem. Soc. Rev.*, 2017, **46**, 3386–3401.

31 M. B. Majewski, A. J. Howarth, P. Li, M. R. Wasielewski, J. T. Hupp and O. K. Farha, Enzyme encapsulation in metal–organic frameworks for applications in catalysis, *CrystEngComm*, 2017, **19**, 4082–4091.

32 Z. Ji, H. Zhang, H. Liu, O. M. Yaghi and P. Yang, Cytoprotective metal–organic frameworks for anaerobic bacteria, *Proc. Natl. Acad. Sci. U. S. A.*, 2018, **115**, 10582–10587.

33 F. Kracke, I. Vassilev and J. O. Krömer, Microbial electron transport and energy conservation – The foundation for optimizing bioelectrochemical systems, *Front. Microbiol.*, 2015, **6**, 575.

34 N. Kornienko, K. K. Sakimoto, D. M. Herlihy, S. C. Nguyen, A. P. Alivisatos, C. B. Harris, A. Schwartzberg and P. Yang, Spectroscopic elucidation of energy transfer in hybrid inorganic–biological organisms for solar-to-chemical production, *Proc. Natl. Acad. Sci. U. S. A.*, 2016, **113**, 11750–11755.

35 J. S. Deutzmann, M. Sahin and A. M. Spormann, Extracellular enzymes facilitate electron uptake in biocorrosion and bioelectrosynthesis, *mBio*, 2015, **6**, 1–8.

36 Z. Luo, K. Zheng and J. Xie, Engineering ultrasmall water-soluble gold and silver nanoclusters for biomedical applications, *Chem. Commun.*, 2014, **50**, 5143–5155.

37 M. Zhou, Z. Lei, Q. Guo, Q. M. Wang and A. Xia, Solvent Dependent Excited State Behaviors of Luminescent Gold(ɪ)–Silver(ɪ) Cluster with Hypercoordinated Carbon, *J. Phys. Chem. C*, 2015, **119**, 14980–14988.

DISCUSSIONS

Biological approaches to artificial photosynthesis: general discussion

Vivek Badiani, Mark Bajada, Matthias Beller, Andrew B. Bocarsly,
Sylvestre Bonnet, Carlota Bozal-Ginesta, Peter Brueggeller,
Julea N. Butt, Flavia Cassiola, Michael Grätzel,
Leif Hammarström, Marta C. Hatzell, Lars J. C. Jeuken,
Burkhard König, Moritz F. Kuehnel, Joshua Lawrence,
Chong-Yong Lee, Marcelino Maneiro, Shelley D. Minteer,
Esther Edwardes Moore, Samuel E. H. Piper, Nicolas Plumeré,
Joost N. H. Reek, Erwin Reisner, Souvik Roy, Jeremy Shears,
Sergii I. Shylin, Han Sen Soo, Andreas Wagner,
Dominik Wielend, Jenny Zhang and Martijn Zwijnenburg

DOI: 10.1039/C9FD90026H

Jenny Zhang opened the discussion of the paper by Shelley Minteer: Was the vision of this study to produce salinity adapted microorganisms in the lab and then release them into the environment? Or was it mainly to try to understand how salinity adaptation can occur in the natural environment, using RNA sequencing to facilitate this?

Shelley Minteer replied: A comprehensive answer is that we are interested in both aspects. Specifically, our vision is that a better understanding of the salinity adaptation strategies will pose the basis to develop photo-bioelectrochemical systems capable of stable and long-term operation in saline environments. Furthermore, elucidating bacterial metabolism and their influence on photo-bioelectrocatalysis will not only allow for more robust and stable devices, but also for the better correlation of current response and changes taking place in the environment. Accordingly, we consider that the deeper understanding of biological processes related to the metabolism of electroactive microorganisms is critical for their successful application in the environment.

Jeremy Shears commented: You mentioned the formation of biofilms. Is this a stress response to the pH of the medium the organisms are exposed to? Do you notice any effects of the biofilm on kinetics? Do you consider them "good" or "bad" from a process perspective? And can you say a little bit more about using alginates as a form of biofilm?

Shelley Minteer responded: In our experiments aimed to the study of salinity adaptation effects on bioelectrocatalysis, we are directly depositing *R. capsulatus* cells on the electrode surface. Since the electrodes are tested right after preparation, no particularly thick biofilm is developed, and we did not investigate biofilm effects on the kinetics of the reactions taking place. However, in view of the in-field application of the technology, biofilm development will play a critical role, as it will facilitate bacterial cells' attachment to the electrode and their stability over long-term operation. We are currently investigating the effects of different quorum sensing autoinducers on *R. capsulatus* biofilm development. Regarding the use of alginate to form artificial biofilms, first, it is important to specify that alginate is one of the three major components of natural biofilms. Based on that, we are unveiling the possibility to prepare the artificial biofilm to enhance *R. capsulatus* salt tolerance, allowing immediate exposure of bacterial cells to solutions characterized by different salinities. It is important to remark that, while this approach has the advantage to facilitate a "prompt adaptation to salinity", its efficacy over long-term operation remains to be validated, thus motivating our interests in elucidating the biological mechanisms of adaptation.

Andrew Bocarsly remarked: Although having a biofilm very close to the electrode interface facilitates electron transfer between the film and the electrode, the biofilm may generate a mass transport limitation on the solution side of the interface. Will this mass transport resistance be a concern?

Shelley Minteer answered: Yes, it is definitely a concern as we improve the performance of the bioelectrodes. Eventually, they will become mass transport limited.

Julea Butt commented: Some reports suggest that bacteria 'in the middle' of biofilms are metabolically inactive, perhaps in stationary phase or dead. I wondered if you have any information on the metabolic state of the cells in your *Ralstonia* biofilms as a function of distance from the electrode?

Shelley Minteer answered: Our biofilms are *Rhodobacter*, but this is an important aspect. As correctly stated, in thick biofilms the cells far from being exposed to the solution will face a scarcity of available substrates, potentially leading to their death. However, for our current experiments the amount of bacterial cells deposited on the electrode surface is quite limited (30 microliters of a 1 g mL^{-1} bacterial cells solution). Moreover, all the obtained electrodes are utilized within a 1–3 h time window from their preparation, and thus, we do not expect a particular influence of dead cells on the obtained bio-photocurrent results. This is indeed a very interesting aspect to be elucidated with future research, and we have ongoing studies aimed at developing new immobilization techniques and redox mediator systems for enhanced operational stability of the biophotoelectrochemical systems, since we have experience with alginate and redox polymers for these applications.[1–3]

1 B. Alkotaini, S. L. Tinucci, S. J. Robertson, K. Hasan, S. D. Minteer and M. Grattieri, Alginate-encapsulated bacteria for hypersaline solutions treatment in microbial fuel cell, *ChemBioChem*, 2018, **19**, 1162–1169.

2 K. Hasan, M. Grattieri, T. Wang, R. D. Milton and S. D. Minteer, Enhanced Bio-electrocatalysis of *Shewanella oneidensis* MR-1 by a Naphthoquinone Redox Polymer, *ACS Energy Lett.*, 2017, 2, 1947–1951.
3 G. Pankratova, D. Pankratov, R. D. Milton, S. D. Minteer and L. Gorton, Following Nature: Bioinspired Mediation Strategy for Gram-positive Bacterial Cells, *Adv. Energy Mater.*, 2019, 9, 1900215.

Marta Hatzell asked: How are you able to control for electrical conductivity differences with the variable salinity wastewaters investigated?

Shelley Minteer replied: Thank you for pointing out this issue. First of all, our experiments were always conducted using a 20 mM MOPS buffer (pH 7) + 10 mM MgCl$_2$ + 50 mM malic acid, adjusting salinity as required. Accordingly, good conductivity of the utilized electrolyte was ensured for all the experiments, despite the different salinities tested. It is correct that an increased conductivity is obtained for the experiments performed at higher salinity (*i.e.* 22 gL^{-1} NaCl). While this would result in an expected current generation increase for a classical electrochemical system, this is not the case for bioelectrochemical systems employing intact bacterial cells, due to the inhibiting effects of high salt content on bacterial cell activity. Finally, it should also be noted that, with the final goal of achieving in-field application for the developed system, our goal is not to strictly control the experimental conditions (which would be not possible during in-field application), but rather to understand how the system behaves when exposed to stressing and variable environmental conditions.

Joshua Lawrence queried: Have you been able to analyse the RNA-Seq data you have collected? If the described horizontal gene transfer mechanism is involved it should be upregulated in salt-adaptation. RNA-Seq data available for other *Rhodobacter* species identified genes associated with salt stress, some of which slowly accumulate over time. Have you considered these as alternative explanations for your observed adaptation to salinity?

Shelley Minteer responded: We are still in the process of using bioinformatics to evaluate all of the data collected, so we can't confirm the mechanism yet.

Marcelino Maneiro commented: I found your approach very interesting. In my opinion, bioremediation of wastewater is crucial. Only 3% of the world's water is fresh water; a mere 0.014% of all water on Earth is both fresh and easily accessible. In this context it is necessary to develop systems which work in saline wastewater environments. My question is about the behaviour of the *R. capsulatus* bacteria in the presence of other living organisms that tend to be present in the wastewater media. How does it interact with them?

Shelley Minteer replied: While considerably fewer research efforts are focused on this particular topic, we consider it of the utmost importance to address the problem of contaminated saline solution release in the environment. The question is very important in view of the in-field application of *R. capsulatus*. Different reports of bioelectrochemical systems operated with mixed consortia biofilms have shown that the presence of various species can enhance both the removal of organic contaminants, as well as the current output of the device,[1,2] as non-electrogenic bacteria can degrade complex molecules to easily degradable compounds that can be utilized by electrogenic bacteria. For the specific case of *R. capsulatus*, we have not yet investigated its performance and electrochemical behavior in the presence of other living organisms that tend to be present in wastewater media. We consider the study of possible interaction between these species (such as members of the *Geobacteraceae* family that has been found on electrode-respiring biofilms formed in wastewater media) and *R. capsulatus* very important, and our future research efforts will be focused in this direction.

1 N. S. Malvankar, J. Lau, K. P. Nevin, A. E. Franks, M. T. Tuominen and D. R. Lovley, Electrical Conductivity in a Mixed-Species Biofilm, *Appl. Environ. Microbiol.*, 2012, **78**, 5967–5971.
2 B. Virdis, D. Millo, B. C. Donose, Y. Lu, D. J. Batstone and J. O. Krömer, Analysis of electron transfer dynamics in mixed community electroactive microbial biofilm, *RSC Adv*, 2016, **6**, 3650–3660.

Jenny Zhang opened the discussion of the paper by Lars Jeuken: What are the main bottlenecks of your systems?

Lars Jeuken responded: The main bottlenecks is that haems in the trans-membrane electron conduit, MtrCAB, have reduction potentials between −0.4 and 0 V *vs.* SHE (at neutral pH). Commonly used hydrogen-evolving catalysts like the DuBois-type nickel catalysts require an overpotential, which means that at neutral pH, a reduction potential below −0.4V is required to reduce the catalyst.

Long-term, bottlenecks can be identified that need to be solved before compartmentalised vesicle systems can be exploited for solar-fuel production. For instance, (1) the transmembrane electron conduit, MtrCAB, is laborious and expensive to purify and (2) MtrCAB requires specific conditions to incorporate into vesicles, which are not always compatible with conditions to encapsulate high levels of catalysts in the lumen of vesicles.

Leif Hammarström remarked: This is an interesting system. You did not mention how the transmembrane electron flow is charge-balanced to allow for continued electron transfer. Do you know what is the charge compensating ion transfer, *e.g.* proton leakage? Do you have any direct data on leakage rates?

Lars Jeuken answered: We hypothesize that charge-balance proceeds *via* proton leakage, but we do not have experimental proof for this. When the dye, Reactive Red 120, in the lumen of MtrCAB-containing vesicles is reduced with dithionite, no enhancement in reduction rates is observed upon the addition of protonophores. The latter suggests that ion or proton leakage is not rate limiting in this system.

Michael Grätzel commented: Transmembrane electron transfer will induce a build up of negative charges in the interior of the vesicle. This will arrest the electron flux unless it is accompanied by proton transfer. The question I have is what is the source of the protons ? Ultimately the source should be water. If the source is not well defined it is not possible to provide a solar to chemical conversion efficiency for the photoreaction.

Lars Jeuken replied: We propose that the transmembrane electron transfer is accompanied by proton transfer. The reduction rate of Reactive Red 120 is rate limited by reductive bleaching of the dye, which is relatively slow and takes place on the minutes to tens-of-minutes timescale. On this timescale, liposomes are somewhat permeable to protons.[1] Control experiments were performed with added protonophore, CCCP, to further increase the proton permeability into the lumen of the liposomes. However, this did not increase the observed rate of reductive bleaching, suggesting that 'base' proton permeability was sufficiently high.

We envision that in future applications, protonophores or other ion-selective channels might need to be added to prevent the build-up of transmembrane chemical or electrochemical gradients.

1 M. Rossignol, P. Thomas and C. Grignon, *Biochim. Biophys. Acta, Biomembr.*, 1982, **684**, 195–199.

Vivek Badiani asked: Are there any methods you have in mind to understand the protein/liposome orientation at the interface? Are there subsequent ideas on how to control this orientation?

Furthermore, do you envision artificial materials to replace the trans-membrane; if so, what types are you considering?

Lars Jeuken responded: Similar questions were raised elsewhere and are more fully answered as part of those discussions. In brief, we currently do not control the orientation of MtrCAB in proteoliposomes, although it is likely that one of the orientations is more favourable as this is often observed when reconstituting membrane proteins in liposomes. Orientation can be monitored with various methods, for instance, protease treatment followed by mass spectrometry analysis.

In the future we are looking to replace the lipid membrane with a polymer membrane or a hybrid polymer–lipid system. Together with a collaborator, we have published hybrid systems which show a much enhanced endurance compared to liposome systems.[1]

1 S. Khan, M. Li, S. P. Muench, L. J. C. Jeuken and P. A. Beales, *Chem. Commun.*, 2016, **52**, 11020–11023.

Chong-Yong Lee asked: In your previous study, you have employed decahaem MtrC which contains 10 haems. Your current work reported 20 haem MtrCAB. Could you please comment on the difference between both systems, and whether the number of haems influences the electron relay functionality?

Lars Jeuken answered: MtrC is a peripheral membrane protein, while MtrCAB is a transmembrane complex. The compartmentalised system requires a trans-membrane complex to transfer electrons across the membrane. Thus, the number of haems does not influence the functionality of the system, but the fact that the haems in MtrCAB transverse the membrane is key. Electron transfer between the haems is known to be much faster (see ref. 1) than the kinetics measured for reduction of Reactive Red 120. Thus, we propose that the number of haems also does not influence the kinetics of the electron conduit MtrCAB.

1 X. Jiang, B. Burger, F. Gajdos, C. Bortolotti, Z. Futera, M. Breuer and Jochen Blumberger, *Proc. Natl. Acad. Sci. U. S. A.*, 2019, **116**, 3425–3430.

Andreas Wagner remarked: I was curious about the future outlook and general concept presented in your paper. If product is produced inside the membrane, it will somehow need to be released as it would otherwise accumulate. As soon as the reduced/oxidised product species would diffuse out, one would expect the same problems with re-oxidation/reduction outside the membrane by an oxidation/reduction catalyst as in a non-separated design approach. What would be the advantage of the membrane in this case?

Lars Jeuken responded: The advantages of compartmentalisation in this system are: (1) to separate oxidation and reduction catalysts, (2) create different chemical environments for reduction and oxidation, and (3) to stabilise the charge separated state by spatial separation of the hole and electron.

Flavia Cassiola commented: Your work is a clever illustration of how the control over organization and assembly should be considered in the design of photosynthetic systems. The transmembrane protein complex MtrCAB forms a long conductive molecular wire, (20 haem in bacteria outer membrane), which was estimated to have a transmembrane electron transfer of 10^3–10^4 electrons per second for the reaction used in your system, according to your paper. Synthesizing such a long molecular wire is not an easy task. Heinz Frei's group has proposed an interesting design (see ref. 1). Do you think the "volume" (as a way to look at MtrCAB as a long conductive molecular wire) that MtrCAB has as a protein can be matched in a synthetic system (not based on isolated proteins)?

1 W. Kim, B. A. McClure, E. Edri and H. Frei, *Chem. Soc. Rev.*, 2016, **45**, 3221–3243.

Lars Jeuken responded: The estimate of 10^3–10^4 electrons per second was determined by White *et al.* (ref. 1). In our work, the transmembrane electron transfer rate is much slower as it is limited by the relatively slow reductive bleaching rate of the dye, Reactive Red 120.

I think the 'length' or 'volume' of the molecular wire can be matched by synthetic systems, although it remains to be seen if synthetic systems can match the electron transfer rate of MtrCAB and the ability of MtrCAB to transfer electrons through a membrane of a vesicle (*i.e.* from a hydrophilic solvent, through a hydrophobic medium and back to a hydrophilic solvent). Besides the work of Frei, interesting work on synthesis conducting wires (some of them transmembrane) is being done by other groups. Some examples are the work of Bazan (ref. 2) and Albinsson (ref. 3). Both conducting polymers and redox polymers have been synthesized, while nanomaterials like carbon nanotubes are also considered.

1 G. F. White, Z. Shi, L. Shi, Z. Wang, A. C. Dohnalkova, M. J. Marshall, J. K. Fredrickson, J. M. Zachara, J. N. Butt, D. J. Richardson and T. A. Clarke, *Proc. Natl. Acad. Sci. U. S. A.*, 2013, **110**, 6346–6351.
2 J. Du, C. Catania and G. C. Bazan, *Chem. Mater.*, 2014, **26**, 686–697.
3 M. Gilberta and B. Albinsson, *Chem. Soc. Rev.*, 2015, **44**, 845–862.

Sylvestre Bonnet remarked: Electrons in molecular wires usually follow a gradient of Gibbs free energy. What about these MtrCAB proteins? Related to the directionality of electron transfer through the membrane, is the Gibbs free energy curve flat, or does it go down from outside to inside?

Lars Jeuken responded: We note that this issue was also addressed in a comment made by Prof. Julea Butt.

The reduction potentials of the MtrCAB haems lie roughly between 0 and −0.4 V *vs.* SHE, but the reduction potentials of the 20 haems in MtrCAB have not been experimentally determined for the individual haems. Earlier simulations (ref. 1) indicated that the free energy profile for electron flow along MtrC has 'up and downs' and no directionality. In recent electronic structure calculations by the group of Blumberger (ref. 2) it was concluded that MtrC can shuttle electrons in both directions with similar efficiency.

1 A. Barrozo, M. Y. El-Naggar and A. I. Krylov, *Angew. Chem., Int. Ed.*, 2018, **57**, 6805–6809.
2 X. Jiang, B. Burger, F. Gajdos, C. Bortolotti, Z. Futera, M. Breuer and Jochen Blumberger, *Proc. Natl. Acad. Sci. U. S. A.*, 2019, **116**, 3425–3430.

Sylvestre Bonnet remarked: How do you control the orientation of the MtrCAB protein when reconstituting the liposomes? What will happen if one does not control this orientation? Does it have a role on the efficacy/rate/occurrence of electron transfer?

Lars Jeuken replied: We do not control the orientation of MtrCAB in proteo-liposomes and it is expected that both orientations are adapted, although it is likely that one of the orientations is more favourable as this is often observed when reconstituting membrane proteins in liposomes. We hypothesize that the orientation of MtrCAB will not influence its ability to shuttle electrons into the liposomes. In a recent electronic structure calculations by the group of Blumberger (ref. 1) it was concluded that MtrC can shuttle electrons in both directions with similar efficiency. Furthermore, although the physiological function of MtrCAB is to shuttle electrons out of the bacterial cell, studies (*e.g.* ref. 2) have shown that the electron transfer direction can be reversed in the bacterial cell.

1 X. Jiang, B. Burger, F. Gajdos, C. Bortolotti, Z. Futera, M. Breuer and Jochen Blumberger, *Proc. Natl. Acad. Sci. U. S. A.*, 2019, **116**, 3425–3430.
2 D. E. Ross, J. M. Flynn, D. B. Baron, J. A. Gralnick and D. R. Bond, *PLoS One*, 2011, **6**, e16649.

Julea Butt commented: I would like to contribute some information relevant to the earlier discussions relating to redox properties of MtrCAB and how its orientation in the vesicle bilayers can be probed. As Lars has indicated, cyclic voltammetry reveals MtrCAB is redox active between approx. 0 and −400 mV *vs.* SHE (ref. 1). Experimental resolution of reduction potentials for individual hemes is challenging. They have very similar optical properties as all hemes are c-type with His/His ligation. In favourable cases information is provided by EPR monitored potentiometric titration as sub-sets of hemes are distinguished by signals determined by the dihedral angle between the ring planes of the axial ligands (ref. 2). Crystal structures are available for the extra-cellular cytochrome, MtrC and its homolog MtrF, which show the hemes as a staggered cross, rather than linear chain. Multiple possible sites of electron exchange between redox partners and the proteins can be identified. The structures also provide a basis for calculation of heme reduction potentials (ref. 3 and 4) and those studies indicate a thermodynamic landscape more akin to a roller-coaster, than a slide. To define the orientation of MtrCAB in the vesicles a number of methods are possible. For example, the externally facing proteins are more accessible to antibodies, protease digestion or dye labelling (ref. 5).

1 R. S. Hartshorne *et al.*, *Proc. Natl. Acad. Sci. U. S. A.*, 2009, **106**, 22169.
2 T. A. Clarke *et al.*, *Proc. Natl. Acad. Sci. U. S. A.*, 2011, **108**, 9384.
3 X. Jiang *et al.*, *Proc. Natl. Acad. Sci. U. S. A.*, 2019, **116**, 3425.
4 H. C.Watanabe *et al.*, *Proc. Natl. Acad. Sci. U. S. A.*, 2017, **114**, 2916.
5 G. F. White *et al.*, *Proc. Natl. Acad. Sci. U. S. A.*, 2013, **110**, 6346.

Jenny Zhang opened the discussion of the paper by Nicolas Plumeré: I read in your paper that you've developed an app for others to use; is this ready for you to show us? Also, can you tell us how this model can be extended to other systems (*e.g.* synthetic systems)?

Nicolas Plumeré responded: Yes, the app to run simulations of the photo-currents and of the concentration gradients can be downloaded from the link given on our webpage. Four different models are always included to choose from. Additional models can be developed on request.

Jenny Zhang asked: When applying the model to synthetic systems, redox polymers are not needed. Is it easier to model this? How does the model differ for a synthetic system that uses a linker to adsorb the photo-harvester to the surface?

Nicolas Plumeré replied: This is correct. If the photosynthesizer can accept electrons directly from the electrode the model is simplifed. The electron mediation by hopping can be omitted. The charge transfer between the photosensitizer and the electrode could be accounted for based on the Butler–Volmer model.

Jenny Zhang remarked: In your model, there are several assumptions. For example, one of the assumptions is for the diffusion coefficient for Y (electron accepter for PSI) to be the same when diffusing through the redox polymer compared to just electrolyte. To what extent is this a valid assumption? What are some other assumptions that still need validation for your model?

Nicolas Plumeré answered: Indeed, the model is based on assumptions. The value of the diffusion coefficient of the electron acceptor in the electrolyte *vs.* its value in the film is a good example. The validity of this particular assumption is true in hydrogel films, which are mostly composed of the electrolyte itself. However if a film with significantly different properties than the solvent itself is used, the model must be adapted with the possibility to use different diffusion coefficients. Other assumptions are, for example, that there is no partition coefficient between the film and the electrolyte, the diffusion in the electrolyte is semi-infinite, and that the photoactive and electron mediating components are homogeneously distributed in the film.

Joshua Lawrence asked: Do you plan on experimentally confirming your model? For example, testing whether differences in diffusion coefficients have the predicted effect described in the paper.

Nicolas Plumeré replied: Agreements between experimental observation and predictions from the model are on-going in our lab. We have already demonstrated that modulation of the kinetic constants of the recombination between the charge carrier and the electrode lead to experimental photocurrents that match both qualitatively and quantitatively the predicted currents. The corresponding manuscript will be submitted soon.

The effect of the diffusion coefficient of the charge carrier is also a very interesting parameter to modulate. We intend to use electrolytes with increasing viscosity to confirm this prediction.

Souvik Roy remarked: I am curious about polymer matrices, can you have a silent matrix in the model where you have direct electron transfer?

Nicolas Plumeré answered: If the redox matrix were silent and if only direct electron transfer was possible, the system would simplify to a monolayer of photosynthetic proteins (since multilayers are not accessible to direct electron transfer). The model does not account for this specific case at the moment but the needed modification for this purpose would be straightforward since it consists of a simplified case of the current model.

Esther Edwardes Moore returned to the discussion of the paper by Lars J. C. Jeuken: Have you considered what catalysts, whether artificial or enzymatic, you might use inside your liposome for fuel synthesis?

Lars Jeuken answered: This question was partly answered in response to an earlier question. The main consideration when selecting appropriate catalysts is that haems in the transmembrane electron conduit, MtrCAB, have a reduction potential between −0.4 and 0 V *vs.* SHE (at neutral pH). −0.4 V *vs.* SHE is thermodynamically sufficient to reduce protons to hydrogen at neutral pH and hence we are focusing on hydrogen-evolving catalysts. Commonly used hydrogen-evolving catalysts like the DuBois-type nickel catalysts require an overpotential, which means that at neutral pH, a reduction potential below −0.4V is required to reduce the catalyst. We are currently considering the use of enzymes (*e.g.* hydrogenases) or Pt nanoparticles, which typically require low overpotentials, although their large size (compared to molecular catalysts) requires technical development to encapsulate them in the lumen of liposomes.

Martijn Zwijnenburg asked: Would leakage of products such as hydrogen out of the liposomes become an issue when using the envisaged biomimetic system for fuel production?

Lars Jeuken answered: In the approach presented in this paper, it is important that the fuel can be exported (either by design or as leak) out of the nano-compartment to prevent a build-up of fuel. In other words, the envisioned role of compartmentalisation in this system is (1) to separate oxidation and reduction catalysts, (2) create different chemical environments for reduction and oxidation, and (3) to stabilise the charge separated state by spatial separation of the hole and electron.

Han Sen Soo commented: The photoreduction rates involving the different light harvesting nanoparticles are fairly different. Are the different rates due to thermodynamic reasons or due to the kinetics of charge transfer? Or could it be due to the charge extraction kinetics? Although the nanoparticles in principle have sufficient potential for the photoreduction, Marcus theory suggests that a larger driving force would increase the rate up to a certain extent and the dye-sensitized TiO_2 systems may happen to have larger driving forces?

Lars Jeuken responded: Our interpretation of the data is that photoreduction rates are determined by kinetics of charge transfer between the light-harvesting nanoparticles and MtrCAB and between MtrCAB and Reactive Red 120 (RR120). The reduction of RR120 by amorphous carbon dots (aCD) is faster in the compartmentalised systems (with MtrCAB) compared to the direct reduction of RR120 by aCD. Additionally, we see that for graphitic carbon dots with core nitrogen doping (g-*N*-CDs) and dye-sensitised TiO_2, MtrCAB is reduced at a higher rate compared to RR120, indicating that RR120 reduction is limited by charge transfer from MtrCAB to RR120.

Jenny Zhang addressed Nicolas Plumeré and Lars Jeuken: One of the biggest issues of using biological components in artificial photosynthesis is stability.

Nicolas, can you extend your model to help us understand, predict or perhaps overcome the issue of stability?

Lars, how is stability an issue for your vesicles? Will the artificial membrane breakdown before the proteins?

Nicolas Plumeré responded: The model could in principle be extended to take into account deactivation processes. If a mechanism for deactivation and the related kinetic parameters are known, they could be integrated into the kinetic scheme to predict the change in photocurrent overtime accordingly. Agreement between experiment and predicted current response would support a hypothesized mechanism.

Lars Jeuken answered: Liposomes and membrane proteins are typically stable on the timescale of hours to days. For future applications, however, stability needs to be extended to weeks or months. Phospholipids with double bonds (such as used in this work) are prone to oxidation, disrupting the hydrophobic core of the lipid bilayer. To stabilise liposomes, it is possible to use (phospho)lipids without double bonds and a typical example is the lipid 1,2-diphytanoyl-sn-*glycero*-3-phosphatidylcholine, which has high chemical and physical stability. An approach we have adapted to stabilise both the lipid membrane and proteins inside the membrane is to make hybrid vesicles of phospholipids and amphilic polymers. In an example with the polymer PBd_{22}-*b*-PEO_{14}, we have recently shown that the functional durability of an oxygen-reducing enzyme can be extended from days/weeks to over a year.[1,2]

1 S. Khan, M. Li, S. P. Muench, L. J. C. Jeuken and P. A. Beales, *Chem. Commun.*, 2016, 52, 11020–11023.
2 R. Seneviratne, S. Khan, E. Moscrop, M. Rappolt, S. P. Muench, L. J. C. Jeuken and P. A. Beales, *Methods*, 2018, 147, 142–149.

Carlota Bozal-Ginesta returned to the discussion of the paper by Shelley D. Minteer: How long do you expect bacteria on the electrode to survive? Do you expect them to grow?

Shelley Minteer replied: There are examples of microbial bioelectrocatalysis in literature continuing for years, but we have only studied them for months, but during that time, they do grow and reproduce.

Sergii I. Shylin asked: In Fig. 5 of the paper (DOI: 10.1039/c8fd00160j), you show a clear difference in biophotocurrent for *R. capsulatus* in fresh water and saline media. In the case of the highest salinity (20 g L^{-1} NaCl), concentration of NaCl is *ca.* 10 times higher than concentration of the supporting electrolyte used in the experiments (10 mM $MgCl_2$). Where does the difference come from? Is it the effect of the electrolyte concentration (*i.e.*, conductivity of the solution) or adaptation of bacteria to saline environments?

Shelley Minteer replied: As correctly noted, there was an increase in bio-photocurrent generation obtained at 20 gL^{-1} NaCl when cells fully adapted to salinity were utilized for the study. It should be noted that when performing the experiments at 20 gL^{-1} NaCl using bacterial cells not adapted to salinity, the bio-photocurrent response was consistently lower (or null) compared to the result obtained in fresh water. Based on this consideration, the response obtained at 20 gL^{-1} NaCl with cells fully adapted to salinity must account for the adaptation of bacterial cells to the increased salinity. It is true, however, that a contribution of increased solution conductivity could also account for part of the enhanced biophotocurrent response obtained with cells fully adapted to salinity.

Souvik Roy asked: Have you tested the substrate scope for your system? Instead of malic acid, can you oxidize a different substrate, such as benzyl alcohol or amines?

Shelley Minteer answered: This is a very important aspect. In our current experiments, we focused on understanding the influence of different salinities on the bioelectrocatalytic properties of *R. capsulatus*, and thus, we performed our study keeping substrate conditions constant, and the substrate scope in the photo-bioelectrochemical system was not investigated. However, the possibility of utilizing a different carbon source as a substrate is of extreme interest, and will be the objective of future research. It is important to note that *R. capsulatus* has a wide substrate scope, being able to utilize lactate, butyrate, propionate, succinate, and malate as a carbon source.[1,2] Furthermore, photocatabolism of nitrophenol and other aromatic compounds has been reported for *R. capsulatus*, opening up its application for the light-driven remediation of contaminated environments in photo-bioelectrochemical systems.[3,4] Since the influence on bioelectrocatalysis of the broad substrate scope has not yet been determined, future research will focus on this aspect.

1 A. Dupuis, M. Chavallet, E. Darrouzet, H. Duborjal, J. Lunardi and J. P. Issartel, The Complex I from *Rhodobacter capsulatus, Biochim. Biophys. Acta, Bioenerg.*, 1998, **1364**, 147–165.
2 M. A. Tichi and R. Tabita, Interactive Control of *Rhodobacter capsulatus* Redox-Balancing Systems during Phototrophic metabolism, *J. Bacteriol.*, 2001, **183**, 6344–6354.
3 R. Blasco and F. Castillo, Light-Dependent Degradation of Nitrophenols by the Phototrophic Bacterium Rhodobacter capsulatus E1F1, *Appl. Environ. Microbiol.*, 1992, **58**, 690–695.
4 C. Sasikala and C. V. Ramana, Biodegradation and Metabolism of Unusual Carbon Compounds by Anoxygenic Phototrophic Bacteria, *Adv. Microb. Physiol.*, 1998, **39**, 339–377.

Souvik Roy queried: How much current enhancement is observed in the absence of malic acid?

Shelley Minteer responded: Due to the current aim of our research, where salinity influence on bioelectrocatalysis was being investigated, we did not focus on optimizing substrate concentration for our experiments. The utilized concentration of 50 mM malic acid was based on recently published works where the influence of different substrates concentrations was studied. Furthermore, it was determined that when no malic acid is present in solution, no photocurrent generation is obtained.[1,2]

1 K. Hasan, K. V. R. Reddy, V. Eßmann, K. Górecki, P. Ó. Conghaile, W. Schuhmann, D. Leech, C. Hägerhäll and L. Gorton, Electrochemical Communication Between Electrodes and *Rhodobacter capsulatus* Grown in Different Metabolic Modes, *Electroanalysis*, 2015, **27**, 118–127.
2 K. Hasan, S. A. Patil, K. Górecki, D. Leech, C. Hägerhälla and L. Gorton, Electrochemical communication between heterotrophically grown *Rhodobacter capsulatus* with electrodes mediated by an osmium redox polymer, *Bioelectrochemistry*, 2013, **93**, 30–36.

Dominik Wielend said: I have a question regarding the benzoquinone you use as electron mediator: In one of your recent publications (ref. 1) you identified benzoquinone to be the best candidate upon several other halogenated benzoquinones as well as one naphthaquinone-derivative.

As benzoquinone is toxic to humans, as well as apparently also to the purple bacteria you mentioned in your Faraday Discussions paper, I wonder if you also considered or tested similar less-toxic compounds like for examples substituted naphtha- or anthraquinone derivatives?

1 M. Grattieri, Z. Rhodes, D. P. Hickey, K. Beaver and S. D. Minteer, *ACS Catal.*, 2019, **9**, 867–873.

Shelley Minteer answered: This is a very important question. As correctly mentioned, p-benzoquinone is a toxic compound, and our initial choice to utilize monomeric quinone-based redox mediators was motivated by the need to clarify the extracellular electron transfer (EET) process for purple bacteria, which was not deeply understood. The monomeric mediators allowed a simple system to study the EET, correlating it to the chemical-physical properties of the utilized mediators; however, the use of these monomeric mediators for the development of the technology on field is not suitable. In view of developing a system where no toxic compounds are introduced in solution, we are currently investigating the photo-bioelectrochemical response of *R. capsulatus* in the presence of a redox polymer. The use of less toxic mediating systems is providing us with the possibility to expand our studies to long-term applications.

Julea Butt remarked: A great thing about using bacteria as a (photo)electro-catalyst is that they can be self sustaining and renewable. *Ralstonia* is a great chassis organism for developing related biotechnology while an alternative, for operating in sea water, might be to use halophilic or halotolerant bacteria. What do you see as the challenges of this alternative approach?

Shelley Minteer responded: The aspect pointed out in this question is critical and of great relevance. First of all, it is correct that, thanks to the capability of halophilic or halotolerant bacteria to grow in a broad range of salinities, they could be applied in bioelectrochemical systems operating in saline conditions. As a matter of fact, we are currently using a halotolerant strain of *Salinivibrio* isolated from the Great Salt Lake to develop bioelectrochemical systems operating in hypersaline environments (salinity higher than 35 g L^{-1}, and for our specific case, higher than 100 g L^{-1}), where the extracellular electron transfer happening at the interface of bacteria–electrode surface allow enhancing the removal of contaminants while the obtained electrical current can be used to monitor the degradation process.[1–3] Accordingly, the application of these organisms is of high relevance for decontamination and monitoring of extreme environments; however, the major challenge of this alternative approach is that these organisms utilize a consistent amount of energy obtained from the oxidation of organic substrates to sustain their metabolism and to grow in such extremely saline conditions. As a result, lower current outputs are obtained, leading to the scientific challenges of enhancing the extracellular electron transfer while ensuring the stability and the capability to grow the microorganisms in such saline conditions. On the contrary, the possibility to apply the photosynthetic organisms *Rhodobacter capsulatus* in a salinity range up to 35–40 g L^{-1}, common for sea and ocean waters, would provide us with the advantage of using sunlight as the energy source, enhancing the amount of electrons available for the extracellular electron transfer with an electrode surface.

1 M. Grattieri, M. Suvira, K. Hasan and S. D. Minteer, Halotolerant Extremophile Bacteria from the Great Salt Lake for Recycling Pollutants in Microbial Fuel Cells, *J. Power Sources*, 2017, **356**, 310–318.
2 M. Grattieri, N. D. Shivel, I. Sifat, M. Bestetti and S. D. Minteer, Sustainable Hypersaline Microbial Fuel Cells: Inexpensive Recyclable Polymer Supports for Carbon Nanotube Conductive Paint Anodes, *ChemSusChem*, 2017, **10**, 2053–2058.
3 M. Grattieri, D. P. Hickey, B. Alkotaini, S. J. Robertson and S. D. Minteer, Hypersaline microbial self-powered biosensor with increased sensitivity, *J. Electrochem. Soc.*, 2018, **165**(5), H251–H254.

Erwin Reisner opened a general discussion of the papers by Shelley Minteer, Nicolas Plumeré and Lars Jeuken: You have presented elegant examples of bacterium- and protein-based systems, which rely on the controlled flow of electrons at the bio-electrode interface or through vesicle membranes. However, natural photosynthesis does not only generate low potential electrons, but also produces proton gradients, in particular the 'proton pump' Photosystem II, for the synthesis of chemical energy carriers such as ATP. What would be the prospects and potential merits in your point of view of producing proton gradients in compartmentalized artificial or semi-biological hybrid systems for the synthesis of chemical energy carriers?

Shelley Minteer answered: This is an aspect of extreme relevance. Our point of view is that the possibility to use semi-biological, or engineered hybrid systems, for the synthesis of chemical energy carriers would have the prospect to rethink classical synthetic routes. Specifically, new bio-electrosynthetic routes could open up for the preparation of important chemicals. An example is the bio-electrocatalytic production of ammonia without the need for an external ATP supply, or costly ATP regeneration systems. Such approaches would have the potential merit of overcoming cost-limitations of current synthetic and biosynthetic routes, achieving cost-effective light-driven bioelectrosynthetic systems. While several challenges remain to be solved to fully implement such technologies, the positive outcome of such research efforts would drastically foster the field of more sustainable and green chemical synthesis.

Lars Jeuken answered: There are potentials for utilising proton (or electrochemical) gradients, but this depends on the envisioned application of the light-driven system. Two examples can be given to illustrate this. In the first, electrochemical gradients formed by transmembrane electron transfer could be utilised for, for instance, ATP formation and recycling (from ADP) if the lipid vesicles also include F0F1-ATPase. The latter system might find applications in chemical syntheses that use ATP-driven biocatalytic steps. In a second example, light-driven proton pumps (*e.g.* bacteriorhodopsin) could be incorporated in these lipid vesicles and light energy could be used to actively lower the pH inside the liposomes to enhance the formation of solar fuels such as hydrogen.

We note, however, that in both these (and other) examples, the complexity of the system increases, potentially reducing the feasibility of these systems in future devices.

Samuel E. H. Piper returned to the discussion of the paper by Shelley D. Minteer: What is the species specificity of the gene transfer agent discussed in your work? Can you imagine a system where *Rhodobacter capsulatus* carries out adaptation to a particular environment and then transfers that adaptation to a different species of microorganism?

Shelley Minteer replied: Previous studies have shown the Rhodobacter capsulatus gene transfer agent (rcGTA) to be species specific.[1] Since the discovery of the rcGTA, many other alphaproteobacteria have been seen to have similar gene transfer agents.[2] One such example from the Rhodobacterales family marine bacteria Roseovarius nubinhibens and Ruegeria mobilis, produce gene transfer agents that are capable of transferring genetic material among other bacteria and even in different phyla.[3] Regarding the possibility to use R. capsulatus to transfer the adaptation to a different species of microorganisms, unfortunately the R. capsulatus GTA is not known to be capable of interspecies gene transfer.

However, a better understanding of rcGTA role in adaptation to salinity and its influence on bioelectrocatalysis could be impactful for other GTAs and their organisms''' adaptions to high salinity, as well as other environmental stress factors.

1 J. D. Wall, P. F. Weaver and H. Gest, Gene Transfer Agents, Bacteriophages, and Bacteriocins of *Rhodopseudomonas capsulata*, *Arch. Microbiol.*, 1975, **105**, 217–224.
2 A. S. Lang, O. Zhaxybayeva and J. T. Beatty, Gene transfer agents: phage-like elements of genetic exchange, *Nat. Rev. Microbiol.*, 2012, **10**, 472–482.
3 L. D. McDaniel, E. Young, J. Delaney, F. Ruhnau, K. B. Ritchie and J. H. Paul, High Frequency of Horizontal Gene Transfer in the Oceans, *Science*, 2010, **330**, 50.

Peter Brueggeller returned to the discussion of the paper by N. Plumeré: Could you imagine that your kinetic model is also suitable for very fast electron transfer? Especially in the case of dyads and triads, where the electron transfer occurs intramolecularly, the velocity is no longer diffusion controlled. This means that the algorithm changes and has to be adapted to effects like "superexchange" in the presence of a suitable bridge.

Nicolas Plumeré replied: In our kinetic scheme, the electron mediation from electrode to the photosynthetic protein occurs by electron hopping through redox relays such as metal complexes or small redox proteins. Such electron hopping is modelled based on Fick's laws because electron transfer (defined by an apparent diffusion coefficient of the electron) is governed by a redox gradient in analogy to the transport of a freely diffusing molecule. If the photosensitizer is directly immobilized on the electrode surface and intramolecular electron transfer is delivering the charges, the model needs indeed to be adapted.

Mark Bajada asked: If you are aiming for the software to be open source, and to be used by other scientists in the field, why did you make use of MATLAB rather than Python? Also, the solver used (ode15s) is quite slow, did you encounter any difficulties when running the calculations?

Nicolas Plumeré replied: We chose to use MATLAB because it is a widely used commercial software, and it has many ready-to-use, built-in functions, which reduce the development time. Although we agree that using MATLAB is unlikely to result in the fastest possible simulation times, it would nevertheless take a significantly longer time to develop the code when compared to other approaches, and both considerations are important. For applications which are extremely computationally demanding, for example, the use of another approach (such as coding in C++) might be required, and the additional time and effort required in implementing the code would be justified. For the numerical solutions generated in this work, however, the simulation times achieved using MATLAB were satisfactory. Since it is possible to convert the simulation script into a stand-alone app that can be used without having to purchase MATLAB (which was done in this case), the ability to run simulations according to this model is widely accessible despite the fact that the app itself was produced using commercial software.

Jenny Zhang asked: How do you expect students to use the app to enhance their research? Can it give us feedback on what components to change to improve their systems?

Nicolas Plumeré answered: The first versions of the app have been used by students in their research projects as well as in teaching exercises since mid-2018. The initial feedback shows that the app is beneficial in three ways:

1. Simulations yield not only the photocurrent but also the concentration gradients of all components involved in current generations. This means one can identify the reason for bottlenecks in current generation. The time and space dependent observations of the film behavior are of high pedagogic value since they enable to "visualize" the meaning of the equation defining the photo-electrochemical processes.

2. Parameter screening enables to identify suitable parameter space for planning experimental strategies. In particular the screening of the effect of the parameters that contribute both to the photocatalytic processes and to the competing recombination processes are valuable since they are not accessible otherwise.

3. Simulations based on the app are also very useful to validate quantitatively a hypothesized mechanism based on the agreement between the predicted current and the experiment.

Jenny Zhang returned to the general discussion on the papers of Shelley Minteer, Nicolas Plumeré and Lars Jeuken: What should be the next steps for the field (biological approaches for artificial photosynthesis)? What is needed in your specific area that will help to push this field to a new level?

Shelley Minteer replied: Whether subcellular (proteins) or cellular bio-photoelectrocatalysis, better materials are needed for interfacing biological entities with electrode surfaces.[1] However, bioengineering is also needed. Microbial cells and proteins were not designed to directly communicate with electrodes (for the most part), so further work is needed to engineer them for better interactions.[2-4] Finally, stability is always an issue with biological approaches,[4] so self-healing and regeneration systems are needed to improve the stability issues.

1 R. D. Milton, T. Wang, K. L.Knoche and S. D. Minteer, Tailoring Biointerfaces for Electrocatalysis, *Langmuir*, 2016, **32**, 2291–2301.
2 G. Güven, R. Prodanovic, and U. Schwaneberg, Protein engineering – an option for enzymatic biofuel cell design, *Electroanalysis*, 2010, **22**, 765–775.
3 N. Sekar, R. Jain, Y. Yan and R. P. Ramasamy, Enhanced photo-bioelectrochemical energy conversion by genetically engineered cyanobacteria, *Biotechnol. Bioeng.*, 2016, **113**, 675-679.
4 M. J. Moehlenbrock and S. D. Minteer, Extended lifetime biofuel cells, *Chem. Soc. Rev.*, 2008, **37**, 1188–1196.

Lars Jeuken answered: I agree with the answer of Prof. Shelley Minteer to this question. The field would need better materials to interface biological entities with inorganic materials, be it catalysts, nanomaterials or macro-electrodes. Similarly, the community needs a better understanding of how biological entities interact with materials; this knowledge is required to underpin the informed engineering of microbes and biomacromolecules. Finally, stability of biological systems needs to be optimised. This can be done by using 'self-healing' or regenerative systems (*i.e.*, living microbes). Alternatively, the field of biotechnology is mature and stabilisation of biocatalysts by classical protein engineering is a well established discipline.

Leif Hammarström opened a discussion of the introductory lecture by Matthias Beller: Can you explain more on how these metal single site catalysts work? Should they be described as going through a sequence of oxidation states, and how does the matrix/surface help in stabilizing these transitions? Essentially, can they be described in the same way as molecular catalysts?

Matthias Beller responded: Yes, in principle single metal center catalysts on a surface can be regarded as supported molecular catalysts. Notably, no metal center exists in an isolated form (at least not at ambient conditions) on such a surface, although such materials are often called "single metal atom catalysts". Similar to classic homogeneous catalysts, the metal center is stabilized by coordinating functional groups. However, in contrast to most organometallic complexes, stabilizing functional groups include oxygen and/or nitrogen atoms, while in homogeneous catalysts P-based ligands still prevail.

Joost Reek remarked: Prof Beller started an interesting discussion on batteries *vs.* solar fuel. Of course energy density is a very important factor for mobile applications and as such solar fuels have a big advantage. The question is whether, for stationary functionalities such as households or energy farms, the energy density of the storage material is also important? You can imagine that cheap but large heavy batteries may be sufficient for households, but can we also imagine advantages for solar fuel approaches for these type of applications?

Matthias Beller responded: The energy density for stationary applications is also an important aspect for their practical implementation. In this respect, chemical storage materials such as hydrogen, methane, methanol, or solar fuels are advantageous compared to batteries. Obviously, current technologies for generating these storage materials from renewable energy have to be improved to become cost competitive.

Burkhard König remarked: The vision of a chemical industry collecting carbon dioxide as carbon starting materials directly from the atmosphere is fascinating. What are the big challenges in realizing this? Are the absorber techniques used by current start up companies a good way or do you see alternatives?

Matthias Beller responded: Fully agree. Apart from increasing the efficiency of CO_2 absorption, also performing reactions of CO_2 under ambient conditions could improve the efficiency. In this respect, the tolerance of present abundant gases (oxygen) is challenging. For example, the reduction of CO_2 in the presence of oxygen! For improved photo- and classic thermal reactions the use of multifunctional catalysts might be a solution.
Perhaps we might propose a Schwerpunktprogramm (SPP) of the DFG on this topic?

Moritz F. Kuehnel asked: There is considerable controversy around CO_2 reduction on TiO_2 in the literature, with concerns over observed CO_2 reduction products actually originating from organic contaminants. Can you comment on these concerns, and how you overcame the problems previously encountered in this field?

Matthias Beller replied: I fully agree, this is an important concern. Due to the small amount of reduction products "ultrapure" materials have to be used. We are confident that the observed carbon monoxide stems from carbon dioxide because of labeling studies.

Flavia Cassiola remarked: I agree with your observation that the majority of the work done on artificial photosynthesis has not considered mechanisms for CO_2 capture. The current assumption is that the photosynthesis system will be fed by a concentrated source of pure CO_2. Except for the biological approaches any reported systems have considered direct air capture of CO_2. What are your thoughts on how CO_2 capture from the environment could be considered in the current proposed devices for artificial photosynthesis? What might we be missing from nature and its ability to capture CO_2 from the environment, which is a very diluted and impure source of CO_2?

Matthias Beller responded: To realize carbon dioxide valorization on a large scale and/or to permit for decentralized usage of carbon dioxide, its capture from air is extremely important. In my opinion this problem can only be overcome by cooperative efforts from biology, chemistry and engineering. On the one hand we can be inspired by the natural RuBisCO system; on the other hand artificial receptors based on our understanding of supramolecular binding principles can be used.

Conflicts of interest

There are no conflicts to declare.

Faraday Discussions

PAPER

Photocatalytically active ladder polymers†

Anastasia Vogel, [ID] [a] Mark Forster, [ID] [b] Liam Wilbraham, [ID] [c]
Charlotte L. Smith, [ab] Alexander J. Cowan, [ID] [b]
Martijn A. Zwijnenburg, [ID] [c] Reiner Sebastian Sprick [ID] [a]
and Andrew I. Cooper [ID] *[a]

Received 21st November 2018, Accepted 19th December 2018

DOI: 10.1039/c8fd00197a

Conjugated ladder polymers (cLaPs) are introduced as organic semiconductors for photocatalytic hydrogen evolution from water under sacrificial conditions. Starting from a linear conjugated polymer (cLiP1), two ladder polymers are synthesized *via* post-polymerization annulation and oxidation techniques to generate rigidified, planarized materials bearing dibenzo[b,d]thiophene (cLaP1) and dibenzo[b,d] thiophene sulfone subunits (cLaP2). The high photocatalytic activity of cLaP1 (1307 $\mu mol\ h^{-1}\ g^{-1}$) in comparison to that of cLaP2 (18 $\mu mol\ h^{-1}\ g^{-1}$) under broadband illumination ($\lambda > 295$ nm) in the presence of a hole-scavenger is attributed to a higher yield of long-lived charges (μs to ms timescale), as evidenced by transient absorption spectroscopy. Additionally, cLaP1 has a larger overpotential for proton reduction and thus an increased driving force for the evolution of hydrogen under sacrificial conditions.

Introduction

Clean and sustainable production of hydrogen is one promising strategy for future zero-emission energy supply.[1] In this context, photocatalysis using heterogeneous semiconductors for water splitting has received much attention. Progress has been made in the application of both inorganic[2] and organic semiconductors, the latter triggered by studies on carbon nitride,[3] which have inspired many follow-up studies.[4] The majority of studies focus on half reactions using sacrificial agents to produce either hydrogen or oxygen, but overall water-splitting systems have also been reported that produce both gases.[5] Conjugated polymer semiconductors have gained much attention recently[6] because of their synthetic modularity, the large number of monomers that are available, and the

[a]Department of Chemistry, Materials Innovation Factory, University of Liverpool, Liverpool, UK. E-mail: aicooper@liverpool.ac.uk
[b]Department of Chemistry, Stephenson Institute for Renewable Energy, University of Liverpool, Liverpool, UK
[c]Department of Chemistry, University College London, London, UK
† Electronic supplementary information (ESI) available. See DOI: 10.1039/c8fd00197a

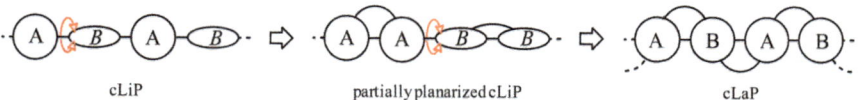

Fig. 1 Graphical representation of conjugated linear polymer (cLiP) with free torsional motions, partially planarized cLiP, and conjugated ladder polymer (cLaP).

resulting tunability in their physical properties. This has triggered the development of a plethora of new types of polymer photocatalysts, including conjugated linear polymers (cLiPs)‡,[7,8] conjugated microporous polymers (CMPs),[9–11] conjugated triazine frameworks (CTFs)[12] and covalent organic frameworks (COFs).[13,14] The modularity of these materials over a wide range of monomer building blocks allows the transfer of photocatalytically active subunits from one class of materials into another. This allows us, in principle, to build structure–property relationships where molecular effects are deconvoluted from solid state packing effects. A complication is that the efficacy of heterogeneous polymer photocatalysts depends on a large number of independent variables, including but not limited to the extent of conjugation and the light absorption cross-section,[15] the residual metal content,[9,11] wettability,[16] thermodynamic driving forces for proton reduction,[15,17] and charge carrier life-times.[18] However, none of these variables have so far been singled out as the most dominant one: instead photocatalytic activity is a complex function of many different interrelated factors, often thwarting attempts to design better catalysts.

This does not mean that there are no viable structural hypotheses for polymer photocatalyst design. For example, previous studies on fluorene-type polymers suggest that partial planarization of poly(p-phenylene) leads to an increase in photocatalytic activity (Fig. 1).[8] Conjugated linear polymers can be further planarized forming a double-stranded polymer; that is, a so-called 'conjugated ladder polymer' (cLaP)[19–21] Ladder polymers restrict the free torsional motion between the monomer units and, in the case of cLaPs (Fig. 1; also **BBL** and **MeLPPP**, Fig. 2B), this leads to a fully coplanar, π-conjugated polymer backbone.[20] Thus, cLaPs tend to exhibit high thermal, optical, and mechanical stability, as well as high resistance to chemical degradation and π-conjugation, along with long exciton diffusion lengths and strong π–π stacking interactions.[19,20] In principle, all of these features are desirable in a polymer photocatalyst. Ladder polymers such as **BBL** (Fig. 2B) have been shown to have high electron mobilities when compared to the conjugated, non-ladderized parent polymer **BBB**.[22] Also, the degree of order within the conjugated ladder polymer **MeLPPP** has been highlighted to be a major contributor to its high charge carrier mobility, which resembles molecular crystals more than conventional, less ordered conjugated polymers.[23]

Conjugated linear polymers containing dibenzo[b,d]thiophene sulfone units (as in **P7**, Fig. 2A) have been shown repeatedly to outperform most other organic photocatalysts.[8,14,24–26] This is attributed to the co-planarization of neighbouring subunits in the polymer, increased hydrophilicity, and strong visible light

‡ In the literature, both conjugated linear and conjugated ladder polymers are abbreviated as cLPs.[20] For the purposes of this paper, we will use the abbreviation cLiP for linear polymers and cLaP for ladderized polymers.

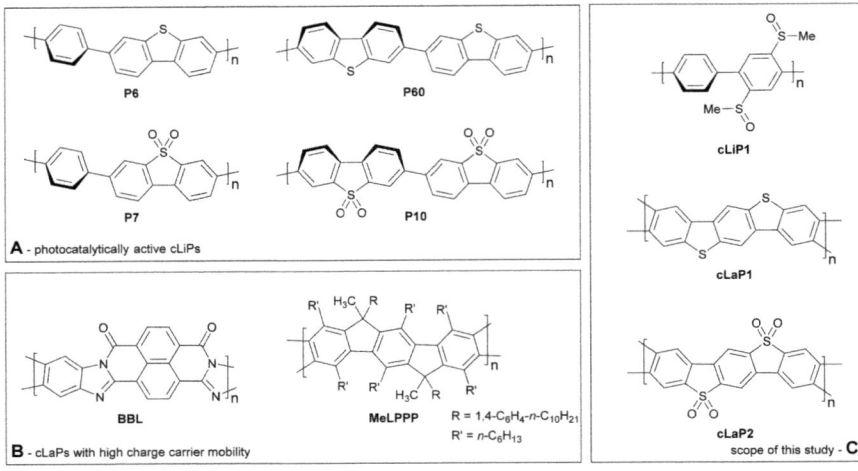

Fig. 2 (a) Photocatalytically active conjugated linear polymers containing dibenzo[b,d] thiophene and dibenzo[b,d]thiophene sulfone building blocks; (b) conjugated ladder polymers with reported high charge carrier mobilities; (c) scope of this study including parent linear conjugated polymer cLiP1 and conjugated ladder polymers cLaP1 and cLaP2.

absorption.[8,14] While there is one report on the use of **BBL** as a photoanode for photoelectrochemical water splitting,[27] no ladder polymers have been reported as bulk powdered photocatalysts for direct water splitting.

Here, we set out to combine structural features derived from highly active linear polymer photocatalysts with the increased conjugation that might be expected for a conjugated ladder polymer. We did this by synthesizing and analyzing a series of related linear (**cLiP1**) and ladder polymers based on dibenzo [b,d]thiophene (**cLaP1**) and dibenzo[b,d]thiophene sulfone (**cLaP2**) units (Fig. 2C), followed by testing these materials for sacrificial photocatalytic water reduction.

Results

Synthesis and characterization

In contrast to the cross-coupling reactions (e.g., Suzuki–Miyaura, Stille or Kumada coupling) that are typically used to yield linear conjugated polymers in one step, ladder polymers are usually synthesized by either (A) a single-step poly-condensation reaction of tetra-functionalized building blocks or (B) polymeriza-tion and subsequent annulation of a bi-functionalized subunit to give a ladderized polymer.[20] For the synthesis of ladder polymers containing dibenzo [b,d]thiophene and dibenzo[b,d]thiophene sulfone, polymerization of an aryl bi-sulfoxide 1 via route B was chosen. Building block 1 was synthesized from 1,4-dibromobenzene (Scheme 1) according to the literature.[28] Polymerization of 1 with 1,4-benzene diboronic acid ester via a Pd(0)-catalyzed Suzuki–Miyaura cross-coupling reaction gave the parent polymer **cLiP1**.[29] An intramolecular ring-closing reaction was then performed using trifluoromethanesulfonic acid (TfOH),[29] and the intermediate polysulfonium salt (**cLaP⁺-Me**) was dealkylated using NEt₄Br to give the ladderized polymer **cLaP1**. Further oxidation of **cLaP1** with H₂O₂ in

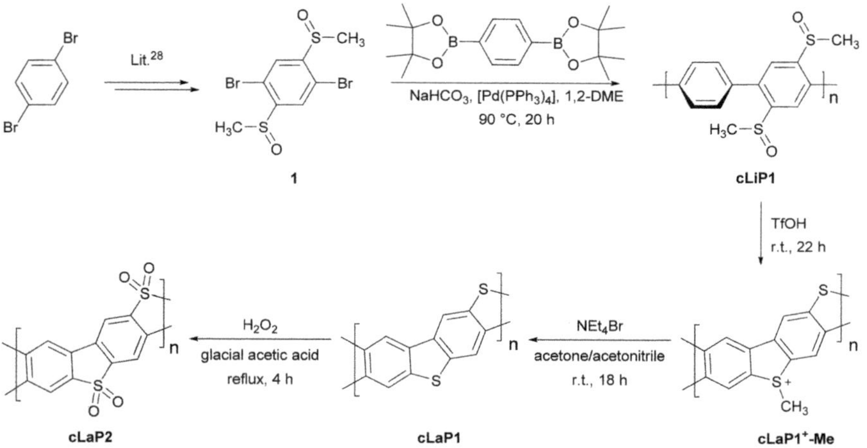

Scheme 1 The synthesis route of conjugated linear polymer cLiP1 and the related conjugated ladder polymers cLaP1 and cLaP2.

glacial acetic acid gives **cLaP2**, which is a ladder-type analogue of the linear dibenzo[b,d]thiophene sulfone polymer (**P10**).[30] ^{1}H{^{13}C} NMR and IR spectra for previously reported compounds (see the Experimental section of the ESI and Fig. S10 and S11†) agree with the literature. It is worth noting that no convenient method is available to quantify the degree of ladderization and degree of dealkylation since all materials were insoluble in common organic solvents. The total insolubility of these materials also precludes molecular weight determination.

Fourier-transform infrared spectroscopy (FT-IR) was used to analyze the insoluble polymers: **cLiP1** has a stretching vibration at 1032 cm^{-1} that can be attributed to the sulfoxide group ($\tilde{\nu}_{S=O}$) and, in the fingerprint region, sharp peaks (837, 747, 674 cm^{-1}) corresponding to wagging C–H and C–C vibrations for 1,4- and 1,2,4,5-substituted benzene subunits are observed (Fig. S10†). Upon ladderization to **cLaP1**, the spectrum is dominated by strong peaks attributed to various stretching vibrations of the triflate anion (635, 1224 cm^{-1} ($\tilde{\nu}_{C-F}$) and 1156 cm^{-1} ($\tilde{\nu}_{O=S=O}$), see also Fig. S11†). The fingerprint region shows a single broad peak at 832 cm^{-1} consistent with a polymer containing 1,2,4,5-substituted benzene subunits. The signals of the triflate anion (compare Fig. S11,† right) indicate that not all of the intermediate sulfonium subunits were demethylated when treated with NEt$_4$Br, despite further attempts to optimize the conditions. Furthermore, the presence or absence of the sulfoxide group (1010 cm^{-1} ($\tilde{\nu}_{S=O}$)) after the ladderization could not be detected due to the overlapping signals of the triflate anion. Thus, no further conclusions on the success rate of the annulation or the presence of defects (non-annulated aryl sulfoxide groups) could be made at this point. Upon oxidation to **cLaP2**, the spectrum shows two strong stretching vibrations (1307 cm^{-1} and 1148 cm^{-1}; symmetric and asymmetric $\tilde{\nu}_{O=S=O}$) indicating the oxidation of the dibenzo[b,d] thiophene moiety to dibenzo[b,d]thiophene dioxide (Fig. S12†). Additionally, no signals associated with the presence of triflate anions were observed. There is also no evidence for over-oxidation nor ring-opening reactions of the polymer to yield sulfonic acid groups.

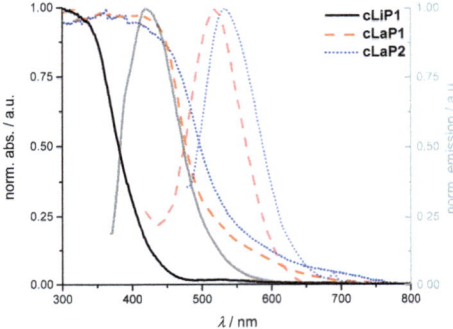

Fig. 3 Normalized UV-vis absorption and photoluminescence emission spectra of **cLiP1** (black, full lines), **cLaP1** (red, dashed lines) and **cLaP2** (blue, dotted lines) from powder samples.

Powder X-ray diffraction patterns (PXRD) show that polymer **cLiP1** has limited long-range order, while **cLaP1** and **cLaP2** are amorphous (Fig. S17–S19†). Scanning electron microscopy (SEM) shows that all of the materials consist of *ca.* 100 nm spheres that are fused together (Fig. S22†). UV-vis and photo-luminescence (PL) spectroscopy were used to probe the optoelectronic properties of **cLiP1**, **cLaP1** and **cLaP2** measured as powders in the solid-state (Fig. 3 and Fig. S1–S9†): polymer **cLiP1** absorbs mostly in the UV region with an absorption edge around 415 nm. Upon ladderization of **cLiP1**, the absorption edge shifts by about 100 nm for **cLaP1**§ and this clearly shows that the system has a higher degree of conjugation as the annulation of neighbouring conjugated units reduces their respective torsion angles close to zero degrees as the system is rigidified (compare Fig. S39†). As expected,[8] the oxidation to **cLaP2** led to only minor changes in the UV-vis spectrum. Polymer **cLiP1** has an estimated optical gap of 3.01 eV, while both ladderized polymers **cLaP1** and **cLaP2** show narrower optical gaps of 2.41 and 2.30 eV (for Tauc plots see Fig. S1–S5†), respectively.

The fluorescence emission spectra in the solid state show the same bath-ochromic shifts as the UV-vis absorption spectra. Moreover, a smaller Stokes' shift and resolved vibronic coupling in the excitation spectrum of **cLaP2** and **cLaP1** compared to **cLiP1** (Fig. S6–S9†) underlines once more the increased rigidity (and symmetry) of the polymer upon ladderization. Time-correlated single photon counting (TCSPC) was used to estimate the lifetime of the excited state in an aqueous suspension (Fig. S32–S34 and Table S3†). Polymer **cLiP1** has a short lifetime of 0.14 ns, which is similar to that of **cLaP1** with 0.21 ns. The dibenzo[*b*,*d*] thiophene sulfone polymer **cLaP2** has the longest lifetime of the materials studied ($\tau_{avg} = 1.71$ ns), as observed previously.[31]

Photocatalytic performance

The photocatalytic activity of these materials for hydrogen evolution from water in the presence of triethylamine (TEA) as a sacrificial electron donor was studied

§ UV-vis and fluorescence spectra for **cLaP1** were recorded using material recovered from experiments for hydrogen evolution as this represents the active catalysts best. A UV-vis spectrum of the as-synthesized **cLaP1** (or rather **cLaP1⁺-Me**) can be found in the ESI.†

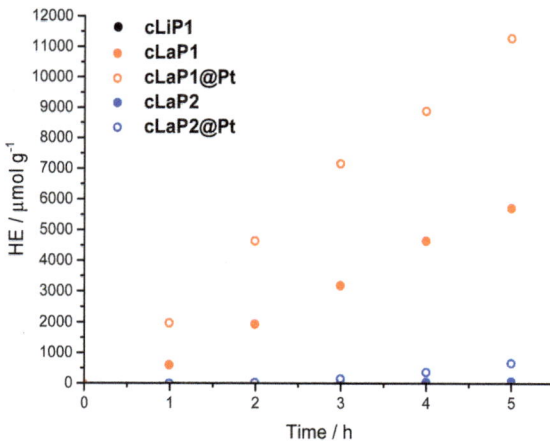

Fig. 4 Hydrogen evolution of **cLiP1**, **cLaP1** and **cLaP2** as well as **cLaP1@Pt** and **cLaP2@Pt** from a H$_2$O/TEA/MeOH mixture under broadband irradiation (300 W Xe light source, $\lambda > $ 295 nm).

under broad-spectrum and visible light irradiation ($\lambda > 295$ nm and $\lambda > 420$ nm; Fig. 4 and Table 1). In addition, methanol was used in the aqueous mixture to enhance miscibility of TEA with water, and to improve wettability of the polymers.[8,33] Polymer **cLiP1** showed very limited activity (15 µmol h^{-1} g^{-1}), even compared to poly(p-phenylene) (232 µmol h^{-1} g^{-1}).[8] Upon ladderization to **cLaP1**, the photocatalytic activity increased dramatically to 1307 µmol h^{-1} g^{-1}. When the material was recovered and used again as a photocatalyst, an increase in the hydrogen evolution rate to over 2000 µmol h^{-1} g^{-1} was observed (Fig. S32†). This can possibly be explained by further demethylation of the catalyst during catalysis by TEA, which reduces the doping levels of the photocatalyst, and a lower effective mass for the recovered catalyst and thus higher HER per gram. This is supported by post-catalysis FT-IR measurements (Fig. S14†): peaks associated with the presence of the triflate counterion are no longer present, and thus a no longer charged polymer species has to be assumed after irradiation. The oxidation of **cLaP1** to **cLaP2**, led to an almost total loss of activity (18 µmol h^{-1} g^{-1}).

All of the materials were tested as synthesized without the addition of any additional metal co-catalyst. However, it has been shown that residual palladium originating from the Suzuki–Miyaura coupling reaction can act as a co-catalyst.[9,34] Inductively coupled plasma mass spectrometry (ICP-MS) measurements show that the residual palladium content decreased from 0.83 wt% for **cLiP1** to 0.38 and 0.36 wt% for **cLaP1** and **cLaP2**, respectively. The lower residual palladium amount might be due to the use of trifluoromethanesulfonic acid and repeated washing of **cLaP1** after the ring-closure reaction. Similarly, the oxidation giving **cLaP2** is performed in acetic acid. The use of acids has been previously shown to decrease the residual palladium loadings of insoluble conjugated microporous polymers.[35] Finally, the photocatalytic activity of **cLaP1** was increased from 317 to 1489 µmol h^{-1} g^{-1} under visible light illumination and from 1307 to 2297 µmol h^{-1} g^{-1} under broadband illumination by *in situ* photodeposition of platinum as

Table 1 Photophysical properties, hydrogen evolution rates and palladium content

Polymer	λ_{edge} [a]/ nm	E_g^{Tauc} [b]/ eV	λ_{em} [c]/ nm	HER ($\lambda > 295$ nm) [d]/ $\mu mol\ h^{-1}\ g^{-1}$	HER ($\lambda > 420$ nm) [d]/ $\mu mol\ h^{-1}\ g^{-1}$	Pd content [e]/ wt%
cLiP1	415	3.01	433	15 ± 2	0^f	0.83
cLaP1	514	2.41	520	1307 ± 26	317 ± 9	0.38
cLaP1@Pt	—	—	—	2297 ± 92	1489 ± 24	0.38 (+1 Pt)^g
cLaP2	548	2.30	534	18 ± 1	0^f	0.36
cLaP2@Pt	—	—	—	272 ± 10	184 ± 7	0.36 (+1 Pt)^g
P60	454	2.68	460	1295 ± 36	641 ± 20	0.49
P60@Pt	—	—	—	1703 ± 102	213 ± 5	0.49 (+1 Pt)^g
P6 (Lit. 8)	448	—	456, 481	1660 ± 12	432 ± 4	0.60
P7 (Lit. 8)	459	—	477	2352 ± 76	1492 ± 32	0.38
P10 (Lit. 26)	473	—	509	—	3260 ± 164	0.40

[a] Absorption onset determined from UV-vis reflectance measurements in the solid state. [b] Optical gap determined from absorption spectra using the Tauc method.[32] [c] Emission peak maximum determined in the solid state. [d] Hydrogen evolution rate determined in H_2O/TEA/MeOH irradiated with a 300 W Xe light source using suitable filters. [e] Palladium content determined via ICP-MS. [f] No hydrogen was detected with five hours of irradiation. [g] Platinum was photodeposited in situ onto the polymer from H_2PtCl_6.

co-catalyst (**cLaP1@Pt**, 1 wt%). Similarly, **cLaP2** showed a 10-fold increase in activity (from 18 to 272 $\mu mol\ h^{-1}\ g^{-1}$) under broadband illumination with platinum as co-catalyst (**cLaP2@Pt**, 1 wt%).

External quantum efficiencies (EQEs) were estimated for **cLaP1** in H_2O/TEA/ MeOH mixtures using monochromatic light and these show that the hydrogen production is indeed photocatalytic (Fig. S31†). At 420 nm an EQE of 1.6% was determined, which increased to 2.8% upon addition of platinum (**cLaP1@Pt**, 1 wt %). This is higher than previously reported for **P1** ($EQE_{420\ nm} = 0.4\%$)[26] and biphenyl-thiophene-co-polymer **P12** ($EQE_{420\ nm} = 1.4\%$)[15] under the same conditions, but lower than that of phenylene-benzothiadiazole-co-polymer **B-BT-1,4** ($EQE_{420\ nm} = 4.0\%$)[7a] in triethanolamine/water mixture loaded with platinum, and phenylene–dibenzo[b,d]thiophene sulfone **P7** ($EQE_{420\ nm} = 7.2\%$)[26] in a H_2O/ TEA/MeOH mixture.

Transient absorption spectroscopy

To try to explain the differences in catalytic activity, we studied the charge carrier lifetimes using transient absorption (TA) spectroscopy. TA has been shown to be an effective tool for studying the formation and lifetime of electron–polaron states in polymer photocatalysts for hydrogen evolution.[26,36] Here we focused on the kinetics of species present on the μs to ms timescales following UV (355 nm) excitation of **cLiP1**, **cLaP1** and **cLaP2** in the presence of a H_2O/TEA/MeOH mixture under a nitrogen atmosphere (Fig. 5). All three materials exhibited transient absorptions between 400–900 nm on the timescale probed, however the amplitude of the TA signal of **cLaP1** was far greater ($\times 5$–10) than those of **cLiP1** and **cLaP2**, indicating an increase in long-lived photogenerated species. The TA spectrum of **cLaP1** contains two distinct absorptions centred at ca. 500 and 630 nm that decay at different rates (Fig. S35 and S36,† $t_{50\%} \sim 18\ \mu s$ (500 nm), 25

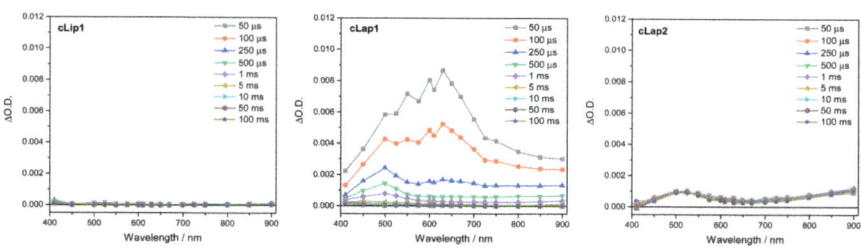

Fig. 5   µs to ms TA spectra (same scale on all y-axes) of **cLiP1**, **cLaP1** and **cLaP2** suspended in a H_2O/TEA/MeOH mixture following excitation with a 355 nm laser (6 ns, 400 µJ cm^{-2}, 0.33 Hz). The large TA features seen with **cLaP1** correlate with the high rate of H_2 evolution.

µs (630 nm) under N_2).¶ In the absence of methanol and TEA the long-lived TA bands at 500 and 630 nm are no longer observed (Fig. 6). The role of TEA has previously been investigated with ultrafast TA spectroscopy[26,36] and it was shown to be required for efficient hole scavenging and for the formation of long-lived electron polarons that are suitable for proton reduction. In the absence of TEA, the excitonic states would be expected to decay rapidly after formation on time-scales faster than those studied here (ps to ns), in line with TCSPC measurements.[37] We therefore propose that the bands at 500 and 630 nm are due to two distinct electron populations within the sample **cLaP1**. Supporting our assignment are TA experiments carried out using oxygen as an electron scavenger. Introduction of O_2 into the system significantly decreases the TA signal at both 500 and 630 nm indicating the removal of electron populations (Fig. S37†). The observation of a long-lived electron signal at 630 nm for **cLaP1** is in line with a recent assignment of an electron polaron state in related dibenzo[b,d]thiophene sulfone linear polymers.[26] However, the assignment of both individual TA features to specific electron populations, potentially related to the presence of some residual sulfonium subunits, is challenging as the transient UV/vis spectra measured contain broad bands and our experimental resolution prevents the observation of fine structure.

Although the magnitudes of the TA signals of **cLaP2** are far smaller than those of **cLaP1** on the µs timescale, the kinetic traces recorded at 500 nm for **cLaP2** do indicate an extremely long-lived ($t_{50\%} = 0.3$ s) photogenerated species. The presence of extremely long-lived charges, potentially with a low thermodynamic driving force, may also be a factor behind the low HER observed for **cLaP2**. To explore this observation further, Pt was added as a co-catalyst in the hope that it may be able to either intercept photoelectrons and prevent trapping or offer suitable catalytic sites to facilitate charge transfer into solution. **cLaP2@Pt** does show an improvement in HER, increasing from 18 µmol h^{-1} g^{-1} to 272 µmol h^{-1} g^{-1} (Table 1), but a comparison of the TA spectra of **cLaP2** and **cLaP2@Pt** shows no clear difference under nitrogen (Fig. S38†), suggesting that the charges observed at 500 nm do not transfer to Pt and that the hydrogen evolution

---

¶ We note that the lifetimes of all of the bands are dependent upon the history of the sample and tend to decrease after prolonged experiments. However, the signal at 500 nm is consistently shorter-lived compared to the signal at 625 nm.

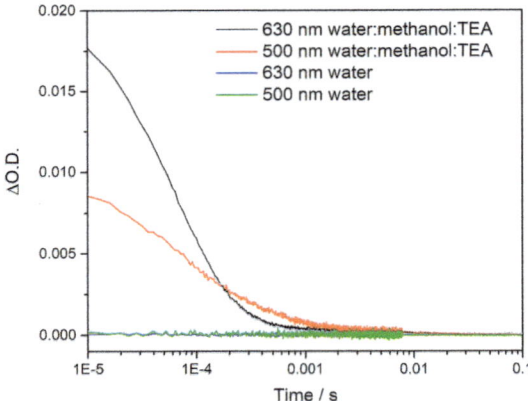

Fig. 6 Kinetic traces recorded at the wavelengths indicated for **cLaP1** in either water alone or a H$_2$O/TEA/MeOH mixture under a nitrogen atmosphere following excitation with a 355 nm (6 ns, 400 μJ cm^{-2}, 0.33 Hz) laser.

catalysed with Pt happens either on a sub 10 μs timescale or that the electron population has absorption features outside of our spectral window. The addition of a Pt co-catalyst also increases the rate of hydrogen evolution in **cLap1** (Table 1). In this case, the increase in HER evolution following Pt addition is accompanied by a significant decrease in the TA magnitudes at both 500 and 630 nm (Fig. S37†) indicating that these previously long-lived charges are now transferring to Pt and on timescales faster than those in our experiment.

Calculations

To gain insights into changes in the thermodynamic driving force for proton reduction and TEA oxidation within the investigated series of polymers, the IP and EA levels as well as optical gaps for varying oligomer lengths (1–9; defined by the number of benzene bi-sulfoxide, thiophene or thiophene dioxide units, respectively) were estimated using a family of recently developed semi-empirical density functional tight-binding methods.[38] Use of such methods, accompanied by a calibration procedure, has been shown to provide accurate optoelectronic properties with accuracy comparable to density functional theory.[39] Different oligomer lengths were tested to ensure that converged values are obtained across both ladder and non-ladder polymer species. Fig. 7 shows the calculated IP and EA values compared to the hydrogen reduction and TEA oxidation potentials at pH = 11.5. We see that each of these polymers, both ladder and non-ladder, can create charge carriers with sufficient thermodynamic driving force to drive the necessary redox chemistry required for hydrogen evolution and TEA oxidation. Both **cLaP1/cLiP1** are predicted to have a larger driving force for proton reduction than **cLaP2**, while **cLaP2/cLiP1** have a larger driving force for overall TEA oxidation to diethylamine and acetaldehyde than **cLaP1**. Assuming that overall TEA oxidation can be effectively described as a combination of two subsequent one-hole transfer-steps to TEA species in solution (I: TEA + h$^+$ → TEA$^+$ → TEA radical (TEAR) + H$^+$ and II: TEAR + h$^+$ + H$_2$O → diethylamine + acetaldehyde + H$^+$), then the fact that **cLaP1** is predicted to have only negligible driving force for the

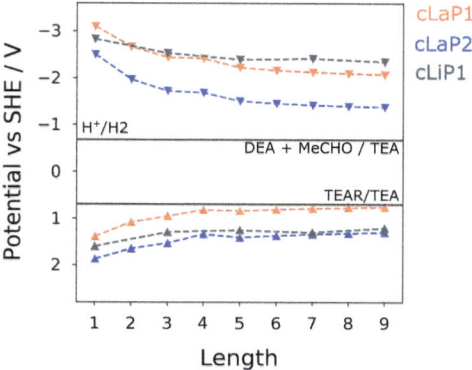

Fig. 7 Predicted IP and EA values (*vs.* SHE) of various (ladder) polymers. IP and EA values have been computed for various oligomer lengths, where 'length' is equal to the number of aromatic rings along the polymer backbone. Hydrogen reduction and TEA oxidation potentials (pH = 11.5) are shown as horizontal lines.

first step, in contrast to **cLaP2**, might suggest that differences in catalytic activity between **cLaP1** and **cLaP2** stems from differences in the driving force for proton reduction. In line with experimental UV-vis absorption spectra, ladder polymers **cLaP1** and **cLaP2** are predicted to have optical gaps that are ~1.0 and ~0.8 eV lower for **cLaP1** and **cLaP2**, respectively, than for **cLiP1** (see Table S4†). Hence, **cLaP1** and **cLaP2** absorb more of the visible spectrum. The larger predicted optical gap of **cLaP1** relative to **cLaP2** (by approximately 0.1 eV) is also in line with experimental spectra.

Discussion

The ladder polymer **cLaP1** outperforms its non-annulated parent polymer, **cLiP1**, significantly, and the optical properties of the material after annulation show a red-shift in the absorption onset. The ability of **cLaP1** to absorb more photons while maintaining a hydrogen reduction driving force at least partially explains its higher photocatalytic activity, especially under filtered visible light. No significant changes in the optical properties were observed upon oxidation of **cLaP1** to **cLaP2**, but the resulting dibenzo[*b,d*]thiophene sulfone material is almost inactive. This low activity of **cLaP2** is surprising since TCSPC shows that **cLaP2** has the longest weighted average lifetime of the exited state; significantly longer than those of **cLiP1** and **cLaP1**. Also, the introduction of dibenzo[*b,d*]thiophene sulfone motifs into other polymer photocatalysts has been reported to give materials with high photocatalytic activity.[8,24,25,31] However, when taking the computationally predicted charge-carrier potentials into account, it becomes clear that **cLaP2** has a reduced overpotential for proton reduction relative to **cLaP1**, while **cLaP1** and **cLaP2** both have a reasonable driving force for TEA oxidation. From a thermodynamic perspective, **cLaP1** thus appears to be the best material in terms of combining thermodynamic driving force with light absorption. This is supported by TA measurements, which show the highest yield of long-lived charges in the case of **cLaP1**. From the TA data, it is clear that there is a direct correlation between the yield of long-lived charges present and the measured hydrogen evolution rate, suggesting that electron polaron states with lifetimes on the μs to ms timescale are required in order for hydrogen evolution to occur. This might rationalize the higher hydrogen evolution yields for **cLaP1**. The greater yield of long-lived photoelectrons may be related to more efficient hole scavenging at early timescales. The catalytic activity of the materials does not correlate with the residual palladium content, but it is unclear whether the threshold for an effect of the residual palladium on the photocatalytic performance has been reached.[9,34] Since the TA spectrum of **cLaP2** shows a persistent long-lived feature, which could be attributed to a deep-trapped charge (Fig. 5, right), we loaded *in situ* **cLaP1** and **cLaP2** with a platinum co-catalyst (1 wt%). In both cases, we observed an increase in photocatalytic performance, but there was no evidence for higher charge carrier yields in the TA spectrum of the platinized **cLaP2**.

In summary, the differences in photocatalytic activity in the series of **cLiP1**, **cLaP1** and **cLaP2** can be rationalized by comparison of charge carrier lifetimes, light absorption, and thermodynamic driving forces. Compared with related linear polymers, such as **P6/P60** and **P7/P10** (Fig. 2A), the idea of extending planarization across the full length of the polymer chain to enhance photocatalytic activity was not realized, at least not without the addition of a co-catalyst. For example, the photocatalytic hydrogen evolution rates for unmodified **cLaP1** (1307 μmol h^{-1} g^{-1}) were found to be similar to those of its linear, non-ladderized analogues **P6** (1660 μmol h^{-1} g^{-1}) and **P60** (1295 μmol h^{-1} g^{-1}). However, when Pt was used as a co-catalyst, **cLaP1@Pt** outperformed **P60@Pt** under both broadband (2297 *vs.* 1703 μmol h^{-1} g^{-1}) and visible light irradiation (1489 *vs.* 213 μmol h^{-1} g^{-1}).

Conclusions

Inspired by photocatalytically active dibenzo[*b,d*]thiophene and dibenzo[*b,d*] thiophene sulfone polymers, we set out to synthesize a new class of ladderized conjugated polymer photocatalysts for photocatalytic evolution of hydrogen from water. Through post-polymerization ladderization, a planarization of the polymer chain and expansion of the π-system could be achieved, as evidenced by the bathochromic shift of the absorption edge. A significant increase in photo-catalytic activity was measured for one ladder polymer (**cLaP1**) while the other (**cLaP2**) remained almost inactive. The difference in photocatalytic activity could be rationalized by analysis of the charge carrier lifetimes *via* TA spectroscopy and comparison of the driving forces derived from calculations. These results suggest that post-polymerization ladderization could be a valuable technique in the preparation of efficient photocatalysts and that ladder polymers containing other photocatalytically active subunits might be considered for future studies.

Conflicts of interest

There are no conflicts to declare.

Acknowledgements

This project has received funding from the European Union's Horizon 2020 research and innovation programme (Marie-Skłodowska-Curie Individual Fellowship to AV) under grant agreement No. 796322. The UK Engineering and

Physical Sciences Research Council (EPSRC) is acknowledged for funding through grants EP/N004884/1 and EP/P034497/1. We thank Catherine M. Aitchison for help with SEM imaging.

Notes and references

1 (a) P. P. Edwards, V. L. Kuznetsov, W. I. F. David and N. P. Brandon, *Energy Policy*, 2008, **36**, 4356–4362; (b) S. Chu and A. Majumdar, *Nature*, 2012, **488**, 294–303; (c) J. M. Ogden, *Phys. Today*, 2002, **55**, 69–75; (d) A. Züttel, A. Borgschulte and L. Schlapbach, *Hydrogen as a Future Energy Carrier*, Wiley, Weinheim, 2011.

2 (a) X. Chen, S. Shen, L. Guo and S. S. Mao, *Chem. Rev.*, 2010, **110**, 6503–6570; (b) Z. Wang, C. Li and K. Domen, *Chem. Soc. Rev.*, 2019, DOI: 10.1039/c8cs00542g; (c) D. Kong, Y. Zheng, M. Kobielusz, Y. Wang, Z. Bai, W. Macyk, X. Wang and J. Tang, *Mater. Today*, 2018, **21**, 897–924.

3 X. Wang, K. Maeda, A. Thomas, K. Takanabe, G. Xin, J. M. Carlsson, K. Domen and M. Antonietti, *Nat. Mater.*, 2009, **8**, 76–80.

4 (a) L. Lin, Z. Yu and X. Wang, *Angew. Chem., Int. Ed.*, 2018, DOI: 10.1002/anie.201809897; (b) W.-J. Ong, L.-L. Tan, Y. H. Ng, S.-T. Yong and S.-P. Chai, *Chem. Rev.*, 2016, **116**, 7159–7329.

5 (a) D. J. Martin, P. J. T. Reardon, S. J. A. Moniz and J. Tang, *J. Am. Chem. Soc.*, 2014, **136**, 12568–12571; (b) G. Zhang, Z.-A. Lan and X. Wang, *Chem. Sci.*, 2017, **8**, 5261–5274; (c) L. Lin, C. Wang, W. Ren, H. Ou, Y. Zhang and X. Wang, *Chem. Sci.*, 2017, **315**, 798; (d) J. Liu, Y. Liu, N. Liu, Y. Han, X. Zhang, H. Huang, Y. Lifshitz, S.-T. Lee, J. Zhong and Z. Kang, *Science*, 2015, **347**, 970–974.

6 (a) G. Zhang, Z.-A. Lan and X. Wang, *Angew. Chem., Int. Ed.*, 2016, **55**, 15712–15727; (b) V. S. Vyas, V. W.-h. Lau and B. V. Lotsch, *Chem. Mater.*, 2016, **28**, 5191–5204.

7 (a) C. Yang, B. C. Ma, L. Zhang, S. Lin, S. Ghasimi, K. Landfester, K. A. I. Zhang and X. Wang, *Angew. Chem., Int. Ed.*, 2016, **55**, 9202–9206; (b) X. Zong, X. Miao, S. Hua, L. An, X. Gao, W. Jiang, D. Qu, Z. Zhou, X. Liu and Z. Sun, *Appl. Catal., B*, 2017, **211**, 98–105; (c) L. Li, R. G. Hadt, S. Yao, W.-Y. Lo, Z. Cai, Q. Wu, B. Pandit, L. X. Chen and L. Yu, *Chem. Mater.*, 2016, **28**, 5394–5399.

8 R. S. Sprick, B. Bonillo, R. Clowes, P. Guiglion, N. J. Brownbill, B. J. Slater, F. Blanc, M. A. Zwijnenburg, D. J. Adams and A. I. Cooper, *Angew. Chem., Int. Ed.*, 2016, **55**, 1792–1796.

9 L. Li, Z. Cai, Q. Wu, W.-Y. Lo, N. Zhang, L. X. Chen and L. Yu, *J. Am. Chem. Soc.*, 2016, **138**, 7681–7686.

10 Y. Xu, C. Zhang, P. Mu, N. Mao, X. Wang, Q. He, F. Wang and J.-X. Jiang, *Sci. China: Chem.*, 2017, **46**, 8574.

11 L. Li and Z. Cai, *Polym. Chem.*, 2016, **7**, 4937–4943.

12 (a) K. Schwinghammer, S. Hug, M. B. Mesch, J. Senker and B. V. Lotsch, *Energy Environ. Sci.*, 2015, **8**, 3345–3353; (b) S. Kuecken, A. Acharjya, L. Zhi, M. Schwarze, R. Schomäcker and A. Thomas, *Chem. Commun.*, 2017, **53**, 5854–5857; (c) L. Li, W. Fang, P. Zhang, J. Bi, Y. He, J. Wang and W. Su, *J. Mater. Chem. A*, 2016, **4**, 12402–12406.

13 (a) L. Stegbauer, K. Schwinghammer and B. V. Lotsch, *Chem. Sci.*, 2014, **5**, 2789–2793; (b) V. S. Vyas, F. Haase, L. Stegbauer, G. Savasci, F. Podjaski, C. Ochsenfeld and B. V. Lotsch, *Nat. Commun.*, 2015, **6**, 8508.

14 X. Wang, L. Chen, S. Y. Chong, M. A. Little, Y. Wu, W.-H. Zhu, R. Clowes, Y. Yan, M. A. Zwijnenburg, R. S. Sprick and A. I. Cooper, *Nat. Chem.*, 2018, **10**, 1180–1189.

15 R. S. Sprick, C. M. Aitchison, E. Berardo, L. Turcani, L. Wilbraham, B. M. Alston, K. E. Jelfs, M. A. Zwijnenburg and A. I. Cooper, *J. Mater. Chem. A*, 2018, **6**, 11994–12003.

16 (a) R. S. Sprick, L. Wilbraham, Y. Bai, P. Guiglion, A. Monti, R. Clowes, A. I. Cooper and M. A. Zwijnenburg, *Chem. Mater.*, 2018, **30**, 5733–5742; (b) Y. Wang, M. K. Bayazit, S. J. A. Moniz, Q. Ruan, C. C. Lau, N. Martsinovich and J. Tang, *Energy Environ. Sci.*, 2017, **38**, 253.

17 C. B. Meier, R. S. Sprick, A. Monti, P. Guiglion, J.-S. M. Lee, M. A. Zwijnenburg and A. I. Cooper, *Polymer*, 2017, **126**, 283–290.

18 (a) O. Elbanna, M. Fujitsuka and T. Majima, *ACS Appl. Mater. Interfaces*, 2017, **9**, 34844–34854; (b) R. Godin, A. Kafizas and J. R. Durrant, *Curr. Opin. Electrochem.*, 2017, **2**, 136–143.

19 U. Scherf, *J. Mater. Chem.*, 1999, **9**, 1853–1864.

20 J. Lee, A. J. Kalin, T. Yuan, M. Al-Hashimi and L. Fang, *Chem. Sci.*, 2017, **8**, 2503–2521.

21 R. G. Jones, J. Kahovec, R. Stepto, E. S. Wilks, M. Hess, T. Kitayama and W. V. Metanomski, in *Compendium of Polymer Terminology and Nomenclature: IUPAC Recommendations 2008*, ed. R. G. Jones, E. S. Wilks, W. V. Metanomski, J. Kahovec, M. Hess, R. Stepto and T. Kitayama, Royal Society of Chemistry, Cambridge, 2009, vol. 16, pp. 318–335.

22 A. Babel and S. A. Jenekhe, *J. Am. Chem. Soc.*, 2003, **125**, 13656–13657.

23 D. Hertel, U. Scherf and H. Bässler, *Adv. Mater.*, 1998, **10**, 1119–1122.

24 Z. Wang, X. Yang, T. Yang, Y. Zhao, F. Wang, Y. Chen, J. H. Zeng, C. Yan, F. Huang and J.-X. Jiang, *ACS Catal.*, 2018, **8**, 8590–8596.

25 C. Dai, S. Xu, W. Liu, X. Gong, M. Panahandeh-Fard, Z. Liu, D. Zhang, C. Xue, K. P. Loh and B. Liu, *Small*, 2018, **14**, 1801839.

26 M. Sachs, R. S. Sprick, D. Pearce, S. A. J. Hillman, A. Monti, A. Y. Guilbert, N. J. Brownbill, S. Dimitrov, X. Shi, F. Blanc, M. A. Zwijnenburg, J. Nelson, J. R. Durrant and A. I. Cooper, *Nat. Commun.*, 2018, **9**, 4968.

27 P. Bornoz, M. S. Prévot, X. Yu, N. Guijarro and K. Sivula, *J. Am. Chem. Soc.*, 2015, **137**, 15338–15341.

28 P. Gao, X. Feng, X. Yang, V. Enkelmann, M. Baumgarten and K. Müllen, *J. Org. Chem.*, 2008, **73**, 9207–9213.

29 A. Haryono, K. Miyatake, J. Natori and E. Tsuchida, *Macromolecules*, 1999, **32**, 3146–3149.

30 (a) K. Kawabata, M. Takeguchi and H. Goto, *Macromolecules*, 2013, **46**, 2078–2091; (b) M. Maisuradze, G. Phalavadishvili, N. Gakhokidze, M. Matnadze, S. Tskhvitaia and E. Kalandia, *Int. J. Org. Chem.*, 2017, **07**, 34–41; (c) H. Gilman and D. L. Esmay, *J. Am. Chem. Soc.*, 1952, **74**, 2021–2024.

31 R. S. Sprick, Y. Bai, C. M. Aitchison, D. J. Woods and A. I. Cooper, *Photocatalytic Hydrogen Evolution from Water Using Heterocyclic Conjugated Microporous Polymers: Porous or Non-Porous?*, accessed 14 November 2018, DOI: 10.26434/chemrxiv.6217451.

32 J. Tauc, *Mater. Res. Bull.*, 1968, **3**, 37–46.

33 R. S. Sprick, B. Bonillo, M. Sachs, R. Clowes, J. R. Durrant, D. J. Adams and A. I. Cooper, *Chem. Commun.*, 2016, **52**, 10008–10011.

34 J. Kosco, M. Sachs, R. Godin, M. Kirkus, L. Francas, M. Bidwell, M. Qureshi, D. Anjum, J. R. Durrant and I. McCulloch, *Adv. Energy Mater.*, 2018, **18**, 1802181.

35 F. Wang, J. Mielby, F. H. Richter, G. Wang, G. Prieto, T. Kasama, C. Weidenthaler, H.-J. Bongard, S. Kegnæs, A. Fürstner and F. Schüth, *Angew. Chem., Int. Ed.*, 2014, **53**, 8645–8648.

36 D. J. Woods, R. S. Sprick, C. L. Smith, A. J. Cowan and A. I. Cooper, *Adv. Energy Mater.*, 2017, **7**, 1700479.

37 X. Weng, Y. Kostoulas, P. M. Fauchet, J. A. Osaheni and S. A. Jenekhe, *Phys. Rev. B*, 1995, **51**, 6838–6841.

38 (*a*) S. Grimme, C. Bannwarth and P. Shushkov, *J. Chem. Theory Comput.*, 2017, **13**, 1989–2009; (*b*) V. Ásgeirsson, C. A. Bauer and S. Grimme, *Chem. Sci.*, 2017, **8**, 4879–4895; (*c*) S. Grimme and C. Bannwarth, *J. Chem. Phys.*, 2016, **145**, 54103.

39 (*a*) L. Wilbraham, E. Berardo, L. Turcani, K. E. Jelfs and M. A. Zwijnenburg, *J. Chem. Inf. Model.*, 2018, **28**, 2450–2459; (*b*) I. Heath-Apostolopoulos, L. Wilbraham and M. A. Zwijnenburg, *Faraday Discuss.*, 2019, DOI: 10.1039/c8fd00171e.

Faraday Discussions

PAPER

Computational high-throughput screening of polymeric photocatalysts: exploring the effect of composition, sequence isomerism and conformational degrees of freedom†

Isabelle Heath-Apostolopoulos, Liam Wilbraham ⓘD
and Martijn A. Zwijnenburg ⓘD *

Received 9th November 2018, Accepted 7th December 2018
DOI: 10.1039/c8fd00171e

We discuss a low-cost computational workflow for the high-throughput screening of polymeric photocatalysts and demonstrate its utility by applying it to a number of challenging problems that would be difficult to tackle otherwise. Specifically we show how having access to a low-cost method allows one to screen a vast chemical space, as well as to probe the effects of conformational degrees of freedom and sequence isomerism. Finally, we discuss both the opportunities of computational screening in the search for polymer photocatalysts, as well as the biggest challenges.

Introduction

Starting with the original work from Fujishima and Honda on the photo-electrolysis of water[1] using a TiO_2 photoanode, hydrogen evolution and water splitting photocatalysis generally involves the use of an inorganic semiconductor as a photoelectrode or photocatalyst. In the 1980s, other Japanese researchers[2,3] demonstrated that conjugated polymers could drive the evolution of hydrogen from aqueous solutions containing various sacrificial electron donors. Carbon nitride was the first polymeric material reported to evolve both hydrogen and oxygen under illumination in the presence of a sacrificial electron/hole donor[4] and was later shown to perform overall water splitting.[5-7] Recently, conjugated polymer photoanodes were also shown to be able to oxidise water as part of a photoelectrochemical cell.[8]

While a much less mature technology than use of inorganic semiconductors, organic polymer photocatalysts offer some very attractive features. In contrast to their inorganic counterparts, polymeric photocatalysts are generally based on the most abundant of elements, C, H, N, S, O; though some polymers are, for the

Department of Chemistry, University College London, 20 Gordon Street, London WC1H 0AJ, UK. E-mail: m. zwijnenburg@ucl.ac.uk

† Electronic supplementary information (ESI) available. See DOI: 10.1039/c8fd00171e

moment, synthesised using less abundant metal catalysts. By way of co-polymerisation, the chemical space of possible polymers is also very large and, as a result, polymer properties are easily and systematically tuneable.[9] A large number of polymers have now been reported to act as photocatalysts, including linear polymers[9–14] quasi-amorphous polymer networks,[12,15–25] *e.g.* conjugated microporous polymers (CMPs), and crystalline polymer networks, *e.g.* crystalline organic frameworks (COFs).[26–29] However, as yet only a minuscule fraction of the relevant chemical space has been explored. As an illustration, ~600 distinct monomers for Suzuki or Stille coupling are readily commercially available, a number which could give rise to ~600 linear homo-polymers, 360 000 ordered binary co-polymers, 200 000 000 ordered ternary co-polymers *etc.* In contrast, probably only on the order of hundred linear polymers have as yet been studied as polymer photocatalysts in the open literature.

As the chemical space of potential polymers is orders of magnitude too large to explore by experiment alone, we have developed computational approaches to predict promising polymers to study in the lab, as well as to rationalise observed activities of synthesized polymers. Our original approach[30] was based on density functional theory (DFT) calculations and allowed us to predict, amongst other things: the electron affinity (EA) of a polymer, which controls the thermodynamic driving force for proton reduction; a polymer's ionisation potential (IP), which controls the thermodynamic driving force for water or sacrificial electron donor oxidation (see Fig. 1); as well as a polymer's optical gap, which controls the wavelength below which light is absorbed. The polymer is modelled as a single-polymer strand embedded in a dielectric continuum that models the environment of the polymer, which, for polymers near the polymer–water interface, is dominated by water. We successfully used this approach to rationalise variation in activities for a significant number of polymers,[9,10,14,18,24,29,31] including *e.g.* the effect of co-polymerisation,[9] and successfully validated it against experimental IP/EA data from the literature.[32]

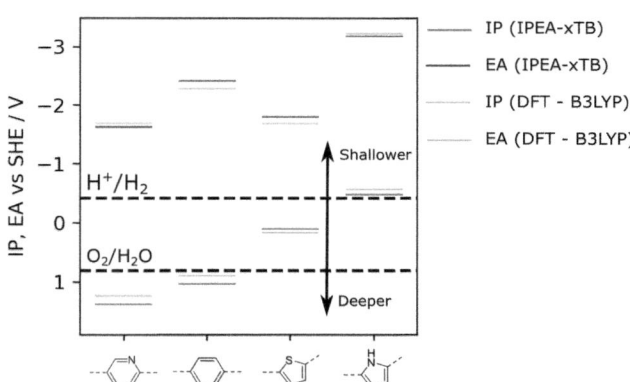

Fig. 1 Examples of IP and EA values calculated for selected polymers. For comparison, values reported were obtained either *via* DFT (B3LYP) or the semi-empirical tight-binding approach used in this work as part of the high-throughput screening workflow (IPEA-xTB). For the ideal photocatalyst, the IP and EA values should straddle the water oxidation and hydrogen reduction potentials, respectively, which are reported here at pH = 7, such as for poly(pyridine) (left). B3LYP data taken from ref. 9 and 14.

However, even DFT calculations are too slow to systematically explore chemical space. To address this issue, we recently developed an approach[33] based on semiempirical tight-binding calculations using the (GFN/IPEA/sTDA)-xTB methods,[34-36] which, after a calibration procedure, gives results that are comparable with DFT at a fraction of the computational cost.

Here we use a series of examples to illustrate the power of our new semi-empirical approach. These include not only a small-scale example of the screening of a co-polymer chemical space for photocatalysts but also screens for the effect on the (co-)polymer properties of (i) different arrangements of monomeric units along the co-polymer chain, sequence isomerism, and (ii) conformerism. In a similar vein to composition, the large number of possible structures resulting from the different possible arrangements of monomeric units in co-polymers and conformation of long polymer chains renders DFT-based methodologies intractable and the sampling of such degrees of freedom is only possible, at this time, using the kind of semi-empirical approach discussed here.

Methodology & computational workflow

As outlined in Fig. 2, the workflow involves multiple steps. Starting from a simplified molecular-input line-entry system (SMILES)[37] representation of each monomer unit, polymer structures were assembled using the Supramolecular Toolkit (*stk*),[38,39] a python library, which takes base functionality from RDKit.[40] We restrict polymer chain length in all cases to oligomers containing 12 aromatic rings along the polymer backbone. We have shown previously that oligomer models of this length provide approximately converged properties with respect to oligomer length.[30]

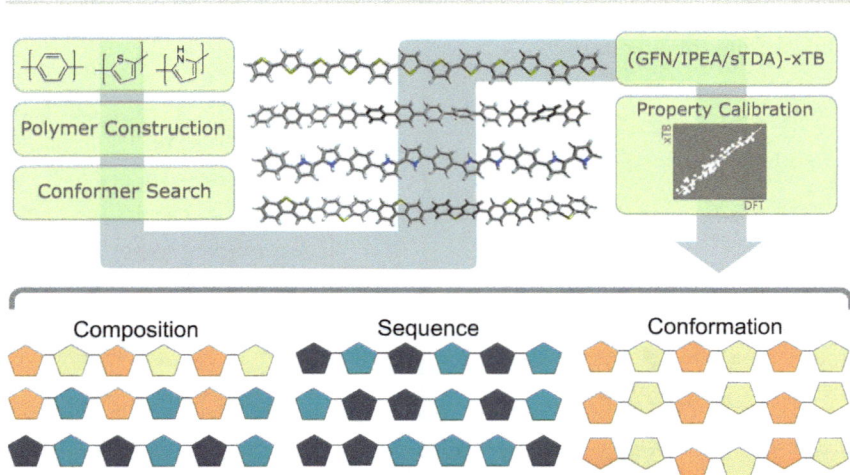

Fig. 2 Schematic of overall high-throughput approach. Starting from 2D representations of monomers (SMILES), 3D polymer models are constructed and undergo a stochastic conformer search. Optoelectronic properties are calculated using the semi-empirical xTB family of methods, which are calibrated to DFT results using a previously-determined linear model. The resulting high-throughput method is used in this work to sample compositional, sequence and conformational degrees of freedom within organic co-polymers (shown schematically using coloured pentagons, bottom).

Conformers for the different oligomer models are generated using a stochastic rather than systematic approach, sampling the conformational space of the polymer randomly using the Experimental-Torsion Distance Geometry with additional basic knowledge (ETKDG) method.[41] Where a single, low-energy conformer is desired, we typically generate 500 conformers per polymer, which undergo a subsequent optimisation and energy ranking procedure using the Merck Molecular Force Field (MMFF)[42] as implemented in RDKit. Where multiple conformers are required, we sample 500 conformers randomly, without energy ranking at the MMFF level. In either case the resulting conformers are subsequently re-optimised using GFN-xTB.[34]

For IP/EA calculations, we use an extension of the parent GFN-xTB method, IPEA-xTB,[36] a differently-parameterised variant of GFN-xTB for the calculation of IP and EA values. For optical gaps, we use the simplified Tamm–Dancoff approach (sTDA)[35,43] applied to orbitals and orbital eigenvalues obtained through xTB (sTDA-xTB).[35] All GFN-xTB calculations were performed using the *xtb* code,[44] while the sTDA results were obtained using the *stda*[45] code. Non-sTDA calculations used the generalised Born surface area solvation model, with the default parameters for water distributed with the *xtb* code. In our previous work,[33] we demonstrated that (TD-)DFT- and (GFN/sTDA)-xTB-derived IP, EA and optical gap values are very strongly, linearly correlated, with very low residual sum of squares values. As a result, we use the simple linear models fitted there to translate the xTB results such that they are maximally comparable to those obtained using our previous (TD-)DFT based approach.

A simple python script,[46] exploiting combinatorics, was used to generate all possible co-polymer sequences at varying co-monomer ratios. For each of the three monomer compositions explored (phenylene–thiophene, phenylene–pyridine & pyridine–thiophene), we generate all possible co-polymer sequences for oligomer length of 12 monomer units and 5 different monomer ratios (*e.g.* phenylene : thiophene in ratios of 1 : 3, 1 : 2, 1 : 1, 2 : 1 and 3 : 1).

Results and discussion

Conformational degrees of freedom

To investigate the sensitivity of the calculated properties to polymer conformation, we calculated IP, EA and optical gap values for 500 randomly generated conformers of four homo-polymers and three co-polymers. Each conformer was optimised using GFN-xTB, with the IP/EA and optical gap values calculated using IPEA-xTB and sTDA-xTB, respectively. Fig. 3 shows the calculated properties for each conformer of each polymer on the *x*-axis and the calculated Boltzmann factor relative to the lowest energy conformer on the *y*-axis. None of the properties calculated are found to be very sensitive to the polymer conformation. In line with previous work by us[33] and others[47] for polymers in the context of organic photovoltaics, the maximum variation of a given property with respect to conformation is generally of the order of 0.1 (e)V. Moreover, the variation for low-energy conformers (Boltzmann factors close to one) is even smaller. While we observe only a weak dependence of IP, EA and optical gap values on polymer conformation, it is possible (and, in some cases, likely) that certain other properties pertinent to photocatalytic water splitting (*e.g.* charge transport, hydrophilicity, absorption intensity) will show a stronger dependence.

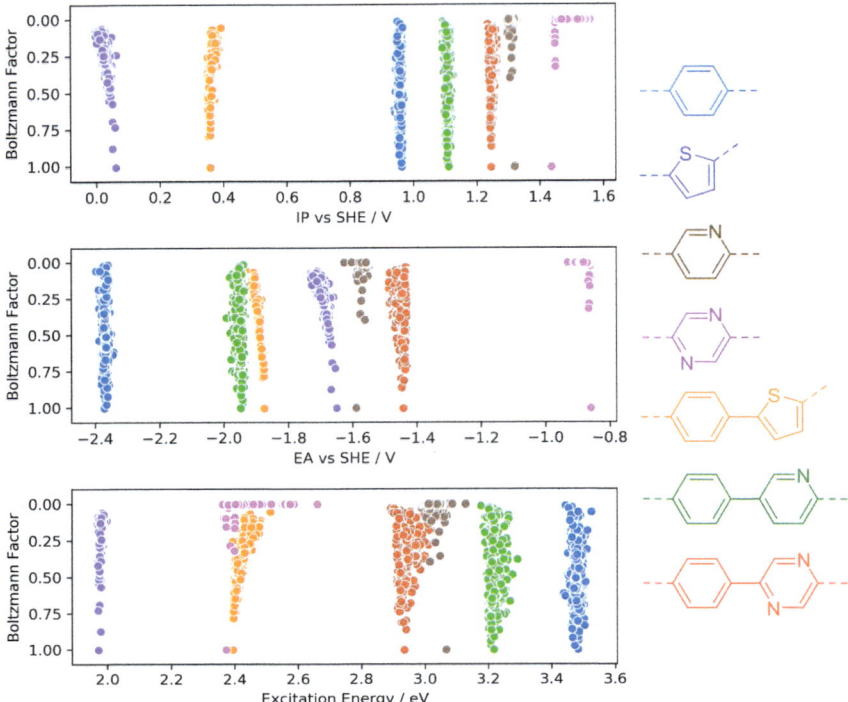

Fig. 3 Effect of conformation on IP, EA and optical gap values for selected homo- and co-polymers. In each case, 500 conformers were randomly generated, and their geometries optimised and properties calculated using (GFN/IPEA/sTDA)-xTB. Coloured chemical structures indicate data shown in the same colour.

From a computational high-throughput screening perspective, the observed low sensitivity to the sampling of conformational degrees of freedom implies that the effect of not finding the true lowest energy conformer on the predicted thermodynamic driving force for proton reduction and water oxidation, as well as on the on-set of light absorption, is only very minor. Hence a minimal conformer search will generally suffice when screening for polymeric photocatalysts. The same weak dependence of IP, EA and optical gap values probably also means that in contrast to chain length and order/disorder in the case of random co-polymers (see below) conformational degrees of freedom do not result in large batch-to-batch variations.

Sequence isomerism and (dis)order

Co-polymers of a fixed overall composition can have a number of distinct sequence isomers, structures with the same overall composition but differing in how the co-monomers are distributed along the polymer chain, *e.g.* the alternating $(AB)_n$ and block A_nB_n isomers. Depending on the synthesis chemistry, either one well-defined sequence isomer or a random mixture is produced experimentally. Being able to predict the properties of one sequence-isomer relative to all others and/or those of a random-mixture is obviously attractive but computationally demanding because of the large number of possible sequence isomers. The calculations, discussed below in more detail, on all the sequence isomers of five different compositions of three potential co-polymer photocatalysts required on the order of 3500 single calculations, where a 'single' calculation involves the structural embedding, conformer search, structure optimisation and calculation of IP, EA and optical gap for each isomer. Hence without an efficient high-throughput procedure, like the one discussed here, this would be a computationally intractable task.

Fig. 4 shows how distributions of properties of sequence isomers of phenylene–thiophene, phenylene–pyridine and pyridine–thiophene co-polymers vary with the co-polymer composition. For each co-monomer ratio, the properties of all combinatorially possible sequence isomers have been calculated, using a fixed oligomer length of 12 monomer units in total. Focussing in the first instance on the mean values of each of the properties (the white central dots in the centre of each of the violins in Fig. 4), we observe in the case of phenylene–thiophene co-polymers that, in line with our more limited sampling in previous work,[9] increasing the thiophene content is predicted to result in progressively shallower IP, deeper EA (see Fig. 1) and lower optical gap values. For phenylene–pyridine co-polymers, increasing the pyridine content results in both deeper IP and deeper EA values, while optical gap values decrease as the fraction of pyridine is increased. When we apply the same analysis to co-polymers of pyridine and thiophene, we predict that, with increasing pyridine content, the IP values become deeper while the EA values remain largely unchanged and the optical gap increases. It should be noted that the change in the mean of a property distribution with composition

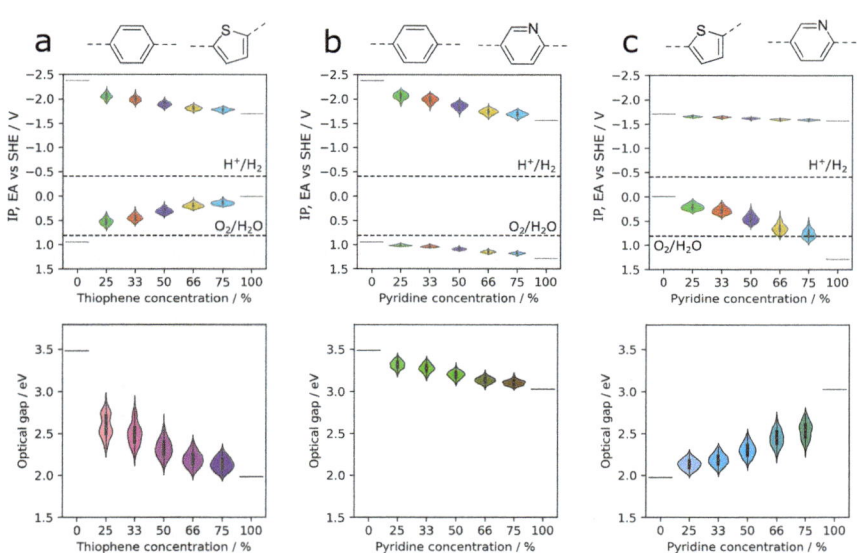

Fig. 4 Distributions of properties (IP, EA and optical gap) of disordered (a) phenylene–thiophene, (b) phenylene–pyridine and (c) pyridine–thiophene co-polymers vary with differing co-monomer ratios. For each ratio, the properties of all possible monomer sequences have been calculated, using a fixed oligomer length of 12 monomer units in total.

can be strongly non-linear. For example, the change in the predicted mean optical gap when going from poly(phenylene) or poly(pyridine) to a co-polymer containing 25% thiophene is much larger than that predicted for going from a thiophene co-polymer containing 25% phenylene or pyridine to poly(thiophene).

Not surprisingly, the extent to which the mean IP, EA or optical gap values of the co-polymers change with composition appears to be linked to the difference in a given property between the corresponding homopolymers. When the difference is large the change in the mean value of that property with composition is also large for the corresponding co-polymer, see e.g. the change in optical gap value with composition for the phenylene–thiophene and pyridine–thiophene co-polymers. Conversely, when the difference in homopolymer properties is small the variation in the mean value of that property is also small, see e.g. the change with composition for the mean IP values of phenylene–pyridine and mean EA values of pyridine–thiophene co-polymers. More surprisingly, the difference in homopolymer properties also appears to control the overall variation in a given property for the different sequence isomers. For example, for phenylene–thiophene co-polymers, optical gap values can vary as much as 0.8 eV between different sequence isomers. In contrast, the small difference between the IP values of poly(phenylene) and poly(pyridine) leads to a variation of less than 0.1 V between sequence isomers of the corresponding co-polymer.

Fig. 5 shows how the 'degree of segregation' of co-monomers influences the overall co-polymer properties. Here we measure the 'degree of segregation' by considering the number of equivalent neighbouring monomeric units for a given sequence isomer. Specifically, this leads to a descriptor which lies between 0 (no identical neighbours, fully alternating) and 1 (only 2 monomers have a neighbour which is non-identical, fully segregated into a block of monomer A and a block of monomer B). For each property (IP, EA, and optical gap) we see that, when fully segregated, the properties of the co-polymer are most similar to those of the corresponding homopolymer with either the shallowest IP, deepest EA or lowest optical gap. Focussing in on the phenylene–thiophene system, finally, for which there is experimental data for (pseudo-)random 1 : 1 co-polymers available in the literature,[9] the fact that the perfectly alternating structure is predicted to have a larger optical gap than the mean value of the predicted optical gap distribution for this composition is in line with the fact that experimentally the (pseudo-)random materials have a smaller optical gap than their alternating counterpart.

In the context of photocatalytic water splitting, Fig. 4 and 5 make clear how the exact co-monomer sequence can influence the relevant properties of a co-polymer. Further, it illustrates how control over co-monomer sequence and hence the sequence isomer produced can be strongly beneficial, especially in terms of the optical gap, even if it cannot always be achieved experimentally.

Overall composition

As an illustration of how our xTB semiempirical approach may be applied to exhaustively screen co-polymer compositions, a library of 10 simple monomer units was combined combinatorially to construct a library of 55 co-polymers (see Fig. 6). The monomer pool contains examples with significantly varying electronic properties, ranging from particularly electron-poor (e.g. pyridine, diazine) to electron-rich monomers (thiophene, pyrrole). Focussing in on co-polymers

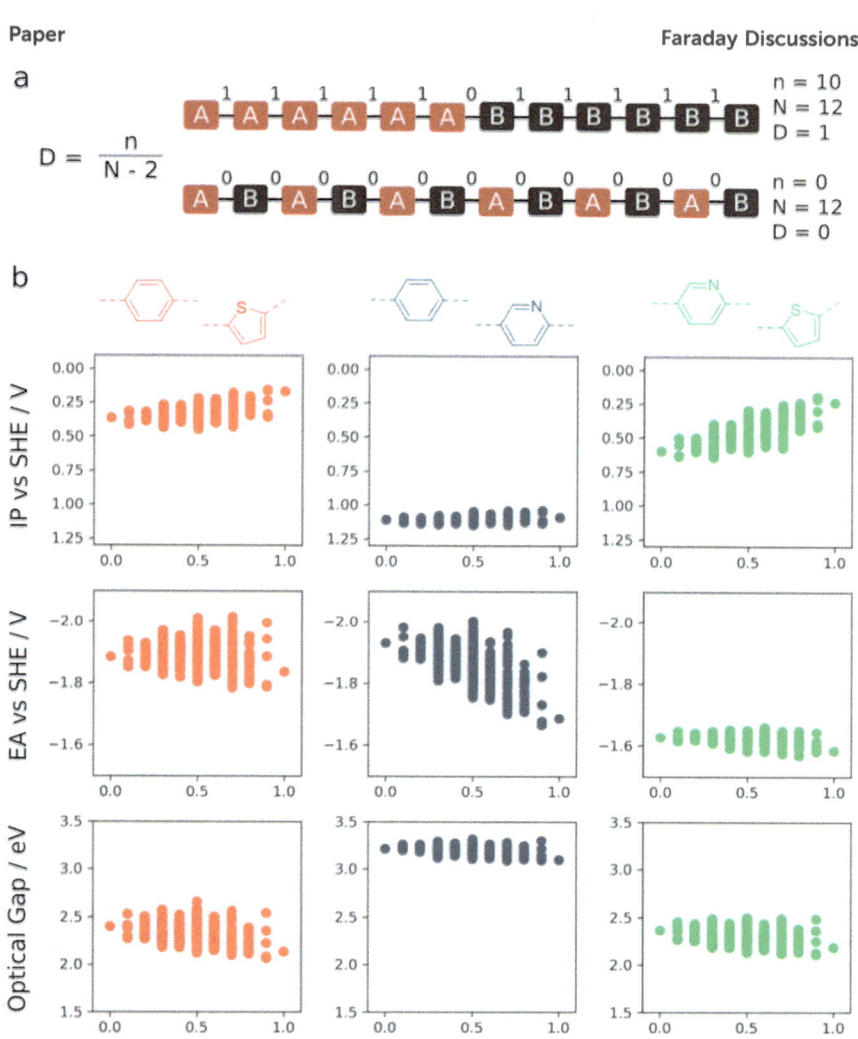

Fig. 5 (a) Schematic of how the 'degree of segregation' (*D*) is measured: the number of equivalent neighbouring monomer units within a given sequence isomer (*n*), divided by the total number of monomer units (*N*) minus 2. This metric spans values between 0 (fully alternating) and 1 (fully segregated into a block of monomer A and a block of monomer B). (b) Illustration of how the 'degree of segregation' of co-monomers influences the overall co-polymer properties (IP, EA and optical gap) for 1 : 1 co-polymers.

containing either thiophene or pyrrole (Fig. 6b), we observe that the incorporation of such electron-rich monomers is predicted to lead to co-polymers with inherently shallow IP and EA values. At the same time, the optical gap values of such materials are low compared to those of the total co-polymer population screened. While this latter property is conducive to water splitting applications – resulting in a greater rate of photon absorption – shallow IP values mean that the thermodynamic driving force for the oxidation of water and, to a lesser extent, sacrificial electron donors such as triethylamine, is largely absent. Conversely, co-

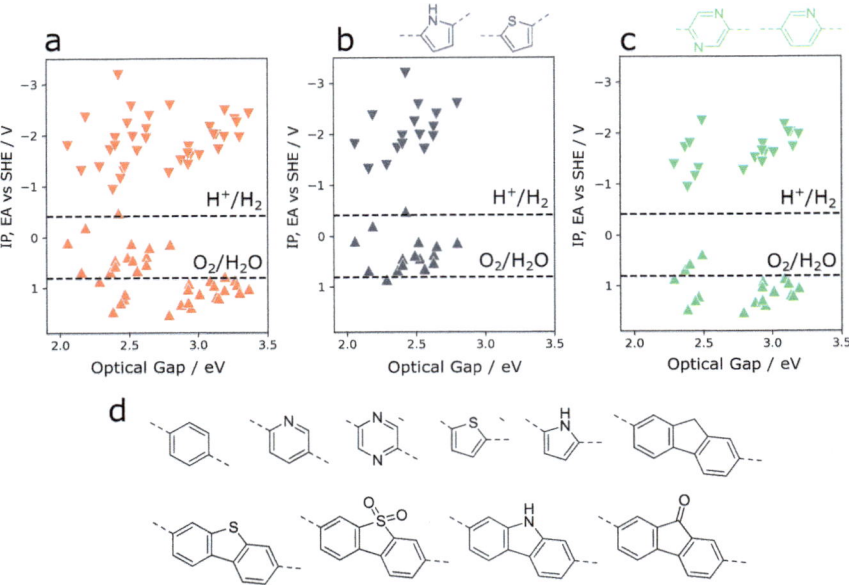

Fig. 6 (a) Binary co-polymer screening results obtained by exhaustive combination of a small library of simple monomers. (b) Results for a subset of co-polymers containing electron-rich monomers (thiophene, pyrrole). (c) Results for a subset of co-polymers containing electron-poor monomers (pyridine, diazine). (d) Monomer library used to produce points shown in (a), of which points in (b) and (c) are subsets. Upward (downward) triangles indicate IP (EA) values.

polymers containing electron-poor monomeric units (pyridine, diazine, see Fig. 6d) are predicted to generally have deep(er) IPs – an attractive property for water oxidation – while retaining the necessary thermodynamic driving force for proton reduction. On the other hand, as a result of these significantly more positive IP potentials, many of these co-polymers also show the widest optical gaps of the overall co-polymer population. Essentially, these two extremes highlight the central challenge of optimising activity[9,48] and high-throughput screening for water splitting photocatalysts – balancing the trade-off between adequate light absorption and thermodynamic reduction and oxidation driving forces. In this context, the high-throughput method described here provides a means of screening very large numbers of co-polymers – even beyond simple binary compositions – in a search for a material with an ideal balance between driving force and light absorption.

Perspective

As demonstrated above using an xTB-based semi-empirical screening approach one can rapidly screen thousands to tens of thousands of (co-)polymers with an accuracy that is comparable to that which could be obtained using DFT. As such one can consider orders of magnitude more polymers than it is possible to screen experimentally, even using robotic synthesis and characterisation platforms. While not sufficient to sample even all possible binary co-polymers based on commercially available monomers, it does become possible, for example, to screen families of co-polymers that share a common monomer, and suggest the best hundred or so for experimental follow-up work – something we are currently actively pursuing together with our experimental collaborators. Studying larger search spaces probably still requires a transition from semi-empirical methods to a machine learning approach. Semi-empirical methods are still useful here in terms of generating the large amount of data required for training such models.

The same considerations also apply when screening sequence isomers or conformers. As we have shown above, the former can be a useful tool to understand what could be achieved experimentally if one could control the exact polymer sequence. It can also be useful to understand the properties of true random co-polymers, especially if it could be combined with some weighting for how likely a particular sequence isomer is to form, *e.g.* by applying Boltzmann weighting using the GFN-xTB total energies.

Perhaps the biggest challenge will be to go beyond considering only IP, EA and optical gap values to also include, for example, transport properties, in the screening of overall composition space or sequence isomerism. While there is no fundamental constraint on doing so, and this indeed has been attempted before for hole transport in the context of polymers for organic photovoltaics,[49] only intramolecular contributions to transport, which do not depend on knowing the intermolecular structure of materials, can be rapidly screened for. Related to this, if transport is indeed relevant it probably makes most sense to compare predictions for materials with experimentally similar particle sizes and hence similar path lengths for transport. Something that is probably difficult to realise in practice. Similar considerations also apply for other experimental variables, such as the concentration of intentionally-added metal co-catalyst and/or leftover noble metal content from polymer synthesis routes.

Conclusions

Besides successfully demonstrating the utility of our low-cost computational workflow for screening polymer photocatalysts, we also demonstrated that conformational degrees of freedom have little influence on optoelectronic properties of polymers that are pertinent to their photocatalytic activity. The ionisation potential and electron affinity of a polymer, which control the thermodynamic driving force for proton reduction and water oxidation, respectively, as well as the polymer's optical gap are predicted to typically change by less than 0.1 (e)V in between low-energy conformers. We have also shown that sequence isomerism in (binary) co-polymers can lead to large variations in these properties between different sequence isomers. This is helpful in understanding the properties of random co-polymers relative to their ordered counterparts, as well as suggesting that synthetic control of the polymer sequence beyond simple alternating co-polymers might be beneficial in optimising a polymer's photocatalytic activity. Finally, we found that, in line with what we knew from more limited previous work, introducing electron-rich co-monomers is predicted to consistently result in co-polymers with small optical gaps but low or negligible thermodynamic driving force for water oxidation, while introducing electron-poor co-monomers is predicted to have the opposite effect.

Conflicts of interest

There are no conflicts to declare.

Acknowledgements

We thank Catherine Aitchison, Yang Bai, Dr Enrico Berardo, Prof. Andrew I. Cooper, Dr Kim Jelfs, Dr Christian Meier, Dr Reiner Sebastian Sprick and Lucas Turcani for useful discussion. The UK Engineering and Physical Sciences Research Council (EPSRC) is acknowledged for funding (EP/N004884/1). I. H. A. thanks University College London for a PhD stipend.

References

1 A. Fujishima and K. Honda, *Nature*, 1972, **238**, 37.
2 S. Yanagida, A. Kabumoto, K. Mizumoto, C. Pac and K. Yoshino, *J. Chem. Soc., Chem. Commun.*, 1985, 474–475.
3 T. Shibata, A. Kabumoto, T. Shiragami, O. Ishitani, C. Pac and S. Yanagida, *J. Phys. Chem.*, 1990, **94**, 2068–2076.
4 X. Wang, K. Maeda, A. Thomas, K. Takanabe, G. Xin, J. M. Carlsson, K. Domen and M. Antonietti, *Nat. Mater.*, 2009, **8**, 76–80.
5 Y. Sui, J. Liu, Y. Zhang, X. Tian and W. Chen, *Nanoscale*, 2013, **5**, 9150–9155.
6 J. Liu, Y. Liu, N. Liu, Y. Han, X. Zhang, H. Huang, Y. Lifshitz, S.-T. Lee, J. Zhong and Z. Kang, *Science*, 2015, **347**, 970–974.
7 L. Lin, C. Wang, W. Ren, H. Ou, Y. Zhang and X. Wang, *Chem. Sci.*, 2017, **8**, 5506–5511.
8 P. Bornoz, M. S. Prévot, X. Yu, N. Guijarro and K. Sivula, *J. Am. Chem. Soc.*, 2015, **137**, 15338–15341.
9 R. S. Sprick, C. M. Aitchison, E. Berardo, L. Turcani, L. Wilbraham, B. M. Alston, K. E. Jelfs, M. A. Zwijnenburg and A. I. Cooper, *J. Mater. Chem. A*, 2018, **6**, 11994–12003.
10 R. S. Sprick, B. Bonillo, R. Clowes, P. Guiglion, N. J. Brownbill, B. J. Slater, F. Blanc, M. A. Zwijnenburg, D. J. Adams and A. I. Cooper, *Angew. Chem., Int. Ed.*, 2016, **55**, 1792–1796.
11 D. J. Woods, R. S. Sprick, C. L. Smith, A. J. Cowan and A. I. Cooper, *Adv. Energy Mater.*, 2017, **7**, 1700479.
12 C. Yang, B. C. Ma, L. Zhang, S. Lin, S. Ghasimi, K. Landfester, K. A. I. Zhang and X. Wang, *Angew. Chem., Int. Ed.*, 2016, **55**, 9202–9206.
13 X. Zong, X. Miao, S. Hua, L. An, X. Gao, W. Jiang, D. Qu, Z. Zhou, X. Liu and Z. Sun, *Appl. Catal., B*, 2017, **211**, 98–105.
14 R. S. Sprick, L. Wilbraham, Y. Bai, P. Guiglion, A. Monti, R. Clowes, A. I. Cooper and M. A. Zwijnenburg, *Chem. Mater.*, 2018, **30**, 5733–5742.
15 Z. Zhang, J. Long, L. Yang, W. Chen, W. Dai, X. Fu and X. Wang, *Chem. Sci.*, 2011, **2**, 1826–1830.
16 S. Chu, Y. Wang, Y. Guo, P. Zhou, H. Yu, L. Luo, F. Kong and Z. Zou, *J. Mater. Chem.*, 2012, **22**, 15519–15521.
17 Z. A. Lan, Y. Fang, Y. Zhang and X. Wang, *Angew. Chem., Int. Ed.*, 2018, **57**, 470–474.

18 R. S. Sprick, J. X. Jiang, B. Bonillo, S. Ren, T. Ratvijitvech, P. Guiglion, M. A. Zwijnenburg, D. J. Adams and A. I. Cooper, *J. Am. Chem. Soc.*, 2015, **137**, 3265–3270.

19 J. Bi, W. Fang, L. Li, J. Wang, S. Liang, Y. He, M. Liu and L. Wu, *Macromol. Rapid Commun.*, 2015, **36**, 1799–1805.

20 K. Schwinghammer, S. Hug, M. B. Mesch, J. Senker and B. V. Lotsch, *Energy Environ. Sci.*, 2015, **8**, 3345–3353.

21 R. S. Sprick, B. Bonillo, M. Sachs, R. Clowes, J. R. Durrant, D. J. Adams and A. I. Cooper, *Chem. Commun.*, 2016, **52**, 10008–10011.

22 L. Li, W. Y. Lo, Z. Cai, N. Zhang and L. Yu, *Macromolecules*, 2016, **49**, 6903–6909.

23 L. Li, Z. Cai, Q. Wu, W. Y. Lo, N. Zhang, L. X. Chen and L. Yu, *J. Am. Chem. Soc.*, 2016, **138**, 7681–7686.

24 C. B. Meier, R. S. Sprick, A. Monti, P. Guiglion, J. S. M. Lee, M. A. Zwijnenburg and A. I. Cooper, *Polymer*, 2017, **126**, 283–290.

25 S. Kuecken, A. Acharjya, L. Zhi, M. Schwarze, R. Schomäcker and A. Thomas, *Chem. Commun.*, 2017, **53**, 5854–5857.

26 L. Stegbauer, K. Schwinghammer and B. V. Lotsch, *Chem. Sci.*, 2014, **5**, 2789–2793.

27 V. S. Vyas, F. Haase, L. Stegbauer, G. Savasci, F. Podjaski, C. Ochsenfeld and B. V. Lotsch, *Nat. Commun.*, 2015, **6**, 8508.

28 P. Pachfule, A. Acharjya, J. Roeser, T. Langenhahn, M. Schwarze, R. Schomäcker, A. Thomas and J. Schmidt, *J. Am. Chem. Soc.*, 2018, **140**, 1423–1427.

29 X. Wang, L. Chen, S. Y. Chong, M. A. Little, Y. Wu, W.-H. Zhu, R. Clowes, Y. Yan, M. A. Zwijnenburg, R. Sebastian Sprick and A. I. Cooper, *Nat. Chem.*, 2018, **10**, 1180–1189.

30 P. Guiglion, C. Butchosa and M. A. Zwijnenburg, *J. Mater. Chem. A*, 2014, **2**, 11996–12004.

31 C. Butchosa, P. Guiglion and M. A. Zwijnenburg, *J. Phys. Chem. C*, 2014, **118**, 24833–24842.

32 P. Guiglion, A. Monti and M. A. Zwijnenburg, *J. Phys. Chem. C*, 2017, **121**, 1498–1506.

33 L. Wilbraham, E. Berardo, L. Turcani, K. E. Jelfs and M. A. Zwijnenburg, *J. Chem. Inf. Model.*, 2018, **28**, 2450–2459.

34 S. Grimme, C. Bannwarth and P. Shushkov, *J. Chem. Theory Comput.*, 2017, **13**, 1989–2009.

35 S. Grimme and C. Bannwarth, *J. Chem. Phys.*, 2016, **145**, 054103.

36 V. Ásgeirsson, C. A. Bauer and S. Grimme, *Chem. Sci.*, 2017, **8**, 4879–4895.

37 D. Weininger, *J. Chem. Inf. Comput. Sci.*, 1988, **28**, 31–36.

38 L. Turcani, E. Berardo and K. E. Jelfs, *J. Comput. Chem.*, 2018, **39**, 1931–1942.

39 *Supramolecular-toolkit*, https://github.com/supramolecular-toolkit/stk, accessed 30 October, 2018.

40 *The RDKit Documentation*, http://www.rdkit.org/docs/, accessed 30 October, 2018.

41 S. Riniker and G. A. Landrum, *J. Chem. Inf. Model.*, 2015, **55**, 2562–2574.

42 T. A. Halgren, *J. Comput. Chem.*, 1996, **17**, 490–519.

43 C. Bannwarth and S. Grimme, *Comput. Theor. Chem.*, 2014, **1040–1041**, 45–53.

44 *xtb – An extended tight-binding semi-empirical program package*, https://www.chemie.uni-bonn.de/pctc/mulliken-center/software/xtb/xtb, accessed 30 October, 2018.

45 sTDA – A simplified Tamm-Dancoff density functional approach for electronic excitation spectra, https://www.chemie.uni-bonn.de/pctc/mulliken-center/software/stda/stda, accessed 30 October 2018.

46 Sequence-generator, https://github.com/ZwijnenburgGroup/sequence-generator, accessed 6 November 2018.

47 N. E. Jackson, B. M. Savoie, K. L. Kohlstedt, T. J. Marks, L. X. Chen and M. A. Ratner, *Macromolecules*, 2014, **47**, 987–992.

48 P. Guiglion, C. Butchosa and M. A. Zwijnenburg, *Macromol. Chem. Phys.*, 2016, **217**, 344–353.

49 N. M. O'Boyle, C. M. Campbell and G. R. Hutchison, *J. Phys. Chem. C*, 2011, **115**, 16200–16210.

Faraday Discussions

PAPER

Visible light-driven water oxidation with a ruthenium sensitizer and a cobalt-based catalyst connected with a polymeric platform†

Zeynep Kap[a] and Ferdi Karadas (iD) [*ab]

Received 5th November 2018, Accepted 11th December 2018

DOI: 10.1039/c8fd00166a

A facile synthesis for a photosensitizer–water oxidation catalyst (PS–WOC) dyad, which is connected through a polymeric platform, has been reported. The dyad assembly consists of a ruthenium-based chromophore and a cobalt–iron pentacyanoferrate coordination network as the water oxidation catalyst while poly(4-vinylpyridine) serves as the bridging group between two collaborating units. Photocatalytic experiments in the presence of an electron scavenger reveal that the dyad assembly maintains its activity for 6 h while the activity of a cobalt hexacyanoferrate and $Ru(bpy)_3^{2+}$ couple decreases gradually and eventually decays after a 3 h catalytic experiment. Infrared and XPS studies performed on the post-catalytic powder sample confirm the stability of the dyad during the catalytic process.

Introduction

Photocatalytic water splitting has been an attractive and promising research topic over the last two decades due to its potential contribution to sustainable and renewable energy development.[1] The main objective with water splitting is to convert solar light into chemical energy and concurrently to produce hydrogen and oxygen. Since the demanding four-electron process of water oxidation is considered as the bottleneck of water splitting, research efforts have been centered on developing efficient assemblies for light-driven water oxidation catalysis.

In general, a photosensitizer (PS), which absorbs sunlight to create holes and electrons, collaborates with a water oxidation catalyst (WOC) to drive the water oxidation reaction in the presence of an electron scavenger. Recently dyads, in which the molecular PS and WOC are covalently coordinated to each other with

---

[a]Department of Chemistry, Bilkent University, 06800 Ankara, Turkey. E-mail: karadas@fen.bilkent.edu.tr
[b]UNAM-Institute of Materials Science and Nanotechnology, Bilkent University, 06800 Ankara, Turkey

† Electronic supplementary information (ESI) available: UV-Vis, FTIR, XPS, XRD, SEM, EDX characterizations, and details of photocatalytic studies. See DOI: 10.1039/c8fd00166a

a suitable linker, have been constructed to develop dye-sensitized photo-electrochemical cells (DSPECs) to enhance the electron transfer and charge transport between molecular units and the semiconductor.[2-5] Several Ru(II) based PS–WOC dyad assemblies have recently been coated on TiO_2 to build dye-sensitized DSPECs with promising faradaic efficiencies for O_2 evolution.[6,7] The catalytic efficiency of dyad systems was investigated in a homogeneous system as well. In a study by Thummel et al., a Ru–Ru dyad assembly showed a TON of 134 under 6 h illumination, which is much higher than its analogous inter-molecular system with a TON of 6.[3] A follow-up study by Thummel et al. showed a TON of 68 under 1 h light illumination in the presence of sodium persulfate at pH 5.3.[8] Sun et al. also prepared different PS–WOC assemblies, incorporating a ruthenium diimine chromophore and a ruthenium-based catalyst.[9] The TON of the assembly was found to be 38 while the separate system showed a TON of 8. In the majority of the dyad systems, ruthenium-based units have been preferred as both a WOC and a PS due to their strong light absorption, long excited state lifetimes, and high efficiencies.[10] Implementing earth-abundant components, particularly for the catalytic site, still remains a significant challenge due to synthetic limitations.

The selection of a proper bridging group is one of the critical parameters for the design of an efficient dyad. Polymeric platforms have also been used for this purpose.[10-14] Several studies indicate that enhanced catalytic efficiency observed on polymeric dyad systems is due to a hopping mechanism along the chain, which results in an intra-assembly electron/hole transfer.[4,13,14] Waters et al. re-ported that an electrode-bound helical peptide PS–WOC assembly has a 10-fold improvement in its catalytic activity compared to its analogous homogeneous system.[12] It has been shown that intra-assembly electron transfer is a key parameter to enhance efficiency and for aligning the distance between units for optimum electron transfer rates.[12,13] Hisaeda et al. emphasized that an assembly with a polymer linkage can also efficiently work even under diluted conditions by fixation of each functional group in the same polymeric unit, thus providing a close distance for electron transfer.[15] In the presence of a polymeric support, stability of the system is also expected to increase by preventing photodecom-position of the photosensitizer.[15,16]

In this study, we present a novel heterogeneous PS–WOC dyad by using poly(4-vinylpyridine) (P4VP) as a bridging platform between a ruthenium chromophore and cobalt-based Prussian blue analogue (PBA). Cobalt hexacyanometalates have recently been demonstrated as promising water oxidation catalysts due to their high catalytic activities, robustness, and stabilities at a wide range of pH (1 to 13).[17-24] Therefore, the use of a CoFe–PBA as a WOC rather than a Ru-based one will be a step forward in the development of entirely earth abundant dyads. P4VP has recently been used to prepare a Co–Fe coordination polymer for water oxidation electrocatalysis by our group.[25] The study involved the coordination of $Fe(CN)_5$ groups to the pyridyl groups of P4VP yielding a robust precursor for the synthesis of amorphous PBAs. In another study, a Co–P4VP assembly has successfully been prepared and found to be an efficient metallopolymer for water reduction electrocatalysis.[26] Given the successful utilization of P4VP for catalytic applications, herein, we propose a synthetically facile dyad, wherein the ruthenium-based molecular photosensitizer is connected to a Prussian blue type water oxidation catalyst through a P4VP platform. Photocatalytic water oxidation performance has been investigated in comparison with a cobalt-based PBA. Characterization techniques have also been performed to evaluate its stability.

Experimental

Starting materials

cis-Bis(2,2′-bipyridine)dichlororuthenium(II) hydrate (Acros Organics, 97%), poly(4-vinylpyridine) (Sigma-Aldrich, MW $\sim$ 60 000), AgNO$_3$ (Sigma-Aldrich, $\geq$99.0), Na$_2$[Fe^{III}(CN)$_5$NO]·2H$_2$O (Alfa Aesar, 98%), and NaOH (Sigma-Aldrich, 98–100.5%) were used. All the solvents were analytical grade and reagents received were used without any further processing. Millipore deionized water (resistivity: 18 mΩ cm) was used for all experiments that required water.

Synthetic procedures

General procedure for synthesis of [Ru–P4VP]. At room temperature, 700.0 mg (1.445 mmol) cis-[Ru(bpy)$_2$Cl$_2$] and 490.9 mg (2.890 mmol) AgNO$_3$ are mixed in 100 mL methanol according to the modified literature.[27] After 1 h vigorous stirring, the precipitated layer of AgCl was filtered through a Celite® filter and removed. [Ru(bpy)$_2$(H$_2$O)$_2$](NO$_3$)$_2$ filtrate was evaporated by a rotary evaporator. [Ru(bpy)$_2$(H$_2$O)$_2$](NO$_3$)$_2$ was added to the solution of 6-fold molar excess of poly(4-vinylpyridine) which was dissolved in 200 mL 4 : 1 ethanol/water. The mixture was refluxed in the dark for 48 h under constant stirring. Completion of the product was monitored by UV-Vis spectroscopy. The resulting solution was evaporated by a rotary evaporator, dissolved in ethanol, and precipitated by ethyl ether.[28] The precipitate was filtered and rinsed with cold water and ethyl ether. Throughout the article, the abbreviation [Ru–P4VP] will be used for the [Ru(bpy)$_2$(P4VP)$_6$].

General procedure for synthesis of the Fe precursor. Na$_3$[Fe^{II}(CN)$_5$NH$_3$]·3H$_2$O was used as the Fe precursor. According to the procedure in literature with slight modifications,[25,29] 30 g of Na$_2$[Fe^{III}(CN)$_5$NO]·2H$_2$O and 4 g NaOH were mixed in 120 mL of water under constant stirring. Throughout the experiment, the temperature was kept below 10 °C. After obtaining a homogenous solution, 25% (v/v) NH$_4$OH solution was added until saturation, followed by the addition of cold methanol until a yellow color was obtained. The product was recrystallized using NH$_4$OH and CH$_3$OH solutions. After vacuum filtration, the resulting precipitate was dried in a vacuum oven overnight at 25 °C. IR (cm^{-1}): 3300(b), 2135(s), 2009(m), 1642(m), 1621(m), 1257(m), 569(m).

General procedure for synthesis of [Ru–P4VP–Fe]. The [Ru–P4VP] was dissolved in methanol and the precursor was added according to a 1 : 2 Ru/Fe ratio. The solution was kept in the dark under constant stirring for 5 days. Cold [Ru–P4VP–Fe] solution was centrifuged with water three times and the solution was discarded. The complex was dried after washing with acetone in a vacuum desiccator. Throughout the manuscript, the abbreviation [Ru–P4VP–Fe] will be used for the [Ru(bpy)$_2$(P4VP)$_6$]–Fe(CN)$_5$ assembly.

General procedure for synthesis of [Ru–P4VP–CoFe]. Cobalt(II) acetate tetrahydrate was used as the Co precursor. [Ru–P4VP–Fe] and the Co precursor were mixed in a 1 : 1 acetonitrile/water solution. The Co precursor was added according to the 3 : 2 Co/Fe stoichiometric ratio. The solution was kept in the dark under constant stirring for 2 days following evaporation by a rotary evaporator.

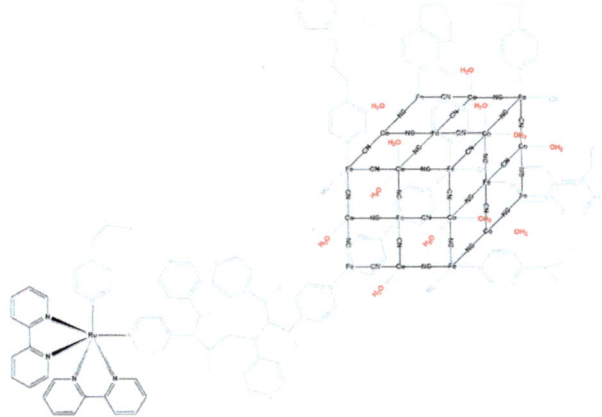

Fig. 1 Proposed structure of the ruthenium chromophore and cobalt-based PBA dyad, incorporating poly(4-vinylpyridine).

Throughout the article, the abbreviation [Ru–P4VP–CoFe] will be used for the $[Ru(bpy)_2(P4VP)_6]$–CoFe(CN)$_5$ assembly. The proposed structure of the assembly is shown in Fig. 1.

Photochemical setup

The oxygen amount was measured with GC (Agilent 7820A, Molesieve GC column (30 m $\times$ 0.53 mm $\times$ 25 μm)) thermostatted at 40 °C which was equipped with a TCD detector thermostatted at 100 °C (Ar as carrier gas). Oxygen evolution was calibrated with a pressure transducer (Omega PXM409-002BAUSBH). The solar light simulator (Sciencetech, SLB-300B, 300 W Xe lamp, AM 1.5 global filter) was calibrated to 1 sun (100 mW cm^{-2}). Experimental setup is shown in ESI† and explained in detail.

Results and discussion

Characterization

Ru(bpy)$_2$Cl$_2$ exhibits two characteristic bands at 526 and 283 nm, which are assigned to metal-to-ligand charge transfer (MLCT) and ligand centered π–π* (LC) transitions, respectively. On the other hand, a blue shift to 465 (with a shoulder at 435 nm) is obtained for [Ru–P4VP] verifying the complex formation, which are in good accordance with absorption profiles of trisbipyridyl–ruthenium(II) complexes (Fig. 2).[30-32] These bands were also observed for [Ru–P4VP–Fe] and [Ru–P4VP–CoFe], which indicates that the ruthenium ion is surrounded with pyridyl groups in all compounds and that the ruthenium site in [Ru–PVP–CoFe] could serve as a chromophore to utilize visible light (Fig. S1†).

The infrared spectrum of [Ru–P4VP] exhibits two major bands at 1417 cm^{-1} and 1597 cm^{-1}, which are attributed to the C=C$_{ring}$ and C=N$_{ring}$ of pyridyl rings (Fig. S2†).[25,33] An additional band in the 2000–2200 cm^{-1} range is observed for [Ru–P4VP–Fe], which corresponds to the C≡N stretches of the [Fe(CN)$_5$] fragment (Fig. 3). The relatively small peak at 2103 cm^{-1} is a result of the partial oxidation

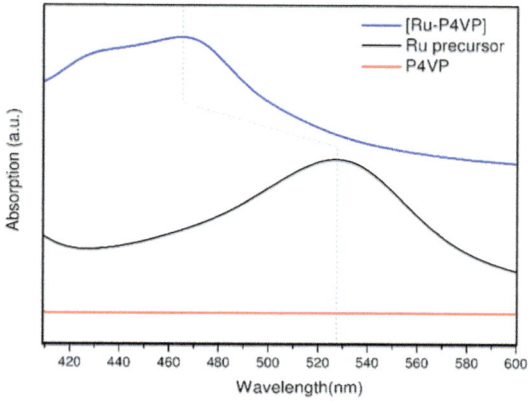

Fig. 2 UV-Vis spectra of P4VP (red), Ru precursor (black), and [Ru–P4VP] (blue) in ethanol.

of the iron sites to Fe^{3+}. The reaction with Co^{2+} leads to a shift to higher frequency due to the formation of the Fe–CN–Co coordination mode.[25,34]

XPS studies indicate an observable change in the binding energy of the Ru $3d_{5/2}$ signal for [Ru-P4VP] compared with that of $Ru(bpy)_2Cl_2$ as a result of the replacement of chloride groups with pyridyl ones. Ru $3d_{5/2}$ signals are considered for comparison due to overlap of the Ru $3d_{3/2}$ and C 1s signals (Fig. 4). Besides, the Ru 3d signals in [Ru-P4VP], [Ru-P4VP-Fe], and [Ru-P4VP-CoFe] samples are similar suggesting no significant changes in the coordination sphere and oxidation state of the ruthenium site. The slight change in the Fe 2p band in [Ru-P4VP-Fe] is attributed to the partial oxidation of $Fe^{3+/2+}$. Two shoulder bands observed at ~711.51 eV and ~725.21 eV in the Fe 2p signals of [Ru-P4VP-Fe] can also be attributed to the aforementioned partial oxidation process (Fig. S3†). Such oxidation is commonly observed in pentacyanoferrate chemistry, and the results are in good agreement with FTIR spectra, which reveals a shoulder band in the cyanide region at 2103 cm^{-1} for [Ru-P4VP-Fe]. Fe 2p signals for [Ru-P4VP-Fe] are

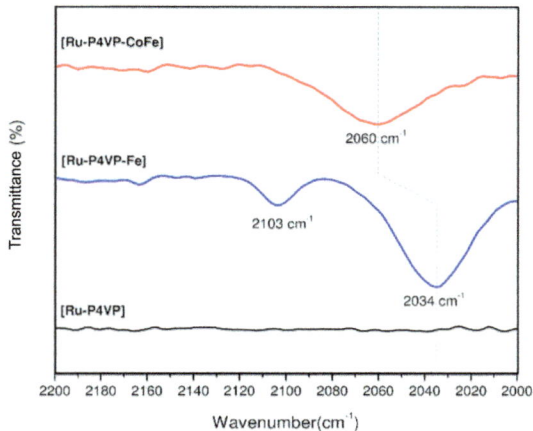

Fig. 3 FTIR spectra of [Ru–P4VP], [Ru–P4VP-Fe], and [Ru–P4VP-CoFe].

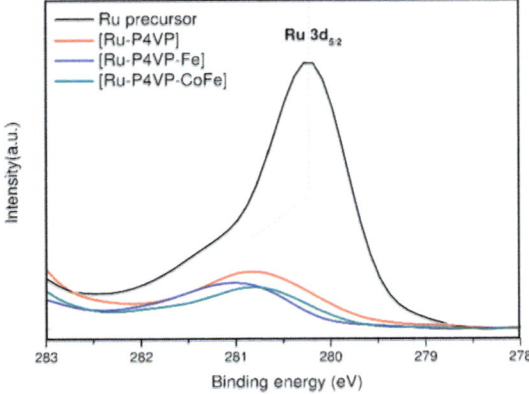

Fig. 4   XPS spectra of the Ru $3d_{5/2}$ signals of [Ru–P4VP–CoFe] (green), [Ru–P4VP–Fe] (blue), [Ru–P4VP] (red), and the Ru precursor (black).

observed at around 708.69 eV and 722.69 eV, which are assigned to Fe $2p_{3/2}$ and Fe $2p_{1/2}$, respectively. The signals of [Ru-P4VP-Fe] and [Ru-P4VP-CoFe] correspond well with those of the Fe precursor. Broad features observed in [Ru-P4VP-CoFe] indicate the presence of multiple oxidation states of iron sites. The N 1s band of [Ru-P4VP] corresponds to the pyridyl groups of P4VP and the bipyridyl groups of the ruthenium fragment (Fig. S4†). A slight shift in the binding energy of [Ru-P4VP] compared with the Ru precursor is attributed to an increase in the electron density of the pyridyl ring because of the π back-bonding interaction between ruthenium and the pyridyl groups of P4VP. A similar response is also observed for [Ru-P4VP-Fe] and [Ru-P4VP-CoFe]. A band observed at higher binding energies reveals the presence of nitrate anions for [Ru-P4VP], which are available to provide charge balance.[35] The cobalt precursor exhibits Co $2p_{3/2}$ and $2p_{1/2}$ peaks at 781.09 eV and 796.96 eV, respectively. Similarly, those of [Ru-P4VP-CoFe] are positioned at 781.01 eV and 796.54 eV, suggesting the presence of cobalt ions with a +2 oxidation state (Fig. 5). Furthermore, the cobalt region of [Ru-P4VP-CoFe]

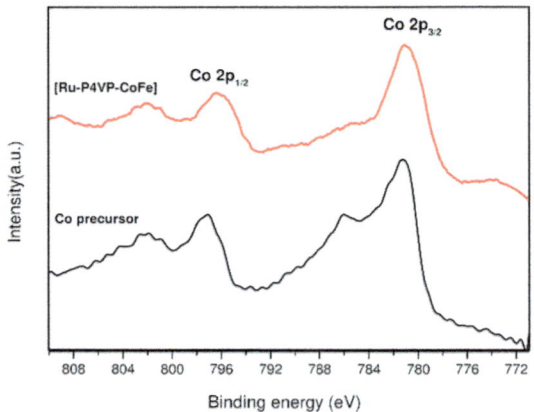

Fig. 5   XPS spectra of the Co 2p signals of the Co precursor, and [Ru–P4VP–CoFe].

contains satellite signals at binding energies approximately 5 eV higher than the principal signals.

XRD analysis conducted on [Ru–P4VP–CoFe] powder reveals characteristic peaks of PB structure. The broad nature of the peaks implies the formation of small PB structures due to the polymeric moiety (Fig. S5†). The structural morphology was also confirmed by SEM studies (Fig. S6†). Cubic structures with particle sizes of around 50 nm were observed. EDX studies also confirm the presence of Ru, Co, Fe, and a small quantity of Na yielding a rough molecular formula of $Na_{0.96}Co_{2.86}[Fe(CN)_5]_{2.13}$–$[P4VP]_6$–$[Ru(bpy)_2]Cl_{2.29}$ (Fig. S7†). It should be noted that an average of two out of six pyridyl groups are estimated to react with $[Ru(bpy)_2]$ fragments[21,22] while the ratio is 1 : 3 for $[Fe(CN)_5]$/pyridyl. SEM studies performed on different regions of the powder sample indicate the lack of a uniform stoichiometric ratio between metal ions. Such a non-uniform distribution can be explained by both the nonstoichiometric nature of PBAs[36] and their integration with a non-uniform Ru–polymer system. Furthermore, the amount of chloride ions is higher in regions where ruthenium is more abundant. A similar trend was also observed for sodium atoms with respect to cobalt and iron atoms, which suggests that $[Ru(bpy)_2]$ and CoFe PBA exhibit a cationic and anionic nature, respectively. Thus, chloride ions are present to provide the charge balance in regions where ruthenium ions are in excess, while Na ions serve a similar purpose for PB structures. Overall, the characterization studies conclude that P4VP is coordinated to both $[Ru(bpy)_2]$ fragments and cubic PB structures.

Catalytic performance

Photocatalytic studies were performed on a suspension solution containing [Ru–P4VP–CoFe] powder and $Na_2S_2O_8$ as the electron scavenger, at pH 7. Photocatalytic experiments were also performed with a previously studied cobalt hexacyanoferrate (labeled as Co–Fe PBA throughout the manuscript) in the presence of a $[Ru(bpy)_3]^{2+}/S_2O_8^{2-}$ couple for comparison under similar conditions.[17] In both experiments, the quantity of O_2 in the gas-tight set-up was measured before and after with gas chromatography. Blank measurements without a catalyst, a chromophore, and an electron scavenger were also carried out under the same conditions.

The experiment for [Ru–P4VP–CoFe] was performed for six cycles with the same batch while the experiment for the $[Ru(bpy)_3]^{2+}$ and Co–Fe PBA couple was performed for three cycles. The $[Ru(bpy)_3]^{2+}$ and Co–Fe PBA curve yields a turnover frequency of 4.5×10^{-4} s^{-1}, which is in good agreement with the previous study.[17] The catalytic activity of $[Ru(bpy)_3]^{2+}$ and Co–Fe PBA system decreases gradually. In the final cycle, the number of moles of O_2 produced reached the value of the blank measurement ($[Ru(bpy)_3]^{2+}$ and Co–Fe PBA without an electron scavenger, Fig. S8†), which is attributed to the decomposition of $[Ru(bpy)_3]^{2+}$ complex under photocatalytic conditions.[17]

The photocatalytic water oxidation performance of PBAs was previously investigated by Galán-Mascarós *et al.* with characterization studies performed on the post-catalytic sample.[17] The origin of the decaying trend was found to be due to the photodecomposition of the Ru chromophore by releasing its pyridyl groups. These groups then poison the catalyst by coordinating to catalytic cobalt sites. On the other hand, [Ru–P4VP–CoFe] maintained its catalytic activity for six cycles. TOF ranges from $3 \times 10^{-4}\ s^{-1}$ to $6 \times 10^{-4}\ s^{-1}$ throughout these cycles. The slight variation in the catalytic performance can be attributed to the change in the morphology of the powder sample during the catalytic process and/or to the rough approximations made for the determination of TOF. For example, all cobalt sites are assumed to be catalytically active in the estimation of TON and TOF for [Ru–P4VP–CoFe]. Given a particle size of 50 nm for cubic-shaped particles obtained by SEM image (Fig. S6†), a rough calculation indicates that only around 3% of the cobalt sites are on the surface and active. Thus, the changes in TOF during each cycle are well within the error range (Fig. 6).

A TON of 11 is obtained after six cycles under a total of 6 h light illumination while the $Ru(bpy)_3^{2+}$ and Co–Fe PBA system achieved only a TON of 2 after three cycles in a 3 h period. The results show that the ruthenium complex in [Ru–P4VP–CoFe] serves as a chromophore similar to $Ru(bpy)_3^{2+}$ and coupling it with a heterogeneous catalyst enhances its stability dramatically (Fig. 7).

Post-catalytic characterization

The stability of the catalyst has been investigated in detail by performing XPS and infrared studies on the post-catalytic powder sample. The suspension solution was filtered, washed several times with distilled water, and dried in a vacuum desiccator to obtain the post-catalytic powder sample.

The XPS analysis of Co 2p and O 1s binding energies in the pristine and post-catalytic samples were carried out for [Ru–P4VP–CoFe]. The spectra of the Co 2p bands exhibit similar Co $2p_{3/2}$, Co $2p_{1/2}$, and satellite bands (Fig. S9†). Moreover, a lack of peaks below 780 eV rules out the decomposition of Co–Fe PBA to a possible catalytically active oxide species.[26] XPS of the O 1s region was conducted to confirm that there is no decomposition of the metal-coordinated clusters which might lead to a mixed metal oxide. Analysis of the O 1s region clearly shows that there is no cobalt oxide species, which typically have binding energies lower than 530 eV, and peaks observed around 531 eV are only due to surface-adsorbed oxygen species (Fig. S10†).[25] Based on the comparison of the Ru $3d_{5/2}$ band of pristine and post catalytic [Ru–P4VP–CoFe], possible formation of RuO_2 can be ruled out (Fig. S11†).[37] It should also be noted that a slight

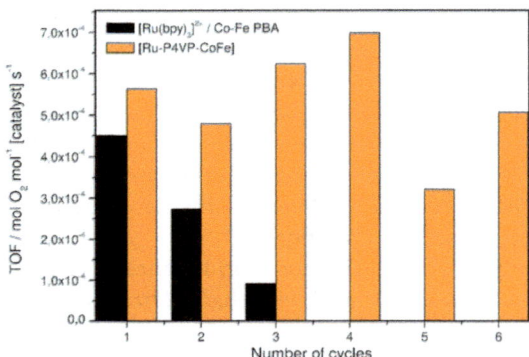

Fig. 6 TOF *vs.* number of cycles comparison of [Ru–P4VP–CoFe] (orange bar) and $[Ru(bpy)_3]^{2+}$/Co–Fe PBA system (black bar). Each cycle duration is 1 h.

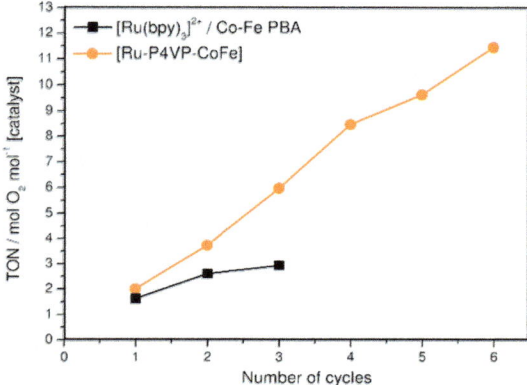

Fig. 7 TON *vs.* number of cycles comparison of [Ru–P4VP–CoFe] (orange, ●) and [Ru(bpy)₃]²⁺/Co–Fe PBA system (black, ■).

broadening in the O 1s peak and a slight shift in the Ru $3d_{3/2}$ peak has been observed for the post-catalytic sample, which could be attributed to the formation of a trace amount of ruthenium oxide species.

The cyanide stretch of the post-catalytic sample shifts to higher wavenumbers, which is attributed to partial oxidation of Co^{2+} to Co^{3+} during photoexcitation (Fig. S12†). As pointed out by Galán-Mascarós *et al.*, this change could also be due to linkage isomerism (CN bond flipping).[17] Furthermore, two major bands of the pristine sample at 1417 cm^{-1} and 1597 cm^{-1}, which are attributed to C=C$_{ring}$ and C=N$_{ring}$ of pyridyl rings, were observed also for the post-catalytic sample (Fig. 8).

Although the [Ru–P4VP–CoFe] assembly showed enhanced activity and stability compared with its uncoordinated analogue system, comparison should also be made with dyad systems reported in the literature. Photocatalytic activity is modest in comparison with the Ru-based dyads reported by Thummel *et al.* (TON of 134 and 68),[3,8] and the trinuclear ruthenium assembly studied by Sun *et al.* (TON of 38).[9]. [Ru–P4VP–CoFe] exhibits a higher TON than a molecular

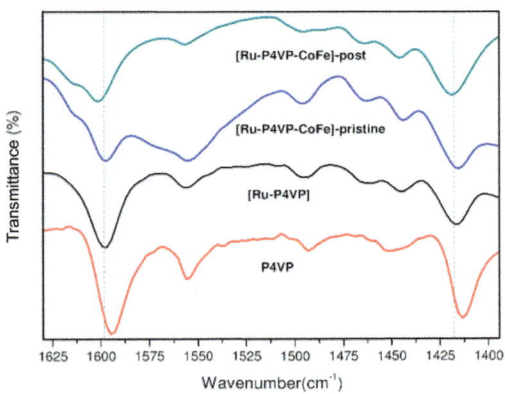

Fig. 8 FTIR spectra of P4VP, [Ru–P4VP], [Ru–P4VP–CoFe], and [Ru–P4VP–CoFe] after six catalytic cycle.

cobalt-based dyad (TON of 5) reported by Sun *et al.*[38] The stability of molecular dyads have generally been investigated with relatively short-term experiments, which are in the order of minutes. This study, however, represents a dyad that maintains its activity during a 6 h photocatalytic experiment. The stability of [Ru–P4VP–CoFe] can be attributed to the inherent robustness of the rigid cyanide network. It is also important to underline that no studies have been reported so far on the re-usability of dyads, and therefore, this study is novel in the line of techniques to analyze catalytic performance of dyads.

Polymeric dyad assemblies presented in the literature are investigated in DSPEC systems where the assembly is anchored to a semiconductor. In such studies, the activities of the dyads are analyzed in different experimental conditions and are reported in terms of current densities and faradaic efficiencies. For this reason, fair comparison of the [Ru–P4VP–CoFe] system with polymeric dyads reported cannot be made.

Conclusions

Overall, a novel heterogeneous PS–WOC dyad, incorporating poly(4-vinylpyridine) as a bridge between a ruthenium chromophore and a cobalt-based PBA, was presented. The structure of each of the intermediate products ([Ru–P4VP] and [Ru–P4VP–Fe]) and that of the final product [Ru–P4VP–CoFe], was monitored and elucidated with infrared, UV-Vis, and XPS studies. EDX studies revealed the formation of a non-stoichiometric compound, wherein the atomic ratio of Ru to Fe atoms varies from 1.87 to 2.56, the average of which yields a rough molecular formula of $Na_{0.96}Co_{2.86}[Fe(CN)_5]_{2.13}$–[P4VP]–$[Ru(bpy)_2]Cl_{2.29}$. SEM and XRD studies indicate the formation of small Prussian blue cubic structures with a size of around 50 nm. Catalytic performance of the dyad was investigated under 1 h light illumination, and oxygen evolution was measured with GC where a pressure transducer was used as a supporting device to sense the pressure during the photocatalytic experiment. The dyad showed slightly higher catalytic activity (a TOF of 5.6×10^{-4} s^{-1}) compared with the relevant multi-component system (a TOF of 4.5×10^{-4} s^{-1}), which could be attributed to the increase in the number of active cobalt sites on the surface due to change in the morphology or improvement in the activity of catalytically active cobalt sites or a combination of both. Moreover, [Ru–P4VP–CoFe] maintained a steady activity for six cycles. Characterization studies performed on the pristine and post-catalytic powder samples confirmed the stability of the assembly under harsh photocatalytic conditions. This result indicates that rigid PB structures serve not only as a water oxidation catalyst but also as a protective layer for the chromophore. Therefore, immobilization of chromophores *via* coordination to cyanide-based frameworks could be a viable approach for the development of robust and active dyad assemblies for photocatalytic water oxidation. The diversity, easy synthesis, and remarkable stability of cyanide-based frameworks make them ideal candidates for the development of dye-sensitized photoanodes for water oxidation.

Conflicts of interest

There are no conflicts to declare.

Acknowledgements

This work is supported by the Scientific and Technological Research Council of Turkey (TUBITAK), grant number 215Z249. Ferdi Karadas thanks TÜBA-GEBİP and BAGEP for young investigator awards.

Notes and references

1 J. L. Dempsey, A. J. Esswein, D. R. Manke, J. Rosenthal, J. D. Soper and D. G. Nocera, *Inorg. Chem.*, 2005, **44**, 6879–6892.

2 J. J. Concepcion, J. W. Jurss, P. G. Hoertz and T. J. Meyer, *Angew. Chem., Int. Ed.*, 2009, **48**, 9473–9476.

3 N. Kaveevivitchai, R. Chitta, R. Zong, M. El Ojaimi and R. P. Thummel, *J. Am. Chem. Soc.*, 2012, **134**, 10721–10724.

4 D. L. Ashford, M. K. Gish, A. K. Vannucci, M. K. Brennaman, J. L. Templeton, J. M. Papanikolas and T. J. Meyer, *Chem. Rev.*, 2015, **115**, 13006–13049.

5 M. Yamamoto, L. Wang, F. Li, T. Fukushima, K. Tanaka, L. Sun and H. Imahori, *Chem. Sci.*, 2016, **7**, 1430–1439.

6 B. D. Sherman, D. L. Ashford, A. M. Lapides, M. V. Sheridan, K. R. Wee and T. J. Meyer, *J. Phys. Chem. Lett.*, 2015, **6**, 3213–3217.

7 B. D. Sherman, Y. Xie, M. V. Sheridan, D. Wang, D. W. Shaffer, T. J. Meyer and J. J. Concepcion, *ACS Energy Lett.*, 2017, **2**, 124–128.

8 L. Kohler, N. Kaveevivitchai, R. Zong and R. P. Thummel, *Inorg. Chem.*, 2014, **53**, 912–921.

9 F. Li, Y. Jiang, B. Zhang, F. Huang, Y. Gao and L. Sun, *Angew. Chem., Int. Ed.*, 2012, **51**, 2417–2420.

10 G. Leem, B. D. Sherman and K. S. Schanze, *Nano Convergence*, 2017, **4**, 37.

11 J. Jiang, B. D. Sherman, Y. Zhao, R. He, I. Ghiviriga, L. Alibabaei, T. J. Meyer, G. Leem and K. S. Schanze, *ACS Appl. Mater. Interfaces*, 2017, **9**, 19529–19534.

12 D. M. Ryan, M. K. Coggins, J. J. Concepcion, D. L. Ashford, Z. Fang, L. Alibabaei, D. Ma, T. J. Meyer and M. L. Waters, *Inorg. Chem.*, 2014, **53**, 8120–8128.

13 S. E. Bettis, D. M. Ryan, M. K. Gish, L. Alibabaei, T. J. Meyer, M. L. Waters and J. M. Papanikolas, *J. Phys. Chem. C*, 2014, **118**, 6029–6037.

14 Z. Fang, A. Ito, H. Luo, D. L. Ashford, J. J. Concepcion, L. Alibabaei and T. J. Meyer, *Dalton Trans.*, 2015, **44**, 8640–8648.

15 H. Shimakoshi, M. Nishi, A. Tanaka, K. Chikama and Y. Hisaeda, *Chem. Commun.*, 2011, **47**, 6548–6550.

16 G. Leem, Z. A. Morseth, K.-R. Wee, J. Jiang, M. K. Brennaman, J. M. Papanikolas and K. S. Schanze, *Chem.–Asian J.*, 2016, **11**, 1257–1267.

17 S. Goberna-Ferrón, W. Y. Hernández, B. Rodríguez-García and J. R. Galán-Mascarós, *ACS Catal.*, 2014, **4**, 1637–1641.

18 E. P. Alsaç, E. Ülker, S. V. K. Nune, Y. Dede and F. Karadas, *Chem.–Eur.J.*, 2018, **24**, 4856–4863.

19 L. Han, P. Tang, Á. Reyes-Carmona, B. Rodríguez-García, M. Torréns, J. R. Morante, J. Arbiol and J. R. Galan-Mascaros, *J. Am. Chem. Soc.*, 2016, **138**, 16037–16045.

20 Y. Yamada, K. Oyama, R. Gates and S. Fukuzumi, *Angew. Chem., Int. Ed.*, 2015, **54**, 5613–5617.

21 Y. Isaka, K. Oyama, Y. Yamada, T. Suenobu and S. Fukuzumi, *Catal. Sci. Technol.*, 2016, **6**, 681–684.

22 Y. Yamada, K. Oyama, T. Suenobu and S. Fukuzumi, *Chem. Commun.*, 2017, **53**, 3418–3421.

23 F. S. Hegner, D. Cardenas-Morcoso, S. Giménez, N. López and J. R. Galan-Mascaros, *ChemSusChem*, 2017, **10**, 4552–4560.

24 Y. Aratani, T. Suenobu, K. Ohkubo, Y. Yamada and S. Fukuzumi, *Chem. Commun.*, 2017, **53**, 3473–3476.

25 M. Aksoy, S. V. K. Nune and F. Karadas, *Inorg. Chem.*, 2016, **55**, 4301–4307.

26 Z. Kap, E. Ülker, S. V. K. Nune and F. Karadas, *J. Appl. Electrochem.*, 2018, **48**, 201–209.

27 A. Devadoss, A. M. Spehar-Délèze, D. A. Tanner, P. Bertoncello, R. Marthi, T. E. Keyes and R. J. Forster, *Langmuir*, 2010, **26**, 2130–2135.

28 C. F. Hogan and R. J. Forster, *Anal. Chim. Acta*, 1999, **396**, 13–21.

29 D. J. Kenney, T. P. Flynn and J. B. Gallini, *J. Inorg. Nucl. Chem.*, 1961, **20**, 75–81.

30 V. Marin, E. Holder and U. S. Schubert, *J. Polym. Sci., Part A: Polym. Chem.*, 2004, **42**, 374–385.

31 B. Durham, J. V. Caspar, J. K. Nagle and T. J. Meyer, *J. Am. Chem. Soc.*, 1982, **104**, 4803–4810.

32 V. Leigh, W. Ghattas, R. Lalrempuia, H. Müller-Bunz, M. T. Pryce and M. Albrecht, *Inorg. Chem.*, 2013, **52**, 5395–5402.

33 N. G. Khaligh, *RSC Adv.*, 2012, **2**, 3321–3327.

34 S. A. V. Jannuzzi, B. Martins, M. I. Felisberti and A. L. B. Formiga, *J. Phys. Chem. B*, 2012, **116**, 14933–14942.

35 V. K. Kaushik, *J. Electron Spectrosc. Relat. Phenom.*, 1991, **56**, 273–277.

36 F. Hegner, I. Herraiz-Cardona, D. Cardenas-Morcoso, N. Lopez, J. R. Galan-Mascaros and S. Gimenez, *ACS Appl. Mater. Interfaces*, 2017, **9**, 37671–37681.

37 D. J. Morgan, *Surf. Interface Anal.*, 2015, **47**, 1072–1079.

38 X. Zhou, F. Li, H. Li, B. Zhang, F. Yu and L. Sun, *ChemSusChem*, 2014, **7**, 2453–2456.

Faraday Discussions

PAPER

Evaluating the impacts of amino acids in the second and outer coordination spheres of Rh-bis(diphosphine) complexes for CO₂ hydrogenation†

Aaron P. Walsh, [iD] ‡§ Joseph A. Laureanti,‡§ Sriram Katipamula,¶
Geoffrey M. Chambers, Nilusha Priyadarshani,‖ Sheri Lense,**
J. Timothy Bays, [iD] John C. Linehan [iD] and Wendy J. Shaw [iD] *

Received 4th November 2018, Accepted 23rd November 2018

DOI: 10.1039/c8fd00164b

To explore the influence of a biologically inspired second and outer coordination sphere on Rh-bis(diphosphine) CO_2 hydrogenation catalysts, a series of five complexes were prepared by varying the substituents on the pendant amine in the $P(Et)_2CH_2N^RCH_2P(Et)_2$ ligands ($P^{Et}N^RP^{Et}$), where R consists of methyl ester modified amino acids, including three neutral (glycine methyl ester (GlyOMe), leucine methyl ester (LeuOMe), and phenylalanine methyl ester (PheOMe)), one acidic (aspartic acid dimethyl ester (AspOMe)) and one basic (histidine methyl ester (MeHisOMe)) amino acid esters. The turnover frequencies (TOFs) for CO_2 hydrogenation for each of these complexes were compared to those of the non-amino acid containing $[Rh(depp)_2]^+$ (depp) and $[Rh(P^{Et}N^{Me}P^{Et})_2]^+$ (NMe) complexes. Each complex is catalytically active for CO_2 hydrogenation to formate under mild conditions in THF. Catalytic activity spanned a factor of four, with the most active species being the NMe catalyst, while the slowest were the GlyOMe and the AspOMe complexes. When compared to a similar set of catalysts with phenyl-substituted phosphorous groups, a clear contribution of the outer coordination sphere is seen for this family of CO_2 hydrogenation catalysts.

Physical and Computational Sciences Directorate, Pacific Northwest National Laboratory, Richland, WA 99352, USA. E-mail: wendy.shaw@pnnl.gov

† Electronic supplementary information (ESI) available. CCDC 1874734. For ESI and crystallographic data in CIF or other electronic format see DOI: 10.1039/c8fd00164b

‡ Authors contributed equally to this work.

§ Current address: Ferro Corporation, Penn Yan, NY 14527, USA.

¶ Current address: Rutgers University, Piscataway, NJ 08854, USA.

‖ Current address: Curium, Maryland Heights, MO.

** Current address: University of Wisconsin-Oshkosh, Oshkosh, WI.

Introduction

Enzymes are able to convert small molecules selectively, efficiently, and with minimal energy input.[1-4] These complex biological catalysts often feature metallic centers within highly tuned coordination environments with intricate second and higher order coordination sphere architectures that facilitate this incredible performance. Research focused on molecular catalysts has strived to achieve the efficacy of enzymes, primarily by modifying the first coordination sphere, the ligands bound to the metal. The second coordination sphere, which is comprised of functional groups which can interact with the metal but aren't bound to the metal, have been instrumental in many of the more recent advancements.[5-7] In spite of these advancements, however, the majority of synthetic catalysts still fall short of enzymatic performance.[8-11] The introduction of outer coordination sphere functionality to small molecule catalysts has proven advantageous in some cases, demonstrating that the bioinspired approach to molecular catalyst design is a viable path to achieve similar catalytic performance.[12-15]

Our previous work with hydrogen oxidation catalysts demonstrated the positive impact of including an outer coordination sphere.[11,16] Using attributes beyond the first and second coordination sphere, we were able to achieve faster TOFs,[17,18] lower overpotentials,[19,20] and electrochemically reversible catalysis while maintaining fast TOFs.[16,21,22] Cooperativity between the first, second, and outer coordination spheres is vital, and the performance is highly sensitive to modest changes possibly due to disruption between intersphere cooperativity.[16,17,23]

In this work, we directly incorporate amino acid analogs into the outer coordination sphere of molecular catalysts for CO_2 hydrogenation. The conversion of CO_2 has potential importance in energy storage and higher value products, as well as in controlling adverse climate effects. The use of second coordination sphere elements in CO_2 reduction has been addressed by us and others. As examples, Fujita and co-workers introduced water soluble ligands allowing the transformation of CO_2 to formate in aqueous solution by iridium complexes, with a protonatable group facilitating CO_2 addition and reduction.[24,25] Kubiak investigated pendant amines in the second coordination sphere using a P_2N_2 framework with rhodium and found that the steric bulk of the pendant amine substituent hindered catalysis.[26] Nilay and co-workers introduced PNP-pincer ligands to iridium and, with the use of ^{13}C-labeled CO_2, showed that CO_2 will directly bind to the iridium metal center with the assistance of a hydrogen bond donor in the second coordination sphere.[27]

Little effort has been focused on the outer coordination sphere of CO_2 hydrogenation catalysts. We provided one such study, investigating a series of $[Rh(P^{Ph}N^{R}P^{Ph})_2]^+$ complexes as a function of the pendant amine substituent.[28] These complexes were extremely sensitive to both a pendant amine in the second coordination sphere and changes in the outer coordination sphere by altering the R group, with TOFs spanning two orders of magnitude. However, the amino acid, amino acid ester, and dipeptide complexes were very slow ($\sim$10 h^{-1}), adding experimental difficulty to the collection of the data. Given the thermodynamics of this system, our expectation was that alkyl-substituted phosphorous groups would result in faster catalysts,[29,30] allowing us to more readily implement and evaluate the impact of the outer coordination sphere on various points in the

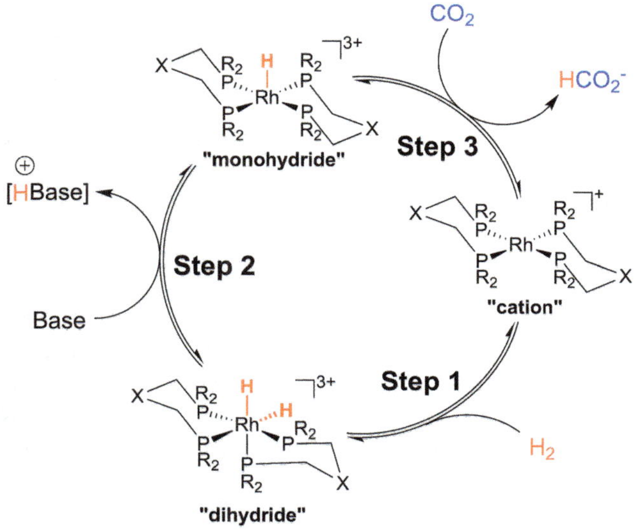

Fig. 1 Proposed mechanism for catalytic hydrogenation of CO_2 by the $[Rh(PN^RP)_2][BF_4]$ catalysts in organic solvents.

proposed catalytic cycle (Fig. 1). Consequently, in this work, we evaluated the analogous $[Rh(P^{Et}N^RP^{Et})_2]^+$ complexes to ultimately allow implementation of a more complex outer coordination sphere. We investigated the role of the pendant amine with just a methyl group, as well as a series of amino acid esters substituted on the pendant amine.

Results

Synthesis of *N,N*-bis(diethylphosphino)amino acid methyl ester (PN^{R}P) ligands and metal complexes

To determine the impact of amino acid residues in the outer coordination sphere on catalytic performance, five new $[Rh(PN^RP)_2][BF_4]$ complexes were prepared with different N-group substituents (Fig. 2), where N^R is N-CH(CH$_2$COOCH$_3$)

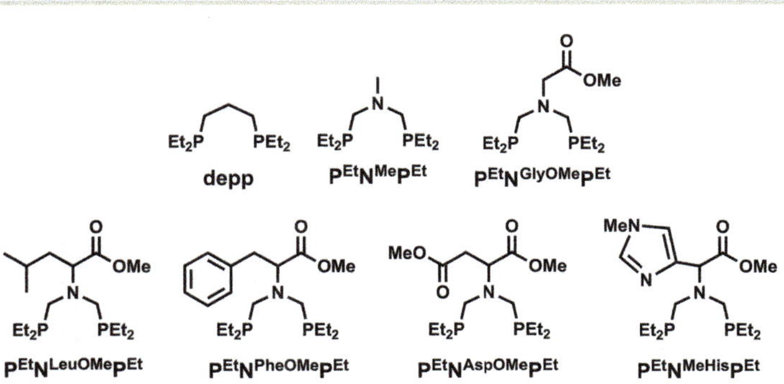

Fig. 2 PNP ligands and their abbreviations.

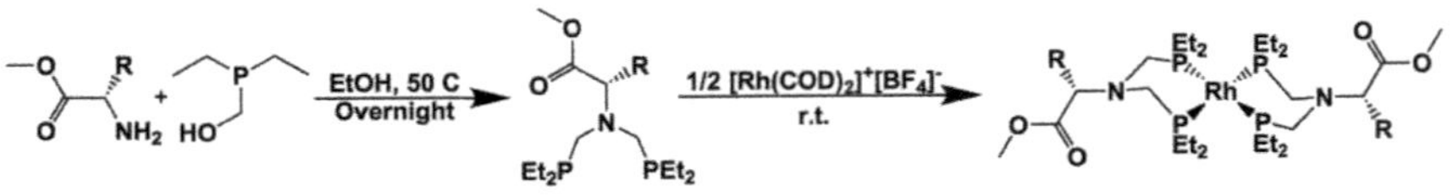

Scheme 1 Overall synthesis of amino acid ester incorporated PNP ligands and metal complexes.

$COOCH_3$ (aspartic acid dimethyl ester, AspOMe), $N\text{-}CH_2COOCH_3$ (glycine methyl ester, GlyOMe), $N\text{-}CH(CH_2CH(CH_3)_2)COOCH_3$ (leucine methyl ester, LeuOMe), $N\text{-}CH(CH_2C_6H_5)COOCH_3$ (phenylalanine methyl ester, PheOMe), and $N\text{-}CH(CH_2\text{-}C_4H_6N_2)COOCH_3$ (histidine methyl ester, MeHisOMe). The simpler $Rh(PN^{Me}P)_2$ and $Rh(depp)_2$ (ref. 31) (depp = diethyl phosphino propane)[32] complexes, were also investigated to probe the mechanism for CO_2 hydrogenation. Neutralized methyl ester protected amino acids were condensed with (hydroxymethyl)diethylphosphine in anhydrous ethanol to produce PN^RP ligands (Scheme 1)[33] generally in good yield (>70%). The addition of $[Rh(COD)_2][BF_4]$ in tetrahydrofuran (THF) to two equivalents of ligand resulted in powders of metal complex generally in good yield (>60%) after precipitation from a concentrated solution. Ligands and complexes were characterized by ^{31}P and ^1H nuclear magnetic resonance (NMR) spectroscopy, and elemental analysis; metal complexes were also

Table 1 ^1H and $^{31}P\{^1H\}$ NMR chemical shifts and J-couplings, and reduction potentials for $[Rh(PN^RP)_2]^+$ under nitrogen and under H_2 in the presence ($[(H)Rh(PN^RP)_2]$) and absence ($[(H)_2Rh(PN^RP)_2]^+$) of Verkade's base. n/a and n/d denote not applicable and not determined, respectively[a]

Complex	$^{31}P\{^1H\}$ NMR (ppm)	$^1J_{RhP}$ (Hz)	^1H NMR hydride (ppm)	$^1J_{RhH}$ (Hz)	$E_{p,a}$ (V vs. Fc)
$[Rh(PN^{PheOMe}P)_2]^+$	4.65	125.7	n/a	n/a	0.23
$[Rh(PN^{Me}P)_2]^+$	5.46	125.2	n/a	n/a	0.11
$[Rh(PN^{MeHisOMe}P)_2]^+$	4.04	125.6	n/a	n/a	0.46
$[Rh(PN^{LeuOMe}P)_2]^+$	4.21	125.2	n/a	n/a	0.3
$[Rh(depp)_2]^+$	5.99	126.6	n/a	n/a	0.14
$[Rh(PN^{AspOMe}P)_2]^+$	4.46	125.5	n/a	n/a	0.35
$[Rh(PN^{GlyOMe}P)_2]^+$	3.87	125.4	n/a	n/a	0.07
$[(H)_2Rh(PN^{PheOMe}P)_2]^+$	17.31, −4.39	a, a	−10.52	145	0.71
$[(H)_2Rh(PN^{Me}P)_2]^+$	18.89, −3.99	a, a	−10.5	145	0.55
$[(H)_2Rh(PN^{MeHisOMe}P)_2]^+$	16.49, −5.51	93.7, a	−10.54	145	0.69
$[(H)_2Rh(PN^{LeuOMe}P)_2]^+$	16.97, −4.51	94.1, a	−10.49	145	0.81
$[(H)_2Rh(depp)_2]^+$	14.92, −6.52	93.4, 82.0	−10.75	145	0.87
$[(H)_2Rh(PN^{AspOMe}P)_2]^+$	17.05, −4.58	a, a	−10.53	145	0.79
$[(H)_2Rh(PN^{GlyOMe}P)_2]^+$	17.93, −4.08	93.9, 82.6	−10.46	145	0.63
$[(H)Rh(PN^{PheOMe}P)_2]$	12.23	139.4	−12.34	140	n/d
$[(H)Rh(PN^{Me}P)_2]$	12.38	139.4	−12.3	140	n/d
$[(H)Rh(PN^{MeHisOMe}P)_2]$	11.95	139.2	−12.37	145	n/d
$[(H)Rh(PN^{LeuOMe}P)_2]$	11.89	139.2	−12.39	145	n/d
$[(H)Rh(depp)_2]$	10.96	139.4	−12.67	145	n/d
$[(H)Rh(PN^{AspOMe}P)_2]$	12.1	139.4	−12.39	145	n/d
$[(H)Rh(PN^{GlyOMe}P)_2]$	12.18	139.2	−12.32	140	n/d

[a] Poorly resolved.

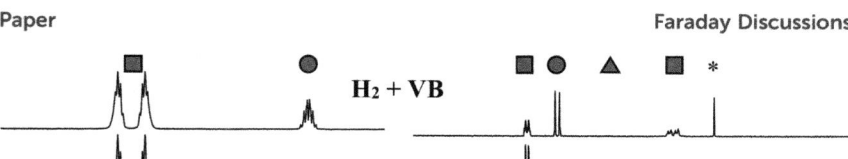

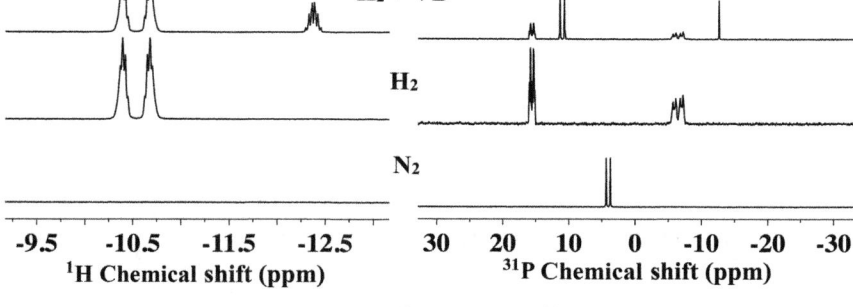

-9.5   -10.5   -11.5   -12.5   30   20   10   0   -10   -20   -30
^{1}H Chemical shift (ppm) ^{31}P Chemical shift (ppm)

Fig. 3 Representative chemical shifts, ^{1}H (left) and ^{31}P{^{1}H} (right), for [Rh(PN^{R}P)$_2$]$^+$ complexes under 1 atm N$_2$ (bottom) or [Rh(PN^{R}P)$_2$]$^+$ complexes under 1 atm H$_2$ in the absence ([(H)$_2$Rh(PN^{R}P)$_2$]$^+$; middle) and presence ([(H)Rh(PN^{R}P)$_2$]; top) of 1.1 equivalents of Verkade's base (VB). Resonances are denoted as follows: squares = dihydride, circles = monohydride, and triangles = cationic species. * denotes protonated Verkade's base.

characterized by electrochemistry as well as single crystal X-ray diffraction. The resulting metal complexes display a characteristic doublet in the ^{31}P{^{1}H} NMR spectra between 3.5 and 6.0 ppm, corresponding to the Rh–P coupling ($^1J_{Rh-P}$ = 125.5 ± 0.5 Hz, Table 1 and Fig. 3, bottom and Fig. S1†) and show dihydrides with the addition of H$_2$ (Fig. 3 and S2†), as expected.

Catalytic hydrogenation of carbon dioxide

All of the complexes evaluated were active for the catalytic hydrogenation of CO$_2$ to formate at room temperature and 34 atm of 1 : 1 CO$_2$: H$_2$ in the presence of Verkade's base (2,8,9-triisopropyl-2,5,8,9-tetraaza-1-phosphabicyclo[3.3.3] undecane) in THF-d_8. The catalytic performance was evaluated using the TOF of each complex (Table 2, Fig. 4). The observed TOFs for the [Rh(PN^{R}P)$_2$]$^+$ catalysts decreased in the following order: NMe > LeuOMe > MeHisOMe > PheOMe >

Table 2 TOF of formate production using [Rh(PN^{R}P)$_2$]$^+$ catalysts. Catalyst loading was 1.0 mM under 34 atm 1 : 1 CO$_2$: H$_2$

[Rh(P^{Et}N^{R}P^{Et})$_2$]$^+$ Catalyst, R=	TOF (h^{-1})	Equivalents Verkade's base for ethyl-substituted	TOF for [Rh(P^{Ph}N^{R}P^{Ph})$_2$]$^{+a}$
NMe	420 (60)	550 (80)	920 (20)^a
MeHisOMe	250 (20)	500	NA
LeuOMe	245 (60)	450 (105)	NA
PheOMe	200 (30)	500 (25)	NA
depp/dppp (no N)	170 (20)	500	150 (50)^a
GlyOMe	100 (20)	520 (10)	11 (4)^a
AspOMe	100 (25)	410 (150)	NA
Gly	NA	NA	120 (10)^a
GlyAlaOMe	NA	NA	12 (7)^a

^{a} Reported in ref. 28.

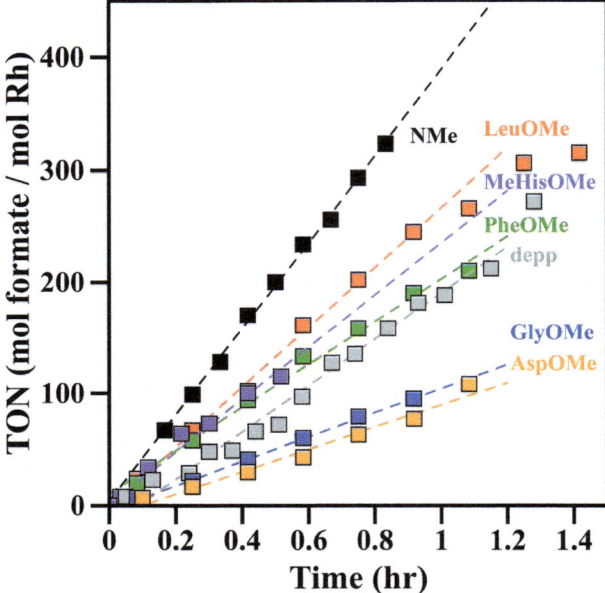

Fig. 4 Representative plots of the production of formate as a function of time, determined for the complexes in this study from operando ^{1}H NMR spectroscopy. Black = NMe, red = LeuOMe, green = PheOMe, purple = MeHisOMe, grey = depp, blue = GlyOMe, and orange = AspOMe. Reaction conditions: $[Rh(PN^{R}P)_2]^+$ (1.0 mM), 1 : 1 CO_2 : H_2, 34 atm, THF-d_8, Verkade's base (515 ± 40 mM).

GlyOMe > AspOMe. Product formation was monitored using operando ^{1}H NMR spectroscopy and quantified using the residual HDO resonance from a capillary insert of D_2O as standard, containing $CoCl_2$ as a relaxation agent to shift the resonance to provide spectral resolution. Catalytic trials exhausted the Verkade's base, as confirmed by the loss of the original $^{31}P\{^{1}H\}$ resonance at 120 ppm and replacement by the protonated Verkade's base resonance at ~-12 ppm in the ^{31}P $\{^{1}H\}$ NMR spectrum.

Thermodynamic studies

Thermodynamic parameters were determined by evaluating the addition of H_2 to form the dihydride and then the addition of base to form the monohydride under lower pressures and concentrations. In all cases for the ethyl-substituted complexes in this study, exposure to 1 atm of H_2 resulted in 100% conversion to the dihydride species, with a representative spectrum shown in Fig. 3 and others shown in Fig. S1 and S2,† and a crystal structure for the NMe complex shown in Fig. 5 with selected distances shown in Table S1.† Upon the addition of Verkade's base, the dihydride is deprotonated to monohydride, with 3–5 equivalents resulting in complete deprotonation (Fig. 3, S1 and S2†). The data with one equivalent of added Verkade's base were used to estimate pK_a values that show an average of 33.3 ± 0.5, and average hydricities of 26.8 ± 0.6 kcal mol^{-1}, assuming a $K_{eq} > 100$ for the H_2 addition step (Table 3).

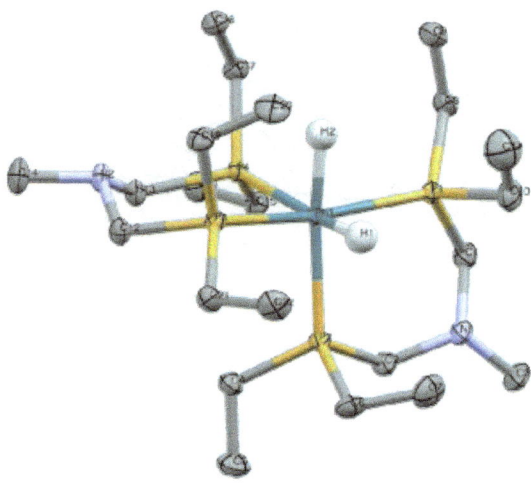

Fig. 5 Crystal structure of the $[(H)_2Rh(P^{Et}N^{Me}P^{Et})_2]^+$ complex shown with ellipsoids at 50% probability. Anions, solvent molecules and hydrogens are not shown for clarity.

Electrochemical analysis of $[Rh(PN^RP)_2][BF_4]$ complexes

Cyclic voltammetry experiments were employed to directly interrogate the electronic properties of the $[Rh(PN^RP)_2]^+$ and $[(H)_2Rh(PN^RP)_2]^{3+}$ species under 1.0 atmosphere of nitrogen or hydrogen, respectively, in a solution of [TBA][BF4] (0.1 M) in THF. The voltammograms were first recorded under a nitrogen atmosphere with $[Rh(PN^RP)_2]^+$ at ~1.0 mM, Fig. 6. Hydrogen was then sparged through the solution, resulting in the rapid formation of the dihydride, as confirmed by the bleaching of the solution from yellow to colorless. Under an atmosphere of nitrogen, one irreversible oxidation was observed that is attributed to the Rh^{III}/ Rh^I couple (*i.e.* oxidation of the 16-electron Rh^I cation to a 14-electron Rh^{III} species).[34] Similar electrochemical behavior was observed under an atmosphere of hydrogen, with the exception of shifts of the anodic peak potential (E_{pa}) of, on average, 0.5 ± 0.025 V in the anodic direction. The oxidation events under hydrogen are attributed to the oxidation of the Rh^{III}-dihydride species to a Rh^I cationic species and are irreversible in all cases.

Table 3 Thermodynamic parameters for $[Rh(P^RN^{R'}P^R)_2]^+$ complexes

Catalyst	pK_a	ΔG_{H^-}
$[Rh(P^{Et}N^{AspOMe}P^{Et})_2]^+$	33.4	25.9
$[Rh(P^{Et}N^{LeuOMe}P^{Et})_2]^+$	33.8	26.1
$[Rh(P^{Et}N^{Me}P^{Et})_2]^+$	32.8	26.7
$[Rh(P^{Et}N^{GlyOMe}P^{Et})_2]^+$	33.3	26.9
$[Rh(P^{Et}N^{PheOMe}P^{Et})_2]^+$	33.7	27.0
$[Rh(P^{Et}N^{MeHisOMe}P^{Et})_2]^+$	33.5	27.3
$[Rh(P^{Et}N^{deppP}P^{Et})_2]^+$	32.7	27.4
$[Rh(P^{Ph}N^{GlyOMe}P^{Ph})_2]^{+a}$	31.7[a]	27.7[a]

[a] Reported in ref. 28.

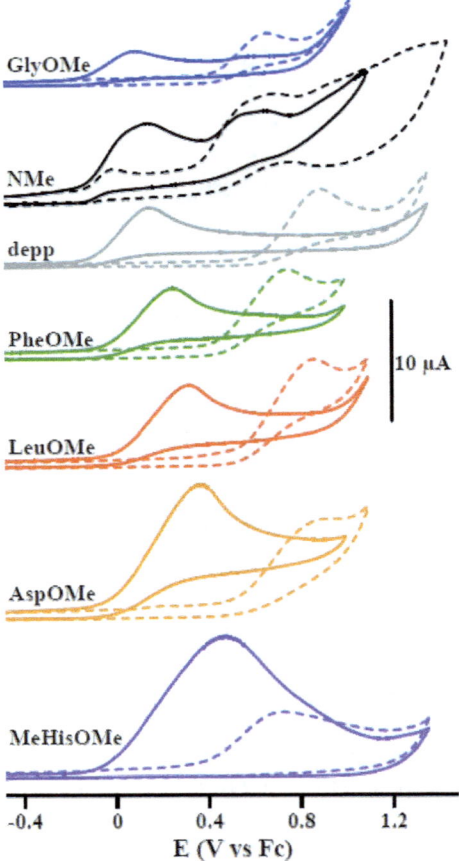

Fig. 6 Cyclic voltammetry of the $[Rh(P^{Et}N^{R}P^{Et})_2]^+$ complexes under N_2 (solid) and H_2 (dashed), recorded in THF at 0.2 V s^{-1}.

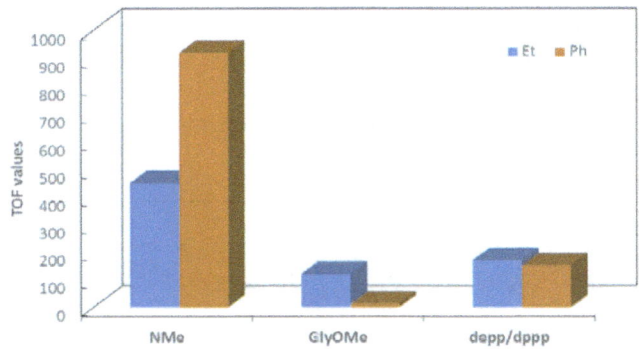

Fig. 7 Observed TOF values of analogous ethyl-substituted (this work) and phenyl-substituted complexes. The different relative rates for complexes substituted differently on the bridge atom (C or N) suggest effects beyond the primary coordination sphere.

Plots of the E_{pa} for the irreversible oxidation of the Rh^{I+} species *vs.* the respective chemical shifts in the $^{31}P\{^1H\}$ NMR produce a linear trend, suggesting that the electron density reported by the ^{31}P NMR reflects the electron density at the metal. Specifically, as electron density is increasingly removed from the rhodium metal and redirected to the phosphorus atoms, a concomitant shift to more positive potentials for the E_{pa} is observed as well as an upfield shift of the resonance in the $^{31}P\{^1H\}$ NMR (Fig. S3†). However, there is no correlation between the observed catalytic CO_2 hydrogenation TOFs and either the E_{pa} or the $^{31}P\{^1H\}$ chemical shift of the $[Rh(PN^RP)_2]^+$, Fig. S4.†

Discussion

General methods for modulating the catalytic activity of homogenous catalysts rely on substitution of the atoms directly coordinating a metal, and varying the electron donating and withdrawing effects of the substituents of these primary coordinating atoms. An alternative strategy is to manipulate the second and outer coordination sphere interactions of the metal, which has been shown to increase the catalytic activity by several orders of magnitude for proton reduction and hydrogen oxidation catalysts, as well as to improve the catalytic efficiency.[7,14–16,25–27,35–41] The influence of the second coordination sphere has been addressed for catalysts for CO_2 reduction,[7,25–27,36,40,41] but less attention has been paid to the influence of the outer coordination sphere. In this work, we investigated the role of a bioinspired second and outer coordination sphere for CO_2 hydrogenation catalysts, where the role of a secondary coordination sphere pendant amine was investigated, and the outer coordination sphere was modulated by substitution of various amino acid ester analogues at the pendant amine.

The TOFs for CO_2 hydrogenation using the seven complexes in this study span a range of four-fold. The pendant amine has a modest impact in these complexes, with an ∼ two-fold increase in TOF for the NMe complex compared to that of the depp complex. For complexes containing a pendant amine, the NMe derivative is the fastest (400 h^{-1}), while the GlyOMe and AspOMe complexes are the slowest (100 h^{-1}). While the NMe complex is the least bulky and is also the fastest, the order of TOFs is not clearly associated with steric bulk, since the GlyOMe complex, which has the smallest side chain, is also the slowest. Furthermore, our combined electrochemical and ^{31}P NMR studies suggest that there is not a pure correlation between the TOF and the electron density at the metal (Fig. S4†). In most of the analyses of the TOF *vs.* another observable, there is often a qualitative correlation with the exception of a couple of points. Unfortunately, the data points that don't correlate are different for each observable. For instance, NMe and AspOMe complexes stand out from the trend for electron density at the metal, while the chemical shift of the ^{31}P resonance is clearly out of line for the depp complex. Overall, the modest change as a function of the outer coordination sphere functional group suggests a minimal impact of the outer coordination sphere when considering this data in isolation.

Influence of the phosphorous substituent

To more fully assess the impact of the outer coordination sphere on this series of complexes, we considered a similar family of complexes that was studied, with phenyl substituents on the phosphorous atom rather than ethyl groups.[28] In that series of complexes, the differences in the TOF as a function of side chain were stark; some of the TOFs are reproduced in Table 2. For instance, the addition of the pendant amine resulted in an order of magnitude increase in the TOF, while the addition of an amino acid slowed the TOF by an order of magnitude, and complexes with the addition of amino acid esters or dipeptides slowed the TOF by two orders of magnitude with respect to the NMe complex. There are two issues of interest to compare. The first is why there was such a significant impact on the TOFs of the phenyl-substituted complexes and not on those of the ethyl-substituted complexes. The second is a comparison of phenyl-substituted and ethyl-substituted complexes with similar functional groups.

The source of the impact on phenyl-substituted complexes

The phenyl-substituted complexes are expected to have slower TOFs based on the high hydricity of the bis-diphosphine substituted Rh complexes as a result of the aromatic substituents.[29,30] The range of TOFs for the ethyl-substituted complexes $(100-400 \text{ h}^{-1})$ was generally faster than that of the phenyl-substituted complexes $(12-920 \text{ h}^{-1})$, with the exception of the phenyl-substituted NMe complex which is at least a factor of two faster than all of the other complexes (Table 2).[28]

The lower hydricity and higher pK_a of the Rh-dihydride in the ethyl-substituted complexes (Table 3) may result in a difference in the rate-limiting step. This possibility was investigated by looking at each step under stoichiometric conditions. In all cases for the ethyl-substituted complexes in this study, exposure to 1 atm of H_2 resulted in 100% conversion to the dihydride species (Fig. S1 and S2†), while the addition of base gave a distribution of dihydride and monohydride. In contrast, the phenyl-substituted complexes added H_2 very poorly. The ΔG_{H_2} values ranged from 0.41 to 1.44 kcal mol^{-1} for the phenyl-substituted complexes containing a pendant amine, representing a <10% conversion upon exposure to 1 atm of H_2.[28] However, upon addition of 3–5 eq. of Verkade's base, complete conversion to the monohydride was observed. Based on these data, we determined that the pK_a values for the ethyl-substituted complexes are slightly higher than that observed for the phenyl-substituted complexes (Table 3), while the hydricities are lower. Both of these observations suggest that it should be harder to deprotonate the dihydride of the ethyl-substituted complexes. Note that no monohydride was observed in the absence of added base in either complex, consistent with our expectation that the pK_a of the pendant amine is too acidic to deprotonate the dihydride species on its own.

The phenyl-substituted complexes were proposed to proceed through the catalytic cycle shown in Fig. 1, with the rate limiting step being the CO_2 to HCO_2^- conversion. Based on the thermodynamic parameters determined for the observations of the ethyl-substituted complexes (Table 3), it is possible that the rate limiting step switches from the CO_2 conversion to formate in the phenyl-substituted complexes (Step 3 in Fig. 1) to deprotonation of the dihydride in the ethyl-substituted complexes (Step 2 in Fig. 1). The collective data from the studies of the phenyl-substituted complexes and the ethyl-substituted complexes imply that the role of the pendant amine and the outer coordination sphere in the phenyl-substituted complexes was to facilitate the addition of CO_2/conversion to formate (Step 3), rather than influencing deprotonation.

Comparison of phenyl-substituted and ethyl-substituted analogs

Comparison of analogous ethyl- and phenyl substituted complexes supports the outer coordination sphere playing a role. In the two sets of complexes, we have three analogous complexes, with the only difference being the substituent on the phosphorous atoms. These are the depp/dppp complexes, the NMe complexes, and the GlyOMe complexes. While we do expect an effect from the primary coordination sphere, if the only effect were from the primary coordination sphere,[29,30] we would expect a similar influence regardless of the second and outer coordination spheres; however, we see differing effects due to the second and outer coordination spheres (Fig. 6), pointing to their clear role. Specifically, in the NMe set of complexes, the TOF of the phenyl-substituted complex is two-times faster than that of the ethyl-substituted complex, while for the GlyOMe complexes, the phenyl-substituted complex is ten-times slower than the ethyl-substituted complex. For the depp/dppp complexes, the TOFs are the same, within error. We would anticipate that the phenyl-substituted complex would always be slower if the primary coordination sphere was the dominant factor;[29,30] instead we observe that they are sometimes slower and sometimes faster, by significantly different amounts, and only with the presence of the pendant amine (Fig. 7). These observations suggest that the phosphorous substituents are not responsible for the observed effect, and consequently implicate the outer coordination sphere in contributing to the catalytic TOF. Computational and experimental studies are ongoing to understand the individual or multiple contributions of the outer coordination sphere.

Conclusion

This series of Rh complexes for CO_2 hydrogenation shows a clear impact of the outer coordination sphere, particularly when compared to the analogous phenyl-substituted complexes. The TOFs of the ethyl-substituted complexes, the complexes that are the focus of this study, span a modest range of a factor of four as a function of amino acid modification. However, when compared to their complementary phenyl-substituted complexes, the TOFs are either $10\times$ slower, $2\times$ faster, or exactly the same, behavior that would not be observed if the primary coordination sphere was the only contribution to the mechanism. The mechanism or mechanisms that are driving this behavior are still under investigation, but these data point to this system being rich for investigation of the effects of the outer coordination sphere.

Methods

General procedures

All experiments were conducted under a nitrogen atmosphere inside a Vacuum Atmospheres glovebox or on a Schlenk line unless otherwise noted. All of the chemicals purchased were of the highest purity commercially available and used as received unless otherwise noted. A NaK alloy was used to dry THF-d_8, which was then distilled under vacuum. Acetonitrile, diethyl ether, dichloromethane, ethanol, and tetrahydrofuran were purified by passage through activated alumina columns in an Innovative Technology, Inc. PS-MD-6 solvent purification system.

UHP (ultra high purity) CO_2, H_2, and gas mixtures were purchased from Oxarc. Polyetheretherketone (PEEK) cells were employed for use with high pressure catalysis using operando NMR spectroscopy, designed and constructed at the Pacific Northwest National Laboratory.[42] All NMR spectroscopy data was collected on a Varian NMRS or Inova operating at 500 MHz ^{1}H frequency. Chemical shift values for ^{31}P$\{^1$H$\}$ NMR spectra were referenced to an external 85% H_3PO_4 standard at 0 ppm. Elemental analysis was performed at the CENTC Elemental Analysis Facility at the University of Rochester. For crystal structure characterization, a suitable crystal was selected, mounted on a MiTeGen MicroMounts pin using Paratone-N oil, and cooled to the data collection temperature (140(2)) K. Data was collected on a Bruker-AXS II CCD diffractometer with 0.71073 Å Mo Kα radiation. Cell parameters were retrieved using Bruker APEX II software,[43] raw data were integrated using SAINTPlus,[44] and absorption correction was applied using SADABS.[45] The structure was solved using direct methods and refined by a least-squares method on F^2 using SHELXS-97 and SHELXL-97,[46] respectively, as well as SHELXL-2018/3 to improve the structure refinement,[47] using the OLEX2 software package as a front end.[48]

Synthetic procedures

General procedure for the neutralization of amino acid methyl esters. Neutralization was conducted in a manner similar to previously described methods.[33] Methyl ester-protected amino acids were purchased as the hydrochloride salt, 1.0 equiv., 5.28 mmol was added to a Morton flask, and dissolved in 20 mL of dichloromethane. Sodium carbonate was added as a solution in water (10.0 equiv., 52.8 mmol), and the resulting mixture was stirred vigorously for 20 minutes. The organic layer was separated and, combined with anhydrous $MgSO_4$ (5.0 g, 40 mmol), and stirred for an additional five minutes to remove any residual water. $MgSO_4$ was then collected over a frit, and the solvent was removed from the organic layer *via* reduced pressure. The final products were isolated as clear, high-boiling oils, and used without further characterization.

General procedure for the synthesis of PN^{R}P ligand. The syntheses of all of the PN^{R}P ligands used in this study were performed in a manner similar to previously described methods.[3] Briefly, diethylphosphinomethanol (0.24 g, 2 mmol) was prepared using a modified preparation from that previously reported.[32] Specifically, diethylphosphine (5.0 g, 55.5 mmol) and paraformaldehyde (1.665 g, 55.4 mmol) were dissolved in 0.05 L of absolute ethanol and the ensuing reaction mixture was stirred overnight at 60 °C under a nitrogen atmosphere. Removal of the solvent under reduced pressure afforded a colorless oil. The resulting alcohol and the neutralized amino acid ester (1 mmol) were dissolved in 25 mL of absolute ethanol in a 50 mL Schlenk flask equipped with a stir bar. The reactions were heated to 50–60 °C for 16 hours under nitrogen, after which the flask was cooled and the solvent was removed under reduced pressure to obtain a colorless oil. Ligands were obtained in >70% yield.

PN^{LeuOMe}P. The leucine methyl ester (LeuOMe) ligand: ^{1}H NMR (500 MHz, THF-d_8): δ 4.12 (m, 1H), 3.61 (s, 3H), 3.08 (dd, 2H), 2.63 (dd, 2H), 1.49 (m, 4H), 1.35 (m, 8H), 1.05 (m, 12H), 0.93 (m, 6H). ^{31}P$\{^1$H$\}$ NMR (202 MHz, THF-d_8): δ −30.49 anal. calcd (found) for $C_{17}H_{37}NO_2P_2$: C, 58.43 (58.6); H, 10.67 (10.92); N, 4.01 (3.72).

$PN^{AspOMe}P$. The aspartic acid methyl ester (AspOMe) ligand: ^{1}H NMR (500 MHz, THF-d_8): δ 3.63 (s, 3H), 3.58 (s, 3H), 2.97 (d, 2H), 2.79 (dd, 1H), 2.58 (ddd, 4H), 1.34 (m, 8H), 1.04 (m, 12H). ^{31}P{^{1}H} NMR (202 MHz, THF-d_8): δ −30.7. Anal. calcd (found) for $C_{16}H_{33}NO_4P_2$: C, 52.59 (52.92); H, 9.1 (9.03); N, 3.83 (3.78).

$PN^{GlyOMe}P$. The glycine methyl ester (GlyOMe) ligand: ^{1}H NMR (500 MHz, THF-d_8): δ 3.62 (s, 3H), 3.35 (s, 2H), 2.92 (s, 4H), 1.37 (q, 8H, J_H = 7.81 Hz), 1.05 t, 12H, J_H = 7.62 Hz. ^{31}P{^{1}H} NMR (202 MHz, THF-d_8): δ −30.9 ppm (s). Anal. calcd (found) for $C_{13}H_{29}NO_2P_2$: C, 53.23 (53.63); H, 9.96 (10.4); N, 4.77 (4.73).

$PN^{PheOMe}P \cdot 1/2H_2O$. The phenylalanine methyl ester (PheOMe) ligand: ^{1}H NMR (500 MHz, THF-d_8): δ 7.22 (m, 5H), 3.78 (m, 1H), 3.63 (s, 3H), 3.16 (q, 4H), 3.01 (dd, 2H), 1.83 (m, 8H), 1.17 (m, 12H). ^{31}P{^{1}H} NMR (202 MHz, THF-d_8): δ 4.88 elem. Anal. calcd (found) for $C_{40}H_{74}N_2O_5P_4$: C, 61.05 (61.36); H, 9.48 (9.75); N, 3.56 (3.43).

$PN^{Me}P$. The methyl substituted amine (Me) ligand: ^{1}H NMR (500 MHz, CD$_3$CN): δ 2.61 (d, 4H), 2.37 (s, 3H), 1.37 (q, 8H), 1.03 (tt, 12H). ^{31}P{^{1}H} NMR (202 MHz, CD$_3$CN): δ −31 (s). Anal. calcd (found) for $C_{11}H_{27}NP_2$: C, 56.15 (55.78); H, 11.56 (11.78); N, 5.95 (5.74).

General procedure for the synthesis of rhodium metal complexes. To a mixture of [(COD)$_2$Rh]BF$_4$ in THF (not soluble, 1 mmol) was added a solution of PN^{R}P ligand (2 mmol) dissolved in THF while stirring. After 1 h, an orange colored solution was obtained. The solvent was removed under vacuum, yielding an orange residue which was re-dissolved in ≥1 mL of THF then added dropwise to −30 °C diethyl ether while stirring vigorously. The rhodium complexes precipitate as yellow solids. The mixture was filtered and dried under vacuum yielding an analytically pure compound. Metal complexes were prepared in >60% yield.

[($PN^{AspOMe}P)_2Rh$]BF$_4$. The AspOMe complex: ^{1}H NMR (500 MHz, THF-d_8): δ 3.87 (t, 2H), 3.71 (s, 6H), 3.63 (s, 6H), 3.07 (dd, 8H), 2.78 (ddd, 4H), 1.84 (m, 16H), 1.19 (m, 24H). ^{31}P{^{1}H} NMR (202 MHz, THF-d_8): δ 2.65 (J_{RhP} = 125.4 Hz). Anal. calcd (found) for $C_{32}H_{66}BF_4$ $N_2O_8P_4$Rh: C, 41.75 (43.48); H, 7.22 (7.51); N, 3.04 (2.86).

[($PN^{GlyOMe}P)_2Rh$]BF$_4$. The GlyOMe complex: ^{1}H NMR (500 MHz, THF-d_8): δ 3.64 (s, 6H), 3.49 (s, 4H), 3.1 (s, 8H), 1.87 (m, 16H), 1.2 (m, 24H). ^{31}P{^{1}H} NMR (202 MHz, THF-d_8): δ 7.04 (J_{RhP} = 125.2 Hz). Anal. calcd (found) for $C_{26}H_{58}BF_4$ $N_2O_4P_4$Rh: C, 40.22 (40.26); H, 7.53 (7.45); N, 3.61 (3.55).

[($PN^{LeuOMe}P)_2Rh$]BF$_4$. The LeuOMe complex: ^{1}H NMR (500 MHz, THF-d_8): δ 3.67 (s, 3H), 3.38 (m, 2H), 3.06 (dd, 8H), 1.84 (m, 16H), 1.65 (m, 4H), 100.52 (m, 2H), 1.19 (m, 24H), 0.92 (m, 12H). ^{31}P{^{1}H} NMR (202 MHz, THF-d_8): δ 2.33 (J_{RhP} = 125.6 Hz). Anal. calcd (found) for $C_{34}H_{74}BF_4$ $N_2O_4P_4$Rh: C, 45.96 (46.81); H, 7.84 (7.98); N, 2.94 (2.76).

[($PN^{PheOMe}P)_2Rh$]BF$_4$. The PheOMe complex: ^{1}H NMR (500 MHz, CD$_3$CN): δ 7.22 (m, 5H), 4.32 (m, 1H), 3.55 (s, 3H), 3.05 (dd, 4H), 2.69 (dd, 2H), 1.34 (m, 8H), 1.01 (m, 12H). ^{31}P{^{1}H} NMR (202 MHz, THF-d_8): δ 2.82 (J_{RhP} = 125.8 Hz). Anal. calcd (found) for $C_{40}H_{70}BF_4N_2O_4P_4$Rh: C, 50.22 (50.47); H, 7.37 (7.28); N, 2.93 (2.81).

[($PN^{Me}P)_2Rh$]BF$_4$. The Me complex: ^{1}H NMR (500 MHz, THF-d_8): δ 2.66 (s, 3H), 2.36 (dd, 4H), 1.78 (m, 8H), 1.13 (m, 12H). ^{31}P{^{1}H} NMR (202 MHz, THF-d_8): δ 3.63 (J_{RhP} = 125.2 Hz). Anal. calcd (found) for $C_{22}H_{54}BF_4N_2P_4$Rh: C, 40.02 (40.07); H, 8.24 (8.34); N, 4.24 (4.15). Crystals of the dihydride complex were prepared by purging a solution in THF with H$_2$. The crystals were clear and prism.

General procedure for catalysis. In a dry, nitrogen glovebox, 300 μL d_8-THF was used to prepare a solution of catalyst (1.0 mM) and Verkade's base (410–550 mM, 2,8,9-triisobutyl-2,5,8,9-tetraaza-1-phosphabicyclo[3.3.3]undecane) at a final volume of 330 μL. A PEEK tube was then charged with this solution and the PEEK cell was fully assembled and isolated before exiting the anaerobic glovebox. Upon attachment to the ISCO high-pressure line, the PEEK cell was evacuated to less than 0.1 torr before refilling with a 1 : 1 mixture of H_2/CO_2 at a constant pressure of 34 atm. Between kinetic measurements the PEEK cell was placed on a vortex mixer to increase the mixing of gas into the solvent. ^{1}H-NMR & ^{31}P{^{1}H}-NMR spectra were recorded every 3–5 minutes between 2–30 minutes. ^{31}P{^{1}H} NMR chemical shifts in THF-d_8 were referenced to protonated Verkade's base at −11.9 ppm in the absence of CO_2, or −12.9 ppm in the presence of CO_2. To quantify the formate production, an external capillary standard ($CoCl_2$, aq., 0.0384 mmol in D_2O) was used inside the PEEK tube and the residual HDO resonance was used as a standard. Turnover frequencies (TOFs) were calculated using the initial slope of the kinetic plots. The TOFs reported are the average of a minimum of three kinetic runs.

Electrochemistry. Cyclic voltammetry experiments were conducted at room temperature in a Vacuum Atmospheres glovebox under 1.0 atmosphere of nitrogen or hydrogen, using THF as the solvent with 0.1 M tetrabutylammonium tetrafluoroborate ([TBA]$^+$[BF4]$^-$) as a supporting electrolyte. The potentiostat employed was a CH Instruments CHI620D electrochemical analyzer, with a conventional three-electrode configuration. A 1.0 mm glassy carbon disk encased in PEEK (Cypress Systems EE040) was used as the working electrode and a glassy carbon rod (Structure Probe, Inc) was used as the counter electrode. Working electrodes were polished with a diamond paste (Buehler) of decreasing size (3.0, 1.0, and 0.25 μm). A pseudo-reference electrode was prepared using a silver wire coated in Cl that was immersed in a solution of THF/[TBA]$^+$[BF4]$^-$ (0.1 M) and separated from the medium using a VyCor frit. Cyclic voltammetry experiments employed a sweep rate of 0.2 V s^{-1}. All of the potentials are referenced to the ferrocenium/ferrocene couple at 0 V.

Conflicts of interest

There are no conflicts of interest.

Acknowledgements

This work was supported by the US Department of Energy (DOE), Office of Science, Office of Basic Energy Sciences (BES), Division of Chemical Sciences, Geosciences & Biosciences. A portion of the work (GMC; ligand synthesis) was supported as part of the Center for Molecular Electrocatalysis, an Energy Frontier Research Center funded by the U.S. DOE, Office of Science, BES. Pacific Northwest National Laboratory (PNNL) is operated by Battelle for the U.S. DOE. Pacific Northwest National Laboratory (PNNL) is a multiprogram national laboratory operated for the DOE by Battelle.

References

1 A. Bachmeier and F. Armstrong, Solar-driven proton and carbon dioxide reduction to fuels—lessons from metalloenzymes, *Curr. Opin. Chem. Biol.*, 2015, **25**, 141–151, DOI: 10.1016/j.cbpa.2015.01.001.

2 T. E. Creighton, *Proteins: Structures and Molecular Properties*, W.H. Freeman and Company, New York, 2nd edn, 1993.

3 R. Wolfenden and M. J. Snider, The depth of chemical time and the power of enzymes as catalysts, *Acc. Chem. Res.*, 2001, **34**, 938–945.

4 R. H. Holm, P. Kennepohl and E. I. Solomon, Structural and functional aspects of metal sites in biology, *Chem. Rev.*, 1996, **96**, 2239–2314.

5 D. L. DuBois, Development of Molecular Electrocatalysts for Energy Storage, *Inorg. Chem.*, 2014, **53**, 3935–3960, DOI: 10.1021/ic4026969.

6 M. Rakowski DuBois and D. L. DuBois, The Roles of the First and Second Coordination Spheres in the Design of Molecular Catalysts for H_2 Production and Oxidation, *Chem. Soc. Rev.*, 2009, **38**, 62–72, DOI: 10.1039/b801197b.

7 E. Fujita, J. T. Muckerman and Y. Himeda, Interconversion of CO_2 and formic acid by bio-inspired Ir complexes with pendent bases, *Biochim. Biophys. Acta, Bioenerg.*, 2013, **1827**, 1031–1038.

8 A. M. Appel, J. E. Bercaw, A. B. Bocarsly, H. Dobbek, D. L. DuBois, M. Dupuis, J. G. Ferry, E. Fujita, R. Hille, P. J. A. Kenis, C. A. Kerfeld, R. H. Morris, C. H. F. Peden, A. R. Portis, S. W. Ragsdale, T. B. Rauchfuss, J. N. H. Reek, L. C. Seefeldt, R. K. Thauer and G. L. Waldrop, Frontiers, opportunities, and challenges in biochemical and chemical catalysis of CO_2 fixation, *Chem. Rev.*, 2013, **113**, 6621–6658.

9 G. Caserta, S. Roy, M. Atta, V. Artero and M. Fontecave, Artificial hydrogenases: biohybrid and supramolecular systems for catalytic hydrogen porduction or uptake, *Curr. Opin. Chem. Biol.*, 2015, **25**, 36–47.

10 W. J. Shaw, The outer-coordination sphere: incorporating amino acids and peptides as ligands for homogeneous catalysts to mimic enzyme function, *Catal. Rev.: Sci. Eng.*, 2012, **54**, 489–550.

11 B. Ginovska-Pangovska, A. Dutta, M. L. Reback, J. C. Linehan and W. J. Shaw, Beyond the active site: the impact of the outer coordination sphere on electrocatalysts for hydrogen production and oxidation, *Acc. Chem. Res.*, 2014, **47**, 2621–2630, DOI: 10.1021/ar5001742.

12 A. Dutta, A. M. Appel and W. J. Shaw, Designing electrochemically reversible H_2 oxidation and production catalysts, *Nat. Rev. Chem.*, 2018, **2018**, 244–252.

13 A. J. P. Cardenas, B. Ginovska, N. Kumar, J. Hou, S. Raugei, M. L. Helm, A. M. Appel, R. M. Bullock and M. O'Hagan, Controlling Proton Delivery through Catalyst Structural Dynamics, *Angew. Chem., Int. Ed.*, 2016, **55**, 13509–13513, DOI: 10.1002/anie.201607460.

14 C. M. Klug, A. J. P. Cardenas, R. M. Bullock, M. O'Hagan and E. S. Wiedner, Reversing the tradeoff between rate and overpotential in molecular electrocatalysts for H_2 production, *ACS Catal.*, 2018, **8**, 3286–3296.

15 F. Schwizer, Y. Okamoto, T. Heinisch, Y. Gu, M. M. Pellizzoni, V. Lebrun, R. Reuter, V. Kohler, J. C. Lewis and T. R. Ward, Artifical metalloenzymes: reaction scope and optimization strategies, *Chem. Rev.*, 2017, **118**, 142–231.

16 A. Dutta, A. M. Appel and W. J. Shaw, Designing electrochemically reversible H_2 oxidation and production catalysts, *Nat. Rev. Chem.*, 2018, **2**, 244–252.

17 A. Dutta, B. Ginovska, S. Raugei, J. A. S. Roberts and W. J. Shaw, OptimizingConditions for Utilization of an H_2 oxidation Catalyst with Outer Coordination Sphere Functionalities, *Dalton Trans.*, 2016, **45**, 9786–9793.

18 A. Dutta, J. A. S. Roberts and W. J. Shaw, Learning from Nature: Arg-Arg Pairing Enhances H_2 oxidation Catalyst Performance, *Angew. Chem., Int. Ed.*, 2014, **53**, 6487–6491.

19 A. Dutta, S. Lense, J. Hou, M. Engelhard, J. A. S. Roberts and W. J. Shaw, Minimal Proton Channel Enables H_2 oxidation and Production with a Water Soluble Nickel-Based Catalyst, *J. Am. Chem. Soc.*, 2013, **135**, 18490–18496.

20 S. Lense, A. Dutta, J. A. S. Roberts and W. J. Shaw, A Proton Channel Allows a Hydrogen Oxidation Catalyst to Operate at a Moderate Overpotential with Water Acting as a Base, *Chem. Commun.*, 2014, **50**, 792–795.

21 A. Dutta, D. L. DuBois, J. A. S. Roberts and W. J. Shaw, Amino acid modified Ni catalyst exhibits reversible H_2 oxidation/production over a broad pH range at elevated temperatures, *Proc. Natl. Acad. Sci. U. S. A.*, 2014, **111**, 16286–16291, DOI: 10.1073/pnas.1416381111.

22 N. Priyadarshani, A. Dutta, B. Ginovska, G. W. Buchko, M. O'Hagan, S. Raugei and W. J. Shaw, Achieving Reversible H_2/H^+ Interconversion at Room Temperature with Enzyme-Inspired Molecular Complexes: A Mechanistic Study, *ACS Catal.*, 2016, **6**, 6037–6049, DOI: 10.1021/acscatal.6b01433.

23 A. Dutta, S. Lense, J. A. S. Roberts, M. L. Helm and W. J. Shaw, The Role of Solvent and the Outer Coordination Sphere on H_2 oxidation Using $[Ni(P^{Cy}_2N^{Pyz}_2)_2]^{2+}$, *Eur. J. Inorg. Chem.*, 2015, **2015**, 5218–5225, DOI: 10.1002/ejic.201500732.

24 E. Fujita, J. T. Muckerman and Y. Himeda, Interconversion of CO_2 and formic acid by bio-inspired Ir complexes with pendent bases, *Biochim. Biophys. Acta, Bioenerg.*, 2013, **1827**, 1031–1038, DOI: 10.1016/j.bbabio.2012.11.004.

25 N. Onishi, S. Xu, Y. Manaka, Y. Suna, W.-H. Wang, J. T. Muckerman, E. Fujita and Y. Himeda, CO_2 Hydrogenation Catalyzed by Iridium Complexes with a Proton-Responsive Ligand, *Inorg. Chem.*, 2015, **54**, 5114–5123, DOI: 10.1021/ic502904q.

26 A. M. Lilio, M. H. Reineke, C. E. Moore, A. L. Rheingold, M. K. Takase and C. P. Kubiak, Incorporation of Pendent Bases into Rh(diphosphine)$_2$ Complexes: Synthesis, Thermodynamic Studies, And Catalytic CO_2 Hydrogenation Activity of $[Rh(P_2N_2)_2]^{(+)}$ Complexes, *J. Am. Chem. Soc.*, 2015, **137**, 8251–8260, DOI: 10.1021/jacs.5b04291.

27 T. J. Schmeier, G. E. Dobereiner, R. H. Crabtree and N. Hazari, Secondary Coordination Sphere Interactions Facilitate the Insertion Step in an Iridium(III) CO_2 Reduction Catalyst, *J. Am. Chem. Soc.*, 2011, **133**, 9274–9277, DOI: 10.1021/ja2035514.

28 J. T. Bays, N. Priyadarshani, M. S. Jeletic, E. B. Hulley, D. L. Miller, J. C. Linehan and W. J. Shaw, The influence of the second and outer coordination spheres on Rh(diphosphine)$_2$ CO_2 hydrogenation catalysts, *ACS Catal.*, 2014, **4**, 3663–3670, DOI: 10.1021/cs5009199.

29 X.-J. Qi, Y. Fu, L. Liu and Q.-X. Guo, Ab initio calculations fo thermodynamic hydricities of transition-metal hydrides in acetonitrile, *Organometallics*, 2007, **26**, 4197–4203.

30 X.-J. Qi, L. Liu, Y. Fu and Q.-X. Guo, Ab Initio Calculations of pKa Values of Transition-Metal Hydrides in Acetonitrile, *Organometallics*, 2006, **25**, 5879–5886, DOI: 10.1021/om0608859.

31 D. L. DuBois, D. M. Blake, A. Miedaner, C. J. Curtis, M. Rawkowski DuBois, J. A. Franz and J. C. Linehan, Hydride Transfer from Rhodium Complexes to Triethylborane, *Organometallics*, 2006, **25**, 4414–4419, DOI: 10.1021/om060584z.

32 C. J. Curtis, A. Miedaner, R. Ciancanelli, W. W. Ellis, B. C. Noll, M. Rakowski DuBois and D. L. DuBois, $[Ni(Et_2PCH_2NMeCH_2PEt_2)_2]^{2+}$ as a Functional Model for Hydrogenases, *Inorg. Chem.*, 2003, **42**, 216–227, DOI: 10.1021/ic020610v.

33 T. M. El Dine, J. Rouden and J. Blanchet, Borinic Acid Catalysed Peptide Synthesis, *Chem. Commun.*, 2015, **51**, 16084–16087.

34 D. Lamprecht and G. J. Lamprecht, Electrochemical oxidation of Rh(I) to Rh(III) in rhodium(I) β-diketonato carbonyl phosphine complexes, *Inorg. Chim. Acta*, 2000, **309**, 72–76.

35 N. P. Boralugodage, R. J. Arachchige, A. Dutta, G. W. Buchko and W. J. Shaw, Evaluating the role of acidic, basic, and polar amino acids and dipeptides on a molecular electrocatalyst for H_2 oxidation, *Catal. Sci. Technol.*, 2017, **7**, 1108–1121, DOI: 10.1039/c6cy02579j.

36 W.-H. Wang, J. F. Hull, J. T. Muckerman, E. Fujita and Y. Himeda, Second-coordination-sphere and electronic effects enhance iridium(iii)-catalyzed homogeneous hydrogenation of carbon dioxide in water near ambient temperature and pressure, *Energy Environ. Sci.*, 2012, **5**, 7923–7926, DOI: 10.1039/c2ee21888g.

37 P. Hosseinzadeh, N. M. Marshall, K. N. Chacon, Y. Yu, M. J. Nilges, S. Y. New, S. A. Tashkov, N. J. Blackburn and Y. Lu, Design of a Single Protein that Spans the Entire 2-V Range of Physiological Redox Potentials, *Proc. Natl. Acad. Sci. U. S. A.*, 2016, **113**, 262–267, DOI: 10.1073/pnas.1515897112.

38 F. Yu, V. M. Cangelosi, M. L. Zastrow, M. Tegoni, J. S. Plegaria, A. G. Tebo, C. S. Mocny, L. Ruckthong, H. Qayyum and V. L. Pecoraro, Protein Design: Toward Functional Metalloenzymes, *Chem. Rev.*, 2014, **114**, 3495–3578, DOI: 10.1021/cr400458x.

39 E. E. Benson, C. P. Kubiak, A. J. Sathrum and J. M. Smieja, *Chem. Soc. Rev.*, 2009, **38**, 89.

40 S. A. Chabolla, C. W. Machan, J. Yin, E. A. Dellamary, S. Sahu, N. C. Gianneschi, M. K. Gilson, F. A. Tezcan and C. P. Kubiak, Bio-inspired CO_2 reduction by a rhenium tricarbonyl bipyridine-based catalyst appended to amino acids and peptidic platforms: incorporating proton relays and hydrogen-bonding functional groups, *Faraday Discuss.*, 2017, **198**, 279–300, DOI: 10.1039/c7fd00003k.

41 T. J. Schmeier, G. E. Dobereiner, R. H. Crabtree and N. Hazari, Secondary Coordination Sphere Interactions Facilitate the Insertion Step in an Iridium(III) CO_2 Reduction Catalyst, *J. Am. Chem. Soc.*, 2011, **133**, 9274–9277, DOI: 10.1021/ja2035514.

42 C. R. Yonker and J. C. Linehan, The use of supercritical fluids as solvents for NMR spectroscopy, *Prog. Nucl. Magn. Reson. Spectrosc.*, 2005, **47**, 95–109.

43 *Bruker APEX 2, version 2010.3-0*, Bruker AXS Inc.: Madison, WI, 2009.

44 *SAINTPlus: Data Reduction and Correction Program*, Bruker AXS Inc.: Madison, WI, 2004, vol. 723a.

45 *SADABS: an Empirical Absorption Correction Program, version 2001/1*, Bruker AXS Inc.: Madison, WI, 2001.

46 G. M. Sheldrick, A short history of SHELX, *Acta Cryst*, 2008, **A64**, 112–122.

47 G. M. Sheldrick, *SHELXL2018*, University of Gottingen, Germany, 2018.

48 O. V. Dolomanov, L. J. Bourhis, R. J. Gildea, J. A. K. Howard and H. Puschmann, OLEX2: a complete structure solution, refinement and analysis program, *J. Appl. Crystallogr.*, 2009, **42**, 339–341.

PAPER

Performance of enhanced DuBois type water reduction catalysts (WRC) in artificial photosynthesis – effects of various proton relays during catalysis†

Wolfgang Viertl,[a] Johann Pann,[a] Richard Pehn,[a] Helena Roithmeyer,[a] Marvin Bendig,[a] Alba Rodríguez-Villalón,[b] Raphael Bereiter,[a] Max Heiderscheid,[a] Thomas Müller,[a] Xia Zhao,[c] Thomas S. Hofer, [ID] [a] Mark E. Thompson, [ID] [d] Shuyang Shi[d] and Peter Brueggeller [ID] *[a]

Received 4th November 2018, Accepted 3rd January 2019

DOI: 10.1039/c8fd00162f

Inspired by natural photosynthesis, features such as proton relays have been integrated into water reduction catalysts (WRC) for effective production of hydrogen. Research by DuBois *et al.* showed the crucial influence of these relays, largely in the form of pendant amine functions. In this work catalysts are presented containing innovative diphosphinoamine ligands: $[M(II)Cl_2(PNP\text{-}C1)]$, $[M(II)(MeCN)_2(PNP\text{-}C1)]^{2+}$, $[M(II)(PNP\text{-}C1)_2]^{2+}$, and $[M(II)Cl(PNP\text{-}C2)]^{+}$ (M = Pt^{2+}, Pd^{2+}, Ni^{2+}, Co^{2+}; PNP-C1 = N,N-bis{(di(2-methoxyphenyl)phosphino) methyl}-N-alkylamine, PNP-C2 = N,N-bis{(di(2-methoxyphenyl)phosphino)ethyl}-N-alkylamine and alkyl = Me, Et, iso-Pr, Bz). Synthetic strategies and detailed characterisation are covered, including ^{1}H-, ^{13}C-, and ^{31}P-NMR analysis, mass spectroscopy and single crystal X-ray diffractometry (XRD). The catalytic properties have been explored by changing the pendant amines and auxiliary methoxy coordination sites, as well as enlarging the ligand backbone. Moreover, confirmed by density functional theory (DFT) calculations based on XRD data *in vacuo* and solvent environment, two very different catalytic cycles are proposed. PNP-C1 shows a classical proton relay, whereas PNP-C2 allows an additional coordination of nitrogen, acting optionally like a pincer. Through new insights into efficiency and stability-increasing influences of proton relays in general, their number per metal centre, an enlarged ligand backbone and the use of solvato instead of halogenido complexes, substantial improvements have been made in catalytic performance over the DuBois *et al.* catalysts and recently self-made WRCs. The turnover

[a]University of Innsbruck, Centrum for Chemistry and Biomedicine, Institute of General, Inorganic and Theoretical Chemistry, Innrain 82, 6020 Innsbruck, Austria

[b]Complutense University of Madrid, Avda. de Séneca, 2, Ciudad Universitaria, 28040 Madrid, Spain

[c]Northwest University, Taibai Road 229, Beilin District, Xi'an 710069, P. R. China

[d]University of Southern California, Irani Chair of Chemistry Department of Chemistry, Chemical Engineering and Materials Science, Seeley G. Mudd Bldg. 3620 McClintock Ave, Los Angeles, CA 90089-1062, USA

† Electronic supplementary information (ESI) available. CCDC 1874808–1874812. For ESI and crystallographic data in CIF or other electronic format see DOI: 10.1039/c8fd00162f

number (TON) related to the single site of cost-efficient nickel WRCs is increased from 11.4 to 637, whereas a corresponding palladium catalyst gives TON as high as 2289.

Introduction

Research in climate change is a young field, even though it has been practiced for several decades. The general consensus in the scientific community is that there is an anthropogenic greenhouse effect. Humans have influenced the climate on Earth and this is directly linked with a proven rise in average temperature and sea level.[1] Politics, economics and society have mostly accepted this fact and due to the urgency of this anthropogenic effect an amazing effort can be noticed in respect of all kinds of green research areas worldwide. An important development was the Paris Agreement with representatives from nearly all countries of the world endorsing it.[2] An important scientific basis for this agreement was the Fifth Assessment Report of the Intergovernmental Panel on Climate Change, where the anthropogenic greenhouse effect was not only proven to be valid, but which also predicted the consequences of further emissions of greenhouse gases by human activity. Artificial photosynthesis has a high potential to decrease greenhouse emissions. Since the one and only inexhaustible source of energy is the sun, artificial photosynthesis is one of the most important research topics in the area of green chemistry.

Three leading voices in the area of water reduction catalysis, hydrogen production and artificial photosynthesis are Fontecave[3-6] at the College de France, Paris, Eisenberg[7-10] at the University of Rochester, NY, USA, and DuBois[11-15] at the Pacific Northwest National Laboratory, WA, USA. In the experiments reported here we follow the direction led by DuBois *et al.* He and his co-workers have focused on diphosphine ligands containing an amine base in the second coordination shell as a proton relay. Homoleptic complexes of this type of diphosphine-amine ligand have been prepared with first-row metals,[16] mainly Ni,[13,17,18] Co,[19,20] Fe[21] and Mn.[22] These complexes have been used mainly as electrocatalysts for proton reduction. Photocatalysts without the integrated proton relay show smaller TON values.[23] The mechanisms of hydrogen production from these catalysts on the molecular level are similar, regardless of whether the primary source of electrons for this reduction step is from electromagnetic irradiation or an applied electrical current.[23] For this reason, electro- and photocatalysts are often exchangeable. Two very interesting nickel complexes of DuBois *et al.* are shown in Fig. 1.[12] One of them (see Fig. 1 left) has a theoretical proton reduction with a turnover frequency (TOF) of more than $100\,000\ \mathrm{s}^{-1}$.

Since 2007 when the active site of the [FeFe] hydrogenase enzyme was clarified by its crystal structure,[24] the role of proton relays, in this case a pendant amine function, has become an important design criterion. Using the knowledge of the structure of the natural hydrogenase enzyme and research results from DuBois *et al.* we synthesised a series of new complexes shown in Fig. 2. The ligand system of DuBois[12] was adapted and modified, leading to two ligand systems PNP-C1 and PNP-C2 (*N*,*N*-bis-[(di-(2-methoxyphenyl)phosphino) methyl]-*N*-alkylamine and *N*,*N*-bis-[(di-(2-methoxyphenyl)phosphino)ethyl]-*N*-alkyl-amine) which were used to build up homo- and heteroleptic complexes

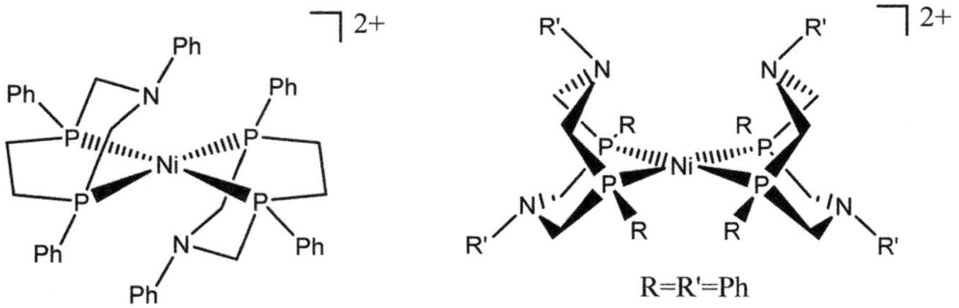

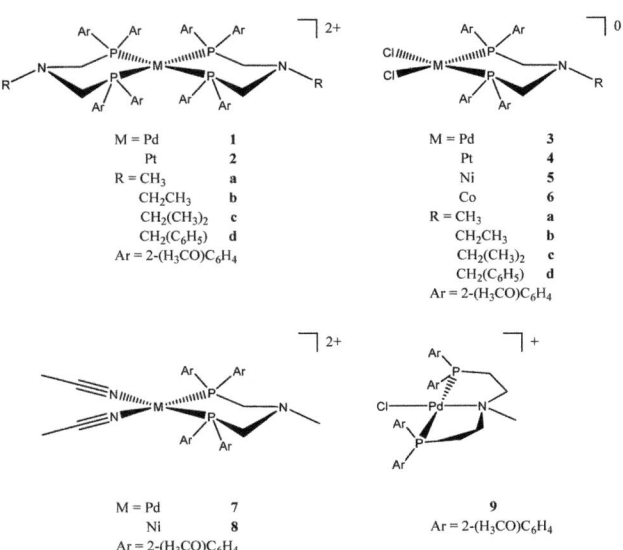

Fig. 1 Molecular catalysts of DuBois *et al.*[12] with a calculated TOF of more than 100 000 s^{-1}.

with Pt, Pd, Ni and Co. The ligands PNP-C1 and PNP-C2 are here newly reported, although analogous compounds with phenyl- instead of 2-methoxyphenyl substitutions on the phosphine site are partially known in literature.[15,25–27] This modification of the phosphine substitutions is a consequence of preliminary works.[28,29] Studies to verify the positive influence of the proton relays in hydrogen production are presented as well as the influence of amine substitution. Moreover hydrogen evolving experiments are discussed using different pH values, catalytically active metals, molar ratios of catalyst and photosensitiser. A comparison of homo-/heteroleptic, solvato/chlorido and PNP-C1/PNP-C2 complexes was also explored. Two of the presented catalysts have been structurally characterized by single crystal XRD data (three more in the ESI†). These structures were subsequently used for DFT calculations which led to two postulated reaction mechanisms.

M = Pd 1
Pt 2
R = CH_3 a
CH_2CH_3 b
$CH_2(CH_3)_2$ c
$CH_2(C_6H_5)$ d
Ar = 2-$(H_3CO)C_6H_4$

M = Pd 3
Pt 4
Ni 5
Co 6
R = CH_3 a
CH_2CH_3 b
$CH_2(CH_3)_2$ c
$CH_2(C_6H_5)$ d
Ar = 2-$(H_3CO)C_6H_4$

M = Pd 7
Ni 8
Ar = 2-$(H_3CO)C_6H_4$

9
Ar = 2-$(H_3CO)C_6H_4$

Fig. 2 Prepared new complexes **1–9** with different ligands (a–d, PNP-C2) and metal centres (Pt, Pd, Ni, Co).

Results and discussion

Influence of pendant amines as proton relays in hydrogen production

To understand the nature of proton relays in more detail we examined the amount of hydrogen produced per catalyst molecule (turnover number: TON) of various homoleptic metal complexes containing two PNP-C1 ligands. For this reason two series of WRCs were synthesised and irradiated with the well-known photo-sensitiser $[Ir(\text{III})(bpy)(ppy)_2](PF_6)$.[4,30–35] The two series of WRCs were $[Pd(\text{II})(PNP-C1-R)_2](PF_6)_2$ **1a–d** and $[Pt(\text{II})(PNP-C1-R)_2](PF_6)_2$ **2a–d** where R stands for methyl-, ethyl-, iso-propyl and benzyl-residues. Despite previous evidence of increased efficiency with proton relays,[17,36–38] a control sample was investigated for both of the series as well. The control sample without pendant amine has the formula $[M(\text{II})(PCCCP)_2](PF_6)_2$ ($M = Pt^{2+}$, Pd^{2+}; PCCCP = 1,3-bis(di(2-methoxyphenyl) phosphino)-n-propane). Hence, this control sample corresponds to the PNP-C1 ligand in which the pendant amine is replaced with a –CH_2– group. The increased efficiency for the catalyst with the proton relay is confirmed with the platinum series (ESI†) and palladium series (see Table 1). Two experiments with different molar ratios of photosensitiser and water reduction catalyst were chosen (PS/WRC = 1 or 25). The low activity of the control sample clearly shows the importance of the proton relay. Regardless of the molar ratio or catalytic active metal, complexes with a diphosphine ligand without proton relay provide less hydrogen (lower TON) in all cases. There is a second notable insight from the palladium series. In comparing the turnover numbers of the different substituted aminodiphosphines there is a clear trend in efficiency. Smaller R-groups give higher hydrogen output. Methylated PNP ligands work the best as proton relays. Interestingly, compound **2d** with a benzyl substitution works slightly better in a molar ratio PS/WRC of 1 than its analogue with iso-propyl. This may be due to an electronic effect of the additional aromatic ring in the second coordination shell. If the ratio is higher (PS/WRC = 25) compound **2d** cannot compete with less sterically bulky groups. A reasonable explanation for the effect of the size of the R-group in the catalysis is related to the fact that an excess of photosensitiser has a positive effect on the electron transfer from photosensitiser to water reduction catalyst. At ratio 1 : 1 an electron transfer occurs rarely compared to ratio 25 : 1. Thus, the catalysts will perform faster with an excess of photosensitiser and produce more hydrogen since the limiting step is no longer the electron transfer. The performance of the WRCs is supported by a flexible ligand backbone which

Table 1 Hydrogen production of the palladium series, dilution ratios PS/WRC of 1 and 25, irradiation with 700 W Hg medium pressure lamp, maximal relative standard deviation 1.5%

Compound	TON	TON
2a	66.8	455.7
2b	54.9	320.0
2c	35.5	268.2
2d	39.8	183.9
Control	11.2	217.4
PS/WRC	1	25
Irradiation time	$t = 21$ h	$t = 25$ h

causes faster catalysis. In total these insights are well in compliance with DuBois *et al.*[39]

The conclusions of these experiments were used for further investigations described in the next sections. The key point there is that pendant amines as proton relays are verified as important in increasing efficiency with compounds utilizing diphosphine ligands. Further experiments were performed exclusively with methylated PNP ligands, due to their higher flexibility and hydrogen outputs.

Hydrogen production of nickel catalyst 5 in respect of pH and WRC/PS ratio

The same irradiation conditions described above were used for $[Ni(\text{II})Cl_2(PNP\text{-}C1\text{-}Me)]$ **5a** (see Table 2). As expected the amount of molecular hydrogen produced per catalyst molecule increases with excess of PS. Thus, the proton transfer is not the rate-limiting step, otherwise the turnover would not rise with the addition of extra equivalents of PS. With a threefold excess of PS, the increase in TON within 5 hours is only about 50% up to 38. Here the cause of this may be a poor oxidation reaction needed to regenerate the catalyst. In the third run, the PS is present in a 25 fold excess and the TON reaches a markedly higher value of 609, *ca.* 25 times higher than when PS/WRC = 1. After 23 hours of irradiating the PS without a WRC present hydrogen is formed with a TON of 6, so the possible contribution of PS to the TON values in Table 2 can be neglected.

Irradiations at different pH values were carried out (see Fig. 3). After 9 hour reaction times, the best result was obtained at a pH of 10, with a TON of 637. By increasing or decreasing the pH, the produced amount of hydrogen drops. Other researchers have reported a number of possible reasons for this behaviour.[40–46] A more basic pH increases the reducing power of the sacrificial donor (accelerating effect), but at the same time the basic conditions decrease the concentration of the proton source needed for hydrogen evolvement (slowing effect). In this case the donor triethylamine (TEA) was always constantly concentrated in the irradiation solution with 10% (v/v), with the pH being adjusted with drops of concentrated hydrochloric acid. Therefore, pH 10 showed the best results concerning the balance of previously mentioned effects. A third influence of the high pH is the higher barrier to form metal hydride bonds and/or the reaction of metal hydrides with further protons.[47] Here again an advantageous balance between the hindrance of bonding reactions within catalysis and the concentration of the necessary sacrificial donor seems to be obtained at pH 10.

Table 2   Hydrogen production by $[Ni(\text{II})Cl_2(PNP\text{-}C1\text{-}Me)]$ **5a** with different molar ratios PS/WRC and within a pH range of 8 to 12.5 during 5 hours, light source 150 W Xe lamp, maximal relative standard deviation 1.5%

Run	Molar ratio PS/WRC	TON	pH	t/h
1	1	24.9	10	5
2	3	38.4	10	5
3	25	608.6	10	5
4	25	176.4	8	5
5	25	381.9	9	5
6	25	430.1	11	5
7	25	327.8	12.5	5

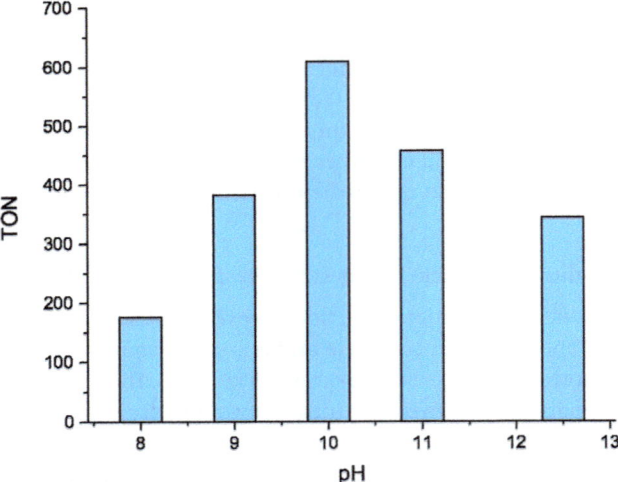

Fig. 3 Hydrogen production of [Ni(ıı)Cl$_2$(PNP-C1-Me)] **5a** with a 25-fold excess of PS, within 9 hours and a pH range of 8 to 12.5, maximal relative standard deviation 1.5%.

Efficiency comparison of catalytically active metals platinum, palladium, nickel and cobalt

After determining the best performing ligand, namely PNP-C1 with methyl substitution and pH condition (10), a study on hydrogen production with different catalytically active metals was carried out. In this study, catalysts **3a–6a** have been investigated. They are all chlorido complexes with one diphosphine ligand, only the transition metal differs. For the irradiation, solvent mixture, irradiation time, sacrificial donor, pH value and light source were kept constant. Platinum **4a** is poorly suited for light-driven hydrogen evolution (see Fig. 4 and ESI†), while palladium **3a** shows the best results. Interestingly, the nickel complex **5a** performs better than the cobalt analogue **6a**, within the first 24 hours. After 24 hours, an extra equivalent of PS was added to the irradiation solutions, after which the efficiency of the cobalt complex **6a** exceeds that of the nickel complex **5a**. This could be explained by either a higher stability of cobalt catalysts and/or some kind of induction period. Compared to similar compounds and photocatalytic experiments there is a reasonable suspicion for the higher stability of the cobalt catalyst during catalysis, especially since cobalt complexes can produce hydrogen even in an aerobic environment[47] (which was also shown for [NiFeSe] hydrogenase[48]). Besides that, the cobalt complex **6a** reaches 68% of the activity of the palladium **3a** analogue after 140 hours, contrary to the nickel complex **5a** which achieves only 45% of palladium's efficiency. Therefore, a very promising approach may be further investigations with cobalt and the presented PNP ligands.

Efficiency comparison of PNP-C1 *versus* PNP-C2, hetero- *versus* homoleptic and chlorido *versus* solvato water reduction catalysts

Enlarging the PNP backbone leads to a more flexible pendant base. The structure of PNP-C2, with an ethylene bridge between the heteroatoms, shows a different

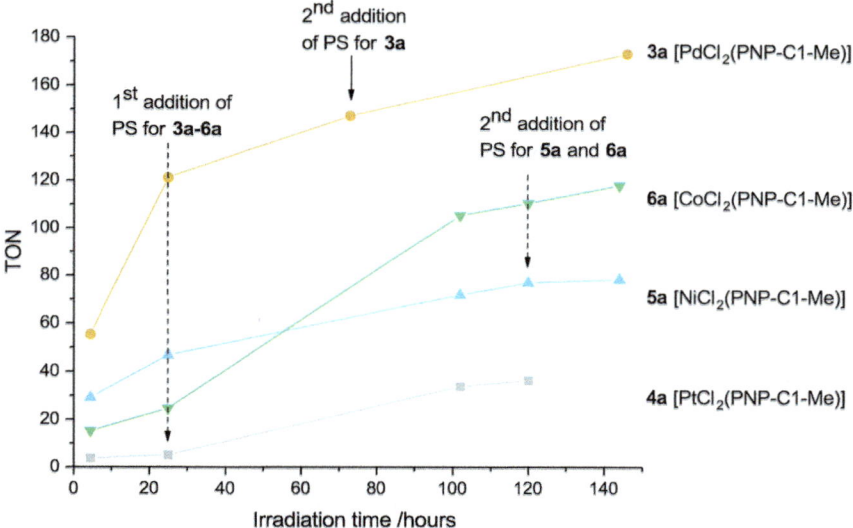

Fig. 4 Time dependent hydrogen production of [M(II)Cl$_2$(PNP-C1-Me)], 3a–6a, with Pt^{2+}, Pd^{2+}, Ni^{2+} and Co^{2+}, an extra equivalent of PS was added after 25 and 120 hours (except for [Pd(II)Cl$_2$(PNP-C1-Me)], 3a, after 25 and 73 hours), irradiation with 150 W Hg medium pressure lamp, maximal relative standard deviation 1.5%.

coordination behaviour, which can be seen in the structural analysis in the following section. PNP-C2 has a coordination behaviour of a pincer ligand.[49–54] The function of the nitrogen as proton relay seems to be deactivated in the pincer coordination mode. DFT calculations suggest an interesting behaviour within the catalysis, where the pendant amine is able to coordinate or deliver protons dependently on the catalytic step. Similar complexes with PNP ligands {[PdCl$_2$(-PNP-C1)] 3a and [PdCl(PNP-C2)](PF$_6$) 9} were synthesised and irradiated. Their performances in hydrogen production are illustrated in Fig. 5. Surprisingly, long-term irradiation gave a higher efficiency with the enlarged ligand backbone of PNP-C2. In this experiment four extra equivalents of photosensitiser were added to both WRCs during 200 hours of irradiation. In the beginning, 3a produces more hydrogen and gives the highest TON for more than 100 hours. In contrast, 9 shows poor performance for the first 100 hours, but then the hydrogen evolution increases dramatically, giving the highest TON out to the 400 hours of the experiment. After 100 hours, the performance of 9 is just starting to increase more dramatically and it eventually outperforms the complex 3a. In the end 3a reaches a TON of 197 after 240 hours whereas 9 achieves 341 turnovers after 263 hours, and 394 turnovers after 357 hours. The TON for 9 is more than 80% than for 3a.

Sequentially, irradiation experiments were repeated with 3a and 9 and the series was expanded with [Pd(PNP-C1)$_2$]$^{2+}$ 1a and [Pd(NCCH$_3$)$_2$(PNP-C1)]$^{2+}$ 7. In Fig. 6 hydrogen production of these four catalysts is shown (1a, 3a, 7 and 9). For this series hydrogen measurements were performed approximately every 40 hours. After each measuring point an extra equivalent of PS was added, 28 times in total, so the entire irradiation time was 1245 hours. Due to the slight ability of the PS to generate hydrogen on its own, a blank sample with no catalyst was

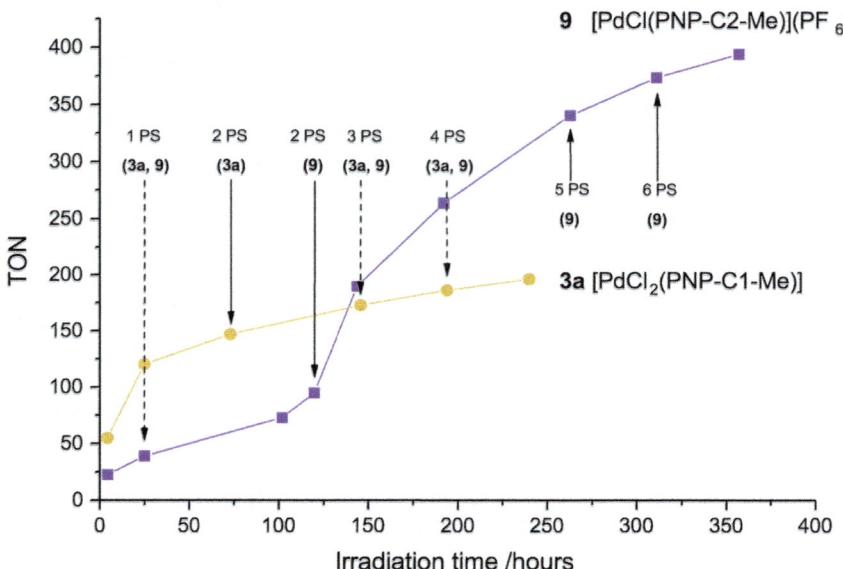

Fig. 5 Time dependent hydrogen production of [Pd(II)Cl₂(PNP-C1-Me)], **3a**, and [Pd(II)Cl(PNP-C2-Me)](PF₆), **9**. An extra equivalent of PS was added after 25, 73, 145 and 193 hours (**3a**), and 25, 120, 145, 193, 263 and 311 hours (**9**), irradiation with 150 W Hg medium pressure lamp, maximal relative standard deviation 1.5%.

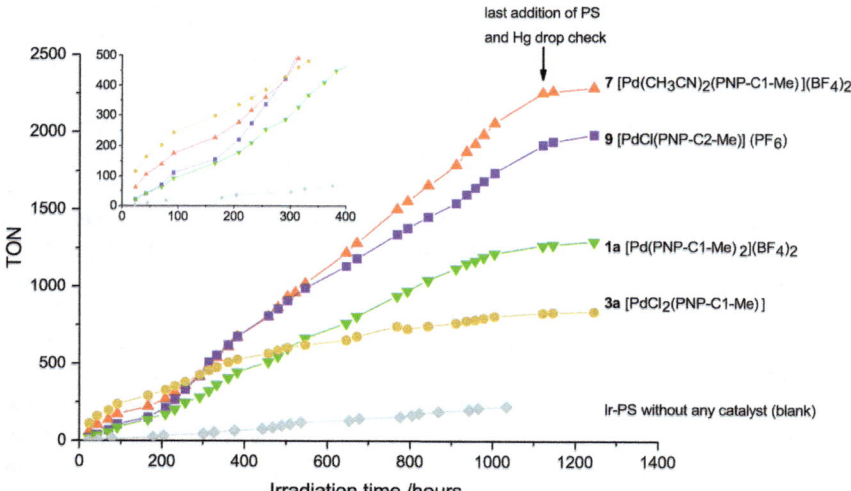

Fig. 6 Time dependent hydrogen production of [Pd(II)(PNP-C1-Me)₂](BF₄)₂, **1a**, [Pd(II)Cl₂(PNP-C1-Me)], **3a**, [Pd(II)(CH₃CN)₂(PNP-C1-Me)](BF₄)₂, **7** and [Pd(II)Cl(PNP-C2-Me)](PF₆), **9**, an extra equivalent of PS was added after each hydrogen measurement, total irradiation time more than 1245 hours, irradiation with 150 W Hg medium pressure lamp; insert in the left top corner shows a zoomed cutout of the first 400 hours of the hydrogen production, maximal relative standard deviation 1.5%.

irradiated as well. As shown in Fig. 6 (diamonds) a linear contribution in hydrogen evolution is obvious due to refilling the PS. Despite this quite small rate which averaged 5.3 turnovers per 24 hours, this contribution is neglected due to the markedly higher rate of the working catalyst. If the PS is excited by the irradiation source, oxidised by catalysts, reduced by the sacrificial donor and cycled again in a continuous manner, one can assume that PS does not have time to produce hydrogen on its own. This supposition is consistent with the significantly higher rate of the system with both the PS and catalyst. To exclude any hydrogen production by Pd nanoparticles, three mercury drops were added after 1121 hours (highlighted in Fig. 6 with a black arrow), which are known to poison metal colloids toward catalytic activity.[55,56] After 1121 hours, the last equivalent of PS was added. The amount of hydrogen produced after mercury addition is the same as the average hydrogen evolution, seen before addition of mercury, eliminating the possibility of colloidal metal particles being responsible for the catalysis.

Compound **3a**, a chlorido complex with short PNP-C1 backbone, and **9**, a chlorido complex with enlarged PNP-C2 backbone, show the same behaviour as discussed above. **3a** starts hydrogen evolution immediately and increases to the highest TON (see Fig. 6 top left). After approximately 300 hours, compound **9** outpaces **3a** and produces hydrogen constantly at quite a high rate, in contrast to **3a**, which appears to degrade over time. By comparing the rates of **3a** and the PS without any catalyst (blank), one can suppose that **3a** is almost completely deactivated after 400 hours of irradiation. However, compound **9** continues working until no additional PS is provided (last addition after 1121 hours). After 1245 hours of catalytic activity, the turnovers for **3a** are 842 and for **9** 1986. This is more than double the amount of hydrogen, corresponding to an efficiency increase of nearly 135%.

Next, three compounds with the short PNP-C1 backbone are compared with each other. **1a** is a homoleptic complex, which is very similar to DuBois' catalysts.[12] **3a** is a chlorido complex and can be seen as half of a DuBois catalyst. Complex **7** is the same as **3a** but with solvent molecules instead of the chlorides, in this case acetonitrile. In the beginning, the half DuBois type catalyst **3a** produces the most hydrogen, followed by the solvato complex **7** and producing the least is the homoleptic complex **1a**. Since there is a conformational change to get from the precatalyst to the actual catalyst, the bulkier homoleptic complex performs worse due to a high reorganisation energy after each catalytic cycle. Most palladium and nickel complexes have a square planar coordination geometry when in the +II oxidation state. By accepting two electrons during catalysis the oxidation state drops to 0, resulting in a conformational change from square planar to tetrahedral.[57] The flexible chlorides can easily change their coordination positions to accommodate this geometry change whereas the bulky PNP-C1 ligand is more rigid. Therefore, the homoleptic DuBois type catalyst must overcome a higher energy barrier to carry out all steps of the catalytic cycle. As a result, compounds **3a** and **7** produce more hydrogen than **1a** at the outset. The two diphosphine ligands stabilise the homoleptic complex **1a**, which leads to a higher stability than **3a**, which is almost completely degraded after 400 hours. Thus, the DuBois type catalyst **1a** can ultimately reach the same TON and overtake the chlorido complex after 500 hours. However, the replacement of chlorides with acetonitrile changes the stability of the catalyst as well. While the chlorido complex **3a** eventually degrades, the solvato complex **7** produces hydrogen continuously at a high rate. In the end, the chlorido complex **3a** achieves a turn-over number of 842, the homoleptic DuBois type complex **1a** reaches 1300 turn-overs and the best result is obtained by the solvato complex **7** with 2289 turnovers.

The enhanced hydrogen production for compound **9**, the chlorido complex with enlarged PNP-C2 backbone, and compound **7**, the solvato complex with shorter PNP-C1 backbone, leads to the suggestion that a solvato complex with the enlarged PNP-C2 ligand may achieve even better results. As a preview one can predict the turnover number of $[Pd(CH_3CN)(PNP-C2-Me)]^{2+}$ by comparing to its PNP-C1 analogue. $[Pd(CH_3CN)_2(PNP-C1-Me)]^{2+}$ **7** (TON = 2289) has a 2.7-times higher efficiency than its chlorido analogue **3a** (TON = 842). Following the same dimension of enhancement, a solvato complex with the enlarged PNP-C2 back-bone could reach a 2.7-times higher hydrogen production than compound **9** $[PdCl(PNP-C2-Me)]^+$, which could be more than 5000 turnovers. This further investigation is planned with complexes based on nickel.

Electrochemical measurements of $[Ni(\textsc{ii})Cl_2(PNP-C1)]$ 5a and $[Pd(\textsc{ii})Cl(PNP-C2)](PF_6)$ 9 in acetonitrile (and water)

We examined the electrochemical properties for a nickel and a palladium catalyst in acetonitrile, namely compounds **5a** and **9**. The cyclic voltammetry (CV) and differential pulse voltammetry (DPV) traces are shown in Fig. 7, with the voltages measured relative to an internal ferrocene reference. Compound **9** is stable to the ferrocene reference, but compound **5a** shows a change in the CV and DPV traces after the ferrocene is added, so the ferrocene couple is not shown in the traces for compound **5a**. Both compounds show weakly reversible $M^{2+/1+}$ and $M^{1+/0}$ reduction waves in anhydrous acetonitrile. While a return wave is not clearly evident in the CV traces, it is visible in the DPV traces, indicating that the reduced species have limited stability. The two reductions for **5a** come at -1.1 and -2.05 V, while those of **9** come at higher potentials of -1.55 and -2.4 V. Oxidation of the two compounds takes place at high potential and is irreversible. When water is added the first and second reduction waves are seen for both compounds prior to the formation of hydrogen, with very little shift in their potentials relative to their values in anhydrous acetonitrile. This indicates that the reduction process in the catalysis is stepwise, *i.e.* $M^{2+} \rightarrow M^{1+} \rightarrow M^0$, and that protonation of the metal centre likely takes place after the second reduction. Compounds **5a** and **9** con-taining PNP-C1-Me and PNP-C2-Me ligands, respectively, have been chosen in order to assess the ability of the different ligands to facilitate proton transfer to the metal center. DFT calculations regarding the potential to catalyze proton transfer indicate that the activation energy to form the hydride is lowered for PNP-C2-Me compared with PNP-C1-Me (*vide infra*). Furthermore, a lower energy barrier occurs for PNP-C2-Me to transfer the second proton and build up dihydrogen. Fig. 7c shows the onset of the catalytic water reduction signal at about -3.0 V for **5a**. This signal is shifted to about -2.5 V for **9** coinciding with the second reduction potential at -2.4 V, which remains as a shoulder (Fig. 7d). This means that the overvoltage for water reduction is reduced in **9** compared with **5a**, thus explaining the better catalytic performance of complexes containing PNP-C2-Me and their enhanced ability to facilitate proton transfer. Therefore, further CV and DPV measurements for complexes containing the two different ligand systems will be carried out to confirm this difference.

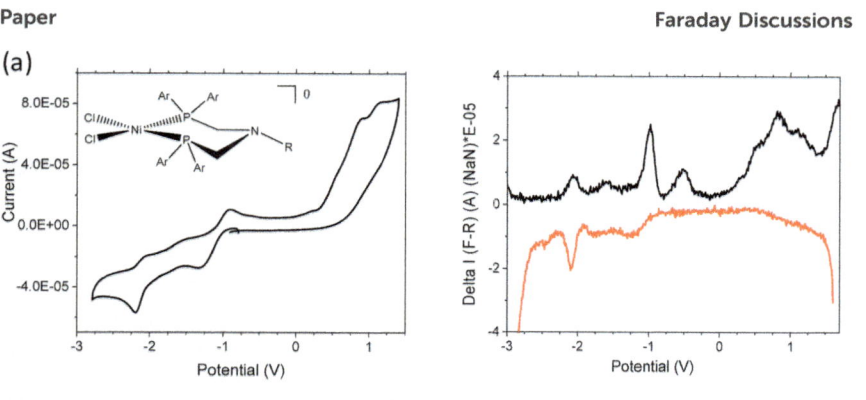

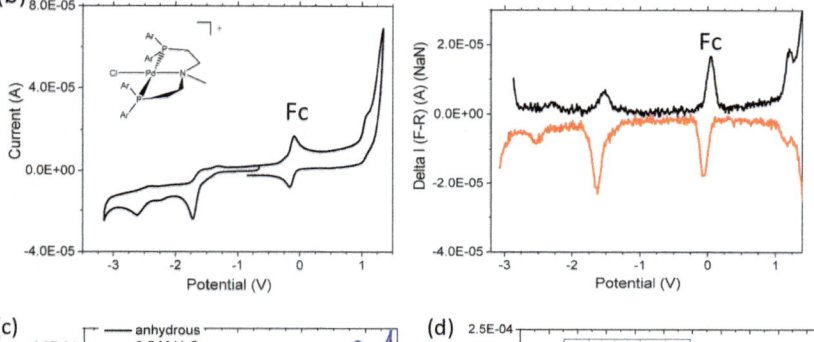

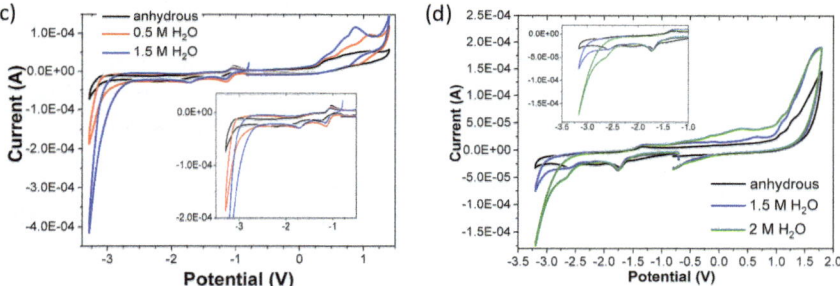

Fig. 7 Electrochemical properties of **5a** (a) and **9** (b) in acetonitrile. Voltages are shown relative to ferrocene/ferrocenium, cyclic voltammograms to the left and differential pulse voltammetry to the right. Electrochemical traces in the presence of water are shown for **5a** in (c) and for **9** in (d).

Structural analysis of catalyst complexes containing at least one PNP ligand by single crystal X-ray diffraction

The molecular structures of [Pd(PNP-C1-Me)$_2$](BF$_4$)$_2$ **1a**, [PdCl$_2$(PNP-C1-Me)] **3a**, [NiCl$_2$(PNP-C1-Me)] **5a**, [NiCl$_2$(PNP-C1-iPr)] **5c** and [PdCl(PNP-C2-Me)](PF$_6$) **9** are presented here (**5a** and **9** are shown in Fig. 8; **1a**, **3a** and **5c** can be found in the ESI†). All structures illustrate the precatalyst with the metal centre in the oxidation state of +II. These precatalysts show a square planar structure concerning the metal centre and coordination partners in all cases, whether the metal is palladium or nickel, the complex is a heteroleptic chlorido or a homoleptic one, the PNP ligand is substituted with a methyl or a sterically more challenging iso-propyl

Fig. 8 ORTEP representations by XRD analysis of [Ni(II)Cl$_2$(PNP-C1-methyl)] (5a, left) and [Pd(II)Cl(PNP-C2-methyl)](PF$_6$) (9, right), anions, hydrogens and solvent molecules omitted for clarity. Selected bond lengths and intramolecular distances [Å]: **5a**: CL1–NI1 2.2242(3), CL2–NI1 2.1964(3), P1–NI1 2.1878(3), P2–NI1 2.1645(3), N1 NI1 3.647(3), O1 NI1 3.437(12), O4 NI1 3.377(7); **9**: Cl1–Pd1 2.3036(8), P1–Pd1 2.2962(8), P2–Pd1 2.3055(8), N1–Pd1 2.086(3), O2 Pd1 3.241(19), O3 Pd1 3.272(16).

group and whether the ligand backbone is short with PNP-C1 or long with PNP-C2. This coordination behaviour fits quite well to literature.[57]

Besides the common square planar structure, the centrosymmetric orientation of two methoxy groups in both structures (5a and 9) is notable. We believe these orientations can form hemilabile coordination bonds within the catalytic cycle. The bond length between oxygen and metal for 5a and 9 range from 3.2 to 3.4 Å. The 2-methoxyphenyl groups reorient to bring the methoxy-oxygens close enough to coordinate to the metal centre, at least in a hemilabile manner. Moreover, the methoxy groups may transfer protons from the pendant amine to the metal centre as well. These two functions are supported by DFT calculations (see following section). Compound 5a shows the usual conformation of complexes with PNP-C1 ligands. The coordination motif with PNP-C1 is maintained with exchanging the amine substitution, the metal centre, the homoleptic or heteroleptic coordination. This ligand coordinates as a diphosphine and the nitrogen atom within the backbone is uncoordinated. Its distance from the metal is 3.65 Å, in contrast to the structure of 9, where the nitrogen metal distance is 2.09 Å. As shown in Fig. 8 the ligand with the enlarged backbone acts as a tridentate ligand. Complexes with the PNP-C2 ligands are known as pincer complexes,[49–54] where coordination bonds occur with all three heteroatoms within the enlarged ligand backbone. Due to a stronger chelate effect, one can assume that compound 9 shows a higher stability than 5a. This stability increase does not affect the precatalyst alone, but all the different catalytic states as well and leads to longer life times and therefore higher hydrogen TONs, as discussed in the previous section. Here, the efficiency increases and the higher turnover number is also strongly justified by a change in the coordination behaviour of the nitrogen atom of the PNP-C2 ligand, as discussed in the following section.

Proposal of catalytic cycle of [Ni(II)(CH$_3$CN)$_2$(PNP-C1)](PF$_6$)$_2$ 8 and [Ni(II)Cl(PNP-C2)](PF$_6$)

The structures of the best performing compounds 7 (the solvato complex with PNP-C1 ligand) and 9 (the chlorido complex with PNP-C2 ligand), discussed in previous sections, were further studied *via* DFT calculations to investigate possible catalytic pathways and propose reasonable explanations for the observed performance enhancement. The computations were carried out by exchanging the palladium metal centre with nickel. The oxidation state of +II for the pre-catalyst was maintained. Below, we postulate the catalytic cycles of the compound $[Ni(\text{II})(CH_3CN)_2(PNP\text{-}C1)](PF_6)_2$ **8** and $[Ni(\text{II})Cl(PNP\text{-}C2)](PF_6)$, which has not been synthesised. The Ni^{2+} complexes show square planar structures in the ground state in all cases in the previous section. The planar structure is well-known in the literature[57] and the following DFT calculations strengthen this postulate, so one can assume the molecule structure is indeed square planar throughout the catalytic cycle. Besides that, we postulate the strong ability of pendant amines to act as proton relays, which has been nicely demonstrated in the solar fuels literature.[11–22,37,58–60]

Two possible reaction mechanisms were calculated for compound **8**. In the first case the catalytic cycle starts with a dissociative proton reduction, where one acetonitrile is removed after reduction of Ni^{+1} to Ni^0 (steps 1 and 2) and before the first protonation (step 3). The other mechanism is an associative one, in which hydride formation takes place in the presence of both acetonitriles. The first mechanism is shown in the ESI† and the associative mechanism, which is energetically more likely and more similar to those proposed by DuBois,[16,36,61–64] is shown in Fig. 9. The proton relay promotes the first protonation (step 3), moving a proton to the reducing metal centre (step 4), the hydride formation (step 5), the second protonation (step 6), followed by passing the second proton to the metal centre (step 7) and formation of dihydrogen (step 8). These steps are quite similar to the mechanisms proposed by DuBois *et al.* Despite that, there are some major

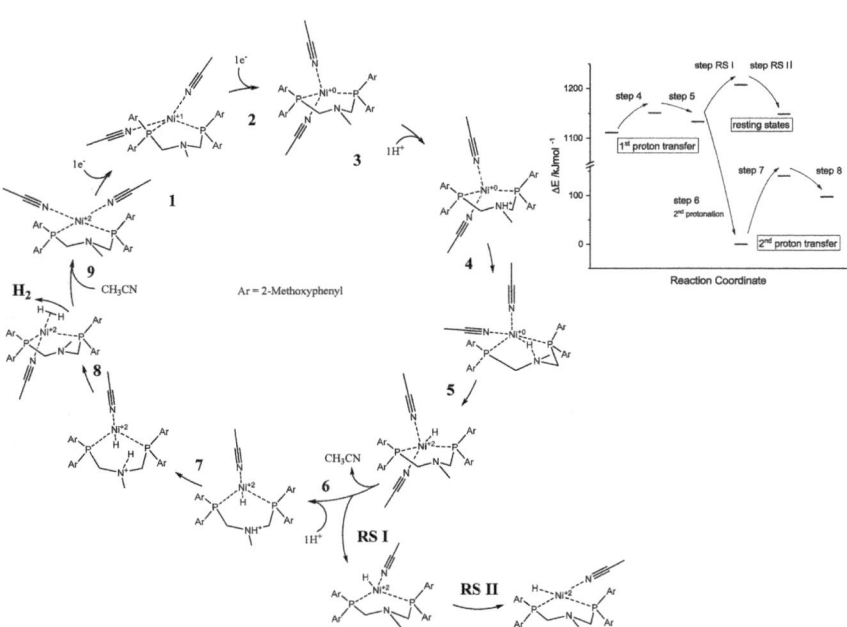

Fig. 9 Calculated catalytic cycle (left) and energy levels including resting state RS in solvent environment (right) of $[Ni(\text{II})(CH_3CN)_2(PNP\text{-}C1\text{-}Me)](PF_6)_2$ **8**.

differences between the catalytic cycle presented in Fig. 9 and DuBois'. Firstly, in the present case a heteroleptic complex is shown, which generates a higher flexibility in geometry changes due to the sterically less challenging acetonitrile ligands. As seen previously, a homoleptic complex cannot provide the same efficiency for hydrogen evolution, because the reorganisation energy of the two diphosphine ligands is too high during conformational changes between square planar and tetrahedral (steps 1 and 2). Secondly, DuBois type catalysts typically consist of four proton relays (see Fig. 1). DuBois and his co-workers published computational studies of the resting states of so-called 'exo' pendant amines.[61] By introducing two or more proton relays within one catalyst molecule, various reaction pathways in the hydrogen movement can take place and decrease the hydrogen production. By using a simpler ligand system, with only one proton relay, and by introducing 2-methoxyphenyl substitutions instead of phenyl, a third aspect has to be taken into account. The methoxy groups can coordinate in a hemilabile fashion to stabilise the metal centre in critical states within the catalytic cycle. Moreover, they can build proton channels, like those in the natural hydrogenase enzymes and support the proton transfer from the pendant amine to the metal centre (shown in Fig. 9 between steps 7 and 8). Finally, we found an additional resting state (steps RS I and RS II) by using ligand systems with backbones like PNP-C1. Here, the first proton reduction takes place, but before a second protonation can occur, the hydride forces the catalyst from an activated tetrahedral geometry to a square planar arrangement as seen for the ground state. To overcome this resting state (Fig. 9, energy levels on the right, step RS II) and return into the catalytic cycle requires 59 kJ mol^{-1} in the solvent environment. This resting state may limit the efficiency of the catalysis. This assumption may be supported in the following by the insights of the catalytic behaviour of $[Ni(\text{II})Cl(PNP-C2)](PF_6)$ and the elimination of this resting state, presented below.

By enlarging the PNP backbone, the coordination chemistry and the mechanism of the catalytic cycle change dramatically (see Fig. 10). As is known for pincer ligands[49-54] with the structure PNP-C2, a tridentate coordination is observed in contrast to the bidentate ligands of the structure PNP-C1. Thus, pincer ligands seem to stabilise complexes more, due to the chelate effect. When receiving electrons from photosensitisers (steps 1 and 2) the metal centre, in this case again nickel, changes its oxidation state (and its Ni–N distance) from +II (1.99 Å) to +I (2.23 Å) to 0 (3.90 Å). During the reduction potential (from excited electrons) one can see a change in the coordination sphere. The nitrogen, formerly coordinating in a pincer mode, disconnects from the metal centre and recovers its function as a proton relay. Otherwise, if there is a protonation before reduction of the precatalyst the tertiary amine becomes quarternary and its coordination ability is lost. In each case the pendant amine can either coordinate or provide protons for reduction depending on the oxidation state of the metal or its own protonation state. After the protonation (step 3) the proton relay can again transfer hydrogen to the metal centre (step 4), which reduces the proton and forms a hydride (step 5). In contrast to the catalytic cycle with the PNP-C1 ligand, in which the energy barrier in solvent environment for this proton reduction is 40 kJ mol^{-1} (Fig. 9, step 4), the activation energy to form the hydride is only 1.0 kJ mol^{-1} here (Fig. 10, step 4; more data about concrete energies including energy diagrams can be found in the

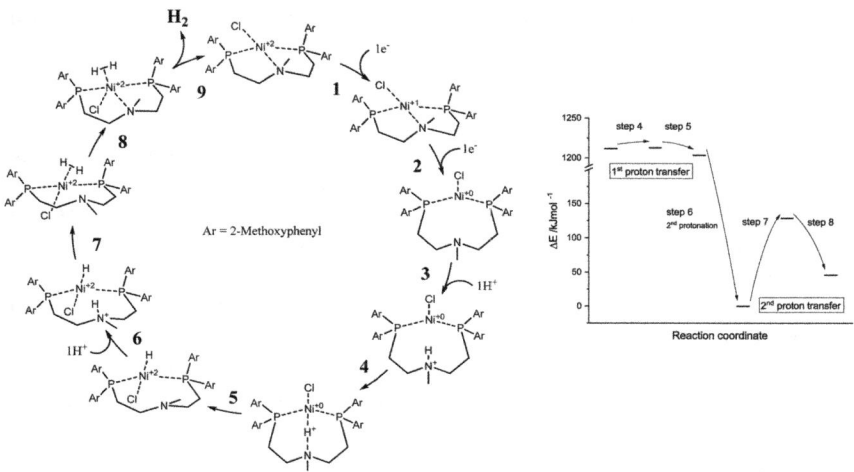

Fig. 10 Calculated catalytic cycle (left) and energy levels in solvent environment (right) of [Ni(II)Cl(PNP-C2-Me)]$^+$.

ESI†). Despite that, there is no possible resting state and therefore no efficiency loss. The nickel centre starts in the ground state as a square planar structure, reaches trigonal planarity after losing the nitrogen and returns to a square planar structure with two phosphorus atoms, one chlorine atom and the newly formed hydride bond. After the second protonation (step 6) there is also a lower energy barrier for PNP-C2 to transfer the second proton and form dihydrogen (Fig. 10, step 7) with a barrier of 129 kJ mol^{-1}, whereas PNP-C1 requires 141 kJ mol^{-1} (Fig. 9, step 7). In the end the complex forms a trigonal bipyramidal conformation (tbp) until molecular hydrogen is released. The deprotonated proton relay is free to coordinate again to the Ni^{+2} metal centre with a coordinative bond length of 2.07 Å. The tbp conformation, formed by two phosphorus atoms, one chlorine atom, one nitrogen atom and coordinated dihydrogen, returns to the square planar structure of the precatalyst after H$_2$ release.

Experimental section

Materials and methods

All preparative work was carried out using standard Schlenk techniques at ambient temperature and pressure, unless otherwise indicated. For the inert gas, Argon 5.0 (Messer) was used. Solvents (Acros Organics/Fisher Scientific, extra dry, with molecular sieve and crown cap) were either degassed by 'freeze–pump–thaw' or by usage of a vacuum pump for at least 15 minutes, dependent on the individual boiling point. Further dehydration was not performed, unless otherwise indicated. Deuterated solvents were purchased from Euriso-Top and used as received. Di-(2-methoxyphenyl)phosphine was prepared by a synthesis strategy of the Brüggeller working group.[65] Other starting materials, such as para-formaldehyde (Sigma-Aldrich), amines, metal salts and bases (Sigma-Aldrich, abcr) were purchased. For filtration, Durapore Membrane Filters (Merck Milli-pore) were used.

Instrumentation

^{1}H, ^{13}C and ^{31}P NMR spectra were recorded on a Bruker Avance DPX 300 spectrometer with internal ^{2}D-lock. Measurements were undertaken at 121.497 MHz for ^{31}P{^{1}H}-NMR (phosphoric acid (85%) as external standard), at 75.476 MHz for ^{13}C{^{1}H}-NMR (calibration based on solvent signal) and at 300.13 MHz for ^{1}H-NMR (calibration based on solvent signal). Mass spectra were recorded on a Finnigan MAT-95 apparatus using MALDI for ionisation and a DAN (1,5-diaminonaph-thalene) matrix. Prior to the XRD measurements, the single crystals were hand-picked using a Leica Wild M10 microscope with vertical illumination and a polarising filter. X-ray diffraction analysis was performed on a Bruker D8 diffractometer with an Incoatec-microfocus-channel, a multi-layer-monochromator and the new CMOS-technology from Bruker (D8 Quest). Cell refinement, data reduction and the empirical absorption correction were performed using Apex II and Saint-V 8.34 A programmes. All structure determination calculations were realised using SHELXTL-NT V6.1 and SHELXL-2014/7. Final refinements on F^2 were done with anisotropic thermal parameters for all non-hydrogen atoms. Irradiation experiments were carried out in a self-made chamber containing water-cooled mercury medium pressure lamps with 150 W and 700 W (TQ 718, $\lambda = 280-700$ nm, $\lambda_{max} = 360$ nm, Heraeus). Hydrogen measurements and chromatograms were recorded on a gas chromatograph (3000 micro-GC, Inficon) with Argon 5.0 as the carrier gas and with a thermal conductivity detector. The calibration window ranged from 1000 to 10 000 ppm concentration of hydrogen. Experimental errors were measured for the calibration gases with hydrogen contents of 1000 ppm and 100 000 ppm. Here, a flask was filled with calibration gas and measured seven times with gas chromatography. The relative standard deviations were 0.4 and 1.5%. Analyses were carried out by usage of EZ IQ programmes.

Electrochemical measurements

Cyclic voltammetry and differential pulsed voltammetry were performed using an VersaSTAT 3 potentiostat. Anhydrous acetonitrile (DriSolv) was used as the solvent under an inert atmosphere, and 0.1 M tetra(n-butyl)ammonium hexa-fluorophosphate (TBAF) was used as the supporting electrolyte. A glassy carbon rod was used as the working electrode, a platinum wire was used as the counter electrode, and a silver wire was used as a pseudoreference electrode. The redox potentials are based on values measured from differential pulsed voltammetry and are reported relative to a ferrocene/ferrocenium (Cp$_2$Fe/Cp$_2$Fe$^+$) redox couple used as an internal reference, while electrochemical reversibility was determined using cyclic voltammetry. The water splitting experiment was carried out by adding deionized water dropwise to the solution.

DFT calculations

In addition to the experimentally recorded gas chromatographs of the hydrogen evolution, DFT simulations were performed by using the Gaussian09 program. Therefore, geometry optimisations (energy minima and transition states) in the gas phase and in the solvent environment were executed. The minima and TS structures were verified using harmonic frequency calculations. For calculations in the gas phase the B3LYP hybrid functional and the basis set 6-31G(d,p) were used. For calculations in the solvent environment the hybrid functional was changed to BP86, but the basis set was the same. The solvent was chosen as in the experimental setup during irradiation. It was a solvent mixture of water and acetonitrile in the ratio 1 : 1. Hence, the relative permittivity was averaged and set as 57.02, while the square of the index of refraction was set as 1.792. Basis structural data was delivered by single crystal X-ray data from compound **5a** and [Pd(II)Cl(PNP-C2-Me)](PF$_6$). To calculate the catalytic cycle of compound **8**, the chlorine atoms of **5a** were exchanged with acetonitriles. Otherwise, within structure **9** the metal centre palladium was exchanged with nickel. For these changes, as well as any visualisations, the programs GausView09 and Mercury 3.9 were used. Computing was performed on LEO3, a high performance compute cluster of the Research Area Scientific Computing at the University of Innsbruck, in operation since September 2011. It consists of 1944 Intel Xeon (Gulftown) compute cores and is equipped with 24 GB RAM per node, *i.e.*, about 4 TB of main memory altogether. The nodes and GPFS storage system are joined by a 40 Gb s^{-1} Infiniband high speed interconnect. In addition, three of the nodes are equipped with Nvidia Tesla M2090 graphics cards and 48 GB of main memory. Additionally, the more time consuming SCRF calculations were alternatively performed on the LEO3E, which is an extension of the existing high performance compute cluster LEO3, financed by the Research Area Scientific Computing, several Institutes of the LFU, and a special grant by the Rector of the LFU. The system consists of 45 nodes with 20 Intel Xeon (Haswell) compute cores each (total 900 cores). Except for two nodes (512 GB) all nodes are equipped with 64 GB RAM, totalling approx. 3.7 TB of RAM. The system has a high-performance low-latency InfiniBand interconnect for MPI communications between nodes and GPFS file system traffic. The GPFS ($SCRATCH) file system has a usable capacity of 54 TB.

Synthesis

The syntheses of the PNP-C1 ligands, their metal complexes [M(II)(PNP-C1)$_2$](BF$_4$)$_2$ and [M(II)Cl$_2$(PNP-C1)$_2$] are all very similar, and the procedures and character-ization data are given for representative examples only. In the ESI† one can find procedure details and data for PNP-C1-methyl, PNP-C2-methyl, [Pd(II)(PNP-C1-methyl)$_2$](BF$_4$)$_2$ **1a**, [Ni(II)Cl$_2$(PNP-C1-methyl)$_2$] **5a**, [Ni(II)(CH$_3$CN)$_2$(PNP-C1-methyl)$_2$](BF$_4$)$_2$ **8** and [Pd(II)Cl(PNP-C2-methyl)$_2$](PF$_6$) **9**. Spectroscopic character-ization data for the remaining ligands and complexes are published elsewhere.[66] The control ligand PCCCP without proton relay was prepared as described in the literature[67] and its complexes [Pd(II)(PCCCP)$_2$](PF$_6$)$_2$ and [Pt(II)(PCCCP)$_2$](PF$_6$)$_2$ are prepared analogously to [M(II)(PNP-C1)$_2$](BF$_4$)$_2$.

Conclusions

Two ligand systems PNP-C1 and PNP-C2 (*N*,*N*-bis-[(di-(2-methoxyphenyl)phos-phino)methyl]-*N*-alkylamine   and   *N*,*N*-bis-[(di-(2-methoxyphenyl)phosphino) ethyl]-*N*-alkylamine) were synthesised and used to form homo- and heteroleptic (chlorido and solvato) complexes with Pt, Pd, Ni and Co. The positive effect of proton relays was confirmed by two series of homoleptic platinum and palladium complexes with PNP-C1-alkyl ligands. Moreover, different substitutions of the pendant amines were examined with the result of the best performing substitution being a methyl group due to its low steric pressure. The best pH conditions for irradiation were found to be at pH 10. Furthermore, different catalytic metals were compared showing the best to be palladium, followed by cobalt (68% of palladium's efficiency) and nickel (45% of palladium's efficiency), ending up with platinum. A comparison in hydrogen evolution of different systems delivered various insights. Heteroleptic complexes produce more hydrogen in the beginning than homoleptic ones, but after some time homoleptic complexes become more efficient at hydrogen production (TON 842 *versus* 1300) due to their lower flexibility within the catalytic cycle and higher stability in long term irradiation experiments. Complexes with chlorines in heteroleptic complexes are less efficient at producing hydrogen in long term experiments; solvato complexes are preferred (TON 842 *versus* 2289). Finally, PNP-2 ligands with an enlarged backbone perform better in proton reduction and hydrogen evolution (TON 842 *versus* 1986), which is consistent with DFT calculations. PNP-C2 ligands combine the properties of stabilising pincer ligands and efficiency increasing proton relays. Taking all these results into account, further investigations on $[Co(\text{II})(CH_3\text{-}CN)(\text{PNP-C2-methyl})](PF_6)$ may be very promising.

Conflicts of interest

There are no conflicts of interest to declare.

Acknowledgements

Financial support from Hypo Tirol Bank, Tyrolean Research Fonds (TWF) and the Tyrolean government (Land Tirol) is gratefully acknowledged. Leopold-Franzens University of Innsbruck supported this work in the form of a doctoral scholarship. Cooperation with VERBUND AG made hydrogen measurements possible. This research was executed as a part of the platform 'materials & nanoscience'. For usage of the supercomputers LEOIII and LEOIIIe we thank the Leopold-Franzens University of Innsbruck and Univ.-Prof. Dr Dr hc. mult. Tilmann Märk. CCDC and FIZ Karlsruhe Deposition Teams are gratefully acknowledged as well as the European Union (European Regional Development Fund, ERDF).

Notes and references

1 R. K. Pachauri and L. A. Meyer, *Climate Change 2014: Synthesis Report. Fifth Assessment Report*, IPCC, Geneva, Switzerland, 2014.

2 J. D. Sutter and J. Berlinger, *Final draft of climate deal formally accepted in Paris*, 2015.

3 P.-A. Jacques, V. Artero, J. Pécaut and M. Fontecave, *Proc. Natl. Acad. Sci. U. S. A.*, 2009, **106**, 20627–20632.

4 P. Zhang, P.-A. Jacques, M. Chavarot-Kerlidou, M. Wang, L. Sun, M. Fontecave and V. Artero, *Inorg. Chem.*, 2012, **51**, 2115–2120.

5 E. S. Andreiadis, P.-A. Jacques, P. D. Tran, A. Leyris, M. Chavarot-Kerlidou, B. Jousselme, M. Matheron, J. Pécaut, S. Palacin, M. Fontecave and V. Artero, *Nat. Chem.*, 2013, **5**, 48–53.

6 V. Artero, M. Chavarot-Kerlidou and M. Fontecave, *Angew. Chem., Int. Ed.*, 2011, **50**, 7238–7266.

7 P. Du, J. Schneider, G. Luo, W. W. Brennessel and R. Eisenberg, *Inorg. Chem.*, 2009, **48**, 4952–4962.

8 Z. Han, W. R. McNamara, M.-S. Eum, P. L. Holland and R. Eisenberg, *Angew. Chem., Int. Ed.*, 2012, **51**, 1667–1670.

9 Z. Han, L. Shen, W. W. Brennessel, P. L. Holland and R. Eisenberg, *J. Am. Chem. Soc.*, 2013, **135**, 14659–14669.

10 Z. Han and R. Eisenberg, *Acc. Chem. Res.*, 2014, **47**, 2537–2544.

11 J. Y. Yang, R. M. Bullock, M. R. DuBois and D. L. DuBois, *MRS Bull.*, 2011, **36**, 39–47.

12 M. L. Helm, M. P. Stewart, R. M. Bullock, M. R. DuBois and D. L. DuBois, *Science*, 2011, **333**, 863–866.

13 M. Rakowski DuBois and D. L. DuBois, *Acc. Chem. Res.*, 2009, **42**, 1974–1982.

14 J. Y. Yang, R. M. Bullock, W. J. Shaw, B. Twamley, K. Fraze, M. R. DuBois and D. L. DuBois, *J. Am. Chem. Soc.*, 2009, **131**, 5935–5945.

15 K. Redin, A. D. Wilson, R. Newell, M. R. DuBois and D. L. DuBois, *Inorg. Chem.*, 2007, **46**, 1268–1276.

16 M. P. Stewart, M.-H. Ho, S. Wiese, M. L. Lindstrom, C. E. Thogerson, S. Raugei, R. M. Bullock and M. L. Helm, *J. Am. Chem. Soc.*, 2013, **135**, 6033–6046.

17 D. L. DuBois and R. M. Bullock, *Eur. J. Inorg. Chem.*, 2011, **2011**, 1017–1027.

18 M. Rakowski DuBois and D. L. DuBois, *Chem. Soc. Rev.*, 2009, **38**, 62–72.

19 G. M. Jacobsen, J. Y. Yang, B. Twamley, A. D. Wilson, R. M. Bullock, M. Rakowski DuBois and D. L. DuBois, *Energy Environ. Sci.*, 2008, **1**, 167.

20 E. S. Wiedner, J. Y. Yang, W. G. Dougherty, W. S. Kassel, R. M. Bullock, M. R. DuBois and D. L. DuBois, *Organometallics*, 2010, **29**, 5390–5401.

21 T. Liu, S. Chen, M. J. O'Hagan, M. Rakowski DuBois, R. M. Bullock and D. L. DuBois, *J. Am. Chem. Soc.*, 2012, **134**, 6257–6272.

22 K. D. Welch, W. G. Dougherty, W. S. Kassel, D. L. DuBois and R. M. Bullock, *Organometallics*, 2010, **29**, 4532–4540.

23 C. M. Strabler, S. Sinn, R. Pehn, J. Pann, J. Dutzler, W. Viertl, J. Prock, K. Ehrmann, A. Weninger, H. Kopacka, L. de Cola and P. Brüggeller, *Faraday Discuss.*, 2017, **198**, 211–233.

24 (*a*) J. C. Fontecilla-Camps, A. Volbeda, C. Cavazza and Y. Nicolet, *Chem. Rev.*, 2007, **107**, 5411; (*b*) P. M. Vignais and B. Billoud, *Chem. Rev.*, 2007, **107**, 4206–4272.

25 C. Klemps, E. Payet, L. Magna, L. Saussine, X. F. Le Goff and P. Le Floch, *Chem. - Eur. J.*, 2009, **15**, 8259–8268.

26 E. Bálint, E. Fazekas, A. Tripolszky, R. Kangyal, M. Milen and G. Keglevich, *Phosphorus, Sulfur Silicon Relat. Elem.*, 2015, **190**, 655–659.

27 Z. Han, L. Rong, J. Wu, L. Zhang, Z. Wang and K. Ding, *Angew. Chem., Int. Ed.*, 2012, **51**, 13041–13045.

28 C. Bianchini, H. M. Lee, A. Meli, W. Oberhauser, F. Vizza, P. Brüggeller, R. Haid and C. Langes, *Chem. Commun.*, 2000, (9), 777–778.

29 C. Bianchini, P. Brüggeller, C. Claver, G. Czermak, A. Dumfort, A. Meli, W. Oberhauser and E. J. Garcia Suarez, *Dalton Trans.*, 2006, (24), 2964–2973.

30 S. Inoue, M. Mitsuhashi, T. Ono, Y.-N. Yan, Y. Kataoka, M. Handa and T. Kawamoto, *Inorg. Chem.*, 2017, **56**, 12129–12138.

31 H. Junge, Z. Codolà, A. Kammer, N. Rockstroh, M. Karnahl, S.-P. Luo, M.-M. Pohl, J. Radnik, S. Gatla, S. Wohlrab, J. Lloret, M. Costas and M. Beller, *J. Mol. Catal. A: Chem.*, 2014, **395**, 449–456.

32 F. Gärtner, B. Sundararaju, A.-E. Surkus, A. Boddien, B. Loges, H. Junge, P. H. Dixneuf and M. Beller, *Angew. Chem., Int. Ed.*, 2009, **48**, 9962–9965.

33 S. Tschierlei, A. Neubauer, N. Rockstroh, M. Karnahl, P. Schwarzbach, H. Junge, M. Beller and S. Lochbrunner, *Phys. Chem. Chem. Phys.*, 2016, **18**, 10682–10687.

34 S. Ramachandra, C. A. Strassert, D. N. Reinhoudt, D. Vanmaekelbergh and L. de Cola, *Z. Naturforsch., B: J. Chem. Sci.*, 2014, **69**, 263–274.

35 M. Mauro, K. C. Schuermann, R. Prétôt, A. Hafner, P. Mercandelli, A. Sironi and L. de Cola, *Angew. Chem., Int. Ed.*, 2010, **49**, 1222–1226.

36 S. Raugei, S. Chen, M.-H. Ho, B. Ginovska-Pangovska, R. J. Rousseau, M. Dupuis, D. L. DuBois and R. M. Bullock, *Chem. - Eur. J.*, 2012, **18**, 6493–6506.

37 S. Raugei, M. L. Helm, S. Hammes-Schiffer, A. M. Appel, M. O'Hagan, E. S. Wiedner and R. M. Bullock, *Inorg. Chem.*, 2016, **55**, 445–460.

38 M. T. Huynh, W. Wang, T. B. Rauchfuss and S. Hammes-Schiffer, *Inorg. Chem.*, 2014, **53**, 10301–10311.

39 M.-H. Ho, M. O'Hagan, M. Dupuis, D. L. DuBois, R. M. Bullock, W. J. Shaw and S. Raugei, *Dalton Trans.*, 2015, **44**(24), 10969–10979.

40 P. Zhang, P.-A. Jacques, M. Chavarot-Kerlidou, M. Wang, L. Sun, M. Fontecave and V. Artero, *Inorg. Chem.*, 2012, **51**, 2115–2120.

41 J. Hawecker, J. M. Lehn and R. Ziessel, *Nouv. J. Chim.*, 1983, **1983**, 271–277.

42 P. Du, K. Knowles and R. Eisenberg, *J. Am. Chem. Soc.*, 2008, **130**, 12576–12577.

43 B. Probst, A. Rodenberg, M. Guttentag, P. Hamm and R. Alberto, *Inorg. Chem.*, 2010, **49**, 6453–6460.

44 T. Lazarides, T. McCormick, P. Du, G. Luo, B. Lindley and R. Eisenberg, *J. Am. Chem. Soc.*, 2009, **131**, 9192–9194.

45 P. Zhang, M. Wang, J. Dong, X. Li, F. Wang, L. Wu and L. Sun, *J. Phys. Chem. C*, 2010, **114**, 15868–15874.

46 B. Probst, C. Kolano, P. Hamm and R. Alberto, *Inorg. Chem.*, 2009, **48**, 1836–1843.

47 A. Call, Z. Codolà, F. Acuña-Parés and J. Lloret-Fillol, *Chem. - Eur. J.*, 2014, **20**, 6171–6183.

48 T. Sakai, D. Mersch and E. Reisner, *Angew. Chem., Int. Ed.*, 2013, **52**, 12313–12316.

49 D. Benito-Garagorri and K. Kirchner, *Acc. Chem. Res.*, 2008, **41**, 201–213.

50 J. Serrano-Becerra and D. Morales-Morales, *Curr. Org. Synth.*, 2009, **6**, 169–192.

51 P. Hermosilla, P. López, P. García-Orduña, F. J. Lahoz, V. Polo and M. A. Casado, *Organometallics*, 2018, **37**, 2618–2629.

52 H. Tanaka, Y. Nishibayashi and K. Yoshizawa, *Acc. Chem. Res.*, 2016, **49**, 987–995.

53 N. Gorgas and K. Kirchner, *Acc. Chem. Res.*, 2018, **51**, 1558–1569.

54 W. Liu, B. Sahoo, K. Junge and M. Beller, *Acc. Chem. Res.*, 2018, **51**, 1858–1869.

55 J. G. Aston and P. Mitacek, *Nature*, 1962, **195**, 70–71.

56 P. Du, J. Schneider, P. Jarosz and R. Eisenberg, *J. Am. Chem. Soc.*, 2006, **128**, 7726–7727.

57 D. Rabinovich, *J. Chem. Educ.*, 2000, **77**, 311.

58 S. Wiese, U. J. Kilgore, M.-H. Ho, S. Raugei, D. L. DuBois, R. M. Bullock and M. L. Helm, *ACS Catal.*, 2013, **3**, 2527–2535.

59 J. M. Darmon, N. Kumar, E. B. Hulley, C. J. Weiss, S. Raugei, R. M. Bullock and M. L. Helm, *Chem. Sci.*, 2015, **6**, 2737–2745.

60 T. A. Tronic, W. Kaminsky, M. K. Coggins and J. M. Mayer, *Inorg. Chem.*, 2012, **51**, 10916–10928.

61 M. O'Hagan, W. J. Shaw, S. Raugei, S. Chen, J. Y. Yang, U. J. Kilgore, D. L. DuBois and R. M. Bullock, *J. Am. Chem. Soc.*, 2011, **133**, 14301–14312.

62 M.-H. Ho, R. Rousseau, J. A. S. Roberts, E. S. Wiedner, M. Dupuis, D. L. DuBois, R. M. Bullock and S. Raugei, *ACS Catal.*, 2015, **5**, 5436–5452.

63 S. Chen, S. Raugei, R. Rousseau, M. Dupuis and R. M. Bullock, *J. Phys. Chem. A*, 2010, **114**, 12716–12724.

64 M. Dupuis, S. Chen, S. Raugei, D. L. DuBois and R. M. Bullock, *J. Phys. Chem. A*, 2011, **115**, 4861–4865.

65 G. Czermak, PhD thesis, University of Innsbruck, 2006.

66 R. Pehn, W. Viertl, M. Bendig, H. Roithmeyer, J. Pann, K. Ehrmann, T. Mueller, H. Kopacka and P. Brüggeller, *Eur. J. Inorg. Chem.*, manuscript in preparation.

67 C. Bianchini, A. Meli, W. Oberhauser, A. M. Segarra, C. Claver and E. J. G. Suarez, *J. Mol. Catal. A: Chem.*, 2007, **265**, 292–305.

Faraday Discussions

PAPER

Photoinduced hole transfer from tris(bipyridine)ruthenium dye to a high-valent iron-based water oxidation catalyst†

Sergii I. Shylin, [ID] *[a] Mariia V. Pavliuk,[a] Luca D'Amario,[ab] Igor O. Fritsky [ID] *[cd] and Gustav Berggren [ID] *[a]

Received 5th November 2018, Accepted 19th December 2018

DOI: 10.1039/c8fd00167g

An efficient water oxidation system is a prerequisite for developing solar energy conversion devices. Using advanced time-resolved spectroscopy, we study the initial catalytic relevant electron transfer events in the light-driven water oxidation system utilizing $[Ru(bpy)_3]^{2+}$ (bpy = 2,2′-bipyridine) as a light harvester, persulfate as a sacrificial electron acceptor, and a high-valent iron clathrochelate complex as a catalyst. Upon irradiation by visible light, the excited state of the ruthenium dye is quenched by persulfate to afford a $[Ru(bpy)_3]^{3+}/SO_4^{•-}$ pair, showing a cage escape yield up to 75%. This is followed by the subsequent fast hole transfer from $[Ru(bpy)_3]^{3+}$ to the Fe^{IV} catalyst to give the long-lived Fe^V intermediate in aqueous solution. In the presence of excess photosensitizer, this process exhibits pseudo-first order kinetics with respect to the catalyst with a rate constant of $3.2(1) \times 10^{10}$ s^{-1}. Consequently, efficient hole scavenging activity of the high-valent iron complex is proposed to explain its high catalytic performance for water oxidation.

Introduction

Climate change, driven by the greenhouse effect and increasing global energy consumption, has become the greatest challenge humanity has ever faced. Most analysts believe that the world's demand for energy will keep growing, providing a strong motivation for research into the development of carbon-neutral and carbon-negative energy sources.[1] In this regard, artificial photosynthesis, harnessing sunlight and water as practically infinite resources, is becoming an attractive way to achieve the sustainable energy goals.[2–4] To date, the development

[a]Department of Chemistry – Ångström Laboratory, Uppsala University, P. O. Box 523, 75120 Uppsala, Sweden. E-mail: sergii.shylin@kemi.uu.se; gustav.berggren@kemi.uu.se

[b]Physics Department, Free University Berlin, Arnimallee 14, 14195 Berlin, Germany

[c]Department of Chemistry, Taras Shevchenko National University of Kyiv, Volodymyrska 64, 01601 Kiev, Ukraine. E-mail: ifritsky@univ.kiev.ua

[d]PBMR Labs Ukraine, Murmanska 1, 02094 Kiev, Ukraine

† Electronic supplementary information (ESI) available. See DOI: 10.1039/c8fd00167g

of efficient water oxidation catalysts remains a challenging task as oxidation of water to dioxygen is a complicated process involving four electron transfer steps coupled with the release of four protons.[5] The classic homogenous photocatalytic water oxidation system consists of a water oxidation catalyst (WOC), a photosensitizer and a sacrificial electron acceptor. Thus, in addition to the challenges associated with bond making and breaking during catalysis, sustainable generation of a long-lived charge-separated state in aqueous solution is required to perform the desired catalytic process.[6,7] Moreover, the photosensitizer must be able to oxidize the WOC, having a redox potential $E_{1/2} > 1.23$ V *versus* the normal hydrogen electrode (NHE). These requirements can be met by ruthenium poly-pyridine complexes, and consequently they have been widely employed as efficient visible light-driven photosensitizers in conjunction with various WOCs.[8–10] Among these complexes, tris(bipyridine)ruthenium ($[Ru(bpy)_3]^{2+}$) is the most common light harvester, with a relatively high oxidation potential $E_{1/2} = 1.26$ V *versus* NHE for the $[Ru(bpy)_3]^{3+}/[Ru(bpy)_3]^{2+}$ redox couple.[11] Due to the long life-time $\tau \sim 10^{-6}$ s of its metal-to-ligand charge transfer (MLCT) excited state, $[Ru(bpy)_3]^{2+}$ may be easily converted to $[Ru(bpy)_3]^{3+}$ under illumination, by electron acceptors such as persulfate.[12] However, there is still room for improving the ability of the WOC to efficiently re-reduce $[Ru(bpy)_3]^{3+}$ to $[Ru(bpy)_3]^{2+}$ and perform sustained water oxidation.

Over the past few decades various heterogeneous WOCs, generally considered more suitable for practical use, and homogeneous analogues, often considered more suitable for mechanistic studies, have been designed. Transition-metal oxides of Groups 7, 8, and 9 were observed early on to have good catalytic properties for oxygen evolution.[13] Among these metals, a privileged position belongs to manganese,[14] as it constitutes the cofactor of photosystem II in living organisms, as well as ruthenium and iridium,[15,16] located diagonally in the periodic table relative to manganese. Oxides of these two noble metals have found their application in commercial proton-exchange membrane electrolyzers due to their stability over a wide pH range, despite their high price and harm to the environment.[17] Like their heterogeneous counterparts, the best-characterized homogeneous WOCs are based on ruthenium and iridium.[18,19] Due to the high costs and low abundance of these metals, there is considerable interest in the use of cheap base metal complexes for water oxidation. To date, a handful of cobalt,[20] manganese,[21,22] copper,[23] nickel[24] and iron compounds have been studied with respect to their potential as WOCs. However, iron, being the most abundant and cheap transition metal, is the least used in molecular WOCs, arguably due to the low stability of iron complexes under oxidative conditions. Only a few molecular iron compounds have been documented as catalysts for chemical,[25–29] electrochemical[30–32] and photochemical water oxidation.[33,34] The most noteworthy iron-based WOCs are TAML (tetraamido macrocyclic ligand) complexes, where enhanced catalyst stability is achieved by utilizing a tetradentate macrocycle, and high-valent iron catalytic intermediates are supported by deprotonated N donor atoms.[25,32,33] We recently reported fast light-driven water oxidation catalyzed by the exceptionally stable iron(IV) cage complex $[Fe^{IV}(L–6H)]^{2-}$, whose structure is depicted in Fig. 1a.[35] In the tris(bipyridine)ruthenium dye photo-oxidant system (Fig. 1b), it shows catalytic performance with a turnover frequency (TOF) of 2.27 s^{-1} and a maximum turnover number (TON) of 365.[36] Its high efficiency has been attributed to the robust clathrochelate ligand that prevents the catalyst from rapid

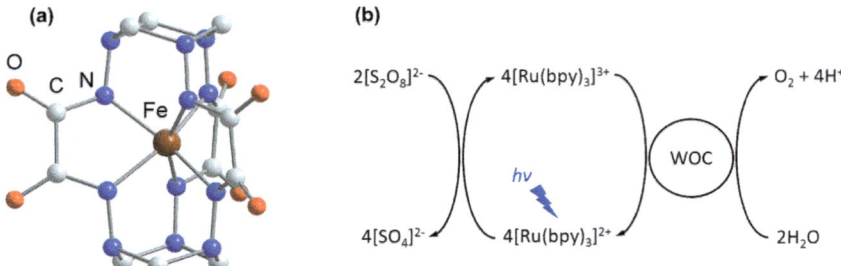

Fig. 1 (a) The molecular structure of the complex anion $[Fe^{IV}(L-6H)]^{2-}$ reported previously.[35] H atoms are omitted for clarity. (b) The photocatalytic cycle of water oxidation to dioxygen by persulfate with $[Ru(bpy)_3]^{2+}$ as the photosensitizer and WOC.

degradation, as has been shown by UV-vis, EPR, Mössbauer spectroscopy, ESI mass spectrometry and DLS studies. The relatively high rate of water oxidation by the catalyst is attributable to its mild water oxidation overpotential of 0.39 V, compared with other iron-based WOCs. In this contribution, we report the hole scavenging activity of the iron(IV) catalyst in the ruthenium dye water oxidation system in order to shed further light on its catalytic efficiency.

Experimental

Synthesis

The general synthesis pathway of iron clathrochelates bearing the $[Fe^{IV}(L-6H)]^{2-}$ complex anion has been described previously.[35] In short, the compound $Na_2[Fe^{IV}(L-6H)]\cdot 2H_2O$ used in the present work as the WOC, was synthesized as follows. $Fe(ClO_4)_3\cdot H_2O$ (0.372 g, 1 mmol, dissolved in 5 ml of water) was added to a warm solution of oxalodihydrazide (0.354 g, 3 mmol, dissolved in 15 ml of water). Then, a solution of NaOH (0.2 g, 5 mmol in 5 ml of water) and an aqueous formaldehyde solution (37%, 0.67 ml, 9 mmol) were added to the resulting mixture. The reaction mixture was stirred for 2 h at room temperature, and then filtered off and the filtrate was removed on a rotary evaporator. The resulting residue was washed with chloroform and ethanol, air dried, and recrystallized from water. The composition of the complex was confirmed by comparing its IR spectrum with that reported previously, and elemental analysis.

Oxygen evolution

Oxygen evolution was detected polarographically using a standard Clark-type oxygraph electrode (Hansatech Instruments) placed in a thermostated cell and separated from the sample solution by an oxygen-permeable Teflon membrane. The signal was recorded every 0.1 s using the Oxygraph software package. An air-saturated aqueous solution was used for calibration of the electrode. All of the experiments were carried out at 20 °C. The cell was purged with argon gas before each experiment, and the solution in the cell (1 ml) was continuously stirred. During light-driven oxygen detection experiments, the buffered solution (borate buffer, 0.1 M, pH 8.0) in the working cell contained $[Ru(bpy)_3](ClO_4)_2$, $Na_2S_2O_8$

and $Na_2[Fe^{IV}(L–6H)]$, the concentrations of which were varied as described in the next section. Visible light LEDs ($\lambda = 450(10)$ nm, 3 W) were used as illumination sources in the photoinduced reactions. For Ru^{III}-induced oxygen evolution, an aqueous solution of $Na_2[Fe^{IV}(L–6H)]$ was added to the freshly prepared solutions of $[Ru(bpy)_3](ClO_4)_3$ (1 mM) in borate buffer (0.1 M, pH 8.0).

Steady-state and time-resolved UV-vis spectroscopy

Steady-state UV-vis spectra were recorded using a Varian Cary 50 spectrometer in a 1 mm cuvette. Time-resolved experiments were carried out using a nanosecond laser spectroscopy setup. The solutions under study were measured in a 1 mm quartz cuvette using a pump-probe methodology. Microsecond transient absorption kinetics were recorded by a Q-switched Nd:YAG laser (Quanta Ray Pro-230), which produced tripled frequency pulses with $\lambda = 355$ nm (13 ns). The laser was coupled with an optical parametric oscillator to obtain the desired wavelength (460 nm) for the pump light. An excitation light power of 30 mJ per pulse was used in all of the experiments. The data were collected using an Edinburgh Instruments LP900 spectrometer equipped with a 450 W Xe lamp used as a probe light source. Light was collected using an Andor CCD camera for TA spectra and a Hamamatsu R928 photomultiplier tube for kinetic traces. Kinetic traces of the optical density at 420 nm (shoulder of the absorption band of $[Ru(bpy)_3]^{2+}$) were studied for solutions containing $[Ru(bpy)_3](ClO_4)_2$ (0.04 mM), $Na_2S_2O_8$ (0.4 mM) and $Na_2[Fe^{IV}(L–6H)]$ (1.0–3.0 µM). For $[Ru(bpy)_3]^{2+}$ luminescence studies, the transient emission spectra and corresponding kinetic traces at 650 nm were recorded (a 580 nm long pass filter was used to block the scattering of pump light).

Dynamic light scattering

DLS experiments were performed using a Zetasizer Nano S scattering system (Malvern Instruments Ltd) that used a uni-phase He–Ne laser (633 nm; 4 mW) working in cross auto-correlation mode. The scattering angle was set to 90° with respect to the incident laser. The intensity correlation curves were analyzed with the Zetasizer family software. The size measurement range was from 0.3 nm to 10 µm.

Results

UV-vis spectroscopy studies

At the outset, we performed a spectroscopy assay of the tris(bipyridine)ruthenium dye photo-oxidant system without the Fe^{IV} catalyst. The UV-vis spectrum of the aqueous solution containing $[Ru(bpy)_3](ClO_4)_2$ (0.05 mM) and $Na_2S_2O_8$ (0.2 mM) recorded in darkness is shown in Fig. 2a (black). It exhibits two overlapping MLCT bands at 420 nm and 455 nm, characteristic of $[Ru(bpy)_3]^{2+}$. Illumination of this assay solution using blue light LEDs ($\lambda \sim 450$ nm) resulted in a rapid loss of color, and in a few seconds nearly 90% of the $[Ru(bpy)_3]^{2+}$ converted to $[Ru(bpy)_3]^{3+}$ (Fig. 2a, red):

$$[Ru(bpy)_3]^{2+} + h\nu \rightarrow [Ru(bpy)_3]^{2+*} \tag{1}$$

$$[Ru(bpy)_3]^{2+*} + S_2O_8^{2-} \rightarrow [Ru(bpy)_3]^{3+} + SO_4^{2-} + SO_4^{\cdot-} \tag{2}$$

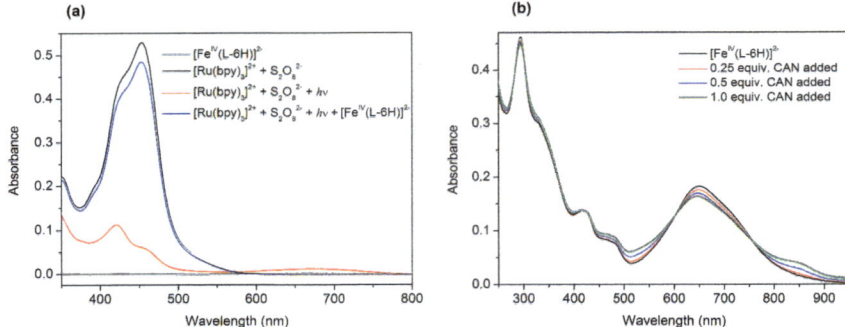

(a)   (b)

Fig. 2   (a) UV-vis spectra demonstrating the photochemical oxidation of $[Ru(bpy)_3]^{2+}$ (0.05 mM) to $[Ru(bpy)_3]^{3+}$ by persulfate (0.2 mM) with the following reduction by the catalyst (0.5 μM). The spectrum of the catalyst (0.5 μM) is given for comparison in grey. (b) UV-vis spectra demonstrating titration of the catalyst $[Fe^{IV}(L-6H)]^{2-}$ (0.02 mM) by CAN.

$$[Ru(bpy)_3]^{2+} + SO_4^{\cdot -} \rightarrow [Ru(bpy)_3]^{3+} + SO_4^{2-} \tag{3}$$

The photochemically generated $[Ru(bpy)_3]^{3+}$ is relatively stable in water, as can be concluded from the follow-up spectroscopic observation. Subsequent addition of 0.01 equivalents of $Na_2[Fe^{IV}(L-6H)]$ (0.5 μM) to this solution resulted in immediate restoration of the yellow color, as practically all of the $[Ru(bpy)_3]^{3+}$ was re-reduced to $[Ru(bpy)_3]^{2+}$ (Fig. 2a, blue). We proposed previously that the catalyst $[Fe^{IV}(L-6H)]^{2-}$ is oxidized by $[Ru(bpy)_3]^{3+}$ to the intermediate state $[Fe^{V}(L-6H)]^{-}$, as a first step in the catalytic process.[36] Alternatively, $[Fe^{IV}(L-6H)]^{2-}$ can be gently oxidized by $(NH_4)_2[Ce(NO_3)_6]$ (CAN), in which case it is possible to monitor the Fe based redox process by UV-vis spectroscopy as CAN does not have absorption bands overlapping with the bands characteristic of $[Fe^{IV}(L-6H)]^{2-}$. After the addition of 0.25–1 equivalents of CAN to $[Fe^{IV}(L-6H)]^{2-}$ in nitric acid (pH 1.5), an increase in absorption is observed around 550 nm and 850 nm at the same time as the intensity of the $[Fe^{IV}(L-6H)]^{2-}$ band at 650 nm decreases (Fig. 2b). The newly formed absorption bands can be assigned to $[Fe^{V}(L-6H)]^{-}$, as has been shown previously using EPR and Mössbauer spectroscopy in a Ru^{III}-induced oxidation assay.[36]

Photochemical and chemical water oxidation

Light-induced water oxidation was demonstrated in an aqueous solution containing $[Fe^{IV}(L-6H)]^{2-}$, $[Ru(bpy)_3]^{2+}$, and $S_2O_8^{2-}$ in borate buffer (pH 8.0) in a cell equipped with a Clark electrode.[36] In this follow-up study, we have optimized the conditions for photochemical water oxidation by varying the concentrations of photosensitizer and sacrificial electron acceptor. The TOF values reach saturation at 2.35 s^{-1}, at concentrations of $[Ru(bpy)_3]^{2+}$ above 0.3 mM and concentrations of $S_2O_8^{2-}$ above 2 mM (Fig. S1 in the ESI†). The evolution of dioxygen during the photocatalytic water oxidation using $[Fe^{IV}(L-6H)]^{2-}$ reaches a plateau after 150–240 s, as the catalytic system becomes inactive after >300 turnovers. Our dynamic light scattering (DLS) studies do not show any traces of nanoparticles after oxygen evolution has stopped (Fig. S2 in the ESI†), it thus appears that the $[Fe^{IV}(L-6H)]^{2-}$ catalyst degrades to other soluble iron complex(-es) unable to perform water oxidation. Still, since it is known that iron hydroxide is catalytically active,[37] we performed control photochemical water oxidation experiments under nearly identical conditions, but using $FeCl_3$ instead of $[Fe^{IV}(L-6H)]^{2-}$ as a precatalyst giving hematite at pH 8.0. When 1 μM $FeCl_3$ is used, oxygen evolution is indeed observed, but the TON and TOF are almost 80% lower than those observed for $[Fe^{IV}(L-6H)]^{2-}$ (Fig. S3†).

The kinetics of water oxidation using the catalyst $[Fe^{IV}(L-6H)]^{2-}$ has been evaluated using the one-electron oxidant $[Ru(bpy)_3](ClO_4)_3$. Addition of $[Fe^{IV}(L-6H)]^{2-}$ (0.2–5 μM) to a solution of $[Ru(bpy)_3]^{3+}$ (1 mM) at pH 8.0 leads to the immediate formation of oxygen that can be detected using the Clark electrode (Fig. 3a). At a catalyst concentration of 1.5 μM, the maximum TON of 45 was reached. As in the photochemical water oxidation studies,[36] the initial rates of water oxidation by $[Ru(bpy)_3]^{3+}$ exhibit a linear dependence on the catalyst concentration (Fig. 3b) with a first-order rate constant of 3.3(1) s^{-1}.

Transient absorption spectroscopy studies

To investigate in detail the light-driven hole transfer from the photo-oxidized ruthenium dye to the iron-based catalyst in aqueous solution, nanosecond transient absorption measurements were performed using the laser pump-probe method. An initial visible-light pump was used to excite $[Ru(bpy)_3]^{2+}$ (eqn (1)) and start a photochemical reaction, and a probe pulse measured the absorbance of the sample solution after different time delays on nano- and micro-second timescales. For the bare ruthenium dye photo-oxidant system consisting of $[Ru(bpy)_3](ClO_4)_2$ (0.04 mM) and $Na_2S_2O_8$ (0.4 mM) without the catalyst, the transient absorption spectra after the pump pulse (460 nm, 30 mJ, 13 ns) resulted in a bleach at 455 nm (Fig. 4a). The intensity of the bleach decreases when the time delay between the pump and probe pulses increases from a few nanoseconds to 1 μs. Moreover, one can observe a shoulder of a new band in the UV region, the

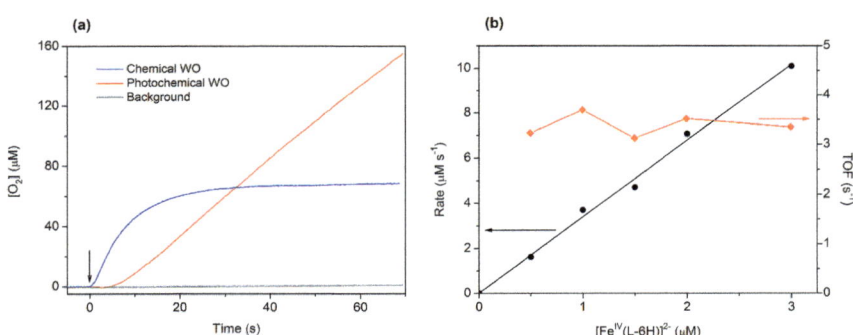

Fig. 3 (a) Traces of oxygen evolution obtained in photochemical (red trace: $[Fe^{IV}(L-6H)]^{2-}$ (1 μM), $[Ru(bpy)_3]^{2+}$ (0.2 mM) and $S_2O_8{}^{2-}$ (2 mM)) and chemical water oxidation experiments (blue trace: $[Fe^{IV}(L-6H)]^{2-}$ (1.5 μM) and $[Ru(bpy)_3]^{3+}$ (1 mM)). Oxygen evolution in the absence of catalyst is given for comparison (grey trace: $[Ru(bpy)_3]^{2+}$ (0.2 mM) and $S_2O_8{}^{2-}$ (2 mM)). The arrow indicates the beginning of the reaction (the start of illumination or addition of the oxidant, respectively). (b) Initial water oxidation rates (circles) and TOF (diamonds) as a function of the catalyst concentration for chemical water oxidation using $[Ru(bpy)_3]^{3+}$ (1 mM).

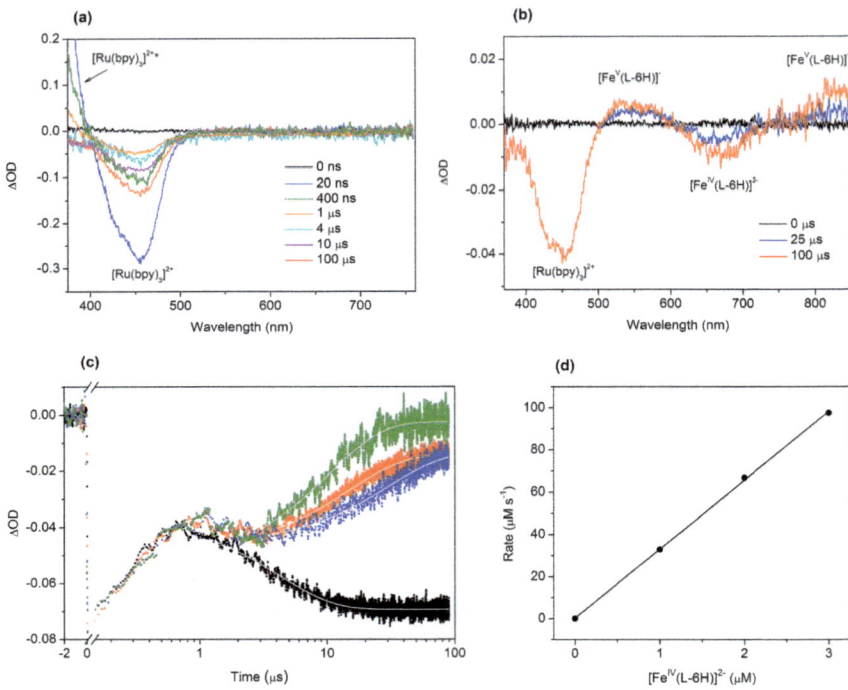

Fig. 4 (a) Transient absorption spectra for the solution containing $[Ru(bpy)_3](ClO_4)_2$ (0.04 mM) and $Na_2S_2O_8$ (0.4 mM) recorded with different time delays after a pump flash. (b) Transient absorption spectra for the solution containing $[Ru(bpy)_3]^{2+}$ (0.04 mM), $S_2O_8^{2-}$ (0.4 mM) and $[Fe^{IV}(L-6H)]^{2-}$ (2 μM) showing the appearance of the bands characteristic of $[Fe^V(L-6H)]^-$. (c) Kinetic traces at 420 nm for the solutions containing $[Ru(bpy)_3](ClO_4)_2$ (0.04 mM), $Na_2S_2O_8$ (0.4 mM) and a variable amount of $[Fe^{IV}(L-6H)]^{2-}$: black trace – 0 μM, blue trace – 1.0 μM, red trace – 2.0 μM, green trace – 3.0 μM. Fits are shown in white. (d) Dependence of the rate of $[Ru(bpy)_3]^{3+}$ reduction by $[Fe^{IV}(L-6H)]^{2-}$ derived from the kinetic traces at 420 nm on the concentration of the catalyst.

intensity of which also decreases on the nanosecond timescale. This transformation can be attributed to the excitation of $[Ru(bpy)_3]^{2+}$ ($\lambda_{abs} = 455$ nm) to the triplet state $[Ru(bpy)_3]^{2+*}$ ($\lambda_{abs} \sim 370$ nm)[38] followed by its relaxation. Due to the low oxidation potential of -0.62 V *versus* NHE,[39] the excited triplet state is oxidatively quenched by $S_2O_8^{2-}$ in aqueous solution yielding $[Ru(bpy)_3]^{3+}$ (eqn (2)). However, we cannot observe generation of $[Ru(bpy)_3]^{3+}$ in the transient absorption spectra directly, as the extinction coefficient for the broad absorption band of $[Ru(bpy)_3]^{3+}$ ($\lambda_{abs} = 670$ nm) is much lower than that for $[Ru(bpy)_3]^{2+}$ ($\lambda_{abs} = 455$ nm) (Fig. 2a). After 1 μs, the intensity of the bleach at 455 nm increases since $[Ru(bpy)_3]^{2+}$ is oxidized by the generated sulfate radical ($SO_4^{\bullet-}$) species (eqn (3)).[40] Under optimized conditions and in the absence of the iron catalyst, $SO_4^{\bullet-}$ quantitatively produces the second equivalent of the oxidized ruthenium species, as ascertained from the transient absorption kinetic traces derived at 420 nm (Fig. 4c, black). The trace reflecting reaction between $[Ru(bpy)_3]^{2+}$ and $SO_4^{\bullet-}$ in the solution can be fitted using exponential decay:

$$\Delta OD = \Delta OD_0 \exp(-t/\tau), \qquad (4)$$

where OD stands for optical density, t is the delay between the pump and probe pulses, and τ is the lifetime of the transient state, resulting in $\tau = 4$ μs. From the OD at 420 nm, the yield of charge-separated products, or cage escape yield, can be calculated using relative actinometry, as previously described.[41] The cage escape yield for the $[Ru(bpy)_3]^{3+}/SO_4^{\bullet-}$ pair is estimated to be 0.75, *i.e.* 75% of the total initial photoinduced electron transfer products are available for subsequent reaction.

The addition of the catalyst $[Fe^{IV}(L-6H)]^{2-}$ to the flash-quench mixture followed by excitation with 460 nm laser pulses results in two bleaches at 455 nm and 650 nm and two bands at 550 nm and ~830 nm observed on the microsecond timescale (Fig. 4b). When comparing these spectral features with the steady-state UV-vis spectra of the catalyst shown in the Fig. 2b and reported previously,[35,36] two positive transient absorbance signals are consistent with the formation of $[Fe^{V}(L-6H)]^-$, while the negative ΔOD signal at 650 nm corresponds to $[Fe^{IV}(L-6H)]^{2-}$:

$$[Fe^{IV}(L-6H)]^{2-} + [Ru(bpy)_3]^{3+} \rightarrow [Fe^{V}(L-6H)]^- + [Ru(bpy)_3]^{2+} \qquad (5)$$

The transient absorption kinetic traces derived at 420 nm for the aqueous solutions containing $[Ru(bpy)_3](ClO_4)_2$ (0.04 mM), $Na_2S_2O_8$ (0.4 mM) and $Na_2[Fe^{IV}(L-6H)]$ (1.0–3.0 μM) are presented in Fig. 4c. The initial growth of the negative ΔOD signal and its following decrease observed between 10 ns and 1 μs reflect the excitation and relaxation of the photosensitizer, respectively, and occur independently of the presence/concentration of the catalyst. The subsequent decrease in the negative ΔOD on the microsecond timescale can be plausibly fitted with a sum of two exponential functions with parameters τ_1 and τ_2, where τ_1 is fixed to 4 μs (derived from eqn (4)). From the fit of the kinetic traces, the pseudo-first order rate constant $k = 3.2(1) \times 10^{10}$ s^{-1} of the reaction between the reduced photosensitizer and the catalyst (eqn (5)) has been extracted (Fig. 4d).

Since the triplet excited state of the photosensitizer, $[Ru(bpy)_3]^{2+*}$, is a very reactive species, one can assume its direct oxidative or reductive quenching by the catalyst, circumventing the persulfate reaction. To investigate this possible scenario, we performed transient emission spectroscopy of the aqueous solution containing $[Ru(bpy)_3](ClO_4)_2$ (0.04 mM) and $Na_2S_2O_8$ (0.4 mM). The photosensitizer was selectively excited by the pump pulse (460 nm), and the luminescence trace at $\lambda_{em} = 650$ nm was recorded (Fig. S4 in the ESI†). When the catalyst $Na_2[Fe^{IV}(L-6H)]$ (2 μM) was added to this mixture, a similar emission trace was obtained. Both traces can be nicely fitted with an exponential decay (eqn (4)) giving lifetimes for the triplet state $[Ru(bpy)_3]^{2+*}$ of 275 ns and 236 ns, respectively. The observed difference suggests that the direct quenching of $[Ru(bpy)_3]^{2+*}$ by $[Fe^{IV}(L-6H)]^{2-}$ is possible but as a minor side process at the given concentrations of catalyst, photosensitizer and persulfate. This phenomenon needs to be investigated in future work.

Discussion

The clathrochelate complex $[Fe^{IV}(L-6H)]^{2-}$, which is indefinitely stable in aqueous solutions at pH 1.0–13.0, acts as a homogeneous catalyst for photocatalytic water oxidation by persulfate with $[Ru(bpy)_3]^{2+}$ as photosensitizer, affording a high TON = 365.[36] In both chemical-driven and light-driven water oxidation, most iron complexes are known to decompose easily into catalytically active iron oxide nanoparticles.[37,42] However, the use of strong acidic media,[27–29] or modification of the working system may prevent formation of FeO_x nanoparticles. For example, surface anchoring,[43] the use of robust polydentate ligands,[33,34] or isolation of a molecular catalyst in a cage of the metal–organic framework (MOF)[44] has been shown to stabilize the molecular structure of transition metal-based catalysts under oxidative conditions. In our work, the iron ion is encapsulated in a clathrochelate cage that allows sustained water oxidation at nearly neutral pH, in both chemical and photochemical assays. To the best of our knowledge, the complex $[Fe^{IV}(L-6H)]^{2-}$ exhibits relatively high catalytic efficiency compared to other molecular iron-based WOCs reported to date (Table 1).

Our transient absorption measurements corroborate the generally accepted scheme of photochemical water oxidation.[12,40] Thus, the excited triplet state of the photosensitizer $[Ru(bpy)_3]^{2+*}$ is quenched by persulfate *via* irreversible electron transfer yielding $[Ru(bpy)_3]^{3+}$ species. The radical anion $SO_4^{\cdot-}$ generated after the quenching reacts with another, non-excited $[Ru(bpy)_3]^{2+}$ molecule yielding a second equivalent of $[Ru(bpy)_3]^{3+}$. The oxidized photosensitizer injects its electron vacancy to the catalyst molecule restoring $[Ru(bpy)_3]^{2+}$, which can give rise to the next cycle of photosensitization and quenching. In the simplified mechanistic model, the hole scavenging reaction must be repeated four times in a row, as the active state of the catalyst must be four-times oxidized to evolve dioxygen from two water molecules. It is usually proposed that a water molecule

Table 1 Comparison of the catalytic performance for selected iron compounds for homogeneous chemical and photochemical water oxidation

Catalyst	Oxidant	pH	TON	TOF (s^{-1})	Ref.
Fe-TAML[a]	CAN	1.0	18	1.3	25
Fe-TAML[a]	CAN	1.0	93	—	27
Fe-TAML[a]	$NaIO_4$	1.0	44	—	27
Fe-TAML[a]	$NaIO_4$	7.0	3	—	27
Fe-TAML[a]	CAN	1.0	17	0.03	33
Fe-TAML[a]	$Ru + S_2O_8^{2-} + h\nu^b$	8.5	220	0.76	33
$[Fe(Pytacn)(OTf)_2]^c$	CAN	0.7	180	0.2	29
$[Fe(Mcp)(OTf)_2]^d$	CAN	0.8	360	0.28	28
$[Fe(Py5OH)Cl]^{-e}$	CAN	1.5	5	0.53	34
$[Fe(Py5OH)Cl]^{-e}$	$[Ru(bpy)_3]^{3+}$	8.0	26.5	2.2	34
$[Fe(Py5OH)Cl]^{-e}$	$Ru + S_2O_8^{2-} + h\nu^b$	8.0	43.5	0.6	34
$[Fe(Py5OH)(MeCN)]^{2-e}$	CAN	1.5	16	0.75	34
$[Fe(Py5OH)(MeCN)]^{2-e}$	$[Ru(bpy)_3]^{3+}$	8.0	7	0.9	34
$[Fe(Py5OH)(MeCN)]^{2-e}$	$Ru + S_2O_8^{2-} + h\nu^b$	8.0	20	0.6	34
$[Fe_2(Hbb)(OMe)(OAc)]^{+f}$	$[Ru(bpy)_3]^{3+}$	7.2	4	0.012	26
$[Fe^{IV}(L-6H)]^{2-}$	$[Ru(bpy)_3]^{3+}$	8.0	45	3.3	This work, 36
$[Fe^{IV}(L-6H)]^{2-}$	$Ru + S_2O_8^{2-} + h\nu^b$	8.0	365	2.27	This work, 36
$FeCl_3^g$	$Ru + S_2O_8^{2-} + h\nu^b$	8.0	63	0.6	This work

[a] Different tetraamido macrocyclic ligand (TAML) complexes were reported (see original publications for details). [b] Photochemical water oxidation using $[Ru(bpy)_3]^{2+}$ as photosensitizer and $S_2O_8^{2-}$ as sacrificial electron acceptor. [c] Pytacn = 1-(2'-pyridylmethyl)-4,7-dimethyl-1,4,7-triazacyclononane; OTf = triflate anion. [d] Mcp = N,N'-dimethyl-N,N'-bis(2-pyridylmethyl)-1,2-cis-diaminocyclohexane; OTf = triflate anion. [e] Py5OH = pyridine-2,6-diylbis(di(pyridin-2-yl)methanol). [f] Hbb = 2,2'-(2-hydroxy-5-methyl-1,3-phenylene)bis(1H-benzo[d]imidazole-4-carboxylic acid). [g] $FeCl_3$ was used as a precatalyst giving iron oxide nanoparticles.

or hydroxide anion adds to the metal center followed by one- or multi-electron oxidation, addition of a second water molecule and then further oxidation.[29,33] In addition, some deviations from the standard scheme are possible, for example direct oxidation of the catalyst by $SO_4^{\cdot -}$. In any case, according to eqn (1)–(3) and (5), two holes are transferred to the catalyst per one absorbed photon; and two photons are needed to drive four-electron oxidation of water to dioxygen in the system, where persulfate is used as the electron acceptor.

One of the factors determining the overall efficiency of the water oxidation system is the rate of the hole transfer reaction shown in eqn (5).[45] The oxidized form of the photosensitizer $[Ru(bpy)_3]^{3+}$ is relatively stable in water, but undergoes irreversible decomposition in the presence of $SO_4^{\cdot -}$ or other reactive species produced during photocatalysis.[40,45,46] Thus, the primary hole scavenging competes with the degradation of $[Ru(bpy)_3]^{3+}$. In certain systems, the loss of photocatalytic activity has been attributed to the decomposition of the photo-sensitizer.[47] Despite the presence of excess amounts of $S_2O_8^{2-}$, the catalysis usually lasts for a few minutes and stops long before the sacrificial electron acceptor is fully consumed owing to decomposition of the light absorber. In the case of the system using $[Fe^{IV}(L–6H)]^{2-}$, the catalyst reduces $[Ru(bpy)_3]^{3+}$ relatively quickly compared to other systems.[40,48,49] Though there are a limited number of reported studies devoted to investigation of the hole scavenging activity of molecular WOCs published to-date, the reported TOF is comparatively high (Table 1). We have also found that addition of a fresh portion of $[Ru(bpy)_3]^{2+}$ to the photocatalytic mixture after >350 turnovers does not reactivate oxygen evolution, indicating that it is not the light absorber, but rather the catalyst, that degrades in the course of water oxidation.

A key challenge for us is to decipher the mechanism of water oxidation cata-lyzed by the clathrochelate complex $[Fe^{IV}(L–6H)]^{2-}$. Our kinetic studies suggest that one molecule of catalyst is involved in oxygen evolution (Fig. 3b). Thus, the typically rate-determining O–O bond formation takes place intramolecularly. According to the generally accepted mechanistic pathway for water oxidation by molecular iron based catalysts, oxidation of an Fe-aqua complex leads to the formation of an oxo-complex Fe=O; then, nucleophilic attack of a second water molecule at the oxo O atom gives rise to the O–O bond.[50] This mechanism is feasible for complexes of Fe, the coordination spheres of which are not saturated (e.g., square planar Fe-TAML complexes). For $[Fe(Py5OH)Cl]^-$ featuring a hex-acoordinated metal ion, it was proposed that the binding mode of the ligand changes during catalysis so that a vacant coordination site opens to form an Fe-aqua complex.[34] However, such a mechanism would not be relevant to $[Fe^{IV}(L–6H)]^{2-}$ with the robust clathrochelate ligand. On the other hand, one might assume a heptacoordinated Fe-aqua intermediate, as was found for Ru-based molecular WOCs.[51] Additionally, we could not completely exclude the outer-sphere electron transfer mechanism, as has been proposed for the related cobalt(II) clathrochelates, which catalyzed water reduction to hydrogen in the $[Ru(bpy)_3]^{2+}$ photo-reductant system.[52] Finally, one could consider an alternative pathway involving α,β-dicarbonyl OH adducts, which could react further to generate intermediate dioxetanes or endoperoxides. This implies that O–O bond formation occurs on the ligand, whereas the iron ion acts as an electron shuttle (since Fe in $[Fe(L–6H)]^{n-}$ may possess oxidation states from +3 till +5, and apparently even till +6 as can be seen in electrochemical experiments).[35] This kind of mechanism was first suggested for the ruthenium "blue dimer", where the O–O bond was proposed to be formed on carbon as a fragment of the four-membered endoperoxide ring C_2O_2.[53] Though we could characterize one of the possible intermediates $[Fe^V(L–6H)]^-$, the mechanism of water oxidation using the clathrochelate catalyst remains elusive. Intermediate species beyond Fe^V seem to be very reactive, and hence have so far not been observable experimentally. To help to understand the catalytic mechanism, future work will include changes to the ligand structure in combination with computational studies.

Summary

The visible light-driven hole transfer from tris(bipyridine)ruthenium dye to the catalyst $[Fe^{IV}(L–6H)]^{2-}$ in an aqueous solution occurs with a pseudo-first-order rate constant $3.2(1) \times 10^{10}$ s^{-1}. To our knowledge, this is the most efficient hole scavenging reported for molecular WOCs in the system utilizing $[Ru(bpy)_3]^{2+}$ as the light absorber and $S_2O_8^{2-}$ as the electron acceptor at nearly neutral pH. Due to the fast hole injection, the oxidized photosensitizer does not accumulate, allowing it to avoid oxidative decomposition. This finding is in harmony with the record high TON reached in photochemical water oxidation by using $[Fe^{IV}(L–6H)]^{2-}$.[36] The efficiency of this catalyst might be further improved by introducing bulky substituents at the methylene groups of the ligand to achieve higher stability. Together with the mechanistic studies, this work is currently ongoing.

Conflicts of interest

There are no conflicts of interest to declare.

Acknowledgements

We would like to thank Prof Leif Hammarström for helpful discussion. SIS acknowledges support provided by the Swedish Institute through individual fellowship (Grant No 23913/2017). IOF acknowledges support from the European Community's Framework Programme (H2020-MSCA-RISE-2017) under grant agreement No 778245.

Notes and references

1 G. A. Jones and K. J. Warner, *Energy Policy*, 2016, **93**, 206–212.
2 L. Hammarström, *Acc. Chem. Res.*, 2015, **48**, 840–850.
3 N. S. Lewis and D. G. Nocera, *Proc. Natl. Acad. Sci. U. S. A.*, 2006, **103**, 15729–15735.
4 L. Hammarström, *Faraday Discuss.*, 2017, **198**, 549–560.
5 J. D. Blakemore, R. H. Crabtree and G. W. Brudvig, *Chem. Rev.*, 2015, **115**, 12974–13005.
6 F. D'Souza and O. Ito, *Chem. Soc. Rev.*, 2012, **41**, 86–96.
7 M. Yamamoto, J. Föhlinger, J. Petersson, L. Hammarström and H. Imahori, *Angew. Chem., Int. Ed.*, 2017, **56**, 3329–3333.
8 D. L. Ashford, M. K. Gish, A. K. Vannucci, M. K. Brennaman, J. L. Templeton, J. M. Papanikolas and T. J. Meyer, *Chem. Rev.*, 2015, **115**, 13006–13049.

9 M. Borgström, N. Shaikh, O. Johansson, M. F. Anderlund, S. Styring, B. Åkermark, A. Magnuson and L. Hammarström, *J. Am. Chem. Soc.*, 2005, **127**, 17504–17515.

10 L. Favereau, A. Makhal, Y. Pellegrin, E. Blart, J. Petersson, E. Göransson, L. Hammarström and F. Odobel, *J. Am. Chem. Soc.*, 2016, **138**, 3752–3760.

11 D. Paul Rillema, G. Allen, T. J. Meyer and D. Conrad, *Inorg. Chem.*, 1983, **22**, 1617–1622.

12 S. Fukuzumi, J. Jung, Y. Yamada, T. Kojima and W. Nam, *Chem.–Asian J.*, 2016, **11**, 1138–1150.

13 A. Harriman, I. J. Pickering, J. M. Thomas and P. A. Christensen, *J. Chem. Soc., Faraday Trans. 1*, 1988, **84**, 2795–2806.

14 M. M. Najafpour, T. Ehrenberg, M. Wiechen and P. Kurz, *Angew. Chem., Int. Ed.*, 2010, **49**, 2233–2237.

15 C. C. L. McCrory, S. Jung, J. C. Peters and T. F. Jaramillo, *J. Am. Chem. Soc.*, 2013, **135**, 16977–16987.

16 T. Audichon, T. W. Napporn, C. Canaff, C. Morais, C. Comminges and K. B. Kokoh, *J. Phys. Chem. C*, 2016, **120**, 2562–2573.

17 W. Smith, *J. Power Sources*, 2000, **86**, 74–83.

18 L. Duan, F. Bozoglian, S. Mandal, B. Stewart, T. Privalov, A. Llobet and L. Sun, *Nat. Chem.*, 2012, **4**, 418–423.

19 J. M. Thomsen, D. L. Huang, R. H. Crabtree and G. W. Brudvig, *Dalton Trans.*, 2015, **44**, 12452–12472.

20 M. V. Pavliuk, E. Mijangos, V. G. Makhankova, V. N. Kokozay, S. Pullen, J. Liu, J. Zhu, S. Styring and A. Thapper, *ChemSusChem*, 2016, **9**, 2957–2966.

21 M. M. Najafpour, G. Renger, M. Hołyńska, A. N. Moghaddam, E.-M. Aro, R. Carpentier, H. Nishihara, J. J. Eaton-Rye, J.-R. Shen and S. I. Allakhverdiev, *Chem. Rev.*, 2016, **116**, 2886–2936.

22 M. V. Pavliuk, V. G. Makhankova, V. N. Kokozay, I. V. Omelchenko, J. Jezierska, A. Thapper and S. Styring, *Polyhedron*, 2015, **88**, 81–89.

23 S. J. Koepke, K. M. Light, P. E. VanNatta, K. M. Wiley and M. T. Kieber-Emmons, *J. Am. Chem. Soc.*, 2017, **139**, 8586–8600.

24 J.-W. Wang, W.-L. Liu, D.-C. Zhong and T.-B. Lu, *Coord. Chem. Rev.*, 2019, **378**, 237–261.

25 W. C. Ellis, N. D. McDaniel, S. Bernhard and T. J. Collins, *J. Am. Chem. Soc.*, 2010, **132**, 10990–10991.

26 B. Das, B. L. Lee, E. A. Karlsson, T. Åkermark, A. Shatskiy, S. Demeshko, R. Z. Liao, T. M. Laine, M. Haukka, E. Zeglio, A. F. AbdelMagied, P. E. Siegbahn, F. Meyer, M. D. Kärkäs, E. V. Johnston, E. Nordlander and B. Åkermark, *Dalton Trans.*, 2016, **45**, 13289–13293.

27 W.-P. To, T. W.-S. Chow, C.-W. Tse, X. Guan, J.-S. Huang and C.-M. Che, *Chem. Sci.*, 2015, **6**, 5891–5903.

28 J. L. Fillol, Z. Codolà, I. Garcia-Bosch, L. Gómez, J. J. Pla and M. Costas, *Nat. Chem.*, 2011, **3**, 807–813.

29 Z. Codolà, I. Garcia-Bosch, F. Acuña-Parés, I. Prat, J. M. Luis, M. Costas and J. Lloret-Fillol, *Chem. - Eur. J.*, 2013, **19**, 8042–8047.

30 M. Okamura, M. Kondo, R. Kuga, Y. Kurashige, T. Yanai, S. Hayami, V. K. K. Praneeth, M. Yoshida, K. Yoneda, S. Kawata and S. Masaoka, *Nature*, 2016, **530**, 465–468.

31 M. K. Coggins, M.-T. Zhang, A. K. Vannucci, C. J. Dares and T. J. Meyer, *J. Am. Chem. Soc.*, 2014, **136**, 5531–5534.

32 S. Pattanayak, D. R. Chowdhury, B. Garai, K. K. Singh, A. Paul, B. B. Dhar and S. S. Gupta, *Chem. - Eur. J.*, 2017, **23**, 3414–3424.

33 C. Panda, J. Debgupta, D. Díaz Díaz, K. K. Singh, S. Sen Gupta and B. B. Dhar, *J. Am. Chem. Soc.*, 2014, **136**, 12273–12282.

34 B. Das, A. Orthaber, S. Ott and A. Thapper, *ChemSusChem*, 2016, **9**, 1178–1186.

35 S. Tomyn, S. I. Shylin, D. Bykov, V. Ksenofontov, E. Gumienna-Kontecka, V. Bon and I. O. Fritsky, *Nat. Commun.*, 2017, **8**, 14099.

36 S. I. Shylin, M. V. Pavliuk, L. D'Amario, F. Mamedov, J. Sá, G. Berggren and I. O. Fritsky, *Chem. Commun.*, 2019, **55**, 3335–3338.

37 X. J. Wu, F. Li, B. B. Zhang and L. Sun, *J. Photochem. Photobiol., C*, 2015, **25**, 71–89.

38 R. S. Khnayzer, V. S. Thoi, M. Nippe, A. E. King, J. W. Jurss, K. A. El Roz, J. R. Long, C. J. Chang and F. N. Castellano, *Energy Environ. Sci.*, 2014, **7**, 1477–1488.

39 P. Diamantis, J. F. Gonthier, I. Tavernelli and U. Rothlisberger, *J. Phys. Chem. B*, 2014, **118**, 3950–3959.

40 G. La Ganga, F. Puntoriero, S. Campagna, I. Bazzan, S. Berardi, M. Bonchio, A. Sartorel, M. Natalic and S. Franco, *Faraday Discuss.*, 2012, **155**, 177–190.

41 M. Ruthkosky, F. N. Castellano and G. J. Meyer, *Inorg. Chem.*, 1996, **35**, 6406–6412.

42 G. Chen, L. Chen, S.-M. Ng, W.-L. Man and T.-C. Lau, *Angew. Chem., Int. Ed.*, 2013, **52**, 1789–1791.

43 B. M. Klepser and B. M. Bartlett, *J. Am. Chem. Soc.*, 2014, **136**, 1694–1697.

44 B. Nepal and S. Das, *Angew. Chem., Int. Ed.*, 2013, **52**, 7224–7227.

45 B. Limburg, E. Bouwman and S. Bonnet, *ACS Catal.*, 2016, **6**, 5273–5284.

46 M. Hara, C. C. Waraksa, J. T. Lean, B. A. Lewis and T. E. Mallouk, *J. Phys. Chem. A*, 2000, **104**, 5275–5280.

47 U. S. Akhtar, E. L. Tae, Y. S. Chun, I. C. Hwang and K. B. Yoon, *ACS Catal.*, 2016, **6**, 8361–8369, and references therein.

48 A. Company, G. Sabenya, M. Gonzalez-Bejar, L. Gomez, M. Clemancey, G. Blondin, A. J. Jasniewski, M. Puri, W. R. Browne, J. M. Latour, L. Que Jr, M. Costas, J. Perez-Prieto and J. Lloret-Fillol, *J. Am. Chem. Soc.*, 2014, **136**, 4624–4633.

49 X. Du, Y. Ding, F. Song, B. Ma, J. Zhao and J. Song, *Chem. Commun.*, 2015, **51**, 13925–13928.

50 I. Gamba, Z. Codolà, J. Lloret-Fillol and M. Costas, *Coord. Chem. Rev.*, 2017, **334**, 2–24.

51 Q. Daniel, P. Huang, T. Fan, Y. Wang, L. Duan, L. Wang, F. Li, Z. Rinkevicius, F. Mamedov, M. S. G. Ahlquist, S. Styring and L. Sun, *Coord. Chem. Rev.*, 2017, **346**, 206–215.

52 I. I. Creaser, L. R. Gahan, R. J. Geue, A. Launikonis, P. A. Lay, J. D. Lydon, M. G. McCarthy, A. W. H. Mau, A. M. Sargeson and W. H. F. Sasse, *Inorg. Chem.*, 1985, **24**, 2671–2680.

53 J. K. Hurst, J. L. Cape, A. E. Clark, S. Das and S. Qin, *Inorg. Chem.*, 2008, **47**, 1753–1764.

Faraday Discussions

PAPER

Light induced formation of a surface heterojunction in photocharged CuWO$_4$ photoanodes†

Anirudh Venugopal [iD] and Wilson A. Smith [iD] *

Received 16th November 2018, Accepted 7th December 2018

DOI: 10.1039/c8fd00179k

Photocharging has recently shown the ability to significantly improve the performance of several metal oxide photoanodes, similar to the enhancements achieved with co-catalysts and passivation overlayers. Herein, we demonstrate the effect of photocharging on CuWO$_4$ photoanodes for the first time, with prolonged AM 1.5 illumination under open-circuit conditions. The photocharging treatment on CuWO$_4$ samples doubled the photocurrent obtained at 1.23 V$_{RHE}$. This enhancement is attributed to the light induced formation of a surface bound copper complex with the solution anion species in the electrolyte. This thin semiconducting copper borate layer forms a heterojunction with the CuWO$_4$, improving the charge separation near the surface and thus suppressing the recombination of charge carriers in the space charge region. The striking similarities in photocharging of different metal oxide semiconductors highlights that the metal oxide semiconductor–electrolyte interface is more complex than previously understood. The formation of this time-dependent light induced surface layer should therefore be considered in all experimental studies on photo-electrochemistry with metal oxide semiconductor photoanodes.

Introduction

A shift to a renewable energy-based society is necessary to contain or limit the catastrophic effects of global climate change.[1] Producing electricity from renewable energy sources could be a way to tackle this problem, but this transition is currently facing a roadblock in dealing with the intermittency of these resources. Current energy storage techniques are not sufficient to tackle the discrepancy between the energy demand and energy production timelines. The viability of a terawatt scale renewable energy society will need this intermittency problem to be dealt with. One way to address this challenge is to (photo-)electrochemically produce commercially valuable chemicals or fuels, that have a good market

Materials for Energy Conversion and Storage (MECS), Department of Chemical Engineering, Faculty of Applied Sciences, Delft University of Technology, Delft 2629HZ, The Netherlands. E-mail: w.smith@tudelft.nl

† Electronic supplementary information (ESI) available. See DOI: 10.1039/c8fd00179k

demand, from the excess renewable electricity. This way the economics of the technology can be balanced, making it self-sustainable and attracting further investments and expansions. (Photo-)electrochemical pathways to produce valuable chemicals such as hydrogen from earth-abundant raw materials, i.e. water, is one such way to do so.[2]

(Photo-)electrochemical water oxidation is an important (half-)reaction of interest within electrochemistry, as it acts as a proton source for more valuable electrochemical (half-)reactions such as hydrogen evolution, CO_2 reduction and ammonia synthesis. Metal oxides have received significant attention in the past as a prospective photoanode material due to their suitable opto-electronic properties, higher stability compared to low bandgap semiconductors and their raw material abundance. However, these metal oxide photoanodes have been severely underperforming, majorly as a result of the complexity involving the multi-electron ($4e^-/4h^+$) oxygen evolution reaction and the existence of different loss mechanisms such as bulk recombination and surface recombination of photo-generated charge carriers at the semiconductor|electrolyte interface (SEI). The latter issue could be dealt with, to an extent, with the addition of catalytic and passivating overlayers such as Co–Pi.[3,4] Unfortunately, these overlayers often result in parasitic light absorption that decrease the overall performance of the composite device.[5]

Recently, the "photocharging" process was introduced as a way to suppress the losses at the SEI of several metal oxide photoanodes ($BiVO_4$, Fe_2O_3, WO_3) without the addition of an overlayer, resulting in an improvement in the overall performance.[6–12] This photocharging treatment led to an increase in the photocurrent and an appreciable cathodic shift in the onset potential in some cases. This was first demonstrated in bismuth vanadate by Trześniewski et al.,[11] improving the performance of $BiVO_4$ tremendously and making its performance comparable to that of $BiVO_4$ with catalytic overlayers. This photocharging treatment was performed by illuminating the metal oxide semiconductor for a prolonged time under open-circuit conditions in a phosphate buffer electrolyte. Li et al.[7] also demonstrated a similar effect on tungsten doped $BiVO_4$, by prolonged illumination of the W : $BiVO_4$ sample in air with UV light, in a procedure they call the UV-curing process. Trześniewski et al.[10] later went on to show the dramatic effect of the [OH^-] concentration in the electrolyte on the degree of suppression of the losses at the SEI, i.e. at more alkaline conditions the photocharging treatment worked better. Liu et al.[9] made use of intensity modulated photoelectron spectroscopy (IMPS) to quantitatively study the kinetics of the photocharged $BiVO_4$ and show that the enhancement from photocharging is both in the bulk and the surface of the $BiVO_4$ photoanode. Favaro et al.[6] used in situ HAXPES to demonstrate that the light-induced suppression of surface losses resulting from the photocharging process in $BiVO_4$, in a phosphate buffer solution, could be a result of the formation of a thin $BiPO_4$ overlayer as a result of the illumination. Li et al.[8] further demonstrated the effect of the UV curing process on WO_3 photoanodes, showing that this light-induced performance enhancement was not confined to only $BiVO_4$. Xie et al.[12] recently demonstrated a similar enhancement on titanium doped hematite photoanodes, by photocharging the Ti : Fe_2O_3 with prolonged illumination under an applied anodic potential.

Copper tungstate ($CuWO_4$) is another n-type metal oxide semiconductor that has been extensively researched for photo-electrochemical water oxidation.[13–18]

The bandgap energy of $\sim$2.25 eV allows it to achieve a theoretical photocurrent density upwards of 10 mA cm^{-2}.[13] The ability of CuWO$_4$ photoanodes to selectively oxidize water over chloride ions in the solution is also an added advantage, opening possibilities for direct water oxidation from sea water.[18] Herein, we demonstrate that copper tungstate also shows an improvement in performance with the photocharging treatment. We show that this photocharging process of CuWO$_4$ results in doubling of the photocurrent of the untreated sample at 1.23 V$_{RHE}$, with only a negligible anodic shift in the onset potential. We also show that this effect is reversible under dark conditions, as also seen with BiVO$_4$. More importantly, these results raise the question of whether the photocharging treatment is universal to all metal oxide semiconductors, and what the implications of this treatment may hold for general SEI's. Gaining more fundamental understanding of the photocharging process will help to gain more insight into the dynamic behaviour of the metal oxide SEI under illumination. In this article, we would like to briefly discuss the features of the photocharging process of CuWO$_4$ and the similarities with the photocharging process of other metal oxide semiconductors. We highlight that the metal oxide semiconductor–electrolyte interface is much more complex than what is previously understood.

Experimental

Preparation of CuWO$_4$ thin films

CuWO$_4$ thin films of $\sim$1.2 μm were deposited onto a conductive FTO coated glass substrate (TEC-15, Hartford Glass Co.) by the spray pyrolysis technique. The precursor solution was prepared by mixing equimolar aqueous solutions of CuSO$_4\cdot$5H$_2$O (99.995% trace metal basis, Sigma-Aldrich) and Na$_2$WO$_4\cdot$2H$_2$O (BioUltra, $\geq$99.0% (T), Sigma-Aldrich), resulting in the precipitation of CuWO$_4$. The solution was further diluted with Milli-Q water to make the concentrations of Cu^{2+} and W^{6+} at 0.01 M in the final precursor solution. The CuWO$_4$ precipitate was dissolved using 25% aqueous NH$_3$ solution (ammonium hydroxide solution, puriss. p.a., reag. ISO, reag. Ph. Eur., $\geq$25% NH$_3$ basis, Sigma-Aldrich) and the final pH of the precursor solution was made up to 11.5. The FTO substrate was cleaned with laboratory grade soap solution followed by Milli-Q water, acetone and isopropanol and eventually dried off using nitrogen gas. This was followed by a 45 minute UV/ozone cleaning step at 60 °C (Novascan UV Ozone cleaner – PSD Pro series). The cleaned substrate was placed on a ceramic heating plate and the plate was slowly heated to 300 °C. The spray deposition was carried out using an automated spray setup, spraying at 1 minute cycles with 1 second spray and 59 seconds delay time, for 100 cycles driven by an overpressure of 0.8 MPa of nitrogen gas. The nozzle substrate distance was fixed at 40 cm and the precursor solution was placed 33 cm below the nozzle and was fed *via* the siphoning effect of the nitrogen flow. After the deposition, the samples were annealed in a tube oven at 500 °C for 5 hours at a ramp rate of 5 °C min^{-1} with an airflow of 80 cm^3 min^{-1}.

Photoelectrochemical measurements

The CuWO$_4$ photoanodes were photoelectrochemically tested using a potentiostat (VersaSTAT 3, Princeton Applied Research) in a three electrode arrangement, with a coiled platinum wire as the counter electrode and an Ag/AgCl (XR300, saturated

KCl and AgCl solution, Radiometer Analytical) reference electrode. The electrolyte used was an aqueous 0.3 M borate buffer (99.97% trace metal basis Boric acid, Sigma-Aldrich) with 0.1 M sodium perchlorate (ACS reagent, $\geq$ 98.0%, Sigma-Aldrich) and the pH of this solution was corrected to 7.5 using a concentrated NaOH (pellets, $\geq$98% Assay, Baker analysed ACS, J.T. Baker, Avantor performance materials) solution. The voltammetric scans were performed at a scan rate of 50 mV s^{-1} unless specified otherwise. A Newport Sol3A Class AAA (type 94023A-SR3) solar simulator, with a 450 W xenon short arc lamp, was used to provide the simulated solar illumination for the photocurrent measurements at the standard AM 1.5 illumination intensity (100 mW cm^{-2}). The samples were illuminated in the back illumination mode. The electrolyte inside the photoelectrochemical cell was constantly stirred using a magnetic stirrer and was purged with nitrogen gas to constantly remove any dissolved gasses in the electrolyte. The incident photon to current efficiency (IPCE) measurements were performed using a 200 W quartz–tungsten halogen lamp coupled with a grating monochromator (Acton SpectraPro 150i) and electronic shutter system (Uniblitz LS6). The illumination intensities of the tungsten halogen lamp were measured using a Newport high performance handheld optical power meter (1919-R) coupled with a UV-enhanced silicon photodetector (918-UV-OD3R). The IPCE measurements were performed at an applied potential of 1.23 V$_{RHE}$. The CuWO$_4$ photoanodes were photocharged by illuminating the sample (with AM 1.5 light) in a photoelectrochemical cell under open-circuit conditions, for prolonged hours. The catalytic efficiency of the CuWO$_4$ photoanodes was tested using methanol as a hole scavenger, by mixing it with the electrolyte used for the voltammetry measurements in a 50% v/v ratio. Using this experiment, we probe the photocurrent generated in the presence of a hole scavenging electrolyte, which assumes near unity transfer of holes to oxidize a compound that is more easily oxidized than water, and use the ratio of these photocurrents to estimate the 'ideal' hole transfer rate if it drives a reaction that does not have a major rate-limiting step. Mott–Schottky plots were obtained for the CuWO$_4$ samples in the dark, in the fixed frequency mode (10 kHz) with a 15 mV amplitude of perturbation. These measurements were performed without any stirring or nitrogen bubbling.

Material characterization

UV-Vis absorption measurements of the CuWO$_4$ samples were obtained using a PerkinElmer Lambda 900 UV/Vis/NIR spectrometer. The measurements were performed inside an integration sphere, in transmittance mode (% T), with a sample tilt of 15 degrees to minimize reflection of the incident beam from the sample. The transmittance data was converted to absorption data (% A) using the formula % $A = 100 - \% \, T$. The measurements were performed between 750 nm and 300 nm, at a scan rate of 250 nm min^{-1} with an integration time of 0.2 s. Absorption coefficient and Tauc plot were calculated from the above obtained transmittance data.

X-ray diffraction (XRD) measurements were performed using a Bruker D8 Advance diffractometer powered with a Co Kα X-ray source ($I = 1.78897$ Å). The samples were scanned at a scan speed of 0.4 seconds per step, with each step being 0.02 degrees, with no rotation.

Scanning electron microscope (SEM) images to study the surface morphology of the $CuWO_4$ samples were obtained using a Joel JSM-6010LA analytical SEM machine. The measurements were performed at a working distance of ~10 mm at an accelerating voltage of 5 kV.

X-ray photoelectron spectroscopy (XPS) measurements were performed using a Thermo Scientific Kα XPS system with an aluminium Kα X-ray source. The measurements were performed with a spot size of 400 μm, pass energy of 50 eV, energy step size of 0.1 eV and a dwell time of 50 ms, along with a flood gun for charge compensation. 10 scans were performed for each of C 1s, O 1s, and W 4f, 300 scans for B 1s, 50 scans for the valence band and 25 scans were performed for the Cu 2p, around their respective binding energies, to obtain the final spectrum. The valence band scans were performed at a pass energy of 100 eV. Measurements were performed for three different spots in each sample and the data was averaged over these three spots, expect for B 1s, where the data was averaged over two different spots. All data presented in this work are corrected for the C 1s peak shifts, using 284.8 eV as the reference point for the C 1s peaks. The data presented in the paper is after background subtraction. A U 2 Tougaard background spectra with a C parameter of −3450 was used in each case. Samples were rinsed with Milli-Q water and dried under N_2 stream, to remove any salt deposits at the surface, before loading to the XPS system.

Results

Voltammetric scans were performed to evaluate the photoelectrochemical performance of untreated (UT) and photocharged (PC) $CuWO_4$ substrates. The potential sweeps were performed between 0 V *vs.* open-circuit potential (OCP) and 1.4 V_{RHE} to obtain the current–voltage plots. Fig. 1 shows the measured photocurrent density of the UT-$CuWO_4$ samples (solid blue line) against the applied potential. A photocurrent density of ~30 μA cm^{-2} was measured at 1.23 V_{RHE} for the UT-$CuWO_4$ sample, with an onset potential of ~0.7 V_{RHE}. The lower slope at potentials less than 0.9 V_{RHE}, compared to the slope of the curve above this potential, would suggest that there could be other non-faradaic processes

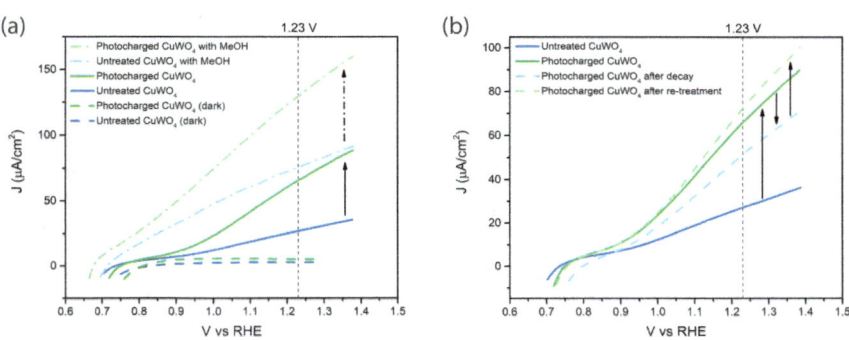

Fig. 1   (a) Voltammetric scans (in light and dark) of UT- and PC-$CuWO_4$ in the back illumination mode. Measurements were performed with and without a hole scavenger. (b) Voltammetric scans demonstrating the decay and re-photocharging of the PC-$CuWO_4$ substrates.

occurring apart from water oxidation at these low potentials. In fact, similar observations were made in other reports on $CuWO_4$ (ref. 13) and WO_3,[19,20] where they attribute it to the de-intercalation of protons or alkali cations from the $CuWO_4$ or WO_3 lattice. After analysing the properties of the UT-$CuWO_4$, these samples were then subjected to the photocharging treatment for 15 hours. Fig. 1(a) also shows the improvement in the photocurrent of the PC-$CuWO_4$ sample (solid green line), compared with the untreated sample (solid blue line). The photocurrent at 1.23 V_{RHE} of the photocharged sample is, at most, two times that of the untreated sample, with only a negligible anodic shift in the onset potential. The increase in photocurrent after the photocharging treatment may be a result of light-induced surface passivation, improvement in the kinetics or catalytic properties, improvement in the bulk charge transport or charge sepa-ration properties, or a combination of some or all of these factors. The latter was true in the case of $BiVO_4$, where a comprehensive set of investigations has been performed since the discovery of this photocharging effect.[6,7,9–11] The evolution of photocurrent with the photocharging time is shown in Fig. S1.†

To get a better understanding of the reason behind the improvement in performance of the $CuWO_4$ samples from photocharging, a hole scavenger was used before and after photocharging and the results were compared to the case without a hole scavenger. Hole scavengers have a much lower theoretical redox potential and, as a result, can be more easily oxidized than water. The catalytic efficiency in the presence of such hole scavengers is assumed to be near 100%, with a complete suppression of all surface losses. Methanol, which is typically used as a hole scavenger in photoelectrochemical studies,[21] was used in a 50% v/v ratio with the aqueous electrolyte to perform the voltammetric scans. Fig. 1(a) also shows the improvement in performance of the UT- and PC-$CuWO_4$ sample, when a hole scavenger is used (dash dot light blue and light green lines, respectively). The performance of the UT-$CuWO_4$ improved when a hole scavenger is used (dash dot light blue line), as expected. This result suggests that the UT-$CuWO_4$ was suffering from surface recombination losses and/or poor water oxidation kinetics. Comparing the performance of UT-$CuWO_4$ with hole scavenger and that of PC-$CuWO_4$ without the hole scavenger suggests that the improvement in perfor-mance of the PC-$CuWO_4$ sample could be a result of improvement in catalytic efficiency. Interestingly, the PC-$CuWO_4$ samples also showed an improvement (dash dot light green line) in photocurrent density across a large potential range when the hole scavenger was used. If the performance enhancement of $CuWO_4$ due to photocharging was purely due to an improvement in catalytic efficiency or suppression of surface recombination losses, the performance of UT- and PC-$CuWO_4$ with hole scavenger would have been similar. Therefore, the improve-ment in performance of PC-$CuWO_4$ compared to UT-$CuWO_4$ in the presence of a hole scavenger suggests that there is a "bulk" component in the performance enhancement due to the photocharging effect, similar to what was found for $BiVO_4$.[9]

The improvement in performance of the PC-$CuWO_4$ was also seen to be reversible under dark conditions. Herein, the sample was first photocharged for 15 hours and then left in the electrolyte under dark OCP conditions for 5 hours. The performance of this "discharged" sample was further evaluated with another voltammetric scan, as seen in Fig. 1(b) (dashed blue line). The sample has indeed discharged, as evident from the decrease in the measured photocurrent. However, the performance of this decayed sample was still better than the untreated $CuWO_4$ sample (solid blue line), suggesting a slow decay process or a permanent modification of the $CuWO_4$ sample. This decaying of the photocharging effect was also seen in $BiVO_4$ by Trześniewski et al.,[11] albeit at a much faster rate at a similar pH, suggesting that there are some similarities and some differences in the mechanism(s) for the photocharging effect in these two materials. The decayed $CuWO_4$ samples could also be re-photocharged in a similar fashion (dashed green line).

To try and understand if this "bulk" component in performance enhancement after photocharging is related to the improvement in optical characteristics of $CuWO_4$, the absorption spectra of $CuWO_4$ before and after photocharging were measured. As seen in Fig. 2(a), the as-deposited $CuWO_4$ samples had an absorption onset of ~550 nm, with a slow increase in the absorption between 550 nm and 420 nm. This characteristic is attributed to an indirect bandgap within $CuWO_4$, which is also seen with some other metal oxide semiconductors.[22] The other characteristic feature in this figure is the absorption above 700 nm. This is attributed to the d–d transitions in copper due to the Jahn–Teller distortion in the copper octahedra (CuO_6) when copper exists in the Cu^{2+} state, as in the case of $CuWO_4$.[23,24] The absorption spectra of the PC-$CuWO_4$ was also measured in a similar fashion and the absorption within the bandgap was seen to be unchanged. Interestingly, the absorption close to the bandgap onset and above the bandgap decreased after the photocharging treatment. This above bandgap absorption could be a result of the combination of multiple factors. One possibility is the change in the absorption onset of the d–d transitions in the Cu 3d orbital. Another is the stray reflection of the incoming light beam during the transmittance measurements, from the complex morphology of the spray deposited $CuWO_4$, while the change in the absorption from any mid-bandgap states originating from the defects within the material may also play a role. To see if this decrease in absorption is related to a change in the optical properties of the material due to the photocharging effect, a Tauc plot was constructed from the absorption data collected before and after photocharging. As seen in Fig. 2(b), the bandgap of the as-deposited and photocharged $CuWO_4$ remained the same at 2.25 eV, similar to other reports in the literature.[13,17]

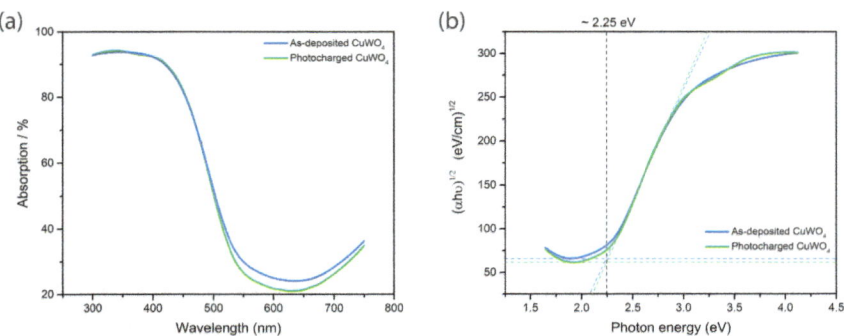

Fig. 2 (a) Absorption spectra of as-deposited and photocharged $CuWO_4$. (b) Tauc plot constructed from the absorption data for the as-deposited and photocharged $CuWO_4$.

Incident photon to current efficiency (IPCE) measurements were performed on the UT- and PC-CuWO$_4$ to determine if this change in absorption at the onset and above the bandgap has altered the wavelength dependent quantum efficiency of CuWO$_4$ due to the photocharging treatment. Fig. 3(a) shows the IPCE of UT- and PC-CuWO$_4$ plotted against the wavelength of the incoming photons. Both the UT- and PC-CuWO$_4$ had a similar onset of the photocurrent, at ~550 nm, suggesting that the change in the absorption near the bandgap onset and above the bandgap does not contribute to the improved performance due to photocharging. The quantum efficiency within the bandgap improved significantly post photo-charging. To understand this improvement, the IPCE data was corrected for the absorption to obtain the absorbed photon to current efficiency (APCE). The APCE reflects the losses due to recombination of the photo-generated carriers within the semiconductor. The APCE, in Fig. 3(b), also shows a similar enhancement due to photocharging of CuWO$_4$, suggesting that the photocharging treatment has resulted in the suppression of recombination losses within the semiconductor.

To investigate if the improvement in performance due to photocharging is a result of some photo-induced morphology change, as suggested in some reports,[7,8] the morphology of UT- and PC-CuWO$_4$ was analysed using scanning electron microscopy (SEM). Fig. 4(a) and (b) show the surface morphologies of the UT-CuWO$_4$ and PC-CuWO$_4$, respectively. The UT-CuWO$_4$ samples were left in the electrolyte under open-circuit conditions for 16 hours without any illumination, whereas the PC-CuWO$_4$ samples were illuminated under the same conditions. A nanoporous surface with an extensive microstructure network is clearly visible in both these images, indicating that the surface has remained mostly intact after the photocharging treatment. This is also shown in Fig. S2(a and b).† To further confirm if the photocharging treatment has led to the formation of any additional active sites at the surface, the electrochemical active surface area (ECSA) of the UT- and PC-CuWO$_4$ was determined by performing cyclic voltammetry scans in the dark, at a series of different scan rates for each of these samples. The double layer capacitance (C_{dl}), obtained from the slope of the current density *versus* scan rate plot, can be directly correlated to the ECSA. In Fig. 5(a), the similar C_{dl} values indicate that there is no major change in the number of active sites in the UT- and PC-CuWO$_4$ samples. The small change in the obtained C_{dl} values could be a result of the error margin in these fitting techniques.

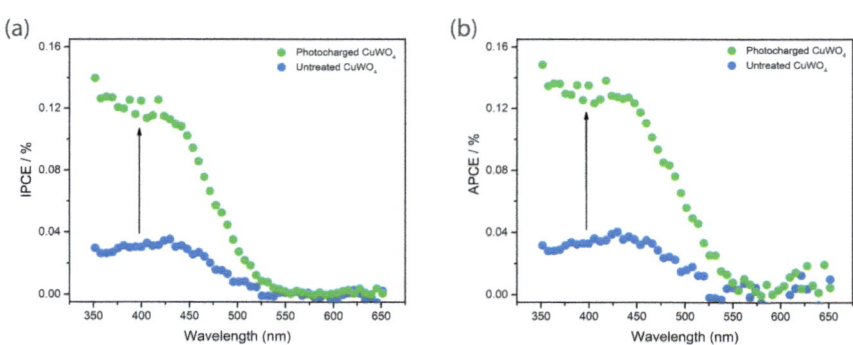

Fig. 3 (a) Incident photon to current efficiency (IPCE) of the UT- and PC-CuWO$_4$ substrates. (b) Absorbed photo to current efficiency (APCE) of the UT- and PC-CuWO$_4$.

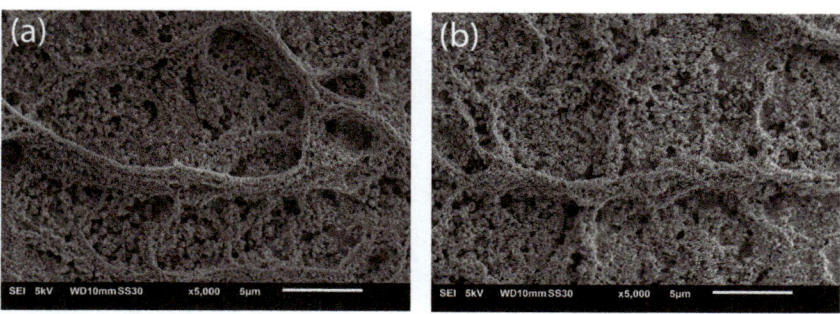

Fig. 4 (a) SEM image of UT-CuWO$_4$ sample. (b) SEM image of PC-CuWO$_4$ sample.

The bulk crystal structure of the UT- and PC-CuWO$_4$ were also analysed with the X-ray diffraction technique. Fig. 5(b) shows the XRD spectra for both the untreated and the photocharged samples. Sharp peaks corresponding to CuWO$_4$ (positions marked in red), agreeing with other reports in the literature,[13,17] indicate a very crystalline thin film of CuWO$_4$. More importantly, no shift in peak positions or a significant increase in the peak intensities were visible after the photocharging treatment, suggesting that the bulk structure has remained unchanged after the photocharging treatment. This is also consistent with the reports on photocharging of BiVO$_4$.[7,11]

To investigate if the chemical composition of the surface has been altered due to the photocharging treatment, the surface composition of the UT- and PC-CuWO$_4$ was analysed using X-ray photoelectron spectroscopy (XPS). Fig. 6(a–c) show the narrow spectra of the Cu 2p, O 1s and B 1s species. The W 4f narrow spectra is shown in Fig. S3(a).† Cu 2p peaks have a signature of four distinct features, when the Cu exists in the Cu^{2+} state.[25,26] The Cu^{2+} satellite peaks (as marked in the figure) are a result of the shake-up transitions in the valence band of copper, due to the photoejection of a core electron.[27] The characteristic Cu 2p$_{3/2}$ peaks for Cu(I)O, Cu(II)O and Cu(II) hydroxide are at 932.18 ± 0.12 eV, 933.76 ± 0.11 eV and 934.67 ± 0.02 eV, respectively[28] and their associated Cu 2p$_{5/2}$ peaks

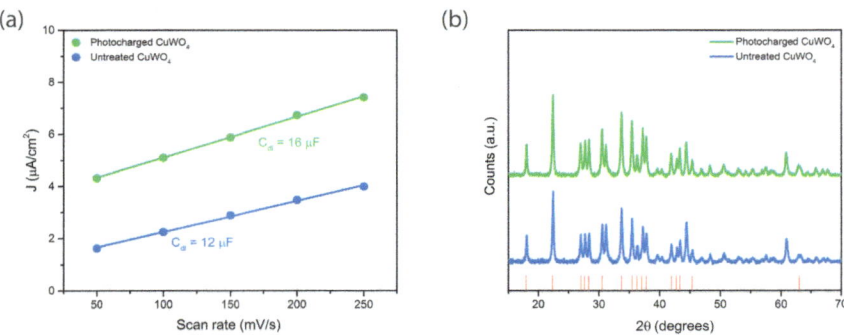

Fig. 5 (a) Electrochemically active surface area (ECSA) plot for UT- and PC-CuWO$_4$, obtained at 1.3 V$_{RHE}$ from performing cyclic voltammetry at different scan rates in the dark. (b) X-ray diffractograms of UT- and PC-CuWO$_4$.

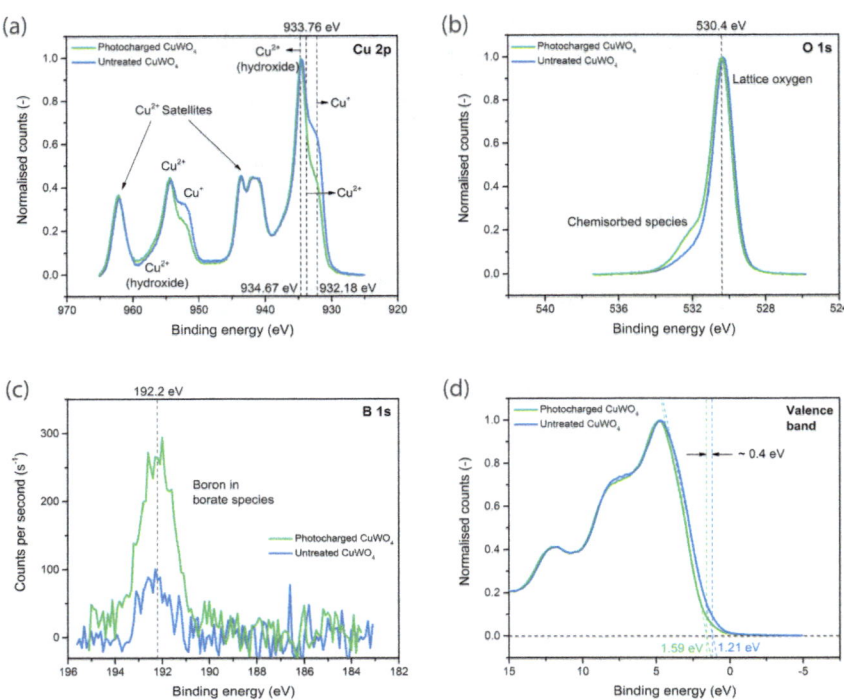

Fig. 6 XPS spectra of (a) Cu 2p, (b) O 1s, (c) B 1s and (d) valence band of UT- and PC-CuWO$_4$.

around 955 eV. The Cu 2p spectra in Fig. 6(a) consists of copper in Cu(I), Cu(II) and Cu(II) hydroxide forms for both the UT- and PC-CuWO$_4$, as evidenced by the complex spectra with peaks between 930 and 940 eV, resulting from the convolution of signals from copper in different forms. Interestingly, the peaks at ~932 eV and ~952 eV diminished after photocharging, which is attributed to Cu$^+$, possibly suggesting that the concentration of Cu$^+$ species at the surface has decreased near the semiconductor–electrolyte interface after photocharging. Similarly, the W 4f and O 1s spectra were also analysed for the UT- and PC-CuWO$_4$. The W 4f$_{7/2}$ has a characteristic peak at 35.8 ± 0.4 eV, when the tungsten is in the +6 oxidation state,[28] with the W 4f$_{5/2}$ doublet peak at ~37.9 ± 0.4 eV. A W 5p$_{3/2}$ peak is also normally visible at ~40.8 eV.[28] The associated lattice oxygen (O 1s) would have a peak at 530.5 ± 0.2 eV.[28] The O 1s peaks could also have a shoulder at higher binding energies. This shoulder peak at ~532 eV is a characteristic of oxygen in the chemisorbed species at the surface.[6,10] All these characteristic peaks for W 4f and O 1s are visible in Fig. S3(a)† and 6(b) for both UT- and PC-CuWO$_4$. The shoulder peak in the O 1s spectra has become more prominent post photocharging, possibly due to the build-up of oxygen in some chemisorbed species during photocharging. Both the O 1s and W 4f spectra also seem to have a minor shift of ~0.1 eV to higher binding energies post photocharging. B 1s spectra was also recorded for both UT- and PC-CuWO$_4$. B 1s has a peak at ~192.2 eV for boron in a borate species.[29,30] This peak is clearly visible in Fig. 6(c) for the PC-CuWO$_4$ sample and less so for the UT-CuWO$_4$. A noisy spectrum even after performing 300 scans indicates that the concentration of this boron species at the surface is low. This data was presented without normalisation due to this low signal to noise ratio. The valence band spectra of the UT- and PC-CuWO$_4$ were measured in a similar fashion, shown in Fig. 6(d). The spectra of the UT- and PC-CuWO$_4$ seem to be similar, except for a noticeable shift of valence band maximum (VBM) to higher binding energies (by ~0.4 eV) for the photocharged sample. The complete valence band spectra is shown in Fig. S3(b).†

To determine if the band edge positions have been altered due to photocharging, Mott–Schottky plots were gathered for the UT- and PC-CuWO$_4$ samples. Additionally, the Mott–Schottky plots were also obtained for the PC-CuWO$_4$ after discharging in the dark OCP conditions, as shown in Fig. 7. The Mott–Schottky plots can be used to determine the flat band potential and the charge carrier densities of CuWO$_4$. The flatband potential can be estimated from the x-intercept of the linear fit to the C_{sc}^{-2} vs. potential plot, while the slope of this linear fit is inversely proportional to the charge carrier density. All three, the UT- and PC-CuWO$_4$ and the discharged sample, had a similar x-intercept of ~0.5 V$_{RHE}$ and similar slopes. This flatband potential value agrees with other reports on spray deposited CuWO$_4$.[14] The small difference, <50 mV, in the x-intercept values could be a factor of the accuracy of the Mott–Schottky method and would fall within the error margin.

Discussion

The photocharging treatment has clearly improved the performance of CuWO$_4$ post photocharging. This was confirmed by both the cyclic voltammetry scans and the IPCE experiments for the UT- and PC-CuWO$_4$. The catalytic efficiency of UT- and PC-CuWO$_4$ was calculated according to the method suggested by Dotan et al.[31] A clear bump in the low potential region of the catalytic efficiency plot (Fig. S4(a)†) confirms the de-intercalation of H$^+$ and Na$^+$ from the CuWO$_4$ lattice,[13,20] as suggested before. The hole scavenger experiments suggested

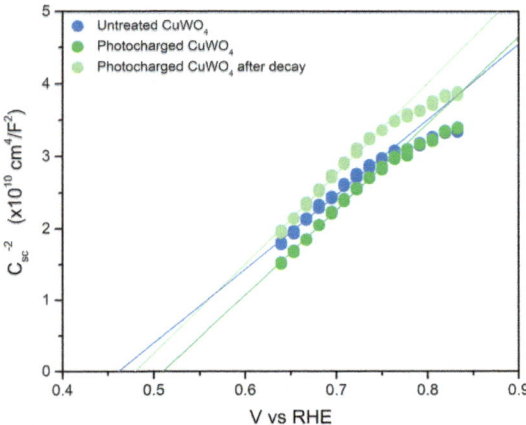

Fig. 7 Mott–Schottky plots of UT-, PC- and discharged CuWO$_4$ samples obtained in the fixed frequency mode in the dark.

a "bulk" enhancement factor from the photocharging effect, together with any surface enhancement. This bulk enhancement post photocharging could be a result of multiple factors; (1) improved light absorption, (2) morphological changes, or (3) improved charge separation in the space charge region. A series of characterisation techniques were performed to better understand this potential bulk enhancement. Absorption spectra of the UT- and PC-$CuWO_4$ showed no major difference within the bandgap. This was confirmed from the Tauc plot which showed a bandgap of ~2.25 eV for both the UT- and PC-$CuWO_4$. In fact, the IPCE plots also had a similar onset for both the UT- and PC-$CuWO_4$. SEM images of the UT- and PC-$CuWO_4$ also didn't show any major difference, with both the micro and nanostructures remaining intact after the photocharging treatment. This was also confirmed by the ECSA plots, which provided similar double layer capacitance values for both the UT- and PC-$CuWO_4$. The XRD data also revealed that the bulk of the $CuWO_4$ remained unchanged post photocharging. The information obtained from the absorption spectra, IPCE plots, SEM images, ECSA experiments and XRD patterns suggest that the bulk enhancement is not a result of any morphological change or improved light absorption due to the photo-charging effect, rather, it could be due to an improved charge separation in the space charge region post photocharging. To verify this, the charge separation efficiency was also calculated as per the method recommended by Dotan et al.,[31] as in Fig. S4(b).† This plot confirms the increase in charge separation efficiency post photocharging.

The chemical composition of the surface of UT- and PC-$CuWO_4$ was analysed using XPS. The data suggested that the surface composition of copper has changed after photocharging. Specifically, there was a decrease in the concentration of Cu^+ at the surface for the PC-$CuWO_4$ samples. Copper predominantly exists in the +2 oxidation state in the bulk lattice of $CuWO_4$. But the surface of such semiconductors could have copper in the +1 oxidation state due to the abrupt termination of the surface. These cations could then act as electron or hole traps at the surface, decreasing the photocurrent obtained from the material. Passivation layers are normally used to suppress such recombination losses at the surface. The decrease in the concentration of Cu^+ at the surface, coupled with the increase in photocurrent after the photocharging treatment, could mean that there is some kind of chemical passivation of these electron traps due to the photocharging effect. In fact, Favaro et al. used in situ HAXPES on $BiVO_4$ and confirmed the light induced formation of a chemical complex between a metal cation within the semiconductor and a solution anion species, which resulted in the "light induced surface passivation".[6] They reported that the formation of this chemical complex is reversible under dark conditions. Similarly, the photo-charging treatment in the case of $CuWO_4$, as in this work, could also form such a complex between the copper atoms at the semiconductor surface and borate or perchlorate ions in the solution. Perchlorate ions in solution are normally considered to be "inert" and are less interactive towards an electrochemical surface compared with a borate anion. Therefore, chances are that the photo-charging treatment could induce the formation of a copper borate complex at the surface, similar to the formation of a $BiPO_4$ complex as in the report of Favaro et al. The B 1s XPS spectra with a peak at ~192.2 eV for the PC-$CuWO_4$ sample, which corresponds to the spectra of boron in a borate form,[29,30] confirms the presence of a borate species at the surface. The O 1s spectra and W 4f spectra did not show a major difference, except for the increase in the peak at 532 eV for the O 1s spectra. As mentioned earlier, this peak is normally attributed to oxygen from chemisorbed species at the surface of the semiconductor. Therefore, this O 1s peak at 532 eV could be due to the oxygen in the borate species at the surface, formed as a result of the photocharging treatment. Similar observations were also made by Favaro et al.[6] and Trześniewski et al.[10] in the case of $BiVO_4$. The valence band XPS spectra for the UT- and PC-$CuWO_4$ also looked similar, expect for a significant shift in the valence band maximum (VBM) by ~0.4 eV. Interestingly though, the Mott–Schottky plots suggests that the CB band position has not changed and is fixed at the 0.5 V_{RHE} post photocharging. The valence band (VB) of $CuWO_4$ is composed of Cu(3d) and O(2p) orbitals.[17,32] There are conflicting reports in the literature about the composition of the conduction band in $CuWO_4$. Some reports suggest that the conduction band (CB) is composed predominantly of W(5d),[17] whereas other reports suggest that it is composed of W(5d) and Cu(3d).[32] It is important to point out here that the penetration depth of a commercial XPS system is ~5 nm from the surface and thus the data represents the chemical composition information of these first five nanometers. Combining the information obtained from the Tauc plot, XRD, Mott–Schottky plot and the XPS data suggests that a heterojunction of two semiconductors could have formed close to the surface, due to the photocharging treatment. The anodic shift of the VBM by ~0.4 eV indicates that a semiconductor with a higher bandgap may have formed at the surface due to the photocharging treatment. Therefore, this heterojunction is composed predominantly of the UT-$CuWO_4$ and a higher bandgap semiconductor at the surface. The unchanged bulk data, as observed in the absorption spectra and XRD data of the UT- and PC-$CuWO_4$ suggest that the thickness of the higher bandgap semiconductor is extremely small, possibly smaller than 5 nm. The origin of this higher bandgap semiconductor is still unclear. The presence of a borate species at the surface post photocharging, as seen in the B 1s XPS spectra, would suggest that this high bandgap semiconductor might be borate-based.

There are reports in the literature about the existence of CuB_2O_4 and $Cu_3B_2O_6$ photocatalysts, both semiconductors with bandgaps of ~3.1 eV and ~2.16 eV, respectively.[29] Therefore, a heterojunction could have formed between the $CuWO_4$ and either of these two semiconductors, close to the surface, due to the photocharging treatment. The improved charge separation in the space charge region, post photocharging, would imply that the band bending at this heterojunction should be favourable for taking the electrons away from the surface and/or drive holes to the SEI for water oxidation. Unfortunately, there are not many reports in the literature on these two copper borate species. Especially, there is no information about the absolute band edge positions and the intrinsic Fermi level positions of these two semiconductors in the literature, to our knowledge. Therefore, further research is needed to identify the exact composition of this photo-induced copper borate complex. The data obtained from the Mott–Schottky plots cannot be used anymore to determine the CB position because of the existence of this heterojunction. Fig. 8 shows an artistic representation of the authors' view of the heterojunction formed, between $CuWO_4$ and copper borate, due to the photocharging process. The improved band bending resulting from this heterojunction improves the charge separation and minimises charge carrier recombination. The formation of this copper complex could be initialised by the

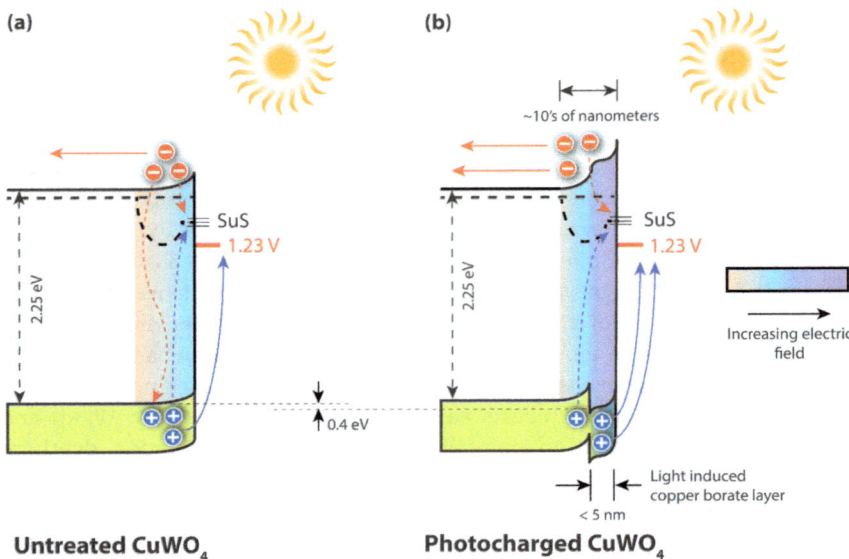

Untreated CuWO$_4$ **Photocharged CuWO$_4$**

Fig. 8 (a) Band diagram representing the band bending in the space charge region for UT-CuWO$_4$. (b) A speculative band diagram showing the additional band bending in the space charge region post photocharging, because of the formation of a heterojunction between CuWO$_4$ and the copper borate complex.

photo-corrosion of CuWO$_4$, similar to the suggestions of Favaro *et al.* on BiVO$_4$.[6] An attempt was made to perform inductively coupled plasma-optical emission spectroscopy (ICP-OES) on the electrolyte samples, post photocharging, to see if any copper or tungsten has leached into the solution in the process of photo-charging (Table S1†). However, the data was too erroneous as the concentrations were too low and hence was too close or below the detection limit. High background noise, especially from the sodium ions in the solution, also made the measurements erroneous. The open-circuit potential measurements (Fig. S5†) during the photocharging process suggests that the formation of this complex reaches an equilibrium, after a few hours of photocharging, as seen from the plateauing of the OCP after a few hours. A similar observation was also made in the case of BiVO$_4$.[10,11] Thus, this film probably does not grow beyond a few nanometers in thickness.

Conclusion

In this work, the effect of photocharging on CuWO$_4$ photoanodes is presented. Prolonged exposure of CuWO$_4$ photoanodes to light under OCP conditions improved the performance of the photoanodes and doubled the photocurrent at 1.23 V$_{RHE}$. This enhancement was seen to be reversible under dark conditions, but at a much slower rate than reports on photocharging of BiVO$_4$.[11] Photo-electrochemical experiments with a hole scavenger suggest that the enhancement in performance is facilitated by a "bulk" factor. Thorough material characterisation suggests that the bulk optical and material properties have not changed post photocharging, and neither does the morphology. It is very striking that all these features are very similar to the reports on photocharging of $BiVO_4$.[10,11]

XPS data suggested a change in the surface chemical composition post photocharging. There was a decrease in the concentration of Cu^+ at the surface after photocharging. Initial indications suggests that this could be due to chemical passivation of trap sites at the surface. An *in situ* HAXPES study on the photocharging of $BiVO_4$ indicated a light-induced formation of $BiPO_4$ layer at the surface of $BiVO_4$ during photocharging in a phosphate buffer.[6] In this article, we speculate on the build-up of a copper borate complex in a similar fashion at the $CuWO_4$ surface due to photocharging. This results in the formation of a heterojunction between $CuWO_4$ and the copper borate complex close to the surface. This proposal is supported by the presence of borate peak in the B 1s spectra for the PC-$CuWO_4$, additional buildup of a peak at 532 eV in the O 1s spectra and a ~0.4 eV anodic shift in the valence band maximum. Copper borate exists in two different forms, CuB_2O_4 and $Cu_3B_2O_6$. Both are proven photocatalysts with a bandgap of ~3.1 eV and 2.16 eV, respectively.[29] Our data suggests that this copper borate layer is extremely thin and does not grow beyond a few nanometers in thickness. The heterojunction formed by these two semiconductors leads to an improved charge separation in the space charge region, close to the surface and thus an improvement in the photocurrent post photocharging. More research is needed to determine the exact composition of this copper borate complex.

This article, along with other reports on photocharging of metal oxide semiconductors ($BiVO_4$, WO_3, Fe_2O_3 (ref. 6–12)), suggests that the metal oxide semiconductor–electrolyte interface is more complex than normally perceived. The effect of this time-dependent light-induced surface layer formation with the electrolyte species at the metal oxide semiconductor surface should be considered in all future works in the field of photo-electrochemistry with metal oxide semiconductors.

Conflicts of interest

There are no conflicts to declare.

Acknowledgements

This work is supported by the NWO VIDI grant awarded to Dr Wilson A. Smith. The authors would like to thank Dr Ibadillah A. Digdaya for having fruitful discussions on the topic, and Kailun Yang and Sanjana Chandrashekar for their help in obtaining the SEM images and for the XRD measurements, respectively.

References

1 An IPCC special report on the impacts of global warming of 1.5 °C above pre-industrial levels and related global greenhouse gas emission pathways, in the context of strengthening the global response to the threat of climate change, sustainable development, and efforts to eradicate poverty, in *2018: Global warming of 1.5 °C*, ed. V. Masson-Delmotte PZ, H. O. Pörtner, D. Roberts, J. Skea, P.R. Shukla, A. Pirani, W. Moufouma-Okia, C. Péan, R. Pidcock, S.

Connors, J. B. R. Matthews, Y. Chen, X. Zhou, M. I. Gomis, E. Lonnoy, T. Maycock, M. Tignor and T. Waterfield, IPCC, 2018.

2 B. J. Sixto Giménez, *Photoelectrochemical solar fuel production*, Springer, Switzerland, 2016.

3 D. K. Zhong, S. Choi and D. R. Gamelin, Near-complete suppression of surface recombination in solar photoelectrolysis by "Co-Pi" catalyst-modified W:BiVO$_4$, *J. Am. Chem. Soc.*, 2011, **133**(45), 18370–18377.

4 T. W. Kim and K.-S. Choi, Nanoporous BiVO$_4$ photoanodes with dual-layer oxygen evolution catalysts for solar water splitting, *Science*, 2014, 1245026.

5 L. Trotochaud, T. J. Mills and S. W. Boettcher, An Optocatalytic Model for Semiconductor–Catalyst Water-Splitting Photoelectrodes Based on *In Situ* Optical Measurements on Operational Catalysts, *J. Phys. Chem. Lett.*, 2013, **4**(6), 931–935.

6 M. Favaro, F. F. Abdi, M. Lamers, E. J. Crumlin, Z. Liu, R. van de Krol, *et al.*, Light-Induced Surface Reactions at the Bismuth Vanadate/Potassium Phosphate Interface, *J. Phys. Chem. B*, 2018, **122**(2), 801–809.

7 T. Li, J. He, B. Pena and C. P. Berlinguette, Curing BiVO$_4$ Photoanodes with Ultraviolet Light Enhances Photoelectrocatalysis, *Angew. Chem., Int. Ed. Engl.*, 2016, **55**(5), 1769–1772.

8 T. Li, J. He, B. Pena and C. P. Berlinguette, Exposure of WO$_3$ Photoanodes to Ultraviolet Light Enhances Photoelectrochemical Water Oxidation, *ACS Appl. Mater. Interfaces*, 2016, **8**(38), 25010–25013.

9 E. Y. Liu, J. E. Thorne, Y. He and D. Wang, Understanding Photocharging Effects on Bismuth Vanadate, *ACS Appl. Mater. Interfaces*, 2017, **9**(27), 22083–22087.

10 B. J. Trześniewski, I. A. Digdaya, T. Nagaki, S. Ravishankar, I. Herraiz-Cardona, D. A. Vermaas, *et al.*, Near-complete suppression of surface losses and total internal quantum efficiency in BiVO$_4$ photoanodes, *Energy Environ. Sci.*, 2017, **10**(6), 1517–1529.

11 B. J. Trześniewski and W. A. Smith, Photocharged BiVO$_4$ photoanodes for improved solar water splitting, *J. Mater. Chem. A*, 2016, **4**(8), 2919–2926.

12 J. Xie, P. Yang, X. Liang and J. Xiong, Self-Improvement of Ti:Fe$_2$O$_3$ Photoanodes: Photoelectrocatalysis Improvement after Long-Term Stability Testing in Alkaline Electrolyte, *ACS Appl. Energy Mater.*, 2018, **1**(6), 2769–2775.

13 D. Bohra and W. A. Smith, Improved charge separation *via* Fe-doping of copper tungstate photoanodes, *Phys. Chem. Chem. Phys.*, 2015, **17**(15), 9857–9866.

14 Y. Gao and T. W. Hamann, Elucidation of CuWO$_4$ Surface States During Photoelectrochemical Water Oxidation, *J. Phys. Chem. Lett.*, 2017, **8**(12), 2700–2704.

15 C. R. Lhermitte and B. M. Bartlett, Advancing the Chemistry of CuWO$_4$ for Photoelectrochemical Water Oxidation, *Acc. Chem. Res.*, 2016, **49**(6), 1121–1129.

16 M. Valenti, D. Dolat, G. Biskos, A. Schmidt-Ott and W. A. Smith, Enhancement of the Photoelectrochemical Performance of CuWO$_4$ Thin Films for Solar Water Splitting by Plasmonic Nanoparticle Functionalization, *J. Phys. Chem. C*, 2015, **119**(4), 2096–2104.

17 J. E. Yourey and B. M. Bartlett, Electrochemical deposition and photoelectrochemistry of CuWO$_4$, a promising photoanode for water oxidation, *J. Mater. Chem.*, 2011, **21**(21), 7651.

18 J. E. Yourey, K. J. Pyper, J. B. Kurtz and B. M. Bartlett, Chemical Stability of $CuWO_4$ for Photoelectrochemical Water Oxidation, *J. Phys. Chem. C*, 2013, **117**(17), 8708–8718.

19 S. H. Baeck, K. S. Choi, T. F. Jaramillo, G. D. Stucky and E. W. McFarland, Enhancement of Photocatalytic and Electrochromic Properties of Electrochemically Fabricated Mesoporous WO_3 Thin Films, *Adv. Mater.*, 2003, **15**(15), 1269–1273.

20 W. Smith, A. Wolcott, R. C. Fitzmorris, J. Z. Zhang and Y. Zhao, Quasi-core-shell TiO_2/WO_3 and WO_3/TiO_2 nanorod arrays fabricated by glancing angle deposition for solar water splitting, *J. Mater. Chem.*, 2011, **21**(29), 10792.

21 M. Valenti, A. Venugopal, D. Tordera, M. P. Jonsson, G. Biskos, A. Schmidt-Ott, *et al.*, Hot Carrier Generation and Extraction of Plasmonic Alloy Nanoparticles, *ACS Photonics*, 2017, **4**(5), 1146–1152.

22 J. K. Cooper, S. Gul, F. M. Toma, L. Chen, Y.-S. Liu, J. Guo, *et al.*, Indirect Bandgap and Optical Properties of Monoclinic Bismuth Vanadate, *J. Phys. Chem. C*, 2015, **119**(6), 2969–2974.

23 J. Ruiz-Fuertes, M. N. Sanz-Ortiz, J. González, F. Rodríguez, A. Segura and D. Errandonea, Optical absorption and Raman spectroscopy of $CuWO_4$, *J. Phys.: Conf. Ser.*, 2010, **215**, 012048.

24 J. E. Yourey, J. B. Kurtz and B. M. Bartlett, Water Oxidation on a $CuWO_4$–WO_3 Composite Electrode in the Presence of $[Fe(CN)_6]_3^{-}$: Toward Solar Z-Scheme Water Splitting at Zero Bias, *J. Phys. Chem. C*, 2012, **116**(4), 3200–3205.

25 O. Y. Khyzhun, T. Strunskus, S. Cramm and Y. M. Solonin, Electronic structure of $CuWO_4$: XPS, XES and NEXAFS studies, *J. Alloys Compd.*, 2005, **389**(1–2), 14–20.

26 Z. Lin, W. Li and G. Yang, Hydrogen-interstitial $CuWO_4$ nanomesh: a single-component full spectrum-active photocatalyst for hydrogen evolution, *Appl. Catal., B*, 2018, **227**, 35–43.

27 M. A. Brisk and A. Baker, Shake-up satellites in X-ray photoelectron spectroscopy, *J. Electron Spectrosc. Relat. Phenom.*, 1975, **7**(3), 197–213.

28 *NIST X-ray Photoelectron Spectroscopy Database [Internet]*, National Institute of Standards and Technology, 2000.

29 J. Liu, S. Wen, X. Zou, F. Zuo, G. J. O. Beran and P. Feng, Visible-light-responsive copper(II) borate photocatalysts with intrinsic midgap states for water splitting, *J. Mater. Chem. A*, 2013, **1**(5), 1553–1556.

30 C. Ong, H. Huang, B. Zheng, R. Kwok, Y. Hui and W. Lau, X-ray photoemission spectroscopy of nonmetallic materials: electronic structures of boron and B_xO_y, *J. Appl. Phys.*, 2004, **95**(7), 3527–3534.

31 H. Dotan, K. Sivula, M. Grätzel, A. Rothschild and S. C. Warren, Probing the photoelectrochemical properties of hematite (α-Fe_2O_3) electrodes using hydrogen peroxide as a hole scavenger, *Energy Environ. Sci.*, 2011, **4**(3), 958–964.

32 M. Lalić, Z. Popović and F. Vukajlović, Electronic structure and optical properties of $CuWO_4$: an *ab initio* study, *Comput. Mater. Sci.*, 2012, **63**, 163–167.

PAPER

Mechanistic insights into C2 and C3 product generation using Ni₃Al and Ni₃Ga electrocatalysts for CO₂ reduction†

Aubrey R. Paris (ID)* and Andrew B. Bocarsly (ID)

Received 15th November 2018, Accepted 5th December 2018

DOI: 10.1039/c8fd00177d

Thin films of Ni₃Al and Ni₃Ga on carbon solid supports have been shown to generate multi-carbon products in electrochemical CO₂ reduction, an activity profile that, until recently, was ascribed exclusively to Cu-based catalysts. This catalytic behavior has introduced questions regarding the role of each metal, as well as other system components, during CO₂ reduction. Here, the significance of electrode structure and solid support choice in determining higher- *versus* lower-order reduction products is explored, and the commonly invoked Fischer–Tropsch-type mechanism of CO₂ reduction to multi-carbon products is indirectly probed. Electrochemical studies of both intermetallic and non-mixed Ni–Group 13 catalyst films suggest that intermetallic character is required to achieve C2 and C3 products irrespective of carbon support choice, negating the possibility of separate metal sites performing distinct yet complementary roles in CO₂ reduction. Furthermore, Ni₃Al and Ni₃Ga were shown to be incapable of generating higher-order reduction products in D₂O, suggesting a departure from accepted mechanisms for CO₂ reduction on Cu. Additional routes to multi-carbon products may therefore be accessible when developing intermetallic catalysts for CO₂ electroreduction.

Introduction

As atmospheric CO_2 becomes an increasingly significant global challenge, the development of efficient and selective electrocatalysts active in CO_2 reduction draws continuous attention. One goal for these catalysts is to find ways to generate higher-order, highly reduced products, effectively constructing C–C bonds from CO_2 subunits. Unfortunately, catalysts capable of electrochemically transforming CO_2 into multi-carbon products are rare; aside from Cu, bimetallic species have shown the greatest promise in this area.[1-3]

Department of Chemistry, Princeton University, Princeton, New Jersey 08544, USA. E-mail: aparis@princeton.edu

† Electronic supplementary information (ESI) available. See DOI: 10.1039/c8fd00177d

Models[4-6] and even machine learning algorithms[7,8] have been implemented to guide the prediction of new bimetallic catalysts capable of reducing CO_2. Many of these calculations incorporate factors such as thermodynamic requirements for reduction of a CO intermediate, likelihood of surface restructuring, impact of exposed crystal faces, and scaling relations between the adsorption energies for CO and protonated reduction intermediates.[9] While these factors are important, other characteristics of both mono- and bimetallic systems, such as catalyst morphology,[10-13] spatial arrangement of the component metals,[14,15] and solid support material,[16] have been shown experimentally to impact product distribution, selectivity, and efficiency. Furthermore, an understanding of mechanisms employed by bimetallic systems—beyond the identification of a common CO intermediate—is lacking. Modeling, machine learning, and overall electrode design could be better informed by systematically exploring these catalytic factors, particularly for bimetallic species that are or may be capable of generating multi-carbon reduction products.

One class of CO_2-reducing bimetallic catalysts that benefits from such an analysis is comprised of Ni–Group 13 (i.e., Ni–G13) intermetallics. Thin films of Ni_3Al[3] and several Ni_aGa_b species (e.g., NiGa, Ni_3Ga, and Ni_5Ga_3)[1,16] supported on carbon have drawn interest due to their unusual ability to electrochemically reduce CO_2 to C2 and C3 products, including 1-propanol, acetone, ethane, and ethanol. While the multi-carbon nature of these products might suggest a mechanism similar to Cu-mediated CO_2 reduction, the true means by which Ni–G13 materials facilitate C–C bond formation is largely unknown. Questions remain regarding the role of each metal in catalysis, such as whether the electrocatalytic layer is composed of an intermetallic compound or two separate metal phases. Further questions include the impact of factors such as carbon support and whether a mechanism resembling Fischer–Tropsch chemistry—often suggested for Cu catalysts[17-19]—is at work. A better understanding of these factors in the context of Ni–G13 catalysts could aid in the prediction and development of additional non-Cu-containing electrocatalysts having the ability to promote C–C bond formation.

Here, we study Ni_3Al on glassy carbon (i.e., Ni_3Al/GC) and Ni_3Ga on highly oriented pyrolytic graphite (i.e., Ni_3Ga/HOPG) to determine the influence of electrode structure (i.e., intermetallic compounds versus non-mixed metals; choice of solid support) in achieving multi-carbon products during CO_2 electroreduction. Our results support the importance of intermetallic character in attaining the desired multi-carbon products, suggesting that the individual metal components do not facilitate separate, stepwise reduction events en route to C–C bonds. Furthermore, the likelihood of C2 and C3 product formation via a Fischer–Tropsch-like process is indirectly assessed via experiments altering the isotopic composition of the electrolyte. In the end, the data reported herein raise questions as to whether Ni–G13 electrocatalysts utilize this commonly invoked mechanism, indicating that another pathway may be possible for achieving Cu-like products on non-Cu-containing catalysts.

Experimental

Ni_3Al and Ni_3Ga were prepared on glassy carbon (GC), highly oriented pyrolytic graphite (HOPG), and reticulated vitreous carbon (RVC) solid supports as described previously *via* the thermal reduction of metal nitrate salts under forming gas atmosphere.[1,3,16] Non-mixed samples (*i.e.*, 3Ni–Al and 3Ni–Ga) were synthesized in an identical fashion, except the precursor metal nitrate solutions were not combined prior to drop-casting. Instead, thin strands of paraffin wax were wound tightly around the center of the carbon support to create a $\sim$3 mm-high physical barrier separating the support into halves; the Ni(II) and Al(III) or Ga(III) nitrate solutions were then drop-casted on separate sides of the barrier while maintaining an overall 3 : 1 ratio of metals. The paraffin was removed prior to furnace treatment to generate a surface interface where a Ni salt was juxtaposed with either an Al or Ga salt.

Materials characterization was performed to confirm that the Ni_3Al and Ni_3Ga intermetallic species were formed in the mixed salt samples, as well as to verify that the metals had remained isolated in the non-mixed samples. The catalyst material was analyzed by powder X-ray diffraction (XRD), following careful removal of the films from their carbon supports, using a Bruker D8 Advance diffractometer with 0.083° step size and CuKα radiation. Surface compositions of the intact electrodes were examined by X-ray photoelectron spectroscopy (XPS) using a Thermo Fisher K-Alpha instrument with a 20 eV pass energy and 50 ms dwell time. XPS spectra were analyzed with the Thermo Scientific Avantage Data System and CasaXPS software. All peaks were referenced to adventitious carbon at 284.8 eV. Scanning electron microscopy (SEM) images were obtained using a FEI XL30 FEG-SEM equipped with an EVEX EDS detector for acquisition of XRD-complementing energy-dispersive X-ray spectroscopy (EDX) data. SEM and EDX data were collected using a 5 keV electron beam and $\sim$12 mm working distance.

Electrolysis experiments were set up by fitting the working electrode, Ag/AgCl reference electrode, and Pt mesh counter electrode (within a gas dispersion tube) into the gas-tight ports of a custom electrochemical cell. As previously described,[3,16] the film-on-carbon working electrode was held using an alligator clip attached to Cu wire, which was threaded through a glass tube sealed with insulating epoxy. The electrolyte, stirred during experimentation, consisted of 0.1 M K_2SO_4 (in H_2O or D_2O, depending on the experiment) buffered with $KHCO_3$ to achieve the desired pH values following CO_2 saturation. Electrolyses were conducted using CH Instruments 760 and 1140 potentiostats until 50 coulombs of charge had passed, and measurements following these experiments failed to yield significant changes in solution pH.

Product analysis was accomplished by sampling the headspace and electrolyte by gas chromatography and ^{1}H-NMR spectroscopy, respectively. CO and ethane were detected using an HP6890 Gas Chromatograph with a Molsieve 5A PLOT capillary column (Agilent), thermal conductivity detector (TCD), and He flow gas. H_2 was detected using an SRI 8610C Gas Chromatograph with a Molsieve column, TCD, and Ar flow gas. Liquid products were analyzed using a Bruker Avance III 500 MHz NMR Spectrometer with a cryoprobe detector. A custom water suppression method permitted direct sampling of electrolyte solutions upon addition of 10 μL of 1,4-dioxane (10 mM) as an internal standard. $^{13}CO_2$ control electrolyses were performed for newly reported catalyst species to confirm product derivation from CO_2, indicated by either peak splitting in the ^{1}H-NMR or gas-phase infrared spectroscopy. H_2 quantification confirmed the charge balance in each experiment discussed herein.

Results and discussion

Whether a heterogeneous, bimetallic CO_2 reduction catalyst consists of an intermetallic species or, more simply, two metals existing in the same electrochemical cell is influenced by the catalytic role of each metal. For example, considering a binary metal catalyst A–B, if metal A reduces CO_2 to CO and metal B selectively reduces CO to a multi-carbon final product, then a heterogeneous mixture of A and B would be expected to perform the catalytic task, while an intermetallic or alloy phase, AB, would potentially effect a different chemistry and product distribution. By association, determining whether the function of a two-metal catalyst depends on alloying could lend insight into the mechanism of CO_2 reduction on bimetallic species. In the case of Ni_3Al/GC and $Ni_3Ga/HOPG$, it is instructive to know whether generation of multi-carbon products is predicated on the presence of an intermetallic compound, as well as what this might mean for the mechanism underlying these catalysts' unique activities.

To examine the importance of electrode structure for these thin film Ni–G13 catalysts, intermetallic and non-mixed electrodes were prepared using a consistent synthetic method based on drop-casting and subsequent heat treatment.[1,3,16] Synthesis of the non-mixed electrodes, referred to as 3Ni–Al/GC and 3Ni–Ga/HOPG, differed from that of the Ni_3Al/GC and $Ni_3Ga/HOPG$ intermetallic species in that the two constituent metals were isolated on the same carbon solid support. This prevented alloying during heat treatment yet permitted drop-casting of the same $3:1$, Ni : G13 stoichiometry featured in the intermetallic films.

The materials characterization of Ni_3Al/GC[3] and $Ni_3Ga/HOPG$,[1,16] thin films having homogeneous compositions, has been reported previously. As shown in Fig. S1,† the non-mixed 3Ni–Al/GC and 3Ni–Ga/HOPG electrodes were striped, as each metal component occupied a distinct portion of the carbon support. Surface characterization data for 3Ni–Al/GC and 3Ni–Ga/HOPG are provided in Fig. 1 and 2, respectively. X-ray photoelectron spectroscopy (XPS) suggests that, much like the parent intermetallics, the surfaces of the Ni, Al, and Ga stripes were either heavily or entirely oxidized. Furthermore, scanning electron microscopy (SEM) of the non-mixed metals shows that, on GC, both stripes were comprised of flat platelets, while the stripes on HOPG were rougher and less uniform in shape. This morphological distribution based on carbon solid support has been reported.[16] A complete summary of characterization data comparing the non-mixed 3Ni–Al/GC and 3Ni–Ga/HOPG electrodes with their parent intermetallics is provided in Table 1.

The electrocatalytic performance of 3Ni–Al/GC and 3Ni–Ga/HOPG was compared to that of Ni_3Al/GC and $Ni_3Ga/HOPG$ using CO_2 electrolysis experiments performed at -1.38 V $vs.$ Ag/AgCl, the potential previously shown to be optimal for CO_2 reduction mediated by the intermetallic species.[3,16] Electrolyses were conducted in two-compartment cells using CO_2-saturated, pH 4.5 K_2SO_4 (0.1 M) as the electrolyte and Pt mesh as the counter electrode. Fig. 3 shows the product distribution achieved using 3Ni–Al/GC and 3Ni–Ga/HOPG alongside that achieved using Ni_3Al/GC and $Ni_3Ga/HOPG$. While Ni_3Al/GC electrochemically reduced CO_2 to form a significant quantity of CO alongside 1-propanol, methanol, and even small amounts of ethanol and acetone,[3] the non-mixed 3Ni–Al/GC

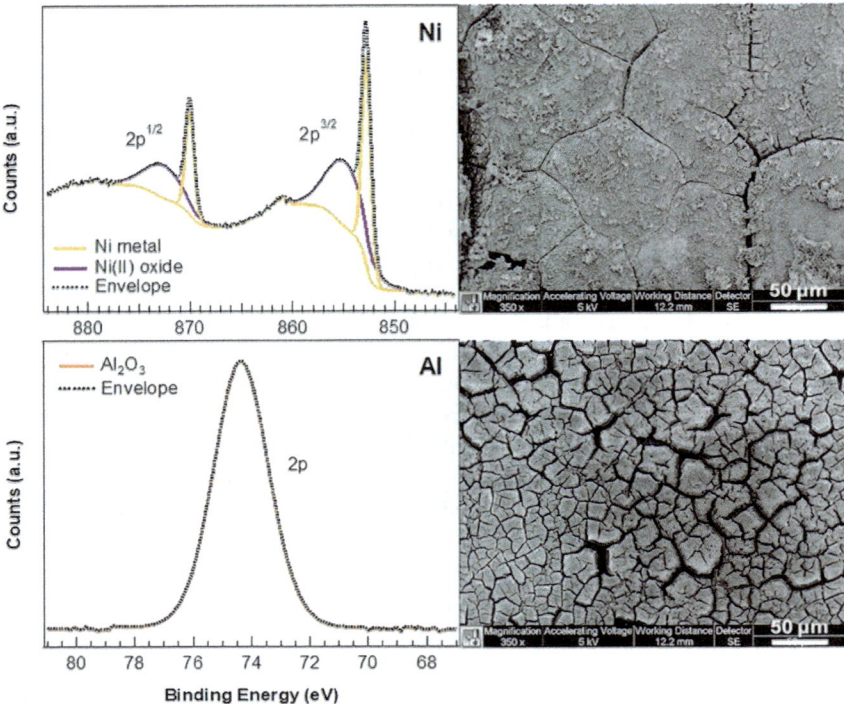

Fig. 1 Surface characterization of 3Ni–Al/GC. XPS data (left) indicate that the surfaces of both the Ni and Al components are heavily oxidized. SEM imaging (right) suggests that both of the metal films on the GC are comprised of relatively flat platelets. SEM images were acquired using a 5 keV electron beam.

electrode only generated small quantities of CO and formate. Similarly, Ni_3Ga/HOPG's small faradaic efficiency for ethane[16] was supplanted by an even smaller quantity of formate when using 3Ni–Ga/HOPG. H_2 evolution, which competes with CO_2 reduction in aqueous electrolyte, accounted for the remaining charge, reaching faradaic efficiencies of approximately 94% and 98% for 3Ni–Al/GC and 3Ni–Ga/HOPG, respectively (Fig. S2†). In all of the subsequent experiments discussed here, charge balance was achieved by quantifying the H_2 produced.

Aside from indicating that, for these Ni-G13 catalysts, intermetallic character is required to achieve higher-order products from CO_2, these electrolysis experiments begin to answer a standing question about how Ni–G13 species facilitate CO_2 reduction. If the Ni and G13 metals had separate roles—facilitating discrete tasks in the stepwise reduction of CO_2 to multi-carbon species—one would have expected the multi-carbon products to be generated even without alloying of the metals, since both metals would have been available to fulfill their respective roles. This assertion is further supported by experiments in which Ni/GC, Al/GC, Ni/HOPG, and Ga/HOPG were synthesized separately, mixed to homogeneity in appropriate 3 : 1 ratios post-furnace treatment, and reapplied to either GC or HOPG prior to electrochemical testing. These scenarios, in which the two metals were in direct, adjacent contact yet were not alloyed, still failed to yield higher-order products. Finally, while surface oxides are consistently present, Table 1

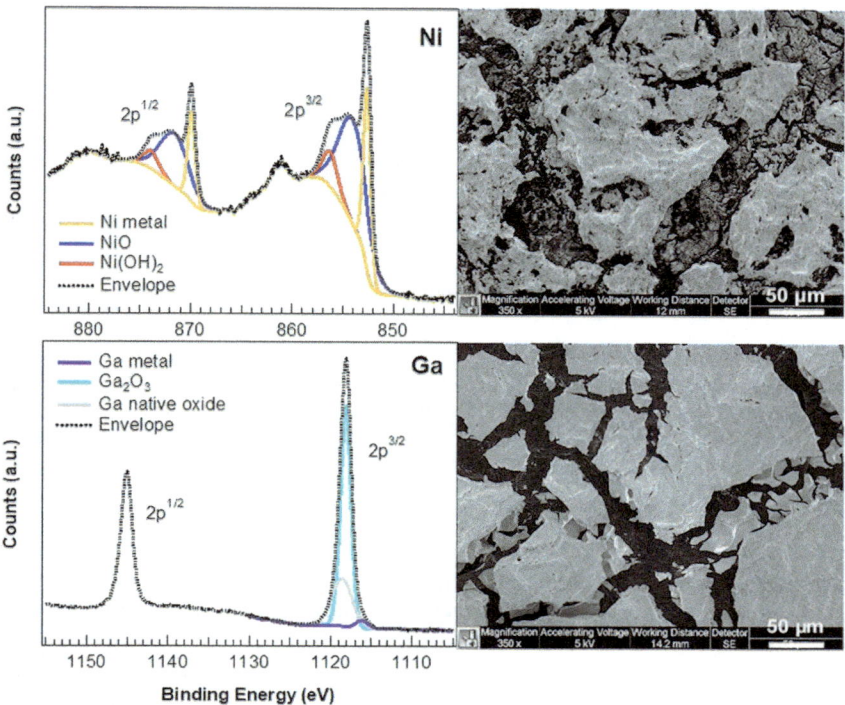

Fig. 2 Surface characterization of 3Ni–Ga/HOPG. XPS data (left) indicate that the surfaces of both the Ni and Ga components are heavily oxidized. SEM imaging (right) shows that the Ni and Ga films on HOPG are made up of rough, non-uniform particles. SEM images were acquired using a 5 keV electron beam.

shows that the surface compositions of the intermetallic and non-mixed electrodes are nearly identical, so variation in surface oxide composition in and of itself cannot explain the differences in catalytic activity. These observations lead to the conclusion that Ni and G13 metals interact synergistically, rather than simply performing complementary tasks, in the CO_2 electroreduction process.

We previously reported that the Ni_3Ga intermetallic system exhibits a solid support dependence during electrochemical CO_2 reduction. Namely, the choice of

Table 1 Summary of materials characterization data comparing Ni_3Al/GC and $Ni_3Ga/HOPG$ intermetallics with non-mixed 3Ni–Al/GC and 3Ni–Ga/HOPG. Bulk composition was determined by a combination of XRD and EDX, while surface composition was obtained from XPS analysis. Data for Ni_3Al/GC and $Ni_3Ga/HOPG$ were reported previously[3,16]

Catalyst	Bulk		Surface	
	Ni	G13	Ni	G13
Ni_3Al/GC		Ni_3Al (cubic)	Ni(II) oxides + Ni	Al_2O_3
3Ni–Al/GC	Ni (cubic)	Al_2O_3 (amorphous)	Ni(II) oxides + Ni	Al_2O_3
$Ni_3Ga/HOPG$		Ni_3Ga (cubic)	Ni(II) oxides + Ni	Ga_2O_3 + Ga
3Ni–Ga/HOPG	Ni (cubic)	Ga_2O_3 (monoclinic)	Ni(II) oxides + Ni	Ga_2O_3 + Ga

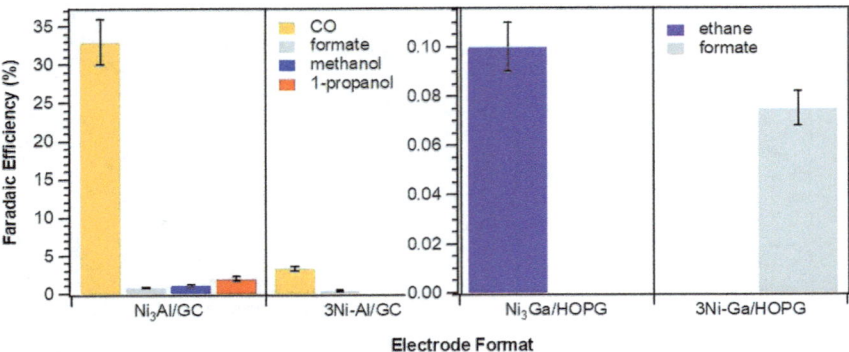

Fig. 3 Distribution of CO_2 reduction products achieved using intermetallic and non-mixed Ni–G13 catalysts on carbon solid supports. Higher-order products could only be generated in the presence of Ni_3Al or Ni_3Ga intermetallic compounds.[3,16] Electrolysis experiments were conducted at -1.38 V vs. Ag/AgCl in pH 4.5, CO_2-saturated K_2SO_4 electrolyte.

carbon support was shown to impact the product distribution and CO_2 reduction pathway, which was largely attributed to the surface character and morphology.[16] This phenomenon introduces the question of whether a solid support dependence might exist for non-mixed Ni–G13 electrocatalysts; one might hypothesize that the ability to generate higher-order products using non-mixed Ni–G13 electrodes might be reinstated by selecting an appropriate support material. Before examining this possibility using the Ni–G13 class of catalysts, it was first necessary to determine whether the product distribution obtained using the Ni_3Al intermetallic relies on solid support choice like the Ni_3Ga analog.

With its product distribution on GC well-characterized,[3] Ni_3Al was synthesized on HOPG and a third support material, reticulated vitreous carbon (RVC; structurally related to GC[20,21]), comprising the three electrode materials previously tested in the Ni_3Ga study. As shown in Fig. 4, XRD analysis indicates that the desired intermetallic compound was synthesized on both carbons, while the morphologies of Ni_3Al/HOPG and Ni_3Al/RVC were consistent with those achieved for Ni_3Ga on the same supports. Specifically, Ni_3Al/RVC was comprised of smooth, platelet structures distributed across the porous carbon framework, whereas Ni_3Al/HOPG's morphology resembled that of 3Ni–Ga/HOPG discussed earlier. XPS analysis of the intermetallic surfaces revealed a likeness to Ni_3Al/GC, as surface-confined Al was comprised of Al_2O_3 and Ni was predominantly—or, in the case of Ni_3Al/RVC, entirely—oxidized prior to CO_2 electrolysis (Fig. S3, S4†).

Electrolysis experiments with the Ni_3Al/RVC and Ni_3Al/HOPG electrodes were performed at -1.38 V vs. Ag/AgCl in pH 4.5, CO_2-saturated K_2SO_4 electrolyte. The results of these experiments, compared to the outcomes achieved using the GC-deposited species, are shown in Fig. 5. While differences in the product faradaic efficiency are recorded for each solid support, the same major reaction products (i.e., CO, 1-propanol, methanol, and formate) are observed in all cases, indicating no change in the product distribution. This result is distinctly different than what has been observed using Ni_3Ga on various carbons, which yielded ethane on HOPG and CO, formate, and trace amounts of methanol on GC and

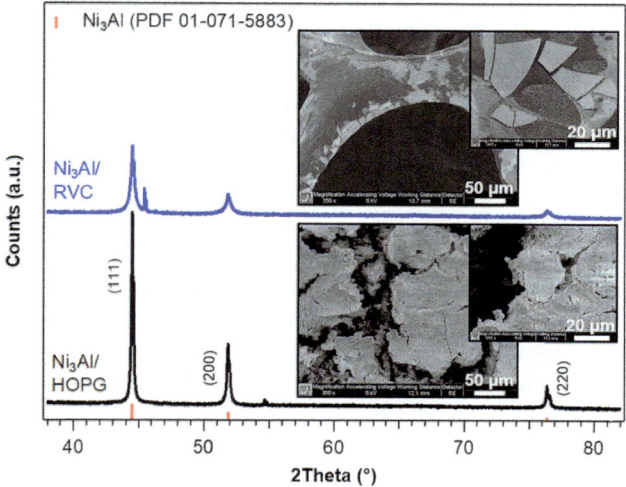

Fig. 4 XRD patterns confirm the synthesis of Ni_3Al on both the RVC (blue; top) and HOPG (black; bottom) solid supports. Unmarked peaks are attributed to carbon. Inset SEM images indicate smooth platelet and rough particle morphologies for the RVC- and HOPG-deposited Ni_3Al, respectively.

RVC.[16] Because Ni_3Al does not exhibit solid support dependence in the intermetallic structure, it was not examined further for this dependence in the non-mixed 3Ni–Al form.

However, given the influence of the solid support on the product distribution attained using Ni_3Ga, the non-mixed 3Ni–Ga analog was prepared and tested on both GC and RVC. This allows for comparison with the 3Ni–Ga/HOPG examined

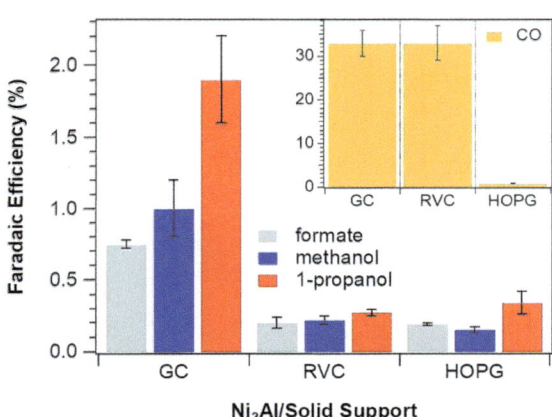

Fig. 5 Faradaic efficiencies for CO_2 reduction products achieved using Ni_3Al films on different carbon solid supports. While the relative efficiencies of products vary, the product distribution remains the same irrespective of support choice. Ni_3Al/GC values were reported previously.[3] Electrolysis experiments were conducted at −1.38 V vs. Ag/AgCl in pH 4.5, CO_2-saturated K_2SO_4 electrolyte.

earlier, as well as the intermetallic species Ni_3Ga/GC and Ni_3Ga/RVC discussed in our previously published work.[16] Complete materials characterization of Ni_3Ga/GC has already been reported, while additional details regarding Ni_3Ga/RVC, indicating similar bulk and surface compositions as the GC variant, are provided in Fig. S5.† Unsurprisingly, characterization revealed that the Ni half of 3Ni–Ga/GC was compositionally and morphologically identical to the Ni half of 3Ni–Al/GC depicted in Fig. 1. For completeness, XRD, XPS, and SEM data for the Ga half of 3Ni–Ga/GC and both halves of 3Ni–Ga/RVC are shown in Fig. S6 and S7.† As was the case for Ni on GC and HOPG, the Ni stripe on RVC was comprised of metallic Ni in the bulk, while the surface featured a mixture of metal and oxides. The Ga stripes on both GC and RVC consisted of monoclinic Ga_2O_3 and were predominantly oxidized at the surface, as well.

Fig. 6 gives the results of 3Ni–Ga/GC and 3Ni–Ga/RVC electrolysis experiments, performed at -1.38 V $vs.$ Ag/AgCl in pH 4.5, CO_2-saturated K_2SO_4, alongside the previously reported results for Ni_3Ga on these solid supports. 3Ni–Ga/GC and 3Ni–Ga/RVC both generated CO as their major product and small amounts of formate and methanol in the electrolyte, in accordance with the carbon-containing products achieved using their intermetallic counterparts. In short, altering the carbon used to support non-mixed Ni and Ga failed to reintroduce the ability to generate higher-order products, an outcome that remains reliant on intermetallic character.

Despite a lack of multi-carbon products, a noteworthy outcome achieved while examining the 3Ni–Ga/RVC electrodes is their high faradaic efficiency for CO ($\sim$52%). To further investigate this result, control electrolyses were performed using electrodes prepared by drop-casting either Ni or Ga onto each carbon support material; the resulting faradaic efficiencies for CO, formate, and methanol are listed in Table S1.† In these experiments, the greatest quantity of CO was

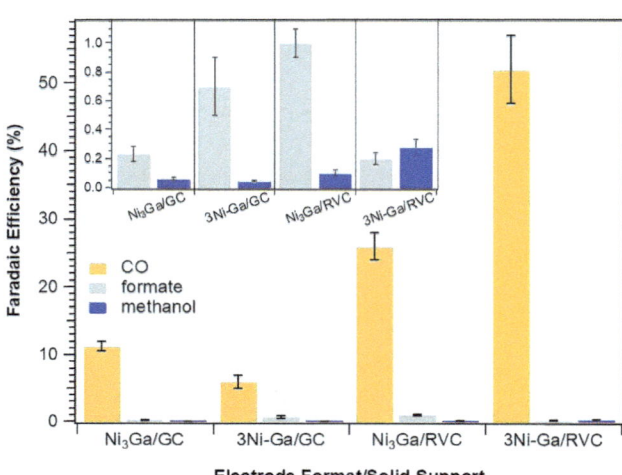

Fig. 6 Faradaic efficiencies for CO_2 reduction products achieved using Ni_3Ga and non-mixed 3Ni–Ga films on GC and RVC solid supports, indicating that the intermetallic and non-mixed species exhibit the same products. Ni_3Ga/GC and Ni_3Ga/RVC values were reported previously.[16] Electrolysis experiments were conducted at -1.38 V $vs.$ Ag/AgCl in pH 4.5, CO_2-saturated K_2SO_4 electrolyte.

produced by the nominal Ga/RVC electrode (actual composition = monoclinic Ga_2O_3/RVC) at ~57% faradaic efficiency (Fig. S8†), outcompeting H_2 evolution. Potential and pH dependence studies using Ga/RVC electrodes suggest that significant quantities of CO can be generated across a range of electrochemical conditions (Fig. S9†). This observation is worthy of note as literature reports of Ga and Ga_2O_3's activity in CO_2 electroreduction are limited. In 1994, Hori *et al.* reported that a Ga electrode could generate CO at a faradaic efficiency of 23% at −1.45 V *vs.* Ag/AgCl in neutral solution,[22] while a more recent report stated that Sn- or Si-doped Ga_2O_3 single crystals could reach a faradaic efficiency of >80% for formate at potentials more negative than −1.8 V *vs.* Ag/AgCl.[23] As such, when considering a combination of faradaic efficiency, overpotential, and chemical stability (Fig. S10†), the monoclinic Ga_2O_3 film on RVC examined here expands— and perhaps improves—upon previous reports of Ga species facilitating two-electron CO_2 reduction to CO.

In any case, a common feature of the most interesting catalyst species studied here (*i.e.*, Ni_3Al/GC and Ni_3Ga/HOPG, which make multi-carbon products, and Ga/RVC, whose CO_2 reduction rivals H^+ reduction) is CO generation capacity. Ga/RVC produces CO at high faradaic efficiencies, while previous reports show that Ni_3Al/GC and Ni_3Ga/HOPG invoke CO as an intermediate en route to higher-order products.[1,3,16] With this in mind, many researchers describe CO_2 electroreduction to highly reduced products, facilitated by Cu-based systems, as a mechanistically Fischer–Tropsch-type process,[21-23] though specific C2 products (*e.g.*, ethylene) have been attributed to different pathways.[24,25] The mechanism by which Ni–G13 intermetallics generate C–C-bonded products is less well-understood. A good first approximation might be determining whether their mechanistic pathways are Fischer–Tropsch-like in nature, thereby testing the prediction that their mechanisms resemble Cu-mediated catalysis.

Fischer–Tropsch catalysis has long been known to exhibit isotope effects when substituting H_2, the key reactant alongside CO, with D_2.[26] To determine whether Ni_3Al/GC or Ni_3Ga/HOPG demonstrate a similar effect, electrolysis experiments were conducted in D_2O (0.1 M K_2SO_4). The faradaic efficiencies for CO, obtained at an operating potential of −1.38 V *vs.* Ag/AgCl, are summarized in Fig. 7, which also includes data obtained using Ni_3Ga/GC (*i.e.*, an intermetallic that generates a modest amount of CO instead of multi-carbon products) and Ga/RVC (*i.e.*, a single-metal species that generates a large amount of CO) for comparison. CO production increases in the presence of D_2O on all of the catalyst species studied, reaching 80% on Ga/RVC. Importantly, Ni_3Al/GC and Ni_3Ga/HOPG fail to generate their higher-order reduction products in D_2O despite the high relative abundance of CO, which serves as the intermediate en route to C2 and C3 products on these electrodes in H_2O-based electrolytes. In fact, Ni_3Ga/HOPG in H_2O has been shown to reduce CO to ethane, and in CO_2 reduction experiments CO is usually only detected in trace quantities;[1,16] in the D_2O experiment described here, CO is quantifiable.

Substituting D_2O for H_2O in these electrochemical systems is clearly debilitating to the Cu-like behavior of Ni–G13 catalysts. In the presence of D_2O, the sustained generation of CO coupled with an inability to reduce it further suggests that CO reduction is rate-limiting, refuting the hypothesis made in our first report on the Ni_3Al/GC system.[4] Furthermore, these results suggest that Ni_3Al/GC and Ni_3Ga/HOPG may not operate *via* a Fischer–Tropsch-like mechanism, despite

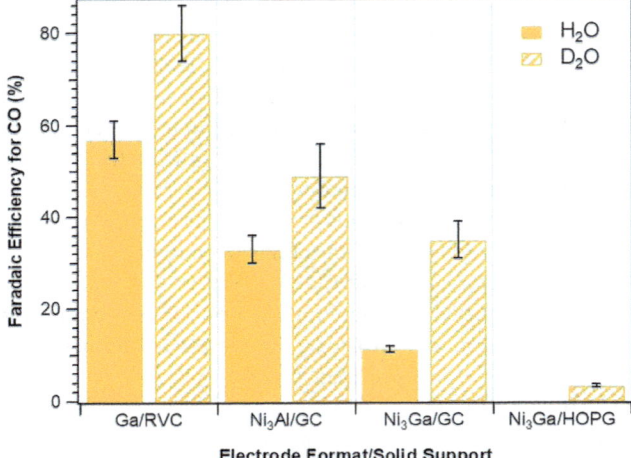

Fig. 7 Faradaic efficiencies for CO achieved using different electrode-support combinations in H_2O- or D_2O-based electrolyte (0.1 M K_2SO_4). The D_2O environment consistently increased CO production and eliminated higher-order product generation where it was previously possible. Electrolysis experiments were conducted at -1.38 V *vs.* Ag/AgCl.

similarities in both reactants and products. H_2/D_2 studies of Fischer–Tropsch catalysis typically reveal an inverse isotopic effect on the order of $r_D/r_H = 1.2$–1.6, where r_D and r_H denote reaction rates in the presence of D_2 and H_2, respectively, though a few catalysts exhibit normal kinetic isotope effects (*i.e.*, $r_D/r_H < 1$).[27,28] To-date, no Fischer–Tropsch catalysts have been reported to fail to generate hydrocarbons altogether when in the presence of D_2 or D_2O.[29] Taken together, these results call into question the suggestion that Ni_3Al/GC and $Ni_3Ga/HOPG$ electrocatalysts facilitate C2 and C3 product formation *via* a traditional Fischer–Tropsch catalytic mechanism.

Conclusions

Here, we have described how the synthesis and electrochemical testing of several Ni–G13 catalyst structures provided insight into the roles of the metals, carbon support, and catalytic mechanism in generating multi-carbon products. These Ni–G13 catalyst systems reduce CO_2 to C2 and C3 species only when in their intermetallic forms, signifying a synergistic role for the two metals and refuting the hypothesis that they perform isolated tasks during catalysis (*e.g.*, CO_2 reduction to CO, CO reduction to higher-order products). Furthermore, the ability to generate higher-order products cannot be rescued by tuning the solid support on which non-mixed metals are deposited. During the experimental process, it was discovered that monoclinic Ga_2O_3 films situated on RVC are effective CO generators, reaching faradaic efficiencies of around 57%. A summary of the product distribution achieved using each electrocatalyst discussed here is provided in Table S2.†

Electrolysis experiments using D_2O-based electrolyte resulted in enhanced CO production for all of the catalysts studied, but the ability of Ni_3Al/GC and $Ni_3Ga/$

HOPG to generate higher-order products was compromised. This extreme isotope effect lends insight into the mechanism of CO_2 reduction on these catalysts, which have been said to exhibit Cu-like CO_2 reduction behavior. Reduction of CO must be rate-limiting given the CO accumulation recorded and, unlike many predictions regarding Cu catalysts, CO_2 reduction to C2 and C3 products appears to proceed *via* a mechanistic route that is distinct from classic Fischer–Tropsch catalysis.

These insights have several key implications for the study of bimetallic catalysts active in CO_2 electroreduction. At present, intermetallic catalysts seem more effective than non-mixed, two-metal species at generating higher-order products from CO_2. Development of non-alloyed systems seeking to combine separate reduction events at distinct metal sites will be even more challenging, as the two metals will need to have compatible electrochemical conditions. Moreover, based on our isotopic experiments, it may be possible to elicit Cu-like CO_2 electroreduction activity without adopting what is presumed to be the dominant CO_2 reduction mechanism on Cu (*i.e.*, Fischer–Tropsch). To the best of our knowledge, the H/D kinetic isotope effect studies described here have not been reported for Cu-based electrocatalysts active in CO_2 reduction. Studies of this type using Cu electrodes could help distinguish critical pathway-product relationships for these popular catalyst materials.

Conflicts of interest

There are no conflicts of interest to declare.

Acknowledgements

The authors acknowledge financial support from the National Science Foundation under grant no. CHE-1800400. In addition, ARP acknowledges funding from the National Science Foundation Graduate Research Fellowship Program under grant no. DGE-1148900. Any opinions, findings, and conclusions expressed in this material are those of the authors and do not necessarily reflect the views of the National Science Foundation.

References

1 D. A. Torelli, S. A. Francis, J. C. Crompton, A. Javier, J. R. Thompson, B. S. Brunschwig, M. P. Soriaga and N. S. Lewis, *ACS Catal.*, 2016, **6**, 2100–2104.
2 R. Kortlever, I. Peters, C. Balemans, R. Kas, Y. Kwon, G. Mul and M. T. M. Koper, *Chem. Commun.*, 2016, **52**, 10229–10232.
3 A. R. Paris and A. B. Bocarsly, *ACS Catal.*, 2017, 7, 6815–6820.
4 H. A. Hansen, C. Shi, A. C. Lausche, A. A. Peterson and J. K. Nørskov, *Phys. Chem. Chem. Phys.*, 2016, **18**, 9194–9201.
5 Z. P. Jovanov, H. A. Hansen, A. S. Varela, P. Malacrida, A. A. Peterson, J. K. Nørskov, I. E. L. Stephens and I. Chorkendorff, *J. Catal.*, 2016, **343**, 215–231.
6 A. A. Peterson and J. K. Nørskov, *J. Phys. Chem. Lett.*, 2012, **3**, 251–258.
7 Z. W. Ulissi, M. T. Tang, J. Xiao, X. Liu, D. A. Torelli, M. Karamad, K. Cummins, C. Hahn, N. S. Lewis, T. F. Jaramillo, K. Chan and J. K. Nørskov, *ACS Catal.*, 2017, 6600–6608.

8 K. Tran and Z. W. Ulissi, *Nat. Catal.*, 2018, **1**, 696–703.

9 Y. Li and Q. Sun, *Adv. Energy Mater.*, 2016, **6**, 1600463.

10 A. Dutta, M. Rahaman, N. C. Luedi, M. Mohos and P. Broekmann, *ACS Catal.*, 2016, **6**, 3804–3814.

11 M. Ma, K. Djanashvili and W. A. Smith, *Angew. Chem., Int. Ed.*, 2016, **55**, 6680–6684.

12 Y. Li, F. Cui, M. B. Ross, D. Kim, Y. Sun and P. Yang, *Nano Lett.*, 2017, **17**, 1312–1317.

13 P. De Luna, R. Quintero-Bermudez, C.-T. Dinh, M. B. Ross, O. S. Bushuyev, P. Todorović, T. Regier, S. O. Kelley, P. Yang and E. H. Sargent, *Nat. Catal.*, 2018, **1**, 103–110.

14 S. Rasul, D. H. Anjum, A. Jedidi, Y. Minenkov, L. Cavallo and K. Takanabe, *Angew. Chem., Int. Ed.*, 2015, **54**, 2146–2150.

15 R. Kortlever, I. Peters, S. Koper and M. T. M. Koper, *ACS Catal.*, 2015, **5**, 3916–3923.

16 A. R. Paris, A. T. Chu, C. B. O'Brien, J. J. Frick, S. A. Francis and A. B. Bocarsly, *J. Electrochem. Soc.*, 2018, **165**, H385–H392.

17 Y. Hori, A. Murata and R. Takahashi, *J. Chem. Soc., Faraday Trans. 1*, 1989, **85**, 2309–2326.

18 Y. Hori, R. Takahashi, Y. Yoshinami and A. Murata, *J. Phys. Chem. B*, 1997, **101**, 7075–7081.

19 R. Kortlever, J. Shen, K. J. P. Schouten, F. Calle-Vallejo and M. T. M. Koper, *J. Phys. Chem. Lett.*, 2015, **6**, 4073–4082.

20 J. Wang, *Electrochim. Acta*, 1981, **26**, 1721–1726.

21 J. M. Friedrich, C. Ponce-de-León, G. W. Reade and F. C. Walsh, *J. Electroanal. Chem.*, 2004, **561**, 203–217.

22 Y. Hori, H. Wakebe, T. Tsukamoto and O. Koga, *Electrochim. Acta*, 1994, **39**, 1833–1839.

23 T. Sekimoto, M. Deguchi, S. Yotsuhashi, Y. Yamada, T. Masui, A. Kuramata and S. Yamakoshi, *Electrochem. Commun.*, 2014, **43**, 95–97.

24 K. J. P. Schouten, Y. Kwon, C. J. M. van der Ham, Z. Qin and M. T. M. Koper, *Chem. Sci.*, 2011, **2**, 1902.

25 K. J. P. Schouten, Z. Qin, E. P. Gallent and M. T. M. Koper, *J. Am. Chem. Soc.*, 2012, **134**, 9864–9867.

26 C. K. Rofer-DePoorter, *Chem. Rev.*, 1981, **81**, 447–474.

27 C. S. Kellner and A. T. Bell, *J. Catal.*, 1981, **67**, 175–185.

28 M. Ojeda, A. Li, R. Nabar, A. U. Nilekar, M. Mavrikakis and E. Iglesia, *J. Phys. Chem. C*, 2010, **114**, 19761–19770.

29 S. Krishnamoorthy, M. Tu, M. P. Ojeda, D. Pinna and E. Iglesia, *J. Catal.*, 2002, **211**, 422–433.

Faraday Discussions

PAPER

Iron phosphate modified calcium iron oxide as an efficient and robust catalyst in electrocatalyzing oxygen evolution from seawater†

Wei-Hsiang Huang and Chia-Yu Lin (ID) *

Received 12th November 2018, Accepted 5th December 2018
DOI: 10.1039/c8fd00172c

Solar fuel generation using seawater as the proton source is fascinating but challenging due to the detrimental chlorochemistry, the lack of active and stable oxygen evolution catalysts operating at seawater pH ($\sim$8) and high turnover conditions. In the present study, iron phosphate modified calcium iron oxide ($CaFeO_x|FePO_4$) modified FTO electrodes were prepared, and their electrocatalytic properties towards the oxygen evolution reaction in both synthetic and natural seawater solutions were investigated. $CaFeO_x|FePO_4$ was prepared by electrodepositing $FePO_4$ onto $CaFeO_x$, prepared by spin-coating and a follow-up annealing process, under a constant applied current for different durations. Mg^{2+}-induced fouling significantly reduces the activity of $CaFeO_x$ and slows down the activation process of $CaFeO_x$, but can be mitigated by surface-modification of $CaFeO_x$ with $FePO_4$. In addition, the presence of additional electrodeposited iron phosphate on the surface of $CaFeO_x$ attenuates the production of corrosive hypochlorite from chloride oxidation. With these unique properties of $FePO_4$, the activated $CaFeO_x|FePO_4$ electrode shows high activity and stability under high turnover conditions, reaching 10 mA cm^{-2} at an overpotential of $\sim$710 mV, with a moderate increase in η ($\sim$70 mV), mainly due to the change in solution pH, over 10 h of electrolysis in phosphate-buffered (0.5 M, pH 7) seawater solution.

Introduction

Direct use of seawater as the proton source for solar fuel generation is of great interest and importance as seawater is the most abundant water source on the Earth, and it avoids competition with the growing demand for fresh water for human activity and reduces the cost for solar fuel generation. In addition, integration of a seawater-based solar-fuel generation device with a solar-fuel cell provides a less energy-intensive approach to producing fresh water from seawater.

Department of Chemical Engineering, National Cheng Kung University, No. 1, University Road, Tainan City 70101, Taiwan. E-mail: cyl44@mail.ncku.edu.tw

† Electronic supplementary information (ESI) available. See DOI: 10.1039/c8fd00172c

Nevertheless, development of an efficient and stable solar-fuel device using seawater is a challenging task due to competing redox reactions and poisoning/fouling effects from impurities in seawater. For example, due to its complicated and kinetically sluggish nature, the desired oxygen evolution reaction (OER) from water oxidation at a flat electrode requires an overpotential (η) of >480 mV to maintain a current density $\geq$ 10 mA cm^{-2}, meeting the requirement of an economic solar-fuel device with >10% solar-to-fuel conversion efficiency[1] at seawater pH (pH $\sim$ 8 (ref. 2)). Such a high η exceeds the thermodynamic requirements of chloride oxidation[3] and directs the current to the kinetically favourable chloride oxidation (e.g., the ratio of exchange current for chlorine production to that for OER is 10^3 to 10^7),[4] which causes the efficiency of solar-fuel devices to deteriorate. In addition, the corrosive chlorine and hypochlorite, produced from chloride oxidation, reduce the stability of the oxygen evolution catalysts (OECs) and thus the durability of the corresponding solar-fuel device.[3]

Operating a solar-fuel device in an alkaline seawater solution, prepared by adding alkali to seawater, has been proposed to mitigate the detrimental chlorochemistry as many Earth-abundant OECs can drive water oxidation at a current density > 10 mA cm^{-2} with η of <400 mV,[5] ensuring the efficient, selective, and stable OER. However, the corrosive nature of alkaline electrolyte necessitates the specialized protection scheme for both the device set-up and the photoelectrodes. In addition, alkaline electrolyte solution is not suitable for fuel generation from photoelectrochemical CO_2 reduction as CO_2 converts to electrochemically inactive carbonates in alkaline solution.[6] Furthermore, the alkali accelerates the precipitation of a passivation oxide layer of indigenous metals in seawater on the electrode surface, which causes the performance of the solar-fuel device to deteriorate. As a result, the development of a highly efficient, robust, and inexpensive OEC that operates in near-neutral (pH $\sim$ 8) seawater is urgently required.

Many Earth-abundant OECs working at neutral or near-neutral pH have been developed recently, including manganese complexes,[7] manganese oxides (MnO_x),[8] self-healing cobalt–phosphate (Co–Pi) and cobalt–borate (Co–Bi),[9] and iron based oxides.[10] Nevertheless, on a flat electrode basis, most of these Earth-abundant material based OECs either require a high η to maintain a current density > 10 mA cm^{-2} or became unstable under a high applied current density. For example, self-healing Co–Pi requires an η of about 480 mV to drive OER at a current density of 1 mA cm^{-2} in phosphate buffer (pH 7), but it becomes unstable at a current density of >5 mA cm^{-2} due to serious chemical dissolution of the active Co species by the accumulated proton from intensive OER and the absence of sufficient Co ions in phosphate buffer for the re-deposition (self-healing) of Co–Pi to compensate for the loss of active Co species.[9c,11] On the other hand, due to the instability of Mn^{3+} species, a precursor for OER, at pH < 9, and the lack of the capability to manage both protons and electrons simultaneously,[8a–c,8e] manganese based OECs often require a much higher η than their nature counterpart, i.e., $CaMn_4O_5$ cluster in photosystem II at neutral pH.

Recently, calcium-containing iron oxide ($CaFeO_x$), which consists of a short-range order γ-Fe_2O_3 domain embedded in an amorphous $CaFe_2O_5$ matrix, was synthesized in our group,[10f] and its application as an electrocatalyst to catalyse oxygen evolution from water oxidation at neutral pH was demonstrated. It was found that during the electrolysis in phosphate buffer the surface calcium leached and redox active iron phosphate ($FePO_4$) formed on the surface of

CaFeO$_x$, which plays a role in mitigating the accumulation of protons generated from OER, thus preventing the chemical dissolution of active iron species by accumulated protons. The calcium-induced structural disorder and the *in situ* formation of iron phosphate were proposed to be the main factors rendering CaFeO$_x$ with higher activity ($\eta = \sim$650 mV@10 mA cm^{-2}) and stability than other calcium free iron oxide based OECs at high turnover conditions.

Encouraged by the high activity and stability of CaFeO$_x$ in neutral (pH 7) aqueous media, we, in the present study, have examined the electrocatalytic properties of CaFeO$_x$ towards seawater oxidation, and improved its activity and stability by surface-modification with FePO$_4$. Through systematic assessments, the beneficial role of FePO$_4$ in improving the activity and stability of CaFeO$_x$|FePO$_4$, by suppressing MgCl$_2$-induced fouling and chloride oxidation, has been discovered. The high activity and stability suggest the applicability of CaFeO$_x$|PO$_4$ for efficient and selective seawater oxidation.

Experimental

General consideration

All of the chemicals used for this work were of analytical grade and were used as received without further purification. Fluorine-doped tin oxide coated glass (FTO; sheet resistance 7 ohm per square, TEC GlassTM 7) was cleaned with a mixture of ammonia, hydrogen peroxide, and deionized water (volume ratio: 1 : 1 : 5) at 70 °C for 30 min and dried under N$_2$ purge before the deposition of catalytic materials. The plating solution, phosphate buffer solution and phosphate-buffered synthetic seawater solution were prepared using deionized water (DIW; 18.2 MΩ cm), whereas the phosphate-buffered seawater solution was prepared using seawater from the sea nearby Yuguang Island (22°58′40″N 120°9′27″E), Tainan city, Taiwan.

Preparation of calcium-containing iron oxide (CaFeO$_x$) thin film modified electrodes

A CaFeO$_x$ thin film modified electrode, designated as FTO|CaFeO$_x$, was prepared by spin-coating an ethanolic solution containing calcium acetate (0.12 M), iron(III) nitrate (0.2 M), and ethanolamine (0.3 M) on to the FTO substrate, followed by an annealing process at 400 °C for 30 min.[10f] Iron phosphate species modified FTO|CaFeO$_x$ electrodes, designated as FTO|CaFeO$_x$|FePO$_{4(t)}$, were prepared by electrochemical deposition at a constant applied current density of 100 μA cm^{-2} for various durations (t) in phosphate buffer (0.1 M, pH 7) containing ferrous sulfate (0.5 mM) under a N$_2$ atmosphere; the oxidation of Fe^{2+} to Fe^{3+} in the presence of phosphate resulted in the precipitation of FePO$_4$ on the surface of FTO|CaFeO$_x$ through eqn (1) & (2):[12]

$$Fe_{(ad)}{}^{2+} + HPO_{4(aq)}{}^{2-} \rightleftarrows FePO_{4(s)} + H_{(aq)}{}^{+} + e^{-} \tag{1}$$

$$Fe_{(ad)}{}^{2+} + H_2PO_{4(aq)}{}^{-} \rightleftarrows FePO_{4(s)} + 2H_{(aq)}{}^{+} + e^{-} \tag{2}$$

Note that to prevent the oxidation of Fe^{2+} ions to Fe^{3+} ions, the phosphate buffer solution was entirely bubbled with nitrogen for at least 30 min prior to the preparation of the plating solution to remove dissolved oxygen before the dissolution of $FeSO_4$.

Preparation of cobalt–phosphate (Co–Pi) thin film modified electrodes

A Co–Pi thin film modified electrode, designated as FTO|Co–Pi, was prepared by electrochemical deposition using a similar synthetic procedure reported previously.[9a] Briefly, Co–Pi was electrodeposited onto the FTO substrate (with an exposed area of ~ 1.0 cm^2) in phosphate buffer (0.1 M, pH 7) containing cobalt nitrate (0.5 mM) at an applied potential of ~ 1.1 V ($vs.$ Ag/AgCl) till a charge of 1.0 C cm^{-2} was passed.

Preparation of electrodeposited iron oxide (e-FeO$_x$) thin film modified electrodes

A FeO$_x$ thin film modified electrode, designated as FTO|FeO$_x$, was prepared by electrochemical deposition using a similar synthetic procedure reported previously.[10c] Briefly, FeO$_x$ was electrodeposited onto the FTO substrate (with an exposed area of ~ 1.0 cm^2) in HEPES buffer (0.1 M, pH 7) containing ferrous sulfate (1.0 mM) using cyclic voltammetry at a scan rate of 50 mV s^{-1} in the potential range between 0.61 to 1.76 V $vs.$ RHE for 25 cycles. Note that to prevent the oxidation of Fe^{2+} ions to Fe^{3+} ions, the phosphate buffer solution was entirely bubbled with nitrogen for at least 30 min prior to the preparation of the plating solution to remove the dissolved oxygen before the dissolution of $FeSO_4$.

Preparation of iridium oxide (IrO$_2$) thin film modified electrodes

An IrO$_2$ thin film modified electrode, designated as FTO|IrO$_2$, was prepared by electrochemical deposition using a similar synthetic procedure reported previously.[13] A plating solution was prepared by adding oxalic acid dehydrate (5 g L^{-1}) and hydrogen peroxide (11.7 g L^{-1}) to an aqueous solution containing iridium(IV) chloride (0.1 M) and mixed for 10 min. After the pH of the plating solution was adjusted to 10.5 by adding K_2CO_3, the plating solution was stirred for 3 days. Thereafter, electrodeposition of IrO$_2$ onto the FTO substrate (with an exposed area of ~ 1.0 cm^2) was carried out by applying a constant current density of 1 mA cm^{-2} for 300 s.

Physical characterization

The amount of Fe^{3+} and Ca^{2+} species in the catalysts of interests and main metal ions (Na^+, Mg^{2+}, Ca^{2+}) in seawater were determined using a Horiba Jobin Yvon JY 2000-2 ICP optical emission spectrometer. The surface composition of the films was verified by X-ray photoelectron spectroscopy (XPS, PHI 5000 VersaProbe system, ULVAC-PHI, Chigasaki, Japan), using a microfocused (100 μm, 25 W) Al X-ray beam, with a photoelectron take-off angle of 45°. The Ar$^+$ ion source for XPS (FIG-5CE) was controlled using a floating voltage of 0.2 kV. The binding energies obtained in the XPS analyses were corrected for specimen charging, by referencing the C 1s peak to 285.0 eV.

Electrochemical characterization

All of the electrochemical measurements were performed in a two-compartment, three-electrode electrochemical cell, separated with a Nafion® 117 film, with a CHI 760 electrochemical workstation (CH Instruments, Inc., USA) at room temperature under a N_2 atmosphere and all potentials (E) were compensated using a built-in IR-compensation program and reported against the reversible hydrogen electrode (RHE) using the equation eqn (3):

$$E \text{ (V } vs. \text{ RHE)} = E \text{ (V } vs. \text{ Ag/AgCl)} + 0.197 + 0.059 \times \text{pH} \tag{3}$$

FTO|CaFeO$_x$, FTO|CaFeO$_x$|FePO$_{4(t)}$, FTO|e-FeO$_x$, and FTO|Co–Pi (all with an exposed area of ~1.0 cm^2) were used as the working electrode connected to a Pt foil counter electrode and an Ag/AgCl$_{sat}$ reference electrode. Linear sweep voltammetry (LSV) at a scan rate of 10 mV s^{-1}, and controlled-current electrolysis (CCE) at an applied current density of 10 mA cm^{-2}, were carried out in phosphate buffer solution (0.5 M, pH 7), phosphate-buffered (pH 7) synthetic seawater, containing phosphate (0.5 M), NaCl (0.36 M) and MgCl$_2$ (0.07 M), and phosphate (0.5 M, pH 7) buffered natural seawater. Note that the phosphate buffer is added as a proton-accepting agent to minimize the change in the solution pH during OER. η is defined as the difference between the applied potential (V $vs.$ RHE) to drive OER at a current density of 10 mA cm^{-2}, and the thermodynamic potential of water oxidation, which is 1.23 V $vs.$ RHE. All of the electrochemical measurements were repeated at least three times.

Oxygen measurement

The amount of O_2 generated from the two-compartment three-electrode electrochemical systems was detected and quantified by headspace gas analysis with an Ocean Optics fluorescence O_2 probe (FOSPOR-R). The O_2 probe was inserted in an anodic compartment through a tightly sealed septum and continuous O_2 readings (O_2 partial pressure) were taken at 1 s intervals throughout the experiment. The three-electrode electrochemical cell was operated with the sequence: at an applied current density of 0 mA cm^{-2} for a period of 30 min (control experiment), followed by 2 h CCE of water at 10 mA cm^{-2} and another 30 min at an applied current density of 0 mA cm^{-2}.

Quantification of hypochlorite

Iodide titration was used for quantification of hypochlorite generated during the CCE experiments in phosphate-buffered synthetic seawater or phosphate-buffered seawater. Right after the CCE experiments, 15 mL of electrolyte was pipetted from the anodic compartment of the two-compartment electrochemical cell to an Erlenmeyer flask. Then 10 mL of freshly prepared KI (0.5 M) was added into the flask under magnetic stirring. Upon a colour change, 10 mL of filtered starch solution was added into the solution to change the solution colour from faint yellow to blue. Finally, the solution was titrated by adding sodium thiosulfate solution (0.01 M) till a change in solution colour from blue to transparent was observed. The amount of sodium thiosulfate reacted ($N_{Na_2S_2O_3}$) was calculated by multiplying the concentration of sodium thiosulfate solution by the volume of sodium thiosulfate solution for titration. The total amount of hypochlorite ($N_{hypochlorite}$) was then calculated by dividing $N_{Na_2S_2O_3}$ by 2 (stoichiometry coefficient) and the volume of electrolyte pipetted for the titration, and multiplied by the total volume of the electrolyte. Two titrations were performed for each CCE experiment, and the average $N_{hypochlorite}$ from the two titrations was taken.

Results and discussion

The effects of magnesium and chloride ions on the stability of FTO|CaFeO$_x$ in neutral conditions were examined using constant-current electrolysis (CCE) at an applied current density of 10 mA cm^{-2}, and the results are shown in Fig. 1. For comparison, the results of CCEs for other state-of-art Earth-abundant OECs, including FTO|Co–Pi and FTO|e-FeO$_x$, are also included in Fig. 1. Note that for a fair comparison, the loading amount of active species, *i.e.*, Co^{2+} (0.37 ± 0.01 μmol cm^{-2}) in FTO|Co–Pi, Fe^{3+} (0.30 ± 0.06 μmol cm^{-2}) in FTO|e-FeO$_x$, and Fe^{3+} (0.28 ± 0.01 μmol cm^{-2}) in FTO|CaFeO$_x$, on these three types of electrodes were kept comparable. It was found that FTO|CaFeO$_x$ exhibited the highest stability under high turnover conditions (10 mA cm^{-2}) compared to the other two state-of-art Earth-abundant materials based OER catalysts, *i.e.*, FTO|Co–Pi and FTO|e-FeO$_x$ in phosphate buffer; the *in situ* formed iron phosphate proton-relay layer in-between the electrolyte and FTO|CaFeO$_x$ surface regulates the local pH, slowing down the accumulation of protons released from the OER and postponing chemical dissolution.[10f] Nevertheless, serious dissolution of CaFeO$_x$ was noticed when the pH of the electrolyte in the working compartment <6.Furthermore, a noticeable decrease in stability of FTO|Co–Pi, FTO|e-FeO$_x$, and FTO|CaFeO$_x$ was observed in synthetic seawater. As these three types of electrodes required an overpotential (η) of >480 mV to maintain a current density of 10 mA cm^{-2} and the local pH decreased during the course of the OER, the oxidation of chloride to corrosive hypochlorite during the CCE experiments became possible. To confirm this, quantification of hypochlorite using iodometric titration was carried out after complete degradation of the catalytic materials, and the results are shown in Table 1. As revealed, different levels of hypochlorite production were noticed in the order FTO|Co–Pi (3.62 ± 1.67 μmole; FE: 2.55 ± 1.27%) > FTO|e-FeO$_x$ (1.12 ± 0.18 μmole; FE: 0.12 ± 0.06%) > FTO|CaFeO$_x$ (0.37 ± 0.25 μmole; FE: 0.02 ±

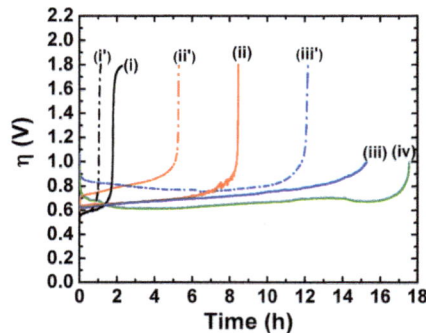

Fig. 1 Overpotential (η) transients, recorded at an applied current density of 10 mA cm^{-2}, of (i, i') FTO|Co–Pi, (ii, ii') FTO|FeO$_x$, (iii, iii') FTO|CaFeO$_x$, and (iv) FTO|CaFeO$_x$|FePO$_{4(t=25)}$ in phosphate buffer (0.5 M, pH 7; curves i, ii, and iii) and phosphate-buffered (pH 7) synthetic seawater solution containing phosphate (0.5 M), NaCl (0.36 M), and MgCl$_2$ (0.07 M) (curves i', ii', iii', and iv) under N$_2$ atmosphere.

Table 1 The amounts of hypochlorite generated from chloride oxidation and the pH values of electrolyte solutions after complete degradation of electrocatalysts in the CCE experiments at an applied current density of 10 mA cm^{-2} in the phosphate-buffered (pH 7) synthetic seawater solution containing phosphate (0.5 M), NaCl (0.36 M), and MgCl$_2$ (0.07 M). The complete degradation of the electrocatalyst is defined as the point when the overpotential to maintain a current density of 10 mA cm^{-2} reached 1.7 V

	Samples			
	FTO|Co–Pi	FTO|FeO$_x$	FTO|CaFeO$_x$	FTO|CaFeO$_x$|FePO$_{4(t=25)}$
Amount of generated OCl$^-$ (μmol)	3.62 ± 1.67	1.12 ± 0.18	0.37 ± 0.25	0
Faradaic efficiency for OCl$^-$ generation (%)	2.55 ± 1.27	0.12 ± 0.06	0.02 ± 0.01	0
pH of electrolyte solution in anodic compartment	6.71 ± 0.00	6.47 ± 0.06	6.00 ± 0.08	5.89 ± 0.07
pH of electrolyte solution in cathodic compartment	7.09 ± 0.00	7.27 ± 0.11	8.28 ± 0.20	8.39 ± 0.93

0.01%). Thus, the formation of corrosive hypochlorite would be an additional factor deteriorating the stability of these electrodes.

On the other hand, FTO|CaFeO$_x$ produced a smaller amount of hypochlorite than FTO|FeO$_x$ and FTO|Co–Pi, which suggests that the kinetics of chloride oxidation at FTO|CaFeO$_x$ is more sluggish than those at the other two electrodes. As the formation of iron phosphate is not possible on FTO|e-FeO$_x$,[10c,10e] the difference in kinetics of chloride oxidation at FTO|CaFeO$_x$ and FTO|e-FeO$_x$ would be attributed to the *in situ* formed iron phosphate. To further examine the effect of iron phosphate, FTO|CaFeO$_x$ surfaces modified with different amounts of iron phosphate, prepared by electrodeposition under a constant applied current density of 100 μA cm^{-2} for different durations (t) (see Experimental section), were subjected to CCE experiments at an applied current density of 10 mA cm^{-2} in the synthetic seawater solution, and the results are shown in Fig. 1 and S1.† Note that the existence of iron phosphate on the FTO|CaFeO$_x$|FePO$_4$ is confirmed by the changes in XPS features (Fig. S2†) after the electrodeposition process, including a positive shift in the binding energy (BE) of the Fe 2p$_{3/2}$ peak (710.4 to 711.2 eV), the appearance of the P 2p$_{3/2}$ peak at 133.7 eV, and the disappearance of the O 1s peak at a BE of 529.5 eV along with a significant increase in the intensity of the O 1s peak at a BE of 531.6 eV.[14] The higher amount of Fe^{3+} (0.31 ± 0.03 μmol cm^{-2}) in FTO|CaFeO$_x$|FePO$_{4(t=25)}$, quantified by ICP measurement, than that in FTO|CaFeO$_x$ suggests that the faradaic efficiency for the deposition of FePO$_4$ is ~100%. The results of the CCE experiments, shown in Fig. 1 and S1,† indicate that FTO|CaFeO$_x$ surface-modified with a suitable amount of iron phosphate, *i.e.*, FTO|CaFeO$_x$|FePO$_{4(t=25)}$, showed enhanced stability in synthetic seawater. The results of iodometric titration (Table 1) show that no hypochlorite can be detected after the CCE experiments, which suggests that the oxidation of chloride is further attenuated in the presence of additional FePO$_4$. The detailed mechanism behind this improved stability by iron phosphate is still under investigation, but somehow it is related to the way in which the iron phosphate formed. In a previous report, iron phosphate was found to gradually form on the $CaFeO_x$ after the activation process, $i.e.$, the period for significant η drop, during the course of the CCE experiment in phosphate containing electrolytes, and as the activation process became longer (curve iii′ in Fig. 1) in the phosphate-buffered synthetic seawater solution, presumably due to the fouling by Mg^{2+} rather than phosphate buffer, its formation rate may not be fast enough to fully cover the surface of the $CaFeO_x$ to slow down the chloride oxidation. In addition, iron phosphate formed at the expense of active iron species, which results in the gradual loss in activity of $CaFeO_x$ and thus gradual increase in overpotential to drive OER at 10 mA cm^{-2}. In contrast, iron phosphate was electrodeposited on $FTO|CaFeO_x|FePO_{4(t=25)}$ before the CCE experiments, the oxidation of chloride can therefore be minimized in the beginning of the CCE experiments. Nevertheless, as iron phosphate itself is not the active species (Fig. S3†), surface modification of $CaFeO_x$ with a greater amount of iron phosphate doesn't further enhance its activity, in terms of η, and stability. The debilitated chloride oxidation at $FTO|CaFeO_x|FePO_{4(t=25)}$ can also be supported by the results of linear sweep voltammetry (Fig. S4†); $FTO|CaFeO_x|FePO_{4(t=25)}$ requires additional overpotential ($\sim$50 mV) to drive a current density of 10 mA cm^{-2} in both phosphate buffer solution (0.5 M, pH 7) containing NaCl (0.5 M) and phosphate-buffered synthetic seawater solution, compared to that required by $FTO|CaFeO_x$. Note that the required overpotential for all of the catalytic materials to maintain a current density increased in the presence of $MgCl_2$ (Fig. 1 and S4†), which could be attributed to the electrode fouling induced by Mg^{2+}. Nevertheless, the initial high overpotential of $FTO|CaFeO_x|FePO_{4(t=25)}$ during the CCE experiment in the phosphate-buffered synthetic seawater solution decreased quickly and reached an η value similar to that obtained in phosphate buffer solution, suggesting that electrode fouling can be quickly mitigated by iron phosphate.

Encouraged by the high stability of $FTO|CaFeO_x|FePO_{4(t=25)}$ in synthetic seawater solutions, our next step was to explore the applicability of $FTO|CaFeO_x|FePO_{4(t=25)}$ in splitting natural seawater. Seawater was collected from the sea nearby Yuguang Island (22°58′40″N 120°9′27″E), Tainan city, Taiwan, and directly used as the solvent to prepare a phosphate-buffered seawater solution for seawater splitting. Fig. 2 shows the chronopotentiograms of FTO|Co–Pi, FTO|e-

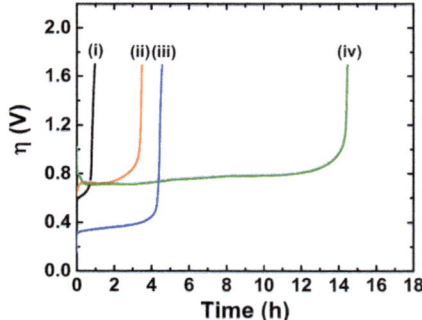

Fig. 2 Overpotential (η) transients, recorded at an applied current density of 10 mA cm^{-2}, of (i) FTO|Co–Pi, (ii) FTO|FeO$_x$, (iii) FTO|IrO$_x$, and (iv) $FTO|CaFeO_x|FePO_{4(25)}$ in phosphate-buffered (0.5 M, pH 7) natural seawater solution under a N_2 atmosphere.

Table 2 The amounts of hypochlorite generated from chloride oxidation, and the pH values of the electrolyte solutions after complete degradation of electrocatalyst in CCE experiments at an applied current density of 10 mA cm^{-2} in the phosphate-buffered (pH 7) natural seawater solution. The complete degradation of the electrocatalyst is defined as the point when the overpotential to maintain a current density of 10 mA cm^{-2} reached 1.7 V

	Samples			
	FTO|Co–Pi	FTO|FeO$_x$	FTO|IrO$_x$	FTO|CaFeO$_x$| FePO$_{4(t=25)}$
Amount of generated OCl$^-$ (μmol)	4.95 ± 0.47	2.33 ± 1.35	0	0
Faradaic efficiency for OCl$^-$ generation (%)	2.80 ± 0.40	0.19 ± 0.07	0	0
pH of electrolyte solution in anodic compartment	6.95 ± 0.02	6.55 ± 0.12	6.66 ± 0.085	6.17 ± 0.03
pH of electrolyte solution in cathodic compartment	7.15 ± 0.04	7.95 ± 0.25	7.78 ± 0.17	9.81 ± 0.07

FeO$_x$, FTO\|IrO$_x$, and FTO\|CaFeO$_x$\|FePO$_{4(t=25)}$ recorded at an applied current density of 10 mA cm^{-2} in the phosphate-buffered seawater solution. As revealed, FTO\|Co–Pi, FTO\|e-FeO$_x$, and FTO\|CaFeO$_x$\|FePO$_{4(t=25)}$ showed lower stability in natural seawater than in synthetic seawater. From the results of iodometric titration, shown in Table 2, it can be found that FTO\|Co–Pi and FTO\|e-FeO$_x$ produced more hypochlorite in the phosphate-buffered seawater solution than in the phosphate-buffered synthetic seawater, which suggests that the pronounced hypochlorite production is one of the main factors further deteriorating the stability of FTO\|Co–Pi and FTO\|FeO$_x$. However, as no detectable hypochlorite and a high FE (99.33 ± 8.04%) for OER (Fig. S5†) were found in the case of FTO\|CaFeO$_x$\|FePO$_{4(t=25)}$, other factors, including impurities in seawater and chemical dissolution by a significant drop in local pH, could play a role in deactivating FTO\|CaFeO$_x$\|FePO$_{4(t=25)}$. Further enhancement in the stability of CaFeO$_x$\|FePO$_{4(t=25)}$, by minimising the change in the local and bulk pH, would therefore be achievable when the batch-type electrolyser in our study is replaced by a flow-type electrolyser. It is interesting to note that FTO\|IrO$_x$ showed the highest activity in terms of η, but it degraded quickly within 5 h of operation in phosphate-buffered seawater electrolyte. As the overpotential of FTO\|IrO$_x$ remained inside the selective overpotential range (<460 mV at pH 7), chloride oxidation at FTO\|IrO$_x$ was not possible, which is in agreement with the results of the iodometric titration (Table 2). As a result, the degradation of FTO\|IrO$_x$ could be mainly attributed to the corrosion.[15]

Conclusions

The applicability of CaFeO$_x$\|FePO$_4$ as an electrocatalyst in efficient and selective seawater splitting was explored in this study. The presence of the additional electrodeposited FePO$_4$ mitigated the MgCl$_2$-induced fouling by facilitating the activation process of CaFeO$_x$\|FePO$_4$, and suppressed the generation of corrosive hypochlorite from chloride oxidation. With these beneficial properties, $CaFeO_x$ modified with an optimal amount of $FePO_4$ (*i.e.*, $CaFeO_x|FePO_{4(t=25)}$) exhibited enhanced stability towards seawater oxidation, outperforming other state-of-art Earth-abundant OECs at high turnover conditions.

Conflicts of interest

There are no conflicts to declare.

Acknowledgements

Financial support from the Ministry of Science and Technology of Taiwan (105-2221-E-006-230-MY2) is gratefully acknowledged.

Notes and references

1 C. C. L. McCrory, S. H. Jung, J. C. Peters and T. F. Jaramillo, *J. Am. Chem. Soc.*, 2013, **135**, 16977–16987.
2 R. Chester and T. D. Jickells, *Marine Geochemistry*, Wiley-Blackwell, New York, 3rd edn, 2012.
3 F. Dionigi, T. Reier, Z. Pawolek, M. Gliech and P. Strasser, *ChemSusChem*, 2016, **9**, 962–972.
4 H. K. Abdelaal, S. M. Sultan and I. A. Hussein, *Int. J. Hydrogen Energy*, 1993, **18**, 545–551.
5 (*a*) Y. Matsumoto, S. Yamada, T. Nishida and E. Sato, *J. Electrochem. Soc.*, 1980, **127**, 2360–2364; (*b*) J. O. M. Bockris and T. Otagawa, *J. Electrochem. Soc.*, 1984, **131**, 290–302; (*c*) J. Suntivich, K. J. May, H. A. Gasteiger, J. B. Goodenough and Y. Shao-Horn, *Science*, 2011, **334**, 1383–1385; (*d*) S. Yagi, I. Yamada, H. Tsukasaki, A. Seno, M. Murakami, H. Fujii, H. Chen, N. Umezawa, H. Abe and N. Nishiyama, *Nat. Commun.*, 2015, **6**, 8249; (*e*) Y. Zhu, W. Zhou, J. Yu, Y. Chen, M. Liu and Z. Shao, *Chem. Mater.*, 2016, **28**, 1691–1697; (*f*) M. Li, Y. Xiong, X. Liu, X. Bo, Y. Zhang, C. Han and L. Guo, *Nanoscale*, 2015, **7**, 8920–8930; (*g*) M. S. Al-Hoshan, J. P. Singh, A. M. Al-Mayouf, A. A. Al-Suhybani and M. N. Shaddad, *Int. J. Electrochem. Sci.*, 2012, **7**, 4959–4973; (*h*) S. Hirai, S. Yagi, A. Seno, M. Fujioka, T. Ohno and T. Matsuda, *RSC Adv.*, 2016, **6**, 2019–2023; (*i*) J. A. Koza, Z. He, A. S. Miller and J. A. Switzer, *Chem. Mater.*, 2012, **24**, 3567–3573; (*j*) Z. Chen, C. X. Kronawitter and B. E. Koel, *Phys. Chem. Chem. Phys.*, 2015, **17**, 29387–29393; (*k*) M. Al-Mamun, X. Su, H. Zhang, H. Yin, P. Liu, H. Yang, D. Wang, Z. Tang, Y. Wang and H. Zhao, *Small*, 2016, **12**, 2866–2871; (*l*) H. Bode, K. Dehmelt and J. Witte, *Electrochim. Acta*, 1966, **11**, 1071–1087; (*m*) F. Song and X. Hu, *Nat. Commun.*, 2014, **5**, 4477.
6 (*a*) Y. Hori and S. Suzuki, *J. Electrochem. Soc.*, 1983, **130**, 2387–2390; (*b*) T. E. Teeter and P. Vanrysselberghe, *J. Chem. Phys.*, 1954, **22**, 759–760; (*c*) P. Vanrysselberghe, G. J. Alkire and J. M. McGee, *J. Am. Chem. Soc.*, 1946, **68**, 2050–2055.
7 (*a*) K. Beckmann, H. Uchtenhagen, G. Berggren, M. F. Anderlund, A. Thapper, J. Messinger, S. Styring and P. Kurz, *Energy Environ. Sci.*, 2008, **1**, 668–676; (*b*) R. Brimblecombe, A. Koo, G. C. Dismukes, G. F. Swiegers and L. Spiccia, *J. Am. Chem. Soc.*, 2010, **132**, 2892–2894; (*c*) M. Yagi and K. Narita, *J. Am. Chem. Soc.*,

2004, **126**, 8084–8085; (*d*) J. Limburg, J. S. Vrettos, H. Y. Chen, J. C. de Paula, R. H. Crabtree and G. W. Brudvig, *J. Am. Chem. Soc.*, 2001, **123**, 423–430; (*e*) J. Limburg, G. W. Brudvig and R. H. Crabtree, *J. Am. Chem. Soc.*, 1997, **119**, 2761–2762; (*f*) M. M. Najafpour and A. N. Moghaddam, *Dalton Trans.*, 2012, **41**, 10292–10297; (*g*) R. Brimblecombe, D. R. J. Kolling, A. M. Bond, G. C. Dismukes, G. F. Swiegers and L. Spiccia, *Inorg. Chem.*, 2009, **48**, 7269–7279; (*h*) R. Brimblecombe, G. F. Swiegers, G. C. Dismukes and L. Spiccia, *Angew. Chem., Int. Ed.*, 2008, **47**, 7335–7338.

8 (*a*) A. Yamaguchi, R. Inuzuka, T. Takashima, T. Hayashi, K. Hashimoto and R. Nakamura, *Nat. Commun.*, 2014, **5**, 4256; (*b*) T. Takashima, K. Hashimoto and R. Nakamura, *J. Am. Chem. Soc.*, 2012, **134**, 1519–1527; (*c*) T. Takashima, K. Hashimoto and R. Nakamura, *J. Am. Chem. Soc.*, 2012, **134**, 18153–18156; (*d*) M. M. Najafpour, K. C. Leonard, F. R. F. Fan, M. A. Tabrizi, A. J. Bard, C. K. King'ondu, S. L. Suib, B. Haghighi and S. I. Allakhverdiev, *Dalton Trans.*, 2013, **42**, 5085–5091; (*e*) A. Ramirez, P. Bogdanoff, D. Friedrich and S. Fiechter, *Nano Energy*, 2012, **1**, 282–289.

9 (*a*) M. W. Kanan and D. G. Nocera, *Science*, 2008, **321**, 1072–1075; (*b*) Y. Surendranath, M. Dinca and D. G. Nocera, *J. Am. Chem. Soc.*, 2009, **131**, 2615–2620; (*c*) D. A. Lutterman, Y. Surendranath and D. G. Nocera, *J. Am. Chem. Soc.*, 2009, **131**, 3838–3839.

10 (*a*) G. Park, Y. I. Kim, Y. H. Kim, M. Park, K. Y. Jang, H. Song and K. M. Nam, *Nanoscale*, 2017, **9**, 4751–4758; (*b*) S. Haschke, Y. L. Wu, M. Bashouti, S. Christiansen and J. Bachmann, *ChemCatChem*, 2015, **7**, 2455–2459; (*c*) M. X. Chen, Y. Z. Wu, Y. Z. Han, X. H. Lin, J. L. Sun, W. Zhang and R. Cao, *ACS Appl. Mater. Interfaces*, 2015, **7**, 21852–21859; (*d*) T. W. Kim and K. S. Choi, *Science*, 2014, **343**, 990–994; (*e*) Y. Z. Wu, M. X. Chen, Y. Z. Han, H. X. Luo, X. J. Su, M. T. Zhang, X. H. Lin, J. L. Sun, L. Wang, L. Deng, W. Zhang and R. Cao, *Angew. Chem., Int. Ed.*, 2015, **54**, 4870–4875; (*f*) H. C. Chiu, W. H. Huang, L. C. Hsu, Y. G. Lin, Y. H. Lai and C. Y. Lin, *Sustainable Energy Fuels*, 2018, **2**, 271–279.

11 A. Minguzzi, F. R. F. Fan, A. Vertova, S. Rondinini and A. J. Bard, *Chem. Sci.*, 2012, **3**, 217–229.

12 (*a*) C. Y. Lin and C. T. Chang, *Sens. Actuators, B*, 2015, **220**, 695–704; (*b*) F. Marken, D. Patel, C. E. Madden, R. C. Millward and S. Fletcher, *New J. Chem.*, 2002, **26**, 259–263.

13 B. S. Lee, S. H. Ahn, H. Y. Park, I. Choi, S. J. Yoo, H. J. Kim, D. Henkensmeier, J. Y. Kim, S. Park, S. W. Nam, K. Y. Lee and J. H. Jang, *Appl. Catal., B*, 2015, **179**, 285–291.

14 (*a*) P. Nagaraju, C. Srilakshmi, N. Pasha, N. Lingaiah, I. Suryanarayana and P. S. S. Prasad, *Appl. Catal., A*, 2008, **334**, 10–19; (*b*) D. H. Yu, C. Wu, Y. Kong, N. H. Xue, X. F. Guo and W. P. Ding, *J. Phys. Chem. C*, 2007, **111**, 14394–14399; (*c*) Y. Wang, Q. Yuan, Q. H. Zhang and W. P. Deng, *J. Phys. Chem. C*, 2007, **111**, 2044–2053.

15 (*a*) H. S. Oh, H. N. Nong, T. Reier, M. Gliech and P. Strasser, *Chem. Sci.*, 2015, **6**, 3321–3328; (*b*) S. H. Hsu, J. W. Miao, L. P. Zhang, J. J. Gao, H. M. Wang, H. B. Tao, S. F. Hung, A. Vasileff, S. Z. Qiao and B. Liu, *Adv. Mater.*, 2018, **30**, 1707261.

Faraday Discussions

PAPER

Fe$_x$Ni$_{9-x}$S$_8$ ($x = 3-6$) as potential photocatalysts for solar-driven hydrogen production?†

David Tetzlaff,[a] Christopher Simon,[bc] Demetra S. Achilleos,[d]
Mathias Smialkowski,[a] Kai junge Puring,[ae] André Bloesser,[bc]
Stefan Piontek,[a] Hatice Kasap,[d] Daniel Siegmund, [ID][e] Erwin Reisner,[d]
Roland Marschall [ID] *[bc] and Ulf-Peter Apfel [ID] *[ae]

Received 12th November 2018, Accepted 18th December 2018

DOI: 10.1039/c8fd00173a

The efficient reduction of protons by non-noble metals under mild conditions is a challenge for our modern society. Nature utilises hydrogenases, enzymatic machineries that comprise iron- and nickel- containing active sites, to perform the conversion of protons to hydrogen. We herein report a straightforward synthetic pathway towards well-defined particles of the bio-inspired material Fe$_x$Ni$_{9-x}$S$_8$, a structural and functional analogue of hydrogenase metal sulfur clusters. Moreover, the potential of pentlandites to serve as photocatalysts for solar-driven H$_2$-production is assessed for the first time. The Fe$_x$Ni$_{9-x}$S$_8$ materials are visible light responsive (band gaps between 2.02 and 2.49 eV, depending on the pentlandite's Fe : Ni content) and display a conduction band energy close to the thermodynamic potential for proton reduction. Despite the limited driving force, a modest activity for photocatalytic H$_2$ has been observed. Our observations show the potential for the future development of pentlandites as photocatalysts. This work provides a basis to explore powerful synergies between biomimetic chemistry and material design to unlock novel applications in solar energy conversion.

Introduction

The generation of dihydrogen (H$_2$) as an environmentally friendly fuel from renewable energy sources is widely regarded as a potential process toward

[a]Inorganic Chemistry I – Bioinorganic Chemistry, Ruhr University Bochum, Universitätsstrasse 150, 44780 Bochum, Germany. E-mail: ulf.apfel@rub.de

[b]Institute of Physical Chemistry, Justus-Liebig-University Giessen, 35392 Giessen, Germany

[c]Physical Chemistry III, University of Bayreuth, 95447 Bayreuth, Germany

[d]Christian Doppler Laboratory for Sustainable SynGas Chemistry, Department of Chemistry, University of Cambridge, Lensfield Road, Cambridge CB2 1EW, UK

[e]Fraunhofer Institute for Environmental, Safety, and Energy Technology UMSICHT, Osterfelder Str. 3, 46047 Oberhausen, Germany

† Electronic supplementary information (ESI) available. See DOI: 10.1039/c8fd00173a

establishing a sustainable energy economy. Notably, about 95% of global H_2 production is currently generated by steam reforming, which is heavily based on limited resources of fossil fuels and is associated with detrimental effects on the Earth.[1]

In contrast to fossil fuel-derived H_2, sunlight-driven processes to produce H_2 such as water splitting *via* photocatalysis or photoelectrochemistry can be regarded as sustainable.[2–5] Commonly, photocatalytic systems for H_2 production consist of a light harvesting semiconductor (SC) material, which is often combined with a H_2 producing co-catalyst.[6] Over the past decades, a wide range of SC materials have been reported and evaluated for photocatalytic H_2 evolution.[7] Of major importance for an effective photocatalyst is a suitable band gap with favourable valence- and conduction bands (VB and CB, respectively). Starting from pioneering work on water splitting using TiO_2,[8] an important class of inorganic materials investigated for photocatalytic H_2 evolution are metal oxides, mainly containing d^0 or d^{10} metal ions (*e.g.* Ti^{IV},[9] Mo^{VI},[10] W^{VI},[11] and Ga^{III},[12] Ge^{IV},[13] Sn^{IV},[14] respectively).[15] However, a major drawback of these materials is their rather limited light absorption and thus poor photocatalytic activity in the visible light spectrum due to a deep VB of oxides. Incorporating the chalcogenides sulphur or selenium into these functional materials was found to have a beneficial influence on their photocatalytic activities, resulting in less positive VB potentials to enable visible light absorption and CB energies suitable for fuel synthesis. Consequently, metal chalcogenides such as ZnS,[16,17] CdS,[18] $CdSe$[19] or multinary metal chalcogenides, *e.g.* alloys of ZnS with Ag or Cu,[15,20,21] have received increasing attention within the past few decades.

In terms of proton reduction using sulphidic minerals, much can be learned from naturally occurring enzymes such as hydrogenases.[22] These sophisticated natural machines enable the efficient formation of H_2 through a combination of a fine-tuned Fe/Ni-sulphide active centre and additional Fe/S-clusters for electron-conductivity through the protein. We have recently reported on the favourable properties of the bulk natural Fe/Ni sulphide mineral pentlandite ($Fe_{4.5}Ni_{4.5}S_8$) as an effective electrocatalyst for the hydrogen evolution reaction.[23–26] The fundamental structural features of this bio-inspired catalyst made from highly abundant and cheap precursors resemble conditions prevalent in metal-sulphide clusters of Fe/Ni-hydrogenases. While being effective as an electrocatalyst, its efficiency as a photocatalyst to drive proton reduction has not yet been reported and is not found in the natural reactivity spectrum of its natural congeners. Since bulk materials are in most cases inactive for photocatalytic conversions, it is well established that material nanostructuring may lead to significant improvement of the photocatalytic performance, due to the fine-tuning of the band gap and, more importantly, kinetic factors such as suppression of electron/hole pair recombination and a larger catalytically active surface area.[15,27]

We herein report synthetic procedures towards nanoparticles of synthetic pentlandites with well-defined Fe : Ni ratios.[24] The obtained materials are characterised by powder-XRD, scanning electron microscopy, UV-Vis and energy dispersive X-ray spectroscopy. To probe the suitability of our materials as photocatalysts for hydrogen evolution, the band gaps and positions of the different $Fe_xNi_{9-x}S_8$ ($x = 3–6$) particles were determined by UV-Vis spectroscopy and Mott–Schottky analysis. Furthermore, for the first time, $Fe_3Ni_6S_8$ is applied as a representative photocatalyst in solar-driven reduction. By presenting this rather surprising and naturally unprecedented activity of pentlandite after nano-structuring, we emphasise the synergetic potential of bioinspired chemistry and materials design.

Experimental section

Materials

All chemimolecular cals were obtained from commercial vendors and used without further purification. Iron, nickel and sulphur were purchased from Sigma-Aldrich with purities of at least 99.98%. $Fe(NO_3)_3 \cdot 9H_2O$ and $Ni(NO_3)_2 \cdot 6H_2O$ (both analytical grade) were purchased from Merck. Citric acid (99%) was purchased from Acros Organics and deionized water was obtained by the water purification system Direct-Q from Millipore. Synthesis of the $[Ni(P_2^{Ph}N_2^{PhCH_2P(O)(OH)_2})_2]$ (**NiP**, molecular DuBois-type Ni proton reduction catalysts) was synthesized as reported previously.[28]

Synthesis of pentlandite nanoparticles by sol–gel approach

$Fe_xNi_{9-x}S_8$ ($x = 3\text{–}6$) nanoparticles were prepared following adopted literature procedures.[29,30] The metal salts, $Fe(NO_3)_3 \cdot 9H_2O$, $Ni(NO_3)_2 \cdot 6H_2O$ (total amount *ca.* 5 g) were dissolved in 15 mL deionized water with the desired stoichiometry of the final product. Citric acid (total metal to citric acid ratio $= 1$) was then added and the solution was kept at 80 °C until a gel was formed. The gel was placed in a tubular furnace which was subsequently purged for 10 min with H_2S/H_2 (15 : 85) at room temperature. Maintaining the H_2S/H_2 flow rate at 40 mL min^{-1}, the temperature was increased up to 300 °C (5 K min^{-1}) and held stable for 1 h. Then, a hydrogen flow was fixed at 70 mL min^{-1}, the temperature was raised to 400 °C (5 K min^{-1}) and held for 3 h. The mixture was then allowed to cool to room temperature while purging with N_2.

Synthesis of pentlandite nanoparticles by ball milling

Bulk $Fe_xNi_{9-x}S_8$ ($x = 3\text{–}6$) was prepared according to a high temperature method recently published by our group.[24] Ball milling of the prepared pentlandite materials was then carried out using the planetary ball mill (Fritsch Pulverisette 7, premium line) with ZrO_2 milling containers (volume: 20 mL) to avoid any potential pollution with iron or nickel. Typically, about 3 g of the materials were milled for 3 min at a speed rate of 700 rpm using ZrO_2 milling balls with a diameter of 10 mm. Next, the materials were additionally milled for 30 min at a speed rate of 1100 rpm after replacing the milling balls by smaller ZrO_2 milling balls with a diameter of 2 mm and adding 4 mL of isopropyl alcohol. The obtained particles were collected by ultracentrifugation (8500 rpm, 10 min).

Physical methods

Powder X-ray diffraction (XRD) was performed on a HUBER powder X-ray diffractometer with Mo-Kα radiation ($\lambda = 0.709$ Å). The reflex positions were converted *via* Bragg's law from Mo to Cu radiation. The crystallite size d of the samples was calculated employing the Debye–Scherrer equation (eqn (1)); k represents the Scherrer constant set to $k = 0.89$, λ the wavelength of the X-ray radiation source and β the full width at half maximum (FWHM) of the diffraction peak at 2θ.

$$d = \frac{k\lambda}{\beta \, \cos(\theta)} \tag{1}$$

The morphology of the samples was studied by scanning electron microscopy (SEM, Jeol JSM 7500F) at a voltage of 5 kV. Elemental mapping of the samples by electron dispersive X-ray spectroscopy (EDX) was also carried out at 5 kV.

Solid state UV/Vis measurements were performed on a Jasco V-670 spectrophotometer by measuring the reflectance of the samples using $BaSO_4$ as a standard. The spectral data were used to plot the absorbance of the samples and estimate the cut-off wavelength *via* two tangents. The band gap energies E were calculated using eqn (2); h is the Planck constant, c the speed of light and $\lambda_{cut\,off}$ the cut-off wavelength.

$$E = \frac{hc}{\lambda_{cutoff}} \tag{2}$$

Mott–Schottky measurements were carried out in the dark utilising a GAMRY Reference 600+ potentiostat. For this purpose, an electrochemical cell with a three-electrode configuration was employed; pentlandite was used as the working electrode, Pt served as the auxiliary electrode, and Ag/AgCl (3 M KCl) as the reference electrode. All experiments were carried out in a 0.5 M H_2SO_4 electrolyte solution, from 1.0 V to −1.0 V *vs.* Ag/AgCl (3 M KCl) (10 mV steps) at 100 Hz. The measured potential was converted to the NHE potential according to $E_{NHE} = E_{Ag/AgCl} + 0.210$ V. The pentlandite working electrode was prepared according to a method recently published by our group.[31]

Photocatalytic measurements

The samples for photocatalytic experiments were prepared by dissolving 2 mg of the photocatalysts $Ni_3Fe_6S_8$ or $NiFe_2O_4$, along with **NiP** (50 nmol) or the precursor for metallic Pt (H_2PtCl_6, 8 wt%) in aqueous EDTA solutions (0.1 M, 3 mL, pH 6) in borosilicate glass vials (7.74 mL). All vessels were purged with N_2 containing 2% CH_4 (internal gas chromatography standard) for 15 min and illuminated with a solar light simulator (Newport Oriel, 100 mW cm^{-2}) equipped with an air mass 1.5 global filter (AM 1.5G). The samples were constantly stirred during irradiation and the temperature was maintained at 25 °C. The amount of H_2 produced was determined by gas chromatography after 48 h of irradiation, by analysing samples of the headspace gas (20 μL) using an Agilent 7890A gas chromatograph, equipped with a 5 Å molecular sieve column and a thermal conductivity detector (TCD).

Results and discussion

In many cases, bulk materials are essentially inactive as photocatalysts. We therefore aimed for a simple and reliable methodology to prepare pentlandite particles with different Fe/Ni ratios in smaller sizes. We anticipated that these nanoparticles may show photocatalytic activity due to improved electron transfer dynamics, short diffusion pathways for charge carriers, and an increased number of catalytic surface sites. Two different synthetic routes were examined. Firstly, a top-down ball milling approach using synthetic bulk pentlandites was investigated. Secondly, we examined a bottom-up sol–gel approach specifically adapted as a novel method for synthesising pentlandites. For the latter approach, a gel prepared from $Fe(NO_3)_3$ and $Ni(NO_3)_2$ was subsequently heated under an atmosphere of H_2S/H_2 (300 K, 1 h) and H_2 (400 K, 3 h). The Fe : Ni ratio within the final product was controlled by utilising the metal precursors in the stoichiometry of the desired product. For both synthetic methods, the different Fe : Ni ratios corresponding to the total formula $Fe_xNi_{9-x}S_8$ ($x = 3$–6) were confirmed by EDX (Table S1 and S2, Fig. S3 and S5†).

The obtained materials were characterised by powder XRD and compared to previously published bulk pentlandite materials containing the same amount of Fe and Ni (Fig. 1).[24] The newly prepared materials reveal similar diffraction patterns compared to $Fe_{4.5}Ni_{4.5}S_8$ suggesting an unaltered overall crystal structure with slightly altered lattice constants of the cubic unit cell (space group of pentlandite: $Fm\overline{3}m$). The particles also remain unaffected by the ball milling process (Fig. 1a). The significantly broadened band pattern of the materials obtained by ball milling as well as the sol–gel process further indicates the successful formation of small crystallites.

While the synthesis of pentlandites $Fe_xNi_{9-x}S_8$ with a high iron to nickel content ($x > 4.5$) by the sol–gel process provided phase pure compounds (Fig. 1b), lower Fe : Ni ratios ($x < 4.5$) resulted in significant formation of Ni_3S_2 (Heazlewoodite) as by-product. Similar behaviour was already observed by Sugaki and Kitakaze showing the preferential formation of Ni_3S_2 at temperatures below 850 °C.[32]

SEM characterisation of the obtained materials revealed pronounced differences between particles derived from ball milling (Fig. 2, top and S2†) and the sol–gel process (Fig. 2, bottom and S4†). While ball milling (Fig. 2, top) results in a broader particle size distribution (1–10 μm), the sol–gel approach (Fig. 2, bottom) yielded smaller particles below 1 μm. However, it is evident that the nanoparticles agglomerated in the course of the catalyst preparation. A crystallite

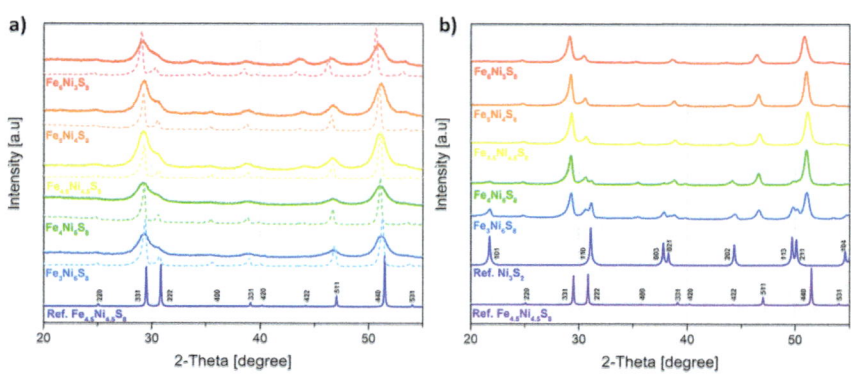

Fig. 1 Powder X-ray diffractograms (Cu Kα) of the samples prepared *via* the (a) ball milling approach and (b) sol–gel process. The dashed lines in (a) represent the diffractograms of the compounds prior to the ball milling process, the solid lines display the diffractograms of the ball-milled samples.

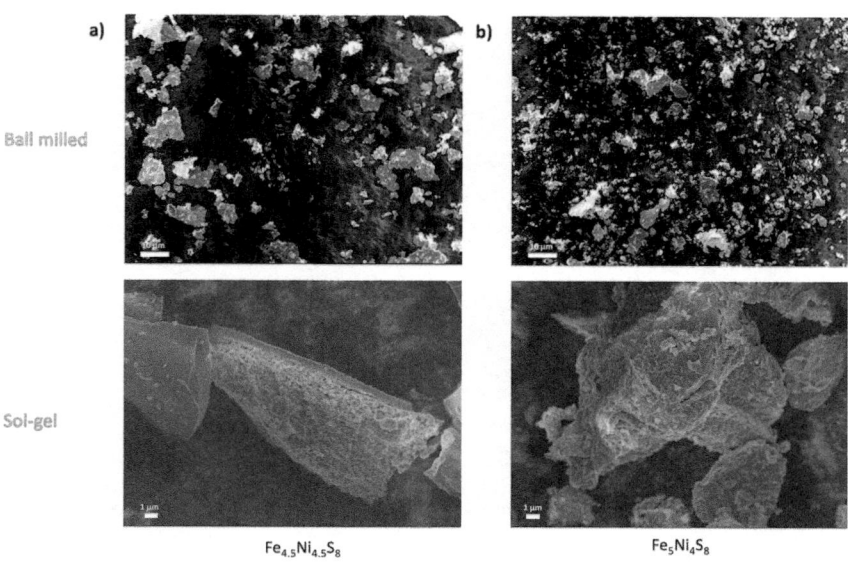

Fig. 2 Comparison of SEM images of (a) $Fe_{4.5}Ni_{4.5}S_8$ and (b) $Fe_5Ni_4S_8$ obtained from ball-milled samples (top) and materials prepared by sol–gel approach (bottom).

size analysis by Debye–Scherrer applied on the 511 reflex (Fig. 1) revealed sizes around 8 nm for the ball-milled samples and approximately 20 nm for the sol–gel samples (Table 1). The apparent discrepancy between this and the SEM analysis is attributed to the higher degree of crystallinity obtained in course of the sol–gel process as well as to the formation of larger crystallites due to extensive sintering processes.

To elucidate the potential of our materials to drive photocatalytic hydrogen production, their electronic band structures were analysed by UV-Vis spectroscopy (Fig. 3a). The measurements reveal a major influence of the pentlandite Fe : Ni ratio on the band gaps of the materials.

For the ball-milled samples, a decrease of the total band gap from 2.39 V ($Fe_3Ni_6S_8$) to 2.02 V ($Fe_6Ni_3S_8$) with an increasing Fe content is observed (Fig. 3c).

Table 1 Comparison of key properties of the presented pentlandite materials

Sample	Crystallite size via PXRD [nm]	Cut-off wavelength [nm]	Band gap [eV]
$Fe_3Ni_6S_8$ (ball-milled)	8.06	519	2.39
$Fe_4Ni_5S_8$ (ball-milled)	7.66	538	2.31
$Fe_{4.5}Ni_{4.5}S_8$ (ball-milled)	8.07	544	2.28
$Fe_5Ni_4S_8$ (ball-milled)	8.77	576	2.16
$Fe_6Ni_3S_8$ (ball-milled)	8.28	616	2.02
$Fe_3Ni_6S_8$ (sol–gel)	18.43	522	2.38
$Fe_4Ni_5S_8$ (sol–gel)	21.50	499	2.49
$Fe_{4.5}Ni_{4.5}S_8$ (sol–gel)	18.06	516	2.41
$Fe_5Ni_4S_8$ (sol–gel)	21.50	529	2.35
$Fe_6Ni_3S_8$ (sol–gel)	15.83	597	2.08

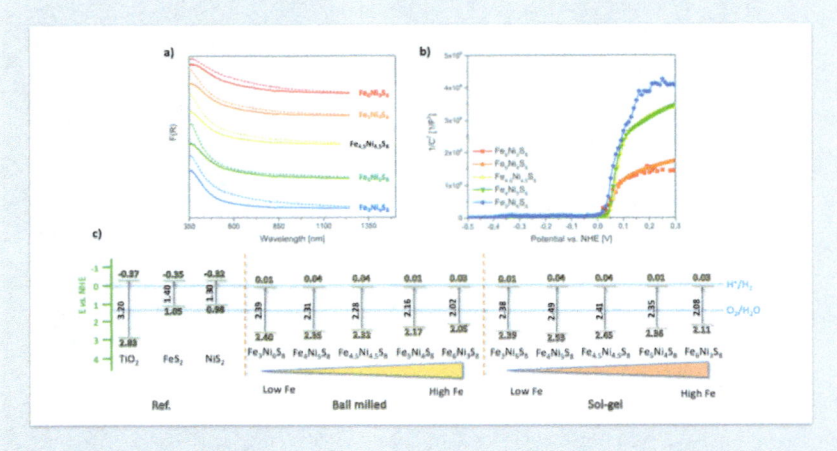

Fig. 3 (a) UV-Vis absorption spectra of the prepared materials (dotted line represents materials from the sol–gel approach and solid line represents ball-milled samples). (b) Mott–Schottky plots of the various ball-milled pentlandite materials measured in 0.5 M H_2SO_4. (c) Energy band diagram representing band gaps and band positions of the presented materials in comparison to TiO_2 and other reference materials.[33]

A similar trend was observed for the sol–gel samples, with the exception of $Fe_3Ni_6S_8$ (2.38 V) which has a slightly narrower band gap compared to $Fe_4Ni_5S_8$ (2.49 V). We assume that this altered behaviour can be reasoned by the small impurities of the material with Ni_3S_2. Comparison between the materials with the same Fe : Ni ratio from the different synthetic attempts reveal that ball-milled pentlandite materials provide smaller band gaps than those samples obtained from the sol–gel synthesis.

A Mott–Schottky analysis was performed in 0.5 M H_2SO_4 with ball-milled and sol–gel-derived materials (Fig. 3b). Since the results for both materials are identical, we subsequently refer to the ball-milled materials for detailed discussion. In all cases, the resulting Mott–Schottky graph has a positive slope indicating n-type conductivity for the pentlandite materials (Fig. S6†). This behaviour is comparable to the common metal chalcogenides FeS_2 and NiS_2 that were reported to show n-type conduction behaviour in 0.5 M H_2SO_4.[34] In addition, the observed flatband potentials of the various pentlandite materials remain comparable across all materials with a potential of 0.01 to 0.04 V *vs.* NHE. Thus, the altered Fe : Ni ratio only has a minor influence on the CB position of the particles, which is slightly below the flatband potential. On the other hand, a clear influence of metal substitution on the VB is apparent, although the VB composition is mainly dictated by sulphur. Thus, the observed different band gap energies are a consequence of altered VB energy caused by metal exchange. Notably, for all samples the CBs are located at approximately the H^+/H_2 redox couple (0 V *vs.* NHE) and might indicate a severely hampered ability to perform photocatalytic H_2 evolution.

We nevertheless decided to conduct photocatalytic experiments with our materials using $Ni_3Fe_6S_8$ as a representative benchmark system and the inverse spinel $NiFe_2O_4$ as a comparable oxide. The potential Fe/Ni-chalcogenide photocatalysts, $Ni_3Fe_6S_8$ and its congener $NiFe_2O_4$ were investigated with a H_2 evolving molecular DuBois-type Ni catalyst[28] (**NiP**, Fig. S1†) or metallic Pt photo-reduced from H_2PtCl_6 (8 wt%) co-catalyst in an aqueous solution of the sacrificial electron donor ethylenediaminetetraacetic acid (EDTA, pH 6) as was previously employed for SC light absorbers such as dye-sensitised TiO_2, nitride or carbon nanodots.[28,35,36]

Subsequently, the $Ni_3Fe_6S_8$ sample was irradiated with simulated solar light (AM 1.5 G, 100 mW cm^{-2}), under N_2 atmosphere (2% CH_4 as internal gas chromatography standard) at 25 °C for 48 h. The H_2 produced was quantified by gas chromatography (GC) upon analysing gas samples of the headspace (20 μL).

Notably, $Fe_6Ni_3S_8$ in conjunction with Pt and **NiP** was found to be a photocatalytic system for H_2 formation. The highest H_2 yield (0.22 ± 0.02 μmol) was observed for the $Ni_3Fe_6S_8$/**NiP** couple after 48 h of irradiation (Fig. 4). This corresponds to a specific activity of 2 μmol H_2 $(g_{photocatalyst})^{-1}$ h^{-1}, which is lower than that of the Pt and **NiP** benchmark systems using carbon dots and carbon nitrides tested under similar conditions. Contrary to the $Fe_6Ni_3S_8$/**NiP** or Pt couple, such systems reveal specific activities up to 7950 μmol H_2 $(g_{photocatalyst})^{-1}$ h^{-1}.[35,37,38] When $Ni_3Fe_6S_8$ was tested with Pt as co-catalyst, 0.07 ± 0.01 μmol H_2 was obtained (Fig. 4), showing the previously reported advantage of using **NiP** over Pt in hybrid photocatalyst systems.[35,38]

To further investigate our material, we performed linear sweep voltammetry experiments of $Fe_xNi_{9-x}S_8$ electrodes in 0.5 M H_2SO_4 in the dark as well as using simulated solar light conditions. (Fig. S7†). The obtained data, however, revealed no significant changes in the current–potential response under both conditions.

Contrary to the sulfidic materials, the oxygen spinel congener, $NiFe_2O_4$ that possesses a band gap of 1.7 ± 0.1 eV (Fig. S8†), failed to produce any H_2 with either **NiP** or Pt under the same experimental conditions (Fig. 4). This behaviour reveals a poorer photocatalytic performance of $NiFe_2O_4$ compared with $Ni_3Fe_6S_8$ and the positive influence of sulphur on the photocatalytic properties upon its incorporation into the structure of the nanoparticles.

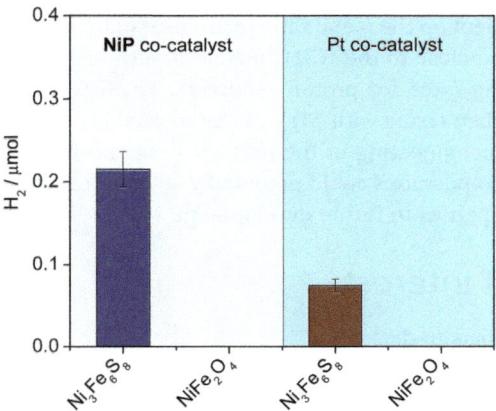

Fig. 4   Solar-driven H_2 evolution using $Ni_3Fe_6S_8$ (2 mg) and $NiFe_2O_4$ (2 mg) nanoparticles as photoabsorbers. **NiP** (50 nmol) or Pt (8 wt%) were employed as co-catalysts and EDTA (0.1 M, 3 mL, pH 6) as the sacrificial electron donor. Conditions: irradiation with simulated solar light (AM 1.5G, 100 mW cm^{-2}, 25 °C) for 48 h under N_2 atmosphere with 2% CH_4.

Within the process of the photocatalytic H_2 generation, tertiary amines such as EDTA and triethanolamine (TEOA) are common sacrificial electron donors in photocatalytic schemes.[39] They are quenched by CB holes to produce N- and C-centred radical cations to ultimately yield a wide range of products such as imi-nodiacetic acid and diethanolamine.[39–42] Minor oxidation products could also derive from the degradation of the ligand framework of the molecular catalyst **NiP** upon quenching the holes of the CB photoexcited state.[42]

The selective exchange of oxygen *vs.* sulphur within mixed-metal Fe/Ni-chalcogenides for photocatalytic applications thus presents a first step in iden-tifying new sustainable photoabsorbers for H_2 production. The fine tuning of the composition/structural properties of such materials by means of smaller particle sizes, crystallinity and different metal compositions will be attempted in the future with the aim of achieving improved photocatalytic performances by a rational band gap design. It is also expected that the incorporation of selenium as heavier chalcogenide could have an influence on the photocatalytic properties of such materials, due to effects on their metallic character and thus their properties as p- or n-semiconductors.

Conclusions

In an attempt to establish pentlandites, $Fe_xNi_{9-x}S_8$ ($x = 3$–6), as potential pho-tocatalysts for solar-driven H_2 production, we realised a versatile preparation method towards pentlandite nanoparticles with varying Fe : Ni content *via* two conceptually different approaches. A novel sol–gel process conveniently yielded particles smaller than in the herein presented alternative ball milling process of bulk pentlandites as determined by SEM measurements. However, pentlandites with a low Fe content ($x < 4.5$) were accompanied with phase impurities from Ni_3S_2 and further optimisation of this synthetic attempt is required in the future. The suitability of these materials as photocatalysts for solar-driven H_2 production was assessed. The herein reported pentlandite materials are n-type semi-conductors with band gaps between 2.08 V and 2.49 V, depending on the Fe : Ni ratio and the particle size. The position of the conduction band was only marginally dependent on the metal ratio in all cases (0.01–0.04 V *vs.* NHE in 0.5 M H_2SO_4) and located close to the H^+/H_2 potential. Although there is limited ther-modynamic driving force for proton reduction, $Fe_3Ni_6S_8$ showed some photo-catalytic activity when tested with **NiP** or Pt as co-catalyst. Band gap tuning of the pentlandites, nanoengineering of the particle sizes and further optimisation of their elemental compositions could potentially result in an improvement in their photocatalytic properties in future development.

Conflicts of interest

There are no conflicts to declare.

Acknowledgements

The authors acknowledge financial support of the Fonds of the Chemical Industry (Liebig grant to U.-P. A.), the Deutsche Forschungsgemeinschaft (Emmy Noether grant to U.-P. A., Cluster of Excellence RESOLV (EXC2033), AP242/2-1; AP242/6-1;

MA 5392/7-1), the Fraunhofer Internal Programs under Grant No. Attract 097-602175 as well as the Christian Doppler Association (Austrian Federal Ministry for Digital and Economic Affairs, the National Foundation for Research, Technology and Development) and OMV.

References

1 S. E. Hosseini and M. A. Wahid, *Renewable Sustainable Energy Rev.*, 2016, **57**, 850–866.

2 W. J. Youngblood, S.-H. A. Lee, K. Maeda and T. E. Mallouk, *Acc. Chem. Res.*, 2009, **42**, 1966–1973.

3 Z. Li, W. Luo, M. Zhang, J. Feng and Z. Zou, *Energy Environ. Sci.*, 2013, **6**, 347–370.

4 L. Steier and S. Holliday, *J. Mater. Chem. A*, 2018, **6**, 21809–21826.

5 Y. Wang, H. Suzuki, J. Xie, O. Tomita, D. J. Martin, M. Higashi, D. Kong, R. Abe and J. Tang, *Chem. Rev.*, 2018, **118**, 5201–5241.

6 S. K. Saraswat, D. D. Rodene and R. B. Gupta, *Renewable Sustainable Energy Rev.*, 2018, **89**, 228–248.

7 T. Yao, X. An, H. Han, J. Q. Chen and C. Li, *Adv. Energy Mater.*, 2018, **8**, 1800210.

8 A. Fujishima and K. Honda, *Nature*, 1972, **238**, 37–38.

9 A. Miyoshi, S. Nishioka and K. Maeda, *Chem.–Eur. J.*, 2018, **24**, 18204–18219.

10 H. Kato, N. Matsudo and A. Kudo, *Chem. Lett.*, 2004, **33**, 1216–1217.

11 P. Dias, T. Lopes, L. Meda, L. Andrade and A. Mendes, *Phys. Chem. Chem. Phys.*, 2016, **18**, 5232–5243.

12 M.-G. Ju, X. Wang, W. Liang, Y. Zhao and C. Li, *J. Mater. Chem. A*, 2014, **2**, 17005–17014.

13 H. Kadowaki, J. Sato, H. Kobayashi, N. Saito, H. Nishiyama, Y. Simodaira and Y. Inoue, *J. Phys. Chem. B*, 2005, **109**, 22995–23000.

14 J. Sato, N. Saito, H. Nishiyama and Y. Inoue, *J. Phys. Chem. B*, 2001, **105**, 6061–6063.

15 S. Y. Tee, K. Y. Win, W. S. Teo, L.-D. Koh, S. Liu, C. P. Teng and M.-Y. Han, *Adv. Sci.*, 2017, **4**, 1600337.

16 G.-J. Lee and J. J. Wu, *Powder Technol.*, 2017, **318**, 8–22.

17 N. Dengo, A. F. De Fazio, M. Weiss, R. Marschall, P. Dolcet, M. Fanetti and S. Gross, *Inorg. Chem.*, 2018, **57**, 13104–13114.

18 Y.-J. Yuan, D. Chen, Z.-T. Yu and Z.-G. Zou, *J. Mater. Chem. A*, 2018, **6**, 11606–11630.

19 F. Qiu, Z. Han, J. J. Peterson, M. Y. Odoi, K. L. Sowers and T. D. Krauss, *Nano Lett.*, 2016, **16**, 5347–5352.

20 H. Kaneko, T. Minegishi, M. Nakabayashi, N. Shibata, Y. Kuang, T. Yamada and K. Domen, *Adv. Funct. Mater.*, 2016, **26**, 4570–4577.

21 I. Tsuji, H. Kato and A. Kudo, *Angew. Chem., Int. Ed.*, 2005, **44**, 3565–3568.

22 F. Möller, S. Piontek, R. G. Miller and U.-P. Apfel, *Chem.–Eur. J.*, 2018, **24**, 1471–1493.

23 B. Konkena, K. junge Puring, I. Sinev, S. Piontek, O. Khavryuchenko, J. P. Dürholt, R. Schmid, H. Tüysüz, M. Muhler, W. Schuhmann and U.-P. Apfel, *Nat. Commun.*, 2016, **7**, 12269.

24 S. Piontek, C. Andronescu, A. Zaichenko, B. Konkena, K. junge Puring, B. Marler, H. Antoni, I. Sinev, M. Muhler, D. Mollenhauer, B. Roldan Cuenya, W. Schuhmann and U.-P. Apfel, *ACS Catal.*, 2018, **8**, 987–996.

25 I. Zegkinoglou, A. Zendegani, I. Sinev, S. Kunze, H. Mistry, H. S. Jeon, J. Zhao, M. Y. Hu, E. E. Alp, S. Piontek, M. Smialkowski, U.-P. Apfel, F. Körmann, J. Neugebauer, T. Hickel and B. Roldan Cuenya, *J. Am. Chem. Soc.*, 2017, **139**, 14360–14363.

26 C. L. Bentley, C. Andronescu, M. Smialkowski, M. Kang, T. Tarnev, B. Marler, P. R. Unwin, U.-P. Apfel and W. Schuhmann, *Angew. Chem., Int. Ed.*, 2018, **57**, 4093.

27 F. E. Osterloh, *Chem. Soc. Rev.*, 2013, **42**, 2294–2320.

28 M. A. Gross, A. Reynal, J. R. Durrant and E. Reisner, *J. Am. Chem. Soc.*, 2014, **136**, 356–366.

29 M. Atif, M. Nadeem and M. Siddique, *Appl. Phys. A: Mater. Sci. Process.*, 2015, **120**, 571–578.

30 I. Bezverkhyy, P. Afanasiev and M. Danot, *J. Phys. Chem. B*, 2004, **108**, 7709–7715.

31 K. junge Puring, S. Piontek, M. Smialkowski, J. Burfeind, S. Kaluza, C. Doetsch and U.-P. Apfel, *J. Visualized Exp.*, 2017, e56087.

32 A. Sugaki and A. Kitakaze, *Am. Mineral.*, 1998, **83**, 133–140.

33 K. C. Christoforidis and P. Fornasiero, *ChemCatChem*, 2017, **9**, 1523–1544.

34 A. M. Huerta-Flores, L. M. Torres-Martínez, E. Moctezuma, A. P. Singh and B. Wickman, *J. Mater. Sci.: Mater. Electron.*, 2018, **29**, 11613–11626.

35 B. C. M. Martindale, G. A. M. Hutton, C. A. Caputo and E. Reisner, *J. Am. Chem. Soc.*, 2015, **137**, 6018–6025.

36 C. A. Caputo, M. A. Gross, V. W. Lau, C. Cavazza, B. V. Lotsch and E. Reisner, *Angew. Chem., Int. Ed.*, 2014, **53**, 11538–11542.

37 B. C. M. Martindale, G. A. M. Hutton, C. A. Caputo, S. Prantl, R. Godin, J. R. Durrant and E. Reisner, *Angew. Chem., Int. Ed.*, 2017, **56**, 6459–6463.

38 H. Kasap, C. A. Caputo, B. C. M. Martindale, R. Godin, V. W. Lau, B. V. Lotsch, J. R. Durrant and E. Reisner, *J. Am. Chem. Soc.*, 2016, **138**, 9183–9192.

39 Y. Pellegrin and F. Odobel, *C. R. Chim.*, 2017, **20**, 283–295.

40 H. S. Chang, G. V. Korshin and J. F. Ferguson, *Environ. Sci. Technol.*, 2006, **40**, 5089–5094.

41 F. Lakadamyali, M. Kato and E. Reisner, *Faraday Discuss.*, 2012, **155**, 191–205.

42 B. C. M. Martindale, E. Joliat, C. Bachmann, R. Alberto and E. Reisner, *Angew. Chem., Int. Ed.*, 2016, **55**, 9402–9406.

Faraday Discussions

PAPER

Distinguishing the effects of altered morphology and size on the visible light-induced water oxidation activity and photoelectrochemical performance of BaTaO$_2$N crystal structures

Mirabbos Hojamberdiev, [ID] *[ab] Kenta Kawashima, [ID] [bc]
Takashi Hisatomi, [ID] [d] Masao Katayama, [ID] [d] Masashi Hasegawa,[a]
Kazunari Domen [ID] [d] and Katsuya Teshima [ID] [be]

Received 8th November 2018, Accepted 28th November 2018

DOI: 10.1039/c8fd00170g

Factors, including crystallinity, morphology, size, preferential orientation, growth, composition, porosity, surface area, *etc.*, can directly influence the optical, charge-separation, charge-transfer and water oxidation and reduction properties of particle-based photocatalysts. Therefore, these factors must be considered when designing high-performance particle-based photocatalysts for solar water splitting. Here, a flux growth method was applied to alter the morphology and size of Ba$_5$Ta$_4$O$_{15}$ precursor oxide crystals using BaCl$_2$, KCl, RbCl, CsCl, KCl + BaCl$_2$ and K$_2$SO$_4$ at different solute concentrations, and the impact of nitridation with and without KCl flux was studied. Specifically, the effects of altered morphology and size on the visible light-induced water oxidation activity and photoelectrochemical performance of the BaTaO$_2$N crystal structures were investigated. Upon nitridation, the samples became porous due to the lattice shrinkage caused by the replacement of 3 O^{2-} with 2 N^{3-} in the anionic network. The BaTaO$_2$N crystal structures obtained by nitridation without KCl flux show higher surface areas than do their counterparts prepared by nitridation with KCl flux because of the formation of porous networks. All of the samples exhibited a high anodic photocurrent upon nitridation without KCl flux compared with those of the samples obtained by nitridation with KCl flux. These findings demonstrate that it is important to specifically engineer photocatalytic crystals to reach their maximum potential in solar water splitting.

[a]Department of Materials Physics, Nagoya University, Furo-cho, Chikusa-ku, Nagoya 464-8603, Japan. E-mail: hmirabbos@gmail.com; hmirabbos@mp.pse.nagoya-u.ac.jp

[b]Department of Environmental Science and Technology, Faculty of Engineering, Shinshu University, 4-17-1 Wakasato, Nagano 380-8553, Japan

[c]Department of Chemistry, The University of Texas at Austin, Austin, Texas 78712, USA

[d]Department of Chemical System Engineering, School of Engineering, The University of Tokyo, 7-3-1 Hongo, Bunkyo-ku, Tokyo 113-8656, Japan

[e]Center for Energy and Environmental Science, Shinshu University, 4-17-1 Wakasato, Nagano 380-8553, Japan

1. Introduction

Solar water splitting using semiconductor photocatalysts is considered an environmentally benign process for the production of renewable, clean and storable hydrogen energy. In addition to the primary requirements (*e.g.*, band edge potentials suitable for overall water splitting, a band gap energy smaller than 3 eV, and stability in the photocatalytic reaction),[1] there are other factors, such as crystal morphology, size, dimension, composition, porosity, and surface area, that can greatly influence the water splitting activity. For instance, as zero-dimensional photocatalysts, nitrogen-doped graphene oxide quantum dots were previously applied for overall water-splitting under visible light illumination, and the p- and n-domains were found to be responsible for the production of H_2 and O_2 gases, respectively.[2] A 100% photon-to-hydrogen production efficiency for the photocatalytic water-splitting reduction half-reaction under visible light irradiation was accomplished by utilization of Pt-tipped CdSe@CdS one-dimensional nanorods with a hydroxyl anion-radical redox couple.[3] By using two-dimensional CdS nanoplatelet-Pt heterostructures, the lifetime of the charge-separated state was extended, and the internal quantum efficiency of H_2 generation exceeded 40% at pH = 8.8–13 and approached unity at pH = 14.7.[4] A three-dimensional plasmonic photoanode was developed using titanium dioxide nanotunnels loaded with gold nanoparticles for water splitting and showed a significant enhancement in photocurrent generation and water splitting compared to that of conventional two-dimensional plasmonic devices.[5] Using Pt-loaded CdS nanoporous structures, an apparent quantum yield of 60% at 420 nm in the visible light production of hydrogen was obtained because the nanopores created within the CdS nanostructures could eliminate the surface recombination of photogenerated holes and electrons.[6]

Perovskite-type transition metal oxynitrides have been regarded as a promising class of inorganic materials (n-type semiconductors), having absorption bands in the wavelength range of 500–750 nm and band gap energies of 1.8–2.5 eV, and they can be used in solar water splitting systems for the production of renewable, storable and clean hydrogen energy. Overall water splitting operable under visible light of up to 600 nm was recently achieved for the first time using a $LaMg_xTa_{1-x}O_{1+3x}N_{2-3x}$ ($x \le 1/3$) solid solution with an amorphous oxyhydroxide surface layer.[7] Very recently, a particulate $BaNbO_2N$ photoanode exhibited a photocurrent of 5.2 mA cm^{-2} at 1.23 V_{RHE} during photoelectrochemical water oxidation under simulated solar irradiation (AM 1.5G), which is the highest yet reported for an oxynitride responsive at wavelengths above 600 nm, due to the enhanced surface crystallinity.[8]

Although there has been significant progress in achieving a higher efficiency of solar-to-hydrogen energy conversion in perovskite-type transition metal oxynitrides, this still needs to be further enhanced for practical applications.[9] One of the possible rationales for the low efficiency of transition metal oxynitride photocatalysts is the presence of a high density of intrinsic defects in their crystal structures, stemming from a long high-temperature nitridation process. Such intrinsic defects act as recombination centers for photogenerated charge carriers, lowering the efficiency.[10] To reduce the density of intrinsic defects and to enhance the photocatalytic water splitting activity of various oxide and oxynitride photocatalysts, we successfully developed a direct flux growth approach and applied the band gap engineering technique of cation doping. Namely, $BaNbO_2$-N,[11] $BaNb_{0.5}Ta_{0.5}O_2N$,[12] $BaTaO_2N$,[13] $SrNbO_2N$,[14] $LaTiO_2N$,[15] W-doped $LaTiO_2N$,[16] $LnTaON_2$ (Ln = La, Pr),[17] $Sr_{1-x}Ba_xW_{1-y}Ta_y(O,N)_3$,[18] $AW(O,N)_3$ (A = Sr, La, Pr, Nd, Eu),[19] and $La_{1-x}Sr_xFe_{1-y}Ti_yO_3$ (ref. 20) crystals were grown by a flux growth method/solid-state reaction and showed an enhanced photocatalytic activity towards photocatalytic water splitting.

As a 600 nm-class photocatalyst, $BaTaO_2N$ has received particular attention due to its smaller band gap ($E_g = 1.8$ eV), suitable band edge positions for visible light-induced water reduction and oxidation, stability in aqueous solutions, and nontoxicity.[13] In this study, we applied a flux growth method to alter the morphology and size of the $BaTaO_2N$ crystal structures and explored the effects of the altered morphology and size on the visible light-induced water oxidation activity and photoelectrochemical performance of the $BaTaO_2N$ crystal structures. The results demonstrate that the water oxidation activity and photo-electrochemical performance of perovskite-type $BaTaO_2N$ crystal structures are greatly influenced by their morphologies and sizes.

2. Experimental

2.1. Preparation

All of the reagents used in this study were purchased from Wako Pure Chemical Industries, Ltd. $BaCO_3$ (96%) and Ta_2O_5 (99.9%) were employed as solutes, while KCl (99.5%), RbCl (95.0%), CsCl (99.0%), $BaCl_2$ (96.0%) or K_2SO_4 (99.0%) was applied as the flux. The solute and flux were manually dry mixed in stoichiometric ratios for 30 min using an agate mortar and pestle. The well-homogenized mixture (10 g) with a 1–50 mol% solute concentration was placed in a platinum crucible with a capacity of 30 cm^3 and closed loosely with a platinum lid, heated at 1000 °C for 10 h at a heating rate of 50 °C h^{-1} and cooled at a cooling rate of 150 °C h^{-1}. The flux-grown $Ba_5Ta_4O_{15}$ crystals were separated from the remaining flux by washing with warm water repeatedly and dried at 100 °C for 12 h. The $BaTaO_2N$ crystal structures were obtained by nitriding the $Ba_5Ta_4O_{15}$ crystals, grown by a flux method using various fluxes and solute concentrations, at 950 °C for 20 h under an NH_3 flow (200 mL min^{-1}), followed by acid treatment (dilute HNO_3), rinsing with water and drying at 100 °C for 12 h. The preparation conditions of the samples are given in Table 1.

2.2. Characterization

The X-ray diffraction (XRD) patterns were acquired with a MiniflexII (Rigaku) diffractometer using Cu Kα radiation ($\lambda = 0.15418$ nm) in the 2θ scan range from 10° to 80° and compared with entries from the ICDD-PDF-2 powder pattern database. The morphology and size of the flux-grown $Ba_5Ta_4O_{15}$ crystals and $BaTaO_2N$ crystal structures were observed using a JSM-7600F field-emission-type scanning electron microscope (JEOL). From SEM images, the average size of the crystals and crystal structures were estimated by measuring at least 300 crystals and their crystal structures using Photo Measure® software (KENIS, Ltd.). High-resolution transmission electron microscopy (HR-TEM) images were captured on an EM-002B microscope (TOPCON) with an accelerating voltage of 200 kV.

Table 1 Preparation conditions, surface areas, photocurrent densities, and O_2 evolution rates of the samples

| Sample name | $Ba_5Ta_4O_{15}$ oxide precursor grown at 1000 °C for 10 h | | $BaTaO_2N$ prepared by nitridation with/without KCl flux at 950 °C for 20 h | Surface area $(m^2\ g^{-1})$ | Photocurrent density at 1.2 V_{RHE} $(mA\ cm^{-2})$ | O_2 evolution (first 3 h)/rate | |
	Flux	Solute concentration (mol%)				µmol	µmol h^{-1}
B10K	$BaCl_2$	10	Yes	5.04	1.19	403.91	134.64
B10			No	6.79	1.48	133.97	44.27
C10K	CsCl	10	Yes	1.41	0.52	56.92	18.97
C10			No	5.62	3.11	142.59	47.53
K1K	KCl	1	Yes	3.30	0.75	100.06	33.35
K1			No	3.53	1.21	132.85	44.28
K10K	KCl	10	Yes	2.89	1.15	77.58	25.86
K10			No	7.62	2.04	76.37	24.86
KB10K	KCl +	10	Yes	2.19	0.97	266.18	88.73
KB10	$BaCl_2$		No	5.09	1.20	112.86	37.62
KS50K	K_2SO_4	50	Yes	6.16	0.71	39.38	13.13
KS50			No	8.58	1.03	68.26	22.75
R10K	RbCl	10	Yes	2.55	0.83	43.13	14.38
R10			No	5.61	1.76	170.55	56.85

Ultraviolet-visible (UV-vis) diffuse reflectance spectra were recorded over a spectral range of 200–800 nm on a V-670 UV-vis-NIR spectrophotometer (JASCO) equipped with an integrating sphere, and the obtained data were converted from reflectance to the Kubelka–Munk (KM) function ($f(R_\infty)$) by the KM method. The structure of the prepared $CoO_x/BaTaO_2N/Ta/Ti$ electrode was examined using a JEM-2800 transmission electron microscope (TEM).

2.3. Photoelectrochemical performance test

For the photoelectrochemical (PEC) measurements, the $BaTaO_2N$ photoanodes were prepared by applying a particle transfer method.[11,21] Briefly, the $BaTaO_2N$ powder loaded with CoO_x (2 wt% Co) nanoparticles was suspended in isopropanol and sonicated for 30 min. Then, the suspension was drop casted onto a 30 mm × 30 mm glass plate followed by drying. Titanium (Ti) and tantalum (Ta) metals were deposited, as a conducting layer, by radio-frequency (RF) magnetron sputtering at 600 °C for 3 h to form a $BaTaO_2N/Ta/Ti$ layer. Subsequently, the layer was peeled off from the glass plate and the excess $BaTaO_2N$ powder (*i.e.*, particles that were loosely attached to the Ta conducting layer) were washed off in distilled water. Finally, the layer was attached to another glass plate using carbon tape and connected to a copper wire by indium solder. The unnecessary part was covered by an epoxy resin.

PEC measurements were conducted using a three-electrode system and potentiostat (HSV-120, Hokuto). The $CoO_x/BaTaO_2N/Ta/Ti$, an Ag/AgCl electrode (in saturated KCl), and a Pt wire were used as the working, reference, and counter electrodes, respectively. The electrolyte used for the measurement was 100 mL of 0.2 M K_2HPO_4 solution at pH = 13 (KOH adjusted). All the measurements were conducted under Ar gas flow. The photocurrent was recorded using linear sweep voltammetry (LSV) with a scan rate of 10 mV S^{-1} under AM 1.5G solar light irradiation (100 mW cm^{-2}, XES-40S1, San-Ei Electric Co., Ltd.).

2.4. Photocatalytic water oxidation test

Photocatalytic water oxidation half-reactions were performed in a side-irradiation-type reactor connected to a closed gas circulation system, equipped with a gas chromatograph (GC-8A, TCD, Ar gas carrier, Shimadzu) and a vacuum pump, under visible light irradiation (300 W Xe lamp with a cutoff filter ($\lambda > 420$ nm)) using 100 mg of $BaTaO_2N$ crystal structures loaded with CoO_x (2 wt% Co) nanoparticles as an O_2 evolution cocatalyst. Additionally, 200 mg of La_2O_3 and 300 mL of 10 mM $AgNO_3$ aqueous solution were employed as a pH buffer and a sacrificial electron scavenger, respectively. Prior to irradiation, the reactor was purged with Ar gas (20 mL min^{-1}) for 2 h to eliminate oxygen and nitrogen (*i.e.*, ambient air). CoO_x nanoparticles were loaded by immersing the $BaTaO_2N$ crystal structures in an aqueous solution of $Co(NO_3)_2 \cdot 6H_2O$ (>99.5%, Wako Pure Chemical Industries, Ltd.), followed by heating at 700 °C for 1 h under an NH_3 flow (200 mL min^{-1}) and reoxidizing at 200 °C for 1 h in air.

3.   Results and discussion

Previously, we attempted to tailor the morphology and size of $BaTaO_2N$ crystals by applying a direct flux growth approach using various fluxes.[13] However, only the KCl flux was favorable for the direct synthesis of cube-like $BaTaO_2N$ submicron-sized crystals. Here, $Ba_5Ta_4O_{15}$ oxide precursors with different morphologies and sizes were therefore grown first by a flux method using various fluxes, and the flux-grown $Ba_5Ta_4O_{15}$ oxide precursors were then subjected to high-temperature nitridation, with and without KCl flux, under flowing NH_3 gas to produce perovskite-type $BaTaO_2N$.

As an example, the XRD patterns of the oxide precursor, synthesized using a 10 mol% solute concentration and KCl flux, before and after high-temperature nitridation under flowing NH_3 gas at 950 °C for 20 h in the presence of KCl flux are shown in Fig. 1. As shown, the XRD pattern of the oxide precursor before high-temperature nitridation is almost identical to that of the hexagonal $Ba_5Ta_4O_{15}$ phase with space group $P\bar{3}m1(164)$ (ICDD PDF# 72-0631). No diffraction peaks belonging to impurity phases are noted, indicating the phase-purity of the flux-grown oxide precursor. The diffraction peaks in the XRD pattern of the sample obtained after high-temperature nitridation can be fully indexed to the perovskite-type cubic $BaTaO_2N$ phase with space group $Pm\bar{3}m$ (no. 221) (ICDD PDF# 84-1748), implying that the high-temperature nitridation under flowing NH_3 at 950 °C for 20 h was sufficient to successfully convert the layered perovskite $Ba_5Ta_4O_{15}$ to perovskite-type $BaTaO_2N$.

Fig. 1b shows the UV-vis diffuse reflectance spectra of the oxide precursor, synthesized using a 10 mol% solute concentration and KCl flux, before and after high-temperature nitridation under flowing NH_3 at 950 °C for 20 h in the presence of KCl flux (as a representative sample). Evidently, the oxide precursor exhibits a strong light absorption in the ultraviolet range due to the band-to-band transition and has an absorption edge at approximately 312 nm, corresponding to an

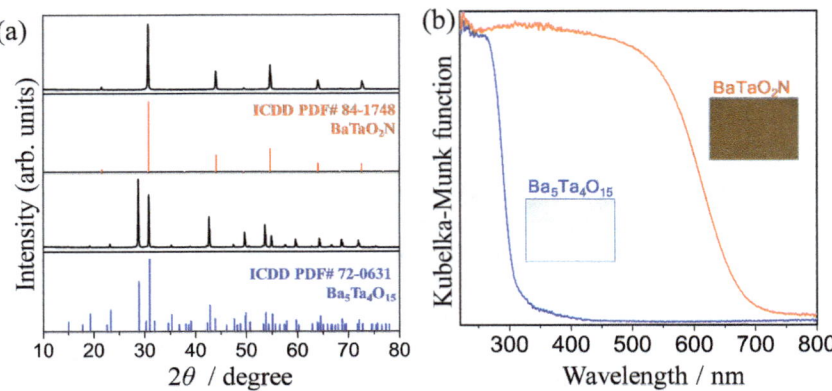

Fig. 1 (a) XRD patterns and (b) UV-vis diffuse reflectance spectra of $Ba_5Ta_4O_{15}$ crystals grown using a 10 mol% solute concentration and KCl flux and $BaTaO_2N$ crystal structures obtained by nitridation of $Ba_5Ta_4O_{15}$ crystals at 950 °C for 20 h in the presence of KCl flux.

optical band gap of 3.97 eV. In contrast, the sample obtained after high-temperature nitridation clearly shows a shift in the absorption edge ($\lambda \approx 665$ nm) toward longer wavelengths in the visible light region due to the energy of the N 2p orbital being higher than that of the O 2p orbitals, and the estimated optical band gap of the sample is approximately 1.86 eV. Additionally, the color of the powdered oxide precursor changes from white to brown after high-temperature nitridation, indicating the successful conversion of hexagonal layered-perovskite oxide $Ba_5Ta_4O_{15}$ to simple cubic perovskite oxynitride $BaTaO_2N$. Absorption beyond the absorption edge, which is generally attributed to the formation of reduced Ta species accompanying anion defects (anion vacancies or O_N anti-site defects), is not observed.

Since no significant differences among the XRD patterns and UV-vis diffuse reflectance spectra of the $Ba_5Ta_4O_{15}$ oxide precursors grown using different fluxes and those of the $BaTaO_2N$ crystal structures obtained by nitridation with and without KCl flux were observed, the samples were further examined by SEM. The SEM images of the $Ba_5Ta_4O_{15}$ oxide precursors and $BaTaO_2N$ crystal structures are compiled in Fig. 2. In the left panels, the SEM images of the $Ba_5Ta_4O_{15}$ oxide precursors are shown. It is apparent that the use of different fluxes and solute concentrations leads to the formation of $Ba_5Ta_4O_{15}$ crystals with various morphologies and sizes. For instance, when the $BaCl_2$ flux with a 10 mol% solute concentration was used, large idiomorphic polyhedral $Ba_5Ta_4O_{15}$ crystals with an average size of 12 μm were formed. Interestingly, primary nanocrystals were selectively attached to the largely exposed surfaces of the polyhedral $Ba_5Ta_4O_{15}$ crystals (Fig. 2a). Plate-like crystals of $Ba_5Ta_4O_{15}$ with well-developed faces and an average size of 0.85 and 2.54 μm were grown using CsCl and KCl fluxes, respectively, with a solute concentration of 10 mol% (Fig. 2b and d). Compared with the CsCl flux, the KCl flux also resulted in the formation of irregular submicron-sized crystals of $Ba_5Ta_4O_{15}$. Similarly, plate-like crystals of $Ba_5Ta_4O_{15}$ with unclear edges and an average size of 380 and 178 nm were grown using the KCl flux with a 1 mol % solute concentration and K_2SO_4 flux with a 10 mol% solute concentration, respectively (Fig. 2c and f). However, when the KCl and K_2SO_4 fluxes with 1 and

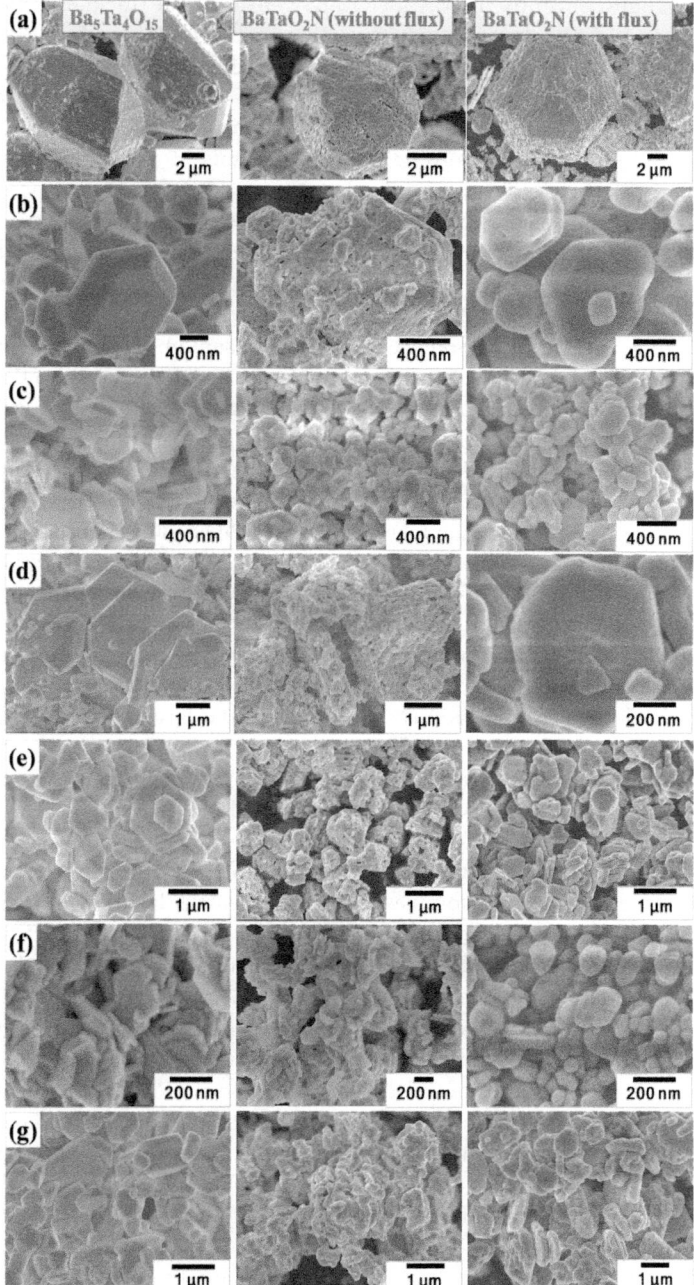

Fig. 2   SEM images of $Ba_5Ta_4O_{15}$ crystals (left column) grown using (a) $BaCl_2$ and a 10 mol % sol. conc., (b) CsCl and a 10 mol% sol. conc., (c) KCl and a 1 mol% sol. conc., (d) KCl and a 10 mol% sol. conc., (e) KCl + $BaCl_2$ and a 10 mol% sol. conc., (f) K_2SO_4 and a 50 mol% sol. conc. and (g) RbCl and a 10 mol% sol. conc. $BaTaO_2N$ crystal structures obtained by nitridation of the flux-grown $Ba_5Ta_4O_{15}$ crystals at 950 °C for 20 h with (right column) and without (middle column) KCl flux.

10 mol% solute concentrations, respectively, were used, the thickness of the plate-like crystals of $Ba_5Ta_4O_{15}$ was substantially reduced in comparison to that obtained with CsCl and KCl (10 mol% solute concentration). When the KCl and $BaCl_2$ fluxes were mixed together, idiomorphic polyhedral $Ba_5Ta_4O_{15}$ crystals with an average size of less than 1 μm were grown along with submicron-sized plate-like crystals (Fig. 2e). Idiomorphic irregular crystals of $Ba_5Ta_4O_{15}$ were grown using the RbCl flux (Fig. 2g). The flux crystal growth is a complex process and needs a thorough investigation to understand the complete mechanism. Here, it is believed that the flux cation size, solute concentration, and flux solubility played important roles in forming the $Ba_5Ta_4O_{15}$ crystals with different morphologies and sizes. In the middle panels, the SEM images of the respective samples after nitridation without KCl flux are represented. As shown, the samples with relatively larger sizes fairly retained the outline of their corresponding oxide precursor crystals (Fig. 2a, b, d, e and g). In contrast, the plate-like crystals of $Ba_5Ta_4O_{15}$, grown using the KCl flux with a 1 mol% solute concentration, completely lost their original shape and turned into coalesced irregular particles (Fig. 2c). The plate-like crystals of $Ba_5Ta_4O_{15}$ grown using the K_2SO_4 flux with a 10 mol% solute concentration fused strongly while keeping their original shape (Fig. 2f). After nitridation, the samples became porous due to the lattice shrinkage caused by the replacement of 3 O^{2-} with 2 N^{3-} in the anionic network, resulting in the structural change from a low-density layered perovskite to a high-density perovskite.[22–24] Again, the samples with a relatively larger crystal size have high numbers of pores (voids) in comparison with the samples of small crystal size because the lattice strain was possibly higher in the larger crystals. Previously, Park and Kim[25] also researched the pore formation in $LaTaON_2$ by nitriding $La_2Ta_2O_7$ particles with different sizes and noted a high pore density for the large precursor oxides. Similarly, Maegli et al.[26] observed that the pore development in $LaTiO_2N$ was more pronounced for $La_2Ti_2O_7$ with a larger particle size, synthesized by a solid-state reaction, than for $La_2Ti_2O_7$ with a smaller particle size, prepared by a polymerized-complex method. In particular, the samples grown using the KCl flux with a 1 mol% solute concentration and the K_2SO_4 flux with a 10 mol% solute concentration possessed no visible pores, presumably due to lower lattice strain. In the right panels, the SEM images of the respective samples after nitridation with KCl flux are shown. It is evident that the nitridation with the KCl flux resulted in the formation of nonporous (dense) $BaTaO_2N$ crystal structures owing to the presence of the flux inducing dissolution and recrystallization steps and facilitating ionic diffusion during nitridation.[25,26] The accumulated lattice strain upon structural transformation was partially released during the dissolution–recrystallization process, hindering the formation of a porous network. Interestingly, only the nitrided samples grown using the $BaCl_2$, CsCl and KCl fluxes with a 10 mol% solute concentration retained the original skeletal morphology of their oxide precursor. Upon nitridation, the thin plate-like crystals of $Ba_5Ta_4O_{15}$ grown using the KCl with a 1 mol% solute concentration and the K_2SO_4 with a 10 mol% solute concentration transformed into submicron-sized idiomorphic polyhedral crystals of $BaTaO_2N$. On the contrary, the idiomorphic crystals of $Ba_5Ta_4O_{15}$ grown using the KCl + $BaCl_2$ and RbCl fluxes altered to plate-like crystals of $BaTaO_2N$ upon nitridation. Note that the morphological and size variations influenced the surface areas of the $Ba_5Ta_4O_{15}$ oxide precursor crystals and $BaTaO_2N$ crystal structures. The $BaTaO_2N$ crystal structures obtained by

nitridation without KCl flux exhibited slightly higher surface areas than their counterparts prepared by nitridation with KCl flux because of the formation of a porous network (Table 1). According to the SEM results, the flux type, solute concentration and nitridation with and without KCl flux were effective at altering the morphologies and sizes of the $Ba_5Ta_4O_{15}$ oxide precursor crystals and $BaTaO_2N$ crystal structures.

Fig. 3 shows the cross-sectional TEM images of a $CoO_x/BaTaO_2N/Ta/Ti$ electrode prepared by a particle transfer method using $BaTaO_2N$ crystal structures obtained by the nitridation of $Ba_5Ta_4O_{15}$ crystals, grown using $BaCl_2$ at a 10 mol% solute concentration, at 950 °C for 20 h without KCl flux. The STEM dark field image of the interfaces (Fig. 3a) clearly shows the average thickness of the $BaTaO_2N$ particle layer to be approximately 2.5 μm and the presence of a porous network among the particles. The EDX elemental mapping images in Fig. 3b–d confirm the 200 nm thickness of the Ta contact layer, which is located between the $BaTaO_2N$ particles and the thicker Ti layer beyond the Ta layer, and the homogenous distribution of barium and tantalum within, and cobalt on the surface of, the $BaTaO_2N$ particles. As shown in Fig. 3e, the primary particles have an average size of less than 200 nm and are in close contact with each other, promoting efficient charge transport. In Fig. 3f, the selected area diffraction pattern, taken with the incident beam along the [001] direction, of the circled region in Fig. 3e reveals the single-crystalline nature of the porous secondary particles.

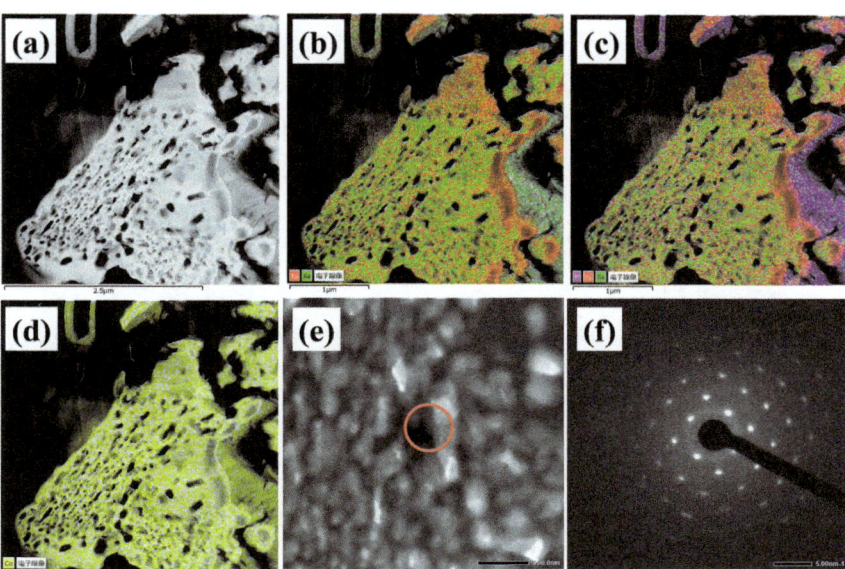

Fig. 3 Cross-sectional TEM images of a $BaTaO_2N/Ta/Ti$ electrode prepared by a particle transfer method using $BaTaO_2N$ crystal structures obtained by nitridation of $Ba_5Ta_4O_{15}$ crystals, grown using $BaCl_2$ and a 10 mol% solute concentration at 950 °C for 20 h without KCl flux: (a) a STEM dark field image of the interfaces, (b–d) EDX element mapping images, (e) a TEM image of $BaTaO_2N$ particles, and (f) a selected area diffraction pattern of the circled region.

Photoelectrochemical (PEC) measurements were performed using the various $BaTaO_2N$ crystal structures obtained by nitridation with and without KCl flux. Fig. 4 shows the current–potential (I–E) curves of the CoO_x-modified $BaTaO_2N/Ta/Ti$ photoanodes under irradiation of AM 1.5G simulated solar light (chopped). Among the samples prepared by nitridation without KCl flux, the $BaTaO_2N$ crystal structures obtained from $Ba_5Ta_4O_{15}$ crystals grown using CsCl at a 10 mol% solute concentration (sample C10) exhibited the highest anodic photocurrent of approximately 3.11 mA cm^{-2} at 1.2 V_{RHE}. The anodic photocurrents of the other samples prepared using different fluxes and solute concentrations decreased in the following order: 2.04 mA cm^{-2} for KCl at a 10 mol% solute concentration (sample KB10) > 1.76 mA cm^{-2} for RbCl at a 10 mol% solute concentration (sample R10) > 1.48 mA cm^{-2} for $BaCl_2$ at a 10 mol% solute concentration (sample B10) > 1.21 mA cm^{-2} for KCl at a 1 mol% solute concentration (sample K1) > 1.20 mA cm^{-2} for KCl + $BaCl_2$ at a 10 mol% solute concentration (sample KB10) > 1.03 mA cm^{-2} for K_2SO_4 at a 50 mol% solute concentration (sample KS50). These

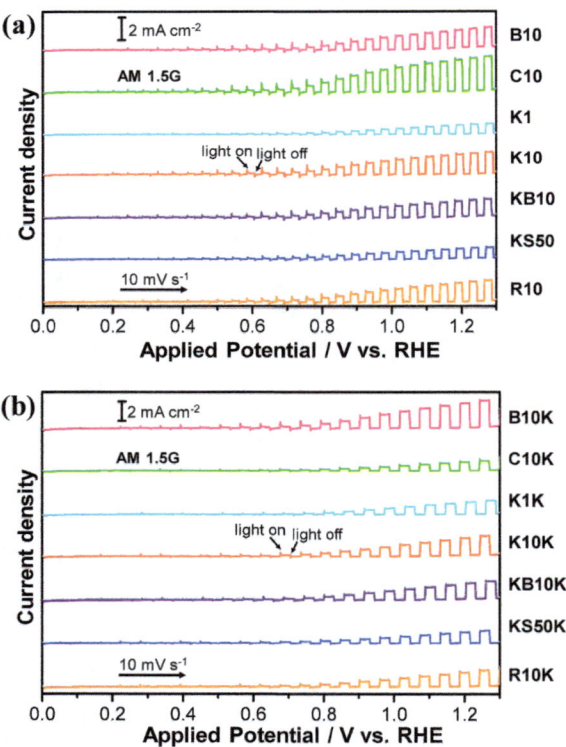

Fig. 4 I–E curves measured under simulated AM 1.5G light of $BaTaO_2N$ crystal structures obtained by nitridation of $Ba_5Ta_4O_{15}$ crystals, grown using $BaCl_2$ and a 10 mol% sol. conc. (pink), CsCl and a 10 mol% sol. conc. (green), KCl and a 1 mol% sol. conc. (light blue), KCl and a 10 mol% sol. conc. (red), KCl + $BaCl_2$ and a 10 mol% sol. conc. (purple), K_2SO_4 and a 50 mol% sol. conc. (blue), and RbCl and a 10 mol% sol. conc. (orange), at 950 °C for 20 h without (a) and with (b) KCl flux. A 0.2 M potassium phosphate aqueous solution adjusted to pH = 13 by adding KOH was used as an electrolyte. The applied potential was swept at +10 mV s^{-1} under intermittent irradiation with a period of 2 s.

results indicate that the morphology, size and porosity are important in influencing the photoelectrochemical performance of $BaTaO_2N$ particles. Notably, the plate-like $BaTaO_2N$ crystal structures (samples C10 and K10) with average sizes of less than 1–2 μm were found to exhibit the highest anodic photocurrents owing to the largely exposed porous surfaces of a similar nature. An increase in the particle size (up to 12 μm) and dimensions (3D) led to a decrease in the photocurrent (sample B10), presumably due to the existence of longer distances for photogenerated electrons and holes, increasing the bulk recombination rate. In contrast, the K1, KB10, KS50 and R10 samples with smaller particle sizes showed lower photocurrents. This is attributed to the surface degradation caused by the excess nitridation of the smaller particles, suggesting that the nitridation time must be determined in advance depending on the particle size of the oxide precursors,[27] the presence of massive interparticle boundaries, and a high surface recombination rate.

Interestingly, the same order for the anodic photocurrents was not observed with the $BaTaO_2N$ crystal structures prepared by nitridation with KCl flux. All of the samples exhibited low photocurrents upon nitridation with KCl flux compared with that of the samples obtained by nitridation without KCl flux. Namely, the highest photocurrents achieved were only 1.15 and 1.19 mA cm^{-2} for the samples prepared using $BaCl_2$ at a 10 mol% solute concentration (sample B10K) and KCl at a 10 mol% solute concentration (sample K10K), which are more than two times lower than that observed above for sample C10. The C10K, K1K, KS50K, R10K, and KB10K samples exhibited photocurrents in the range of 0.52–0.97 mA cm^{-2}. Such a significant difference in the photocurrents can be explained by the presence of a high number of dangling bonds and high surface areas in the samples with porous surfaces prepared by nitridation without KCl flux in comparison with those of the samples with faceted and smooth surfaces prepared by nitridation with KCl flux. These dangling bonds possibly acted as nucleation centers, which led to a good dispersion of CoO_x nanoparticles on the surfaces,[26] while a larger surface area provided many active sites for the reaction.

Previously, Landsmann et al.[28] proposed five rules to design high-performance particle-based photoanodes based on $LaTiO_2N$, demonstrating the importance of a greater surface area, fewer intraparticle boundaries, large cocatalyst particles, extended interparticle interfaces, and micrometer-sized particles. Similar to our results, they demonstrated that the large porous $LaTiO_2N$ particles could show the highest photocurrent compared with that of the dense $LaTiO_2N$ particles synthesized by a NaCl/KCL flux-assisted solid-state reaction.[28] On the contrary, the highest ever reported photocurrent value of 1.5 mA cm^{-2} at 1.23 V_{RHE} was achieved for $BaNbO_2N$ after nitridation of its pre-calcined oxide precursors with NaI flux.[29] Recently, a Ta_3N_5 polyhedron array photoanode with preferential exposure of {001} facets grown using RbCl flux exhibited a photocurrent of 5.6 mA cm^{-2} at 1.23 V_{RHE},[30] whereas nanometer-sized $LaTaON_2$ single crystals grown using LiCl flux could afford a photocurrent density of 5.1 mA cm^{-2} at 1.23 V_{RHE}, which is the highest reported PEC performance for $LaTaON_2$.[31] The above-shown results indicate that the flux method is important to specifically engineer photocatalytic crystals to reach their maximum potential in photoelectrochemical water splitting.

To evaluate the photocatalytic activity of the as-prepared samples, water oxidation half-reaction was performed under visible light in the presence of

AgNO$_3$. The reaction time courses for photocatalytic O$_2$ evolution over the CoO$_x$-loaded BaTaO$_2$N crystal structures, obtained by nitridation with and without KCl flux, under visible light are shown in Fig. 5. During the O$_2$ evolution testing, Ag$^+$ ions were continually reduced into Ag^0 nanoparticles which lower light absorptivity and cover active sites, slightly diminishing their O$_2$ evolution rates over increasing the reaction time (especially ~3–5 h).[16] Hence, to evaluate the photocatalytic activity accurately, the O$_2$ evolution amounts within first 3 h were compared and used to calculate the O$_2$ evolution rates. Overall, most of the BaTaO$_2$N crystal structures obtained by nitridation without flux showed higher O$_2$ evolution rates than their counterparts obtained by nitridation with KCl flux due to a good dispersion of CoO$_x$ cocatalyst nanoparticles and porous network providing many active sites.[15,26] Conversely, only two samples, B10K and KB10K, prepared by nitridation with KCl flux showed higher O$_2$ evolution rates of 134.64 and 88.73 μmol h^{-1}, respectively, among all samples possibly because of their larger particle size, greater exposed surfaces of a similar nature, high crystallinity, reduced surface defect density, and fewer grain boundaries.[32] Although surface area is slightly higher in the samples obtained by nitridation without flux, it might not play a more important role in the water oxidation reaction compared to crystallinity. As Kudo and Miseki[33] pointed out, high crystallinity is more necessary than a high surface area for water splitting because the recombination of photogenerated electrons and holes is a critical issue for uphill reactions. However, K10 and K10K exceptionally showed almost same O$_2$ evolution rates owing to the relatively large difference in surface area which chancels the positive effect of high crystallinity in K10K. As shown in Table 1, the O$_2$ evolution rate over the BaTaO$_2$N crystal structures obtained by nitridation without KCl flux followed the order: 22.75 (KS50) < 24.86 (K10) < 37.62 (KB10) < 44.27 (B10) < 44.28 (K1) < 47.53 (C10) < 56.85 (R10) μmol h^{-1}. In contrast, the O$_2$ evolution rate over the BaTaO$_2$N crystal structures obtained by nitridation with KCl flux did not follow this trend and progressed in a different order: 13.13 (KS50K) < 14.38 (R10K) < 18.97 (C10K) < 25.86 (K10K) < 33.35 (K1K) < 88.73 (KB10K) < 134.64 (B10K) μmol

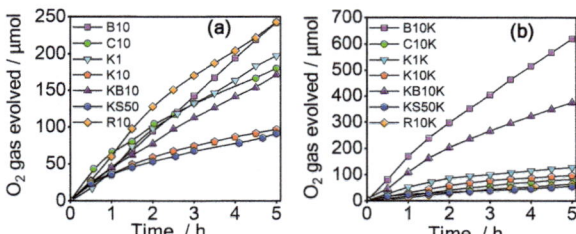

Fig. 5   Reaction time courses for the photocatalytic oxygen evolution over BaTaO$_2$N crystal structures obtained by nitridation of Ba$_5$Ta$_4$O$_{15}$ crystals, grown using BaCl$_2$ and 10 mol% sol. conc. (rectangle), CsCl and 10 mol% sol. conc. (circle), KCl and 1 mol% sol. conc. (down-pointing triangle), KCl and 10 mol% sol. conc. (pentagon), KCl + BaCl$_2$ and 10 mol% sol. conc. (up-pointing triangle), K$_2$SO$_4$ and 50 mol% sol. conc. (hexagon), and RbCl and 10 mol% sol. conc. (diamond), at 950 °C for 20 h without (a) and with (b) KCl flux. Photocatalytic reaction conditions: 100 mg photocatalyst loaded with CoO$_x$ cocatalyst (2 wt% Co); aqueous solution of AgNO$_3$ (10 mM, 300 mL); 200 mg La$_2$O$_3$ (pH buffer); light source − 300 W Xe lamp fitted with a cold mirror (CM-1) and a cutoff filter (λ > 420 nm); a side-irradiation-type reaction vessel was used in this study.

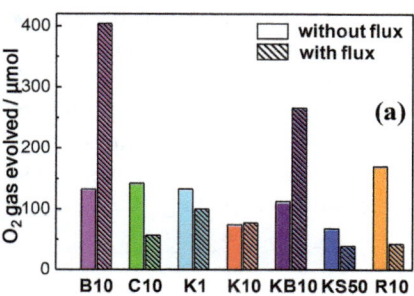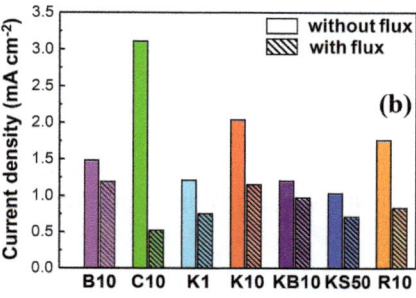

Fig. 6 Comparison of (a) the amount of photocatalytically evolved oxygen (first 3 h) and (b) photocurrent density at 1.2 V_{RHE} of BaTaO$_2$N crystal structures obtained by nitridation of Ba$_5$Ta$_4$O$_{15}$ crystals, grown using BaCl$_2$ and 10 mol% sol. conc. (pink), CsCl and 10 mol% sol. conc. (green), KCl and 1 mol% sol. conc. (light blue), KCl and 10 mol% sol. conc. (red), KCl + BaCl$_2$ and 10 mol% sol. conc. (purple), K$_2$SO$_4$ and 50 mol% sol. conc. (blue), and RbCl and 10 mol% sol. conc. (orange), at 950 °C for 20 h with and without KCl flux.

h^{-1}. Additionally, negligible amounts of N$_2$ gas ($\sim$0–47 µmol) were evolved after 5 h O$_2$ evolution testing due to photocorrosion of the as-obtained oxynitrides. The N$_2$ evolution rates were drastically decreased upon increasing the testing time, implying the somewhat high photostability of the as-obtained oxynitrides.

Similar to their photoelectrochemical performances, the photocatalytic O$_2$ evolution activities of the samples with smaller particle sizes were lower than those of the samples with larger particle sizes (Fig. 6). Again, this can be explained by the presence of interparticle boundaries and surface degradation caused by the excess nitridation of such small particles, giving rise to surface defects. Both factors simultaneously led to an increase in the fast surface recombination rate, which negatively affected the O$_2$ evolution rate. For the samples with larger particle sizes, long-lived electrons greatly induced the four-electron oxidation of water due to the less frequent surface recombination of photogenerated electrons and holes and photocorrosion.[34] As Amano et al.[35] described, the most important factor influencing the photocatalytic water oxidation efficiency of BaTaO$_2$N particles is the number of surviving photogenerated holes, which is dictated by the morphology and size of the photocatalyst.

4.   Conclusions

In summary, the effects of altered morphology and size on the visible light-induced water oxidation activity and photoelectrochemical performance of the BaTaO$_2$N crystal structures were investigated. Upon nitridation, the samples became porous due to the lattice shrinkage caused by the replacement of 3 O^{2-} with 2 N^{3-} in the anionic network. The BaTaO$_2$N crystal structures obtained by nitridation without KCl flux showed a higher surface area than did their counterparts prepared by nitridation with KCl flux because of the formation of a porous network. All of the samples prepared by nitridation without KCl flux exhibited high anodic photocurrents compared with the samples obtained by nitridation with KCl flux due to the high number of dangling bonds that acted as nucleation centers for highly dispersed CoO$_x$ nanoparticles. However, a different trend was observed for the photocatalytic O_2 evolution. The samples with smaller particle sizes exhibited low photoelectrochemical performances and photocatalytic O_2 evolution activities compared to their counterparts with larger particle sizes because of the high number of interparticle boundaries and surface degradation. Although the efficiency achieved in this study for $BaTaO_2N$ is not the highest, the findings clearly demonstrate the importance of controlling the morphology, size, porosity and crystallinity in achieving higher efficiencies for solar water splitting.

Conflicts of interest

There are no conflicts to declare.

Acknowledgements

This research was supported in part by the Japan Technological Research Association of Artificial Photosynthetic Chemical Process (ARPChem).

References

1 K. Maeda and K. Domen, *J. Phys. Chem. C*, 2007, **111**, 7851–7861.
2 T. Yeh, C. Teng, S. Chen and H. Teng, *Adv. Mater.*, 2014, **26**, 3297–3303.
3 P. Kalisman, Y. Nakibli and L. Amirav, *Nano Lett.*, 2016, **16**, 1776–1781.
4 Q. Li, F. Zhao, C. Qu, Q. Shang, Z. Xu, L. Yu, J. R. McBride and T. Lian, *J. Am. Chem. Soc.*, 2018, **140**, 11726–11734.
5 R. Takakura, T. Oshikiri, K. Ueno, X. Shi, T. Kondo, H. Masuda and H. Misawa, *Green Chem.*, 2017, **19**, 2398–2405.
6 N. Bao, L. Shen, T. Takata and K. Domen, *Chem. Mater.*, 2008, **20**, 110–117.
7 C. Pan, T. Takata, M. Nakabayashi, T. Matsumoto, N. Shibata, Y. Ikuhara and K. Domen, *Angew. Chem., Int. Ed.*, 2015, **54**, 2955–2959.
8 J. Seo, T. Hisatomi, M. Nakabayashi, N. Shibata, T. Minegishi, M. Katayama and K. Domen, *Adv. Energy Mater.*, 2018, **8**, 1800094.
9 T. Takata, C. Pan and K. Domen, *Sci. Technol. Adv. Mater.*, 2015, **16**, 033506.
10 S. Chen, T. Takata and K. Domen, *Nat. Rev. Mater.*, 2017, **2**, 17050.
11 M. Hojamberdiev, E. Zahedi, E. Nurlaela, K. Kawashima, K. Yubuta, M. Nakayama, H. Wagata, T. Minegishi, K. Domen and K. Teshima, *J. Mater. Chem. A*, 2016, **4**, 12807–12817.
12 K. Kawashima, M. Hojamberdiev, K. Yubuta, K. Domen and K. Teshima, *J. Energy Chem.*, 2018, **27**, 1415–1421.
13 M. Hojamberdiev, K. Yubuta, J. J. M. Vequizo, A. Yamakata, S. Oishi, K. Domen and K. Teshima, *Cryst. Growth Des.*, 2015, **15**, 4663–4671.
14 K. Kawashima, M. Hojamberdiev, O. Mabayoje, B. R. Wygant, K. Yubuta, C. B. Mullins, K. Domen and K. Teshima, *CrystEngComm*, 2017, **19**, 5532–5541.
15 K. Kawashima, M. Hojamberdiev, H. Wagata, K. Yubuta, J. J. M. Vequizo, A. Yamakata, S. Oishi, K. Domen and K. Teshima, *J. Phys. Chem. C*, 2015, **119**, 15896–15904.
16 K. Kawashima, M. Hojamberdiev, H. Wagata, M. Nakayama, K. Yubuta, S. Oishi, K. Domen and K. Teshima, *Catal. Sci. Technol.*, 2016, **6**, 5389–5396.

17 M. Hojamberdiev, M. F. Bekheet, J. N. Hart, J. J. M. Vequizo, A. Yamakata, K. Yubuta, A. Gurlo, M. Hasegawa, K. Domen and K. Teshima, *Phys. Chem. Chem. Phys.*, 2017, **19**, 22210–22220.

18 K. Kawashima, M. Hojamberdiev, C. Stabler, D. Vrankovic, K. Yubuta, R. Riedel, K. Domen and K. Teshima, *Materials for Renewable and Sustainable Energy*, 2017, **6**, 10.

19 K. Kawashima, M. Hojamberdiev, H. Wagata, E. Zahedi, K. Yubuta, K. Domen and K. Teshima, *J. Catal.*, 2016, **344**, 29–37.

20 M. Hojamberdiev, K. Kawashima, M. Kumar, A. Yamakata, K. Yubuta, A. Gurlo, M. Hasegawa, K. Domen and K. Teshima, *Int. J. Hydrogen Energy*, 2017, **42**, 27024–27033.

21 T. Minegishi, N. Nishimura, J. Kubota and K. Domen, *Chem. Sci.*, 2013, **4**, 1120–1124.

22 M. Hojamberdiev, H. Wagata, K. Yubuta, K. Kawashima, J. J. M. Vequizo, A. Yamakata, S. Oishi, K. Domen and K. Teshima, *Appl. Catal., B*, 2016, **182**, 626–635.

23 M. Hojamberdiev, M. F. Bekheet, E. Zahedi, H. Wagata, J. J. M. Vequizo, A. Yamakata, K. Yubuta, A. Gurlo, K. Domen and K. Teshima, *Dalton Trans.*, 2016, **45**, 12559–12568.

24 K. Kawashima, M. Hojamberdiev, H. Wagata, K. Yubuta, S. Oishi and K. Teshima, *Cryst. Growth Des.*, 2015, **15**, 333–339.

25 N.-Y. Park and Y.-I. Kim, *J. Mater. Sci.*, 2012, **47**, 5333–5340.

26 A. E. Maegli, S. Pokrant, T. Hisatomi, M. Trottmann, K. Domen and A. Weidenkaff, *J. Phys. Chem. C*, 2014, **118**, 16344–16351.

27 M. Kodera, Y. Moriya, M. Katayama, T. Hisatomi, T. Minegishi and K. Domen, *Sci. Rep.*, 2018, **8**, 15849.

28 S. Landsmann, A. E. Maegli, M. Trottmann, C. Battaglia, A. Weidenkaff and S. Pokrant, *ChemSusChem*, 2015, **8**, 3451–3458.

29 M. Kodera, H. Urabe, M. Katayama, T. Hisatomi, T. Minegishi and K. Domen, *J. Mater. Chem. A*, 2016, **4**, 7658–7664.

30 Z. Shi, J. Feng, H. Shan, X. Wang, Z. Xu, H. Huang, Q. Qian, S. Yan and Z. Zou, *Appl. Catal., B*, 2018, **237**, 665–672.

31 J. Zhou, C. Zhou, Z. Shi, Z. Xu, S. Yan and Z. Zou, *J. Mater. Chem. A*, 2018, **6**, 7706–7713.

32 Z. Wang, Y. Inoue, T. Hisatomi, R. Ishikawa, Q. Wang, T. Takata, S. Chen, N. Shibata, Y. Ikuhara and K. Domen, *Nat. Catal.*, 2018, **1**, 756–763.

33 A. Kudo and Y. Miseki, *Chem. Soc. Rev.*, 2009, **38**, 253–278.

34 Y. He, J. E. Thorne, C. H. Wu, P. Ma, C. Du, Q. Dong, J. Guo and D. Wang, *Chem*, 2016, **1**, 640–655.

35 F. Amano, E. Ishinaga and A. Yamakata, *J. Phys. Chem. C*, 2013, **117**, 22584–22590.

DISCUSSIONS

Synthetic approaches to artificial photosynthesis: general discussion

Catherine M. Aitchison, Virgil Andrei, Daniel Antón-García, Ulf-Peter Apfel, Vivek Badiani, Matthias Beller, Andrew B. Bocarsly, Sylvestre Bonnet, Peter Brueggeller, Christine A. Caputo, Flavia Cassiola, Simon T. Clausing, Andrew I. Cooper, Charles E. Creissen, Víctor A. de la Peña O'Shea, Wolfgang Domcke, James R. Durrant, Michael Grätzel, Leif Hammarström, Anna Hankin, Marta C. Hatzell, Ferdi Karadas, Burkhard König, Moritz F. Kuehnel, Sarah Lamaison, Chia-Yu Lin, Marcelino Maneiro, Shelley D. Minteer, Aubrey R. Paris, Ernest Pastor, Chanon Pornrungroj, Joost N. H. Reek, Erwin Reisner, Souvik Roy, Constantin Sahm, Ravi Shankar, Wendy J. Shaw, Sergii I. Shylin, Wilson A. Smith, Katarzyna Sokol, Han Sen Soo, Reiner Sebastian Sprick, Wolfgang Viertl, Anastasia Vogel, Andreas Wagner, David Wakerley, Qian Wang, Dominik Wielend and Martijn A. Zwijnenburg

DOI: 10.1039/C9FD90024A

Moritz F. Kuehnel opened a discussion of the paper by Andrew I. Cooper: What do you know about the HOMO and LUMO localisation on the polymers? I suppose this is strongly affected by introducing heteroatoms such as sulfur, and that this causes the changes observed when oxidising the thiophene to sulfolane moieties. Can you use different heteroatoms to tweak the redox potentials?

Andrew I. Cooper answered: Yes – by changing the heteroatom one can change the catalytic activity, but this can also change a variety of other things such as the surface hydrophilicity and (in some cases) the polymer molecular weight, or, in the case of networks, the surface area. As such, it is often unclear whether these effects come from changes to the redox potentials or a variety of factors.

Moritz F. Kuehnel asked: How much is known about the residual palladium in the polymer? What do you know about its environment, which I assume will depend on functional groups in the polymer backbone, *e.g.* donor groups? I am wondering if the observed differences in activity for different polymers are a result of the different palladium environments, rather than other, more easily determined factors. such as the band gap, *etc.* Do you have any EXAFS data?

Andrew I. Cooper responded: No, no EXAFS data as yet, but the palladium metal can vary in size from a few nanometers (or even palladium clusters) up to 50 nm or so. Ian McCulloch, James R. Durrant and coworkers have shown in ref. 1 that palladium can be catalytically active in such materials at very low concentrations. In their studies, very low palladium concentrations (<40 ppm) have an effect on the photocatalytic hydrogen evolution rate.

1 J. Kosco, M. Sachs, R. Godin, M. Kirkus, L. Francas, M. Bidwell, M. Qureshi, D. Anjum, J. R. Durrant and I. McCulloch, *Adv. Energy Mater.*, 2018, **8**, 1802181.

Virgil Andrei enquired: The polymers employ residual palladium from the Suzuki–Miyaura coupling as the hydrogen evolution co-catalyst. Have you considered using molecular catalysts to expand the scope of the photocatalysis towards CO_2 reduction? The polymers possess aromatic units; therefore, molecular catalysts may be easily attached to the polymeric backbone *via* π–π stacking interactions.

Andrew I. Cooper responded: Yes, we are working on this – for example, by introducing bipyridyl metal binding sites into the polymers.

James R. Durrant commented: I am intrigued that the most active polymer you show appears to be very hydrophobic. For proton reduction photocatalysis I would have expected that some degree of polymer hydrophilicity would be preferred. Could you comment on this?

Andrew I. Cooper responded: Hydrophilicity is an important variable, but it is one such factor – the hydrogen evolution rate is also affected by the band gap, the energy levels, particle size, surface area, and the level of precious metals in the material, which act as co-catalysts. A big difficulty in designing these materials is that no single variable dominates the catalytic activity (see for example: ref. 1).

1 Y. Bai, L. Wilbraham, B. J. Slater, M. A. Zwijnenburg, R. S. Sprick and A. I. Cooper, *J. Am. Chem. Soc.*, 2019, **141**, 9063–9071.

Wendy J. Shaw remarked: I am curious as to whether secondary/tertiary structures (similar to biopolymers) have been investigated and what their role might be in controlling reactivity?

Andrew I. Cooper responded: No, only in the context of crystalline COFs and amorphous analogues, where strong differences in catalytic activity are observed, *e.g.* as in ref. 1. It is hard to control secondary/tertiary structure in these rigid polymers.

1 X. Wang, L. Chen, S. Y. Chong, M. A. Little, Y. Wu, W. Zhu, R. Clowes, Y. Yan, M. A. Zwijnenburg, R. S. Sprick and A. I. Cooper, *Nat Chem.*, 2018, **10**, 1180.

Michael Grätzel commented: I assume that a sacrificial donor was used in the light driven hydrogen evolution process. Could Professor Cooper please provide the chemical structure of the donor? I wonder whether the hydrogen produced under light arises from the reduction of water or from the sacrificial donor. Could Professor Cooper also show the action spectrum for the hydrogen generation to see whether it matches the absorption spectrum of the ladder polymer acting as a sensitizer?

Andrew I. Cooper responded: Yes, we used triethylamine or triethanolamine generally, or in some cases ascorbic acid, but we have also tested non-hydrogen containing donors, such as Na_2S, to exclude the possibility of the hydrogen coming from the donor.

Ravi Shankar opened a discussion of the paper by Martijn A. Zwijnenburg: The screening method that you have applied for polymeric photocatalysts is very interesting. Is there scope to apply or expand this screening method towards other types of catalysts, such as porous catalysts?

Martijn A. Zwijnenburg responded: Our approach is specific for photocatalysts but other groups are working on similar approaches for other material classes, including porous materials.

Burkhard König commented: Conjugated polymers show a specific conjugation length. How many repeating units do you have to consider in your evaluation of polymer photocatalyst properties to describe them well by computational methods?

Martijn A. Zwijnenburg answered: Yes, indeed. In our experience oligomeric models with a length of twelve units, as used in this work, provide approximately converged properties with respect to oligomer length.

Wendy J. Shaw asked: What metrics are the computational results determined by? What makes a good catalyst? If you (and Professor Cooper) do not know what is bad then how do you know that you are not missing something?

Martijn A. Zwijnenburg answered: We know that a good photocatalyst should at least absorb visible light and have sufficient driving force for all desired solution reactions, proton or CO_2 reduction and water or sacrificial electron donor oxidation. What we know less about is what other factors are (the most) important (*e.g.* wettability, reaction kinetics, and charge–carrier and exciton transport), what the best (computational) descriptors for these additional factors are, and the nature of the interplay and trade-offs between the different properties. However, ignoring what we do not fully understand, this still means that by predicting the optical gap and potentials of materials we can rule out many of them by computational screening and only concentrate on the more promising remaining ones experimentally. To paraphrase Frank Westheimer: A couple of months in the laboratory can frequently save a couple of hours of computer time. Another use of (computational) screening in my opinion is to gain an understanding of how photocatalyst properties are interconnected by generating data sets that are large enough to extract inter-property correlations.

Wolfgang Domcke said: The exciton binding energy can be quite large in these polymeric materials and therefore it is an important parameter. Will you come up with specific numbers in the future?

Martijn A. Zwijnenburg responded: While we have not done so here, the exciton binding energy could in principle be approximated from the difference in the fundamental gap, itself defined as the difference between a polymer's ionisation potential and electron affinity, and its optical gap. In previous work (ref. 1) we approximated the exciton binding energy for an exciton in the polymer bulk and near the polymer–water interface using DFT and found values of 1200 and 170 meV, respectively. The fact that even the latter value is much larger than the kT at room temperature (26 meV) suggests that excitons in the polymer photocatalyst likely dissociate on the polymer–water interface or polymer–polymer interface in the case of a heterojunction, with the electron and hole after dissociation ending up in different phases.

1 P. Guiglion, C. Butchosa and M. A. Zwijnenburg, *Macromol. Chem. Phys.* 2016, **217**, 344.

Marcelino Maneiro opened a discussion of the paper by Ferdi Karadas: I think that cyclic voltammetry would shed more light on the characterization of your system, and even on its catalytic mechanism. Have you tried to perform cyclic voltammetry on your Ru–P4VP–CoFe dyad? In relation to the catalytic activity, do you have any evidence that water is oxidized at the cobalt sites?

Ferdi Karadas replied: I agree that cyclic voltammetric studies would definitely help to characterize the assembly in more detail. Such measurements, however, were not performed mainly because the assembly does not dissolve in water and coating this polymeric assembly on an FTO electrode *via* conventional methods was not successful.

The identity of the catalytic site was not investigated in this study. The origin of the catalytic site has been investigated previously and it is well-established in Prussian blue systems that hexacyanometal sites are not catalytically active sites and that it should be the cobalt sites that are coordinated to at least one water molecule. Please check our recent papers (ref. 1 and 2).

1 T. G. U. Ghobadi, E. A. Yildiz, M. Buyuktemiz, S. S. Akbari, D. Topkaya, Ü. Isci, Y. Dede, H. G. Yaglioglu and F. Karadas, *Angewandte Chemie Int. Ed.*, 2018, **57**, 17173.
2 E. P. Alsaç, E. Ülker, S. V. K. Nune, Y. Dede and F. Karadas, *Chem.: Eur. J.*, 2018, **24**, 4856.

Matthias Beller enquired: Following on from the previous question, how sure are you about the stability of this system? Did you compare it with other systems?

Ferdi Karadas replied: The ruthenium chromophore is connected to the Prussian blue (PB)-type water oxidation catalyst in our study. We also performed photocatalytic studies with a regular PB system in the presence of a ruthenium chromophore, which is not connected to PB. Previous studies in the literature as well as our photocatalytic studies and post-catalytic characterization measurements suggest that the connected assembly exhibits a much higher stability.

Michael Grätzel asked: Professor Karadas stated that his $Co(CN)_6$ Prussian blue-type catalyst remained intact during his reaction as a catalyst for the oxidation of water to oxygen. However it appears unlikely that this is the case for the catalyst molecules exposed to water that are actively involved in the oxygen evolution reaction. Rather, they are expected to be converted to cobalt oxides which in turn will catalyze the water oxidation reaction.

Ferdi Karadas answered: Prussian blue analogues (PBAs) are non-oxide based coordination networks, which make them way more stable than molecular catalysts. It should also be noted that the solubility product of PBAs is much lower than those of oxides due to the highly sigma-donating nature of the cyanide ligand, which leads to the formation of a strong M–CN–M' coordination mode. Furthermore, our XPS studies did not reveal any possible formation of oxides. The stability of PBAs have been studied comprehensively in ref. 1 and 2 as well.

1 L. Han, P. Tang, Á. Reyes-Carmona, B. Rodríguez-García, M. Torréns, J. R. Morante, J. Arbiol and J. R. Galan-Mascaros, *J. Am. Chem. Soc.*, 2016, **138**, 16037.
2 E. P Alsaç, E. Ülker, S. V. K. Nune, Y. Dede and Ferdi Karadas, *Chem.: Eur. J.*, 2018, **24**, 4856.

Catherine M. Aitchison queried: What do you think is the rate-limiting step in the system and have you tried to optimise the ratio of the components (*i.e.* sensitizer to polymer to catalyst)? Do you think this would also be important for stability given that you mention that a possible deactivation pathway of the catalyst is poisoning by photosensitizer ligands?

Ferdi Karadas replied: We have not performed any experiments to investigate the rate-limiting step. It could, however, be speculated that the rate-determining step could be the activation of the catalytic cobalt sites given its high overpotential. Within the project, two different assemblies with different stoichiometric ratios of Ru : Fe have been prepared. Since their performances are comparable only one of them was reported, so our preliminary studies do not indicate any correlation between the ratio of the components and the stability or performance of the catalyst.

Andrew B. Bocarsly asked: Following up on Michael Grätzel's question, how certain are you that the outer surface of the material is the catalysis site? Cyanide bridged cyanometalate complexes are known to form microporous structures. Thus, your substrate can penetrate into the interior of the cyanometalate layer. Furthermore, these materials tend to be rather defective with open coordination sites, thus interior catalytic sites may be available. Your materials may have some similarity to the $[Ru(bpz)_3]^{2+}$/cyanometalate layered structure discussed in ref. 1.

1 C. Hidalgo-Luangdilok and A. B. Bocarsly, *Inorg. Chem.*, 1990, **29**, 2894.

Ferdi Karadas answered: Thanks for sharing this paper. We are certain that cobalt sites that are coordinated to at least one water molecule should be catalytically active sites and we have previously shown in ref. 1 that this type of active cobalt site can be increased by incorporating polymers to this chemistry. The polymeric moiety probably increases the number of defective cobalt sites, as suggested, by decreasing the dimensionality of the Prussian blue structure. These defective sites are also known to occur in the interior of the network. We have also performed porosity studies on these polymer–PB hybrid structures (which is not reported in this manuscript), which show that their microporous behavior disappears since the polymeric groups probably block the microchannels. So even if there is a considerable number of active cobalt sites in the bulk of the material, these are not accessible. Overall, polymeric groups decrease the dimensionality of the PB structures at the expense of blocking the microchannels.

1 M. Aksoy, S. V. K. Nune and F. Karadas, *Inorg. Chem.*, 2016, **55**, 4301.

Reiner Sebastian Sprick addressed all of the presenters: When using organic polymers as photocatalysts stability is a concern, particularly when performing water oxidation. Would you be able to comment on this and have you performed experiments to explore this?

Martijn A. Zwijnenburg responded: The stability of polymer photocatalysts when using a sacrificial electron donor (SED) does not seem to be an issue (see ref. 1), however this might be different when oxidising water rather than a SED, especially if the former has sluggish kinetics. Supporting evidence for this hypothesis comes from work in the literature on a polymer consisting of triazine units linked by disulfide bridges (see ref. 2). This material is reported to stably evolve hydrogen in the presence of a SED but to photocorrode when using pure water.

1 R. S. Sprick, B. Bonillo, R. Clowes, P. Guiglion, N. J. Brownbill, B. J. Slater, F. Blanc, M. A. Zwijnenburg, D. J. Adams and A. I. Cooper, *Angew. Chem. Int. Ed.*, 2016, **55**, 1792.
2 Z. Zhang, J. Long, L. Yang, W. Chen, W. Dai, X. Fu and X. Wang, *Chem. Sci.*, 2011, **2**, 1826.

Ferdi Karadas answered: Our post-characterization studies mainly focused on the stability of the metal sites and whether they were transformed to any oxide under photocatalytic conditions. No special method has been performed to explore the stability of the polymer. The visual observation of the powder suspension, however, reveals that the suspension is colorless even after the photocatalytic experiment. This implies that none of the ruthenium fragment went into the solution. Therefore, even if the polymeric chain is broken during the photocatalysis, it was limited and did not lead to disconnection of the chromophore from the catalyst. The maintenance of the activity for six cycles also supports this hypothesis.

Matthias Beller followed this by asking: The sacrificial reductant plays a significant role. We get what we screen for – we screen for the best catalyst that works, *e.g.* triethanolamine, ammonium nitrate. How much can we learn from this for a real system? In the long term? Should we all focus on electrocatalytic reactions?

Martijn A. Zwijnenburg replied: It is undoubtedly true that the use of sacrificial electron donors (SEDs) biases the results and that one in the long term runs the risk of optimising a photocatalyst for oxidising a SED instead of water. In the short term, however, I still see a good case for the use of SEDs in photocatalyst discovery. For example, one would not want to miss a potentially promising photocatalyst system for overall water splitting because in the absence of the ideal co-catalyst water oxidation is too sluggish for any activity to be observed.

Ferdi Karadas answered: Since there are many possible systems to target and since the preparation and analysis of each of these systems is a tedious task, the use of a sacrificial agent is a practical solution for the sake of an efficient use of time. This solution, of course, comes with the expense of moving away from real condition analysis, which could give results beyond the ballpark or lead to wrong directions in certain cases. Thus, researchers should make their own risk analysis to choose the ideal conditions to test their systems.

Christine A. Caputo asked Martijn A. Zwijnenburg: There is a large molecular space to explore computationally. How many experimental validation points do you think you need to provide confidence that your approach is robust?

Martijn A. Zwijnenburg responded: Good question. As, as it stands, we do not predict photocatalytic activity as such but rather the materials properties that underlie it, we typically benchmark our ability to predict those. In terms of the number of validation points, the more the better. This is simple for some properties, *e.g.* the optical gap, as UV-Vis spectra are measured routinely, but harder for other properties, *e.g.* the ionisation potential and electron affinity of polymers, the measurement of which requires very specialized equipment, and as a result values have only been reported for a very limited set of polymers in the literature.

Dominik Wielend said: You have mentioned at the end of your paper that one of your future goals is the prediction of intermolecular properties like charge transport, *etc.* My question is, if you could estimate trends (maybe just qualitatively), which polymer combinations might tend to form intermolecular interactions (either stacking, which might be influenced by a possible bent structure of the polymer, or interactions similar to hydrogen-bonded pigments)?

Martijn A. Zwijnenburg responded: Thanks, that is an interesting suggestion; something we cannot do as yet but might be worth looking into in the future. In general, *a priori* predicting how molecules pack in the solid state is a very difficult problem, especially for amorphous or poorly-crystalline materials, like most polymers. The extent to which a more approximate approach might work will probably be controlled by how strongly a given property depends on the exact details of the packing and/or how common a particular packing motif is.

Joost N. H. Reek asked: With your new computational strategy you can evaluate many compounds, which is great. It is, however, possible that at the end you will have found a polymer with great properties that cannot be synthesised. So do you plan to implement any strategies to also take into account the synthetic availability?

Martijn A. Zwijnenburg answered: As for this work we typically consider commercially available monomers in combination with coupling chemistry that is experimentally known to be quite resilient, we are not so worried about synthetic accessibility here. In general, however, this is a very pertinent point, which several groups are trying to address. A good example from the literature is the work by Gómez-Bombarelli *et al.* on materials for the emissive layer of organic light-emitting diodes (ref. 1).

1 R. Gómez-Bombarelli, J. Aguilera-Iparraguirre, T. D. Hirzel, D. Duvenaud, D. Maclaurin, M. A. Blood-Forsythe, H. S. Chae, M. Einzinger, D.-G. Ha, T. Wu, G. Markopoulos, S. Jeon, H. Kang, H. Miyazaki, M. Numata, S. Kim, W. Huang, S. I. Hong, M. Baldo, R. P. Adams and A. Aspuru-Guzik, *Nat. Mater.* 2016, **15**, 1120.

Joost N. H. Reek posed to Ferdi Karadas: In your approach you are using polymers to which the catalyst and the chromophores are attached in a random fashion. In nature all of the components required for efficient photosynthesis are highly organised and well positioned with respect to one another. Of course it is nice that you show that this random organisation does lead to photo driven water oxidation, but it is unclear what the efficiency is (maybe only a small fraction of the components is sufficiently organised to be active). Can you comment on the efficiency (do you know this?), and do you think we need to develop strategies that allow for better spatial control of the components, for example by using supra-molecular assembly?

Ferdi Karadas replied: I agree that the organisation of the components in an assembly is of crucial importance for better charge separation and efficiency. Interrogation of the efficiency or methods for improving the efficiency, although important, is beyond the scope of this study. It should be noted that the main objective of this proof-of-concept study is only to show that photosensitizers can be connected to a water oxidation catalyst *via* cyanide chemistry and that the chromophore is more stable in such an assembly. Given the results, our recent efforts have now been dedicated to the addition of each of these chromophores *in situ* on a semi-conductor. As suggested, the use of supramolecular assemblies can also provide an alternative synthetic pathway towards this goal.

Simon T. Clausing addressed all of the presenters: Andrew and Martijn presented polymers that are designed to fulfil all roles for photocatalytic hydrogen generation at once: light-harvesting, electron relay, a hydrogen evolving catalyst, and interface to water. It has been noted by Andrew that if you change one parameter to make it more efficient, you "break" another. Would it therefore not be better to focus on developing polymers that can, for example, be wettable light-harvesters with good electron relay functionality, and supply an external catalyst? Or have a catalytically active polymer, but supply an external photosensitizer? Ferdi showed a system like that; is that not an approach that might lead to quicker results, as it is modular by design?

Andrew I. Cooper responded: Yes, this is a very good idea in principle and recently we've been looking at more modular approaches. Putting the modules together again becomes complex, though...

Martijn A. Zwijnenburg replied: Indeed, using a composite rather than one polymeric material as photocatalyst is probably a promising strategy. One could actually argue that the commonly used polymers (inadvertently) loaded with metal nanoparticles are already examples of such composites, as well as the polymer–polymer and polymer–oxide heterojunctions reported in the literature. Moving from single materials to composites, however, could be a double-edged sword. The same complexity that makes it possible to independently optimise processes/properties will also make the workings of a composite photocatalyst more difficult to study and understand.

Ferdi Karadas added: As in natural photosynthesis, I believe that the preparation of a multifunctional assembly, wherein each component (chromophore, catalyst, donor, acceptor, *etc.*) is assigned for a special task, could be a viable approach for better efficiency. However, the field still needs new assemblies and proof-of-concept studies to explore its boundaries and establish its foundations. So even assemblies with lower efficiencies could help us learn more about this process.

Ernest Pastor addressed Martijn A. Zwijnenburg: Do you have any thoughts on the mechanistic role of the scavenger in these systems? For example in CO_2 reduction with molecular complexes it is proposed that the TEA or TEOA can participate in the mechanism at several points as you can form very reactive radicals. This could also affect the stability of the compounds.

Martijn A. Zwijnenburg answered: My hypothesis is that, in the case of TEA, the mechanism involves outer-sphere electron transfer from TEA in the first step and its singly-oxidised deprotonated counterpart in the second step. The latter is indeed very reactive in the sense that it is strongly reductive (predicted potential < -2.5 V *vs.* SHE at pH 11.5, the likely pH of a TEA solution, compared with $+0.7$ V for the first oxidation step and -0.7 V for the overall oxidation of TEA under the same conditions, see ref. 1 and 2). From a thermodynamic perspective this suggests that the first oxidation step will act as an effective kinetic barrier that needs to be overcome for overall oxidation of TEA to occur and that the second oxidation step is likely to be very fast. Beyond this we currently have very little mechanistic insight, something that hopefully will change in the near future.

1 R. S. Sprick, C. M. Aitchison, E. Berardo, L. Turcani, L. Wilbraham, B. M. Alston, K. E. Jelfs, M. A. Zwijnenburg and A. I. Cooper, *J. Mater. Chem. A*, 2018, **6**, 11994.
2 M. Sachs, R. S. Sprick, D. Pearce, S. A. J. Hillman, A. Monti, A. A. Y. Guilbert, N. J. Brownbill, S. Dimitrov, X. Shi, F. Blanc, M. A. Zwijnenburg, J. Nelson, J. R. Durrant and A. I. Cooper, *Nat. Commun.*, 2018, **9**, 4968.

Andreas Wagner asked Andrew I. Cooper: Is the difference in the stacking of the polymers dependent on their hydrophobicity? Could this influence the number of active sites and thereby explain differences in reactivity? Have you tried to use, *e.g.* circular dichroism to study the "folding" or stacking of these polymers in solution?

Andrew I. Cooper answered: We have not used circular dichroism. These polymers are mostly semi-crystalline at best; other materials such as covalent organic frameworks are much more crystalline (as shown in ref. 1) and also porous.

1 X. Wang, L. Chen, S. Y. Chong, M. A. Little, Y. Wu, W.-H. Zhu, R. Clowes, Y. Yan, M. A. Zwijnenburg, R. S. Sprick and A. I. Cooper, *Nature Chemistry*, 2018, **10**, 1180.

Erwin Reisner addressed all of the speakers: Our community is still mostly optimising photocatalytic half-reactions using sacrificial electron donors and acceptors, which causes the problems just discussed and they also need to be ultimately removed to catalyse useful closed redox cycles. Also, single light absorber photocatalyst systems have lower theoretical solar-to-fuel conversion efficiencies than dual light absorber systems and work by Domen, Kudo, Abe and others has previously shown that photocatalytic systems using semiconductor particle pairs can be constructed with suitable reversible redox shuttles. I would therefore like to hear your opinion on shifting the emphasis of our efforts to optimising our half-reactions using *reversible* redox shuttles rather than *irreversible* sacrificial donors. Optimising photocatalytic half-reactions with reversible donors is more challenging as it also introduces issues with back-reaction, but would in principle allow the coupling of individually optimised half-reactions *via* a reversible mediator to produce a functional closed redox system.

Andrew I. Cooper responded: I think this is absolutely the right way forward. As previously mentioned at the meeting, "you get what you screen for". I think there is an urgent need to move away from sacrificial reagents, unless (perhaps) these are coupled to biome schemes, waste degradation, *etc.*, and hence additively useful.

Martijn A. Zwijnenburg replied: I agree, two materials coupled together in Z-scheme by a reversible redox mediator is a strategy worth pursuing. From a computational perspective, this is something we could readily screen for using the methodology described in our contribution. In the defence of sacrificial electron donors, use of them has allowed Prof. Cooper's group and us to demonstrate that hydrogen evolution activity is quite a common property of conjugated polymers, something I suspect few people would have expected to be the case five years ago.

Erwin Reisner continued: Following on from the previous discussion on avoiding sacrificial reagents and developing closed redox cycles, I'd be curious to hear from the delegates about suitable redox shuttles (such as those developed by Domen, Abe, Kudo *et al.*), and recent progress on coupling semiconductor powders with reversible mediators for solar fuel synthesis. This topic may require more attention to allow progress using suspension/solution systems.

Joost N. H. Reek answered: I fully agree on this and we are currently looking into this. In fact, it is the central topic of poster 22 presented by Didjay Brugman.

Michael Grätzel commented: The groups of Professor Anders Hagfeldt (now at EPFL) and Professor Licheng Sun at Uppsala University have developed Cu(ı) complexes acting as redox shuttles that produce very high photo-voltages exceeding 1.2 V in dye sensitized solar cells. The ligands of these complexes are engineered to conserve a nearly tetrahedral configuration upon oxidation to the corresponding Cu(ıı) complexes. These systems mimic the natural redox relays present in copper proteins which are coordinatively bound to histidine moieties and show very high electron exchange rates.

Víctor A. de la Peña O'Shea enquired: The use of polymers in the photocatalytic reactions is a very important challenge, where different reaction media, electron donors, illumination sources and many other parameters have been used to improve the photocatalytic activity. What is your opinion about the redox chemical reactions that occur over the catalytic surfaces? Are the active sites well defined?

Andrew I. Cooper answered: We do not know much about the specific active sites at present – it is possible that specific functional groups play a catalytic role, but as yet there are no mechanistic studies to confirm that.

Martijn A. Zwijnenburg added: Understanding the mechanism by which polymer photocatalysts can evolve hydrogen and/or oxygen is, in my opinion, the next big challenge in the field. Not much is known with certainty in this area. One can hypothesize that for polymers loaded with palladium and platinum nano-particles, either added intentionally or as the remains of the catalysts used to synthesize the polymers, the mechanism involves electron transfer to these nanoparticles, which then reduces the protons and evolves the hydrogen. Indeed there is experimental evidence in ref. 1 that, at least for some polymer photo-catalysts, removing any traces of palladium results in the disappearance of any hydrogen evolution activity. However, there also reports in the literature that appear to suggest that certain polymers might be able to evolve hydrogen in the absence of any noble metal.

1 J. Kosco, M. Sachs, R. Godin, M. Kirkus, L. Francas, M. Bidwell, M. Qureshi, D. Anjum, J. R. Durrant and I. McCulloch, *Adv. Energy Mater.*, 2018 **8**, 1802181.

Wolfgang Viertl opened a discussion of the paper by Wendy J. Shaw: How strong do you estimate the influence of the hemilabile coordinations from the outer coordination sphere to the metal centre to be? Do you think that oxygen from the amino acid residues can coordinate to the metal hydride or the metal centre itself without any additional hydride? Are any stabilisation effects considered here?

Wendy J. Shaw answered: Given the likely flexibility of the ligand, it is unlikely that the COOH or COOMe groups are binding to the active site. In a related version of the catalyst, which has significantly limited structural flexibility, we think we see this under very restricted conditions. Although we have not seen any evidence by NMR or electrochemistry at this stage, it is possible that they are interacting with the hydride and this is something under further investigation.

Constantin Sahm said: I noticed that high CO_2 pressures were used during the catalysis. I was wondering why that is and if the catalysts presented here would work under ambient CO_2 pressure as well?

Wendy J. Shaw responded: They do add CO_2 at 1 atm CO_2 and convert it to $HCOO^-$ under stoichiometric conditions. The higher pressures of CO_2 are used to facilitate the rate of catalysis for measurable kinetics in a reasonable timeframe.

Moritz F. Kuehnel asked: Do you think the observed differences between the ethyl and phenyl-containing complexes are to do with changes in the ligand flexibility?

Wendy J. Shaw answered: My expectation is that both ligands would have similar flexibility. However, it is possible that the Ph-substituted ligands would have less access to the active site than the Et-substituted ones. That said, the Ph-substituted complexes had a larger impact from the outer coordination sphere than those that are Et-substituted, so it does not correlate well (or is at least an anti-correlation). It is something to consider, though, as we continue to understand this system.

Leif Hammarström enquired: When decorating catalyst with proton transfer acid/base groups in the secondary sphere to accelerate hydride formation, the assumption is that reprotonation of base is not rate-limiting. We have seen an example in Fe_2 azadithiolate where protonation of the aza nitrogen is much slower than diffusion controlled. Do you have more data on the topic? Do you see any evidence for or against proton transfer limitations?

Wendy J. Shaw answered: While we do not think that is the role of the pendant amine here, with our H_2 oxidation/production $Ni(P_2N_2)^{2+}$ catalysts we definitely saw enhancements with additional proton relays, providing evidence that a single proton relay is not always enough. We think this is due to the hydrophobic groups on the rest of the molecule limiting the transfer of the proton to the solvent. Additional functional groups, such as carboxyl groups, positioned correctly, significantly enhance proton transport into and out of the molecule. This has resulted in lower overpotentials and faster rates.

Shelley D. Minteer remarked: There seems to be an increasing number of researchers using protein enzymes as biological inspiration for catalysts. From a big picture perspective, are researchers also using nucleic acid enzymes (*i.e.* deoxyribozymes) for biological inspiration? It seems as though they are smaller and it is easier to control the outer coordination sphere than in complex proteins.

Wendy J. Shaw responded: Yes, I believe Yi Lu at UIUC is taking this approach. A particular advantage is DNA origami, where DNA can be put in very specific locations. Two disadvantages are the limited number of functional groups compared to proteins and also the sensitivity of DNA to high salts, which would limit their relevance in electrochemical systems.

Souvik Roy queried: Compared to the PNP ligands, how do the Rh–P_2N_2 complexes perform towards CO_2 hydrogenation?

Wendy J. Shaw responded: Cliff Kubiak and co-workers evaluated Rh–P_2N_2 complexes for CO_2 hydrogenation in ref. 1. He characterized 5 different catalysts. All were active for CO_2 hydrogenation, but were slower than $Rh(depe)_2$ under the conditions used. They suggested that this was due to steric blocking due to the P_2N_2 ligands.

1 A. M. Lilio, M. H. Reineke, C. E. Moore, A. L. Rheingold, M. K. Takase and C. P. Kubiak, *J. Am. Chem. Soc.*, 2015, **137**, 8251.

Souvik Roy asked: Protonation of CO_2 to generate formate seems to be the rate-determining step. Have you tried attaching charged residues to the pendant amines in the second coordination sphere, such as guanidine or imidazolium?

Wendy J. Shaw replied: Yes, we tried to make the arginine (guanidinium group) complex, but were not able to. As a side note, the synthetic challenges we have speak to the earlier question on the synthetic intensity of many of these approaches. We made a Me–His, but had to protect the His due to solubility issues. We did consider making a lysine variant (with a free amine), but there is ample evidence that in the presence of CO_2 this will result in a carbamate. We had less trouble with the Ph-complexes and solubility, so revisiting those complexes with positively charged groups would make sense, although our current plans are to use a structured, model protein and alter the charge within the context of that scaffold.

Daniel Antón-García opened a discussion of the paper by Peter Brueggeller: When you compare the activities of the different metal centers for the chlorido complexes shown in Fig. 4 of your paper (DOI: 10.1039/c8fd00162f), have you checked whether all of the complexes maintain their molecular integrity? Or do they decompose to form nanoparticles capable of hydrogen evolution?

Peter Brueggeller replied: Comparable to the case of Fig. 6 of our paper (DOI: 10.1039/c8fd00162f), where the mercury drop check is depicted in the diagram, this test has also been carried out for the chlorido complexes of Fig. 4. The amount of hydrogen production is the same as that without mercury, indicating that no nanoparticles are responsible for the observed hydrogen evolution. Furthermore, especially for palladium, it has been shown in ref. 1 that only protected nanoparticles lead to reasonable hydrogen production.

1 J. Prock, S. Salzl, K. Ehrmann, W. Viertl, R. Pehn, J. Pann, H. Roithmeyer, M. Bendig, H Kopacka, L, Capozzoli, W. Oberhauser, G. Knör and P. Brüggeller, *ChemPhotoChem*, 2018, 2, 271.

Daniel Antón-García commented: In the case of the Pt and Co chlorido complexes shown in Fig. 4 of your paper (DOI: 10.1039/c8fd00162f), there seems to be an induction period of 24 h after which the activity increases. How can you explain this if there is no change in the molecular structure of the catalyst?

Peter Brueggeller replied: This can be explained by the exchange of chlorido ligands for solvent molecules. Fig. 6 of our paper (DOI: 10.1039/c8fd00162f) shows that solvato complexes perform better than their chlorido analogues. This exchange is certainly dependent on the kind of metal used. In Fig. 4 the Pt and Co chlorido complexes show an induction period of 24 h, since this is the time needed for the production of the solvato complexes in these cases.

Christine A. Caputo remarked: In your study you extended the PNP ring size, and this had the effect of changing the coordination of the ligand on the metal centre, and ultimately fundamentally changed the mechanism of the catalytic reaction. Have you thought much about this change and how it impacts the results you observed?

Peter Brueggeller responded: Our PNP ligands have been tailored to make different coordination properties possible. Thus for the PNP–C1 ligands coordination of the nitrogen atom is impossible. This makes it feasible for the function as proton relay. By contrast, the nitrogen atom in the PNP–C2 ligands leads to two fused five-membered rings upon coordination. At first glance this should inhibit the proton relay function. However, the catalytic performance of the PNP–C2 ligands is better than that of their PNP–C1 counterparts. DFT calculations show that a different mechanism is responsible for this amazing behaviour, which could be applicable also for other pincer-type ligands.

Moritz F. Kuehnel posed: In your long-term experiments you are continuously adding additional Ir photosensitiser. What are its decomposition products? When you did control experiments, you only used the Ir complex without an HER co-catalyst, and you nevertheless observed some H_2 formation. From my own experience, it is vital to consider that photosensitisers can form HER catalysts not only from their own decomposition products, but also with composition products of the HER catalyst. Have you performed control experiments in which you combined the Ir photosensitiser with *e.g.* only the ligand used for the HER co-catalyst?

Peter Brueggeller responded: It is well known in our laboratory that the Ir photosensitiser alone produces a minor amount of hydrogen as indicated in the paper. For this kind of photosensitiser this is known in the literature. Ir nanoparticles are believed to be responsible for this effect (*e.g.* see M. Beller's work on copper iodide using Ir photosensitisers in ref. 1). However, we have performed amalgam tests in order to study the influence of nanoparticles. Since there was no drop in activity, the observed HER is dominated by molecular species. We did not use Ir as the metal for our PNP catalysts. I agree that this is a good idea to check whether water reducing catalysts (WRC) based on Ir are active catalysts.

1 H. Junge, Z. Codolà, A. Kammer, N. Rockstroh, M. Karnahl, S.-P. Luo, M.-M. Pohl, J. Radnik, S. Gatla, S. Wohlrab, J. Lloret, M. Costas and M. Beller, *J. Mol. Catal. A: Chem.*, 2014, **395**, 449.

Christine A. Caputo said: Going back to the steric effect, there is likely a difference in the functionality of the various groups on your pendant amines and in your paper you attribute this to simply an electronic effect, however you cannot discount the steric effect necessarily. Molly O'Hagan at Pacific Northwest National Lab showed by NMR studies that long alkyl chains on the amines in the P_2N_2 type ligands caused them to move more slowly the bigger the R group (ref. 1). I think this is an important effect that is underappreciated.

1 A. J. P. Cardenas, B. Ginovska, N. Kumar, J. Hou, S. Raugei, M. L. Helm, A. M. Appel, R. M. Bullock and M. O'Hagan, *Angew. Chem., Int. Ed.*, 2016, **55**, 13509.

Peter Brueggeller answered: I agree, this is certainly a so-called stereo-electronic effect as described by P.W.N.M. van Leeuwen. Different R groups change the basicity of the pendant amines and their accessibility for protons due to steric constraints. This has been confirmed by DFT calculations. Slower movement of the protons is of course also a disadvantage. Quite clearly, in our study there is a beneficial effect of small R groups. This important effect has been confirmed in the meantime by preliminary experiments, where R is now H.

Michael Grätzel opened a discussion of the paper by Sergii. I Shylin: Dr Shylin used a sacrificial acceptor, *i.e.* peroxodisulfate, to irreversibly photo-oxidize $Ru(\text{ii})(bipy)_3$ to $Ru(\text{iii})(bipy)_3$. The latter in turn was used to oxidize water to oxygen using an $Fe(\text{iv})$ complex as a molecular catalyst. I was wondering whether the time resolved laser photolysis experiments revealed any changes in the transient absorption that could be attributed to the conversion of the $Fe(\text{iv})$ to the $Fe(\text{v})$ complex.

Sergii I. Shylin replied: On the microsecond timescale, we observe a negative absorption peak at 650 nm corresponding to the $Fe(\text{iv})$ complex and two positive absorption peaks at 550 nm and 830 nm coming from the $Fe(\text{v})$ complex, seen in Fig. 4b in the paper (DOI: 10.1039/c8fd00167g).

Michael Grätzel remarked: I wonder whether Dr Shylin could propose a mechanism for the formation of oxygen from the reaction of the $Fe(\text{v})$ complex with water.

Sergii I. Shylin answered: The active species should be four-times oxidized to evolve oxygen from two water molecules. The $Fe(\text{v})$ complex characterized in our paper is only a one-electron oxidized species, and we suggest that it is oxidized further. Intermediates beyond $Fe(\text{v})$ appear to be too active, and hence are not observed in the steady-state spectra. Currently we are working on calculations of possible active species involved in oxygen evolution.

James R. Durrant enquired: To drive water oxidation, your catalyst is likely to need to undergo four oxidations. Do you have any evidence that the electron transfer kinetics change with the oxidation state of the catalyst?

Sergii I. Shylin replied: At the moment we know that the overall oxygen evolution undergoes first-order kinetics, and that the initial electron transfer step is a first-order reaction as well. A detailed investigation of the following catalytic steps would certainly provide important additional insights into the reaction kinetics. However, in our flash photolysis experiment we excite only a small amount of photosensitizer and are thus restricted to monitoring the initial hole transfer step. Observation of hole transfer events at the second (or later) step will require quantitative transformation of the catalyst to a desired intermediate prior to the flash, something which has so far proven difficult to achieve.

Sylvestre Bonnet remarked: In Fig. 3 in your paper (DOI: 10.1039/c8fd00167g) you show that photochemical water oxidation works at a constant rate for more than 60 s. In the experimental part you mention an experiment of 140 s. What happens after that time? Do you see the photocatalytic WO stopping? If you do, is the catalyst or the photosensitiser becoming inactive, or both? Did you try adding more photosensitiser or more catalyst to see if photocatalysis would resume?

Sergii I. Shylin responded: Water oxidation stops after 150–240 s. We have found that the iron complex degrades during catalysis due to the decrease in the pH ($2H_2O \rightarrow O_2 + 4H^+ + 4e^-$). We cannot reactivate water oxidation after these 150–240 s by adding more buffer or base, but the TON can be improved by using more concentrated buffer from the beginning.

Erwin Reisner continued: I have a question regarding the water oxidation mechanism involving the Fe catalyst. You have discussed a possible mechanism for the water oxidation catalysis involving a hepta-coordinate Fe–H_2O species and refer to previously reported Ru-based molecular water oxidation catalysts. However, Ru catalysts such as those reported by Sun and others involve a larger metal ion and a strained equatorial ligand environment, which enables water coordination. The Fe catalyst is well encumbered by the clathrochelate ligand and it is not clear how an aqua ligand would bind. Also, you mention an alternative water oxidation mechanism on the ligand – could you please elaborate on this further and provide more details (maybe also in the context of precedent literature or other evidence)?

Sergii I. Shylin responded: Indeed, the clathrochelate complex discussed in our paper is extremely rigid, so that the formation of the Fe=O oxo-species is questionable unless the integrity of the complex is lost. We are inclined to assume that O–O bond formation may occur on a ligand, whereas the metal center is involved indirectly and facilitates accumulation of holes. A mechanism involving ligand was proposed for the ruthenium "blue dimer" in ref. 1 (see Scheme 4), where an O–O bond is formed on carbon as a fragment of the four-membered endoperoxide ring, C_2O_2. We acknowledge that the mechanism of water oxidation using the clathrochelate complex remains elusive and is not discussed in our paper. Currently we are working on the modification of clathrochelate ligands in order to stabilize the active species (to make it 'less active'). In parallel, calculation studies of the supposed intermediates are ongoing.

1 J. K. Hurst, J. L. Cape, A. E. Clark, S. Das and C. Qin, *Inorg. Chem.*, 2008, **47**, 1753.

Christine A. Caputo enquired: At the end of your manuscript you suggest that you could modify the ligand by adding a methyl group, and hypothesized that this would provide the complex with more stability. Is the synthetic approach to making the proposed methylated molecule quite straightforward?

Sergii I. Shylin replied: The iron(IV) clathrochelate complex spontaneously assembles in aqueous media from low-valent iron salts, oxalodihydrazide and formaldehyde in the presence of atmospheric oxygen. It is a simple one-pot template reaction that can be reproduced even in a minimally equipped school laboratory. The use of other aldehydes instead of formaldehyde would lead to modified clathrochelates, *e.g.* starting from iron(III), oxalodihydrazide and acetaldehyde, the methylated complex can be obtained.

Christine A. Caputo addressed a question to Wendy J. Shaw, Peter Brueggeller and Sergii I. Shylin: In many of these papers, the structure activity relationships rely on the synthesis of a library of ligands and complexes, which is a significant effort and often the synthetic modification is quite challenging. My question to the panel is: is the approach and the synthetic effort worth the payoff?

Peter Brueggeller responded: In this context it is worthwhile to learn from nature also with respect to coordination chemistry. Nature has invented porphyrin ligands during evolution. From a preparative point of view these ligands would be very expensive when produced by chemists. On the other hand these ligands combine chelate and macrocyclic effects, which are absolutely necessary for the complexation of metals like magnesium. Also, for artificial photosynthesis one can postulate a simple rule: inexpensive metals like 3d metals lead to a more difficult coordination chemistry and there is the need for sophisticated ligands. Since at the end the costs for a whole artificial device count, the synthetic effort is worth the payoff, especially when earth-abundant metals come into play.

Wendy J. Shaw replied: Yes. In order to get to a more sustainable future, we need to be able to develop better catalysts. From everything we know, the best way to do this is to mimic features of enzymes, not necessarily structurally but functionally. This requires good synthetic capabilities to test the proposed principles. For an ultimate application, the synthesis will need to be scalable, high yield and reasonably straight forward, but getting to a solution will require many syntheses, including homogeneous, heterogeneous, and biological. Investing in synthetic capabilities is essential to develop the catalysts we need to answer the fundamental questions we have.

Leif Hammarström commented: Regarding photocatalytic water oxidation or hydrogen production, where the rate of gas evolution is plotted as a function of the irradiation time, the observed rate does not directly reflect the rate of any catalyst steps. Instead, it is equal to the rate of photon absorption times the quantum yield for product formation. Quantum yields are typically low, so most of the reaction is recombination or side reactions. Thus, a comparison of rates between different systems does not necessarily give information on the rate of the catalytic steps.

Peter Brueggeller responded: In our photocatalytic HER experiments the Ir photosensitiser was clearly the less stable species, when compared with the WRC. However, the photosensitiser alone produces very poor HER results. This means that side reactions involving the photosensitisers can be neglected. Also the production of reactive nanoparticles has been ruled out by amalgam tests, thus excluding also these side reactions. Furthermore, we have studied a whole series of WRC, where only the metals differ in otherwise identical systems. So why not make the different metals used responsible for the observed HER? The proposed different catalytic cycles for the PNP–C1 and PNP–C2 ligands are confirmed by DFT calculations. However, I agree that these cycles are related to the forward reactions and indeed recombination reactions could also differ.

Sergii I. Shylin responded: I agree with your comment. In our photocatalytic experiments, we used excess amounts of photosensitizer $[Ru(bpy)_3]^{2+}$. Hence, we assumed that the overall performance of the photocatalytic system was determined by the catalyst. In addition, we investigated the kinetics of water oxidation using the one-electron oxidant $[Ru(bpy)_3]^{3+}$ (prepared separately) in darkness. In both the photochemical and chemical experiments, the initial rates of oxygen evolution exhibited a linear dependence on the catalyst concentration.

Michael Grätzel continued: As a follow up of Prof. Hammarström's question, the point is that the $[Ru(bpy)_3]^{3+}$ may not have a high enough redox potential to drive the oxidation of Fe(v) to Fe(vi) On the other hand the peroxosulfate anion radical formed as an intermediate from the one electron reduction of peroxodisulfate does have a high enough redox potential to drive this reaction. To rule out this possibility one would have to use a one electron oxidant such as Ce(iv) to affect the water oxidation.

Sergii I. Shylin replied: That is true. We cannot completely exclude the impact of the sulfate radical on oxygen evolution in photochemical water oxidation. However, we have shown that $[Ru(bpy)_3]^{3+}$ prepared separately does evolve oxygen from water when added to the Fe(iv) catalyst. Hence, $[Ru(bpy)_3]^{3+}$ is able to oxidize the catalyst beyond Fe(v). Alternatively, we could use ceric ammonium nitrate, but it would have required an acidic pH, something which we would like to avoid.

Anastasia Vogel directed a question to Wendy J. Shaw: Considering the interaction of the outer coordination sphere with both the active centre as well as the solvent, do you expect to see a solvent effect on the series of presented TONs/TOFs? If so, may there be an optimum pairing of solvent and outer coordination sphere residues?

Wendy J. Shaw responded: Within a given solvent, I expect to see limited solvent effects. The solvent primarily affects the hydricity, so comparisons between solvent are the most meaningful. It is possible that as we go to a larger protein structure, the protein effectively shields the solvent in different ways, resulting in an effective solvent effect, but I would attribute this more to the scaffold than to the solvent. There is likely an optimum combination of solvent effects and scaffold effects, but ultimately I believe that being able to precisely control where particular atoms are, solvent or scaffold, will yield the best catalysts.

Anastasia Vogel continued: Following up on the previous question, would you expect such a solvent effect on TONs/TOFs due to the interaction with the outer coordination sphere to be larger than other changes (*e.g.* changed mass transport due to different polarities of solvents)?

Wendy J. Shaw responded: I think there is that possibility. In the end, the rate-determining step for any given reaction will dictate what effects the solvent and the scaffold have. In this case, within a given solvent, there is no evidence of an overriding impact beyond controlling the hydricity.

Joost N. H. Reek asked: I really like your approach of controlling reactivity using the second coordination sphere, installing functional groups that are also playing important roles in enzyme cavities. Looking at the mechanism that you propose it seems that the CO_2 activation (step 3) is RDS. (1) Do you have more detailed insight into how this CO_2 activation occurs?

(2) Mark Roberts demonstrates with his porphyrin based catalyst that the secondary interaction with CO_2 is important. In the CO_2 activation step, negative charge accumulates on oxygen which means that cations or hydrogen bonds may stabilise the TS and thus speed up this reaction. You have not included such functional groups in your library. Have you considered functional groups that have hydrogen donors? Simple amine bonds may be sufficient.

Wendy J. Shaw answered: Yes we considered amines, but due to the formation of carbamates, we have not pursued it. The Me–His system is not charged and we were unable to synthesize the arginine system. It is a great thought though and one we are now pursuing in the context of a structured protein scaffold. Regarding the CO_2 addition for the Ph complexes, previous work by Linehan *et al.* in ref. 1 using a Co metal center calculated either a mechanism in which CO_2 bound, or one in which the CO_2 reacted directly with the hydride. The results were within a few kcal mol^{-1}, so either option for the mechanism could be equally likely. We were hoping that the Rh system would provide more distinction. Unfortunately, with the Et complexes, extensive computational DFT studies do not provide agreement with the experimental studies, specifically the deprotonation step appears to be much easier in the calculations than is observed experimentally. We are continuing to apply more advance computational studies to address this, but at this stage, we do not have a definitive answer on the mechanism.

1 M. S. Jeletic, M. T. Mock, A. M. Appel and J. C. Linehan, *J. Am. Chem. Soc.*, 2013, **135**, 11533.

Joost N. H. Reek remarked: The ligands that you use have a lot of flexibility which could potentially lead to different conformations of these ligands around the metal center. It could be that the functional groups that you install on your ligands may influence the distribution of conformations How can you distinguish between real secondary sphere effects and an indirect effect in which the functional groups just changes the ligands conformations? Could phosphorus NMR be of any help?

Wendy J. Shaw answered: Great observation – there is no doubt that the PNP ligands are moving relative to the metal center. This is further supported by ample studies on the P_2N_2 ligands which do show structural restriction. We do not see any difference by ^{31}P-NMR and have done many low temperature experiments to investigate this. While we do not see any direct correlation with mobility, even in the P_2N_2 rings which are much more motionally restricted, it is still difficult to create interactions which slow these motions. This can be done with groups that can interact with each other (aryl rings, or very long chains which can slow chair-to-boat interconversions). In this case we don't have anything large enough to slow these interconversions. The aryl rings could potentially interact to lower the PNP mobility, but there is no evidence that this is correlated to the TOF. It is worth considering, and within the protein scaffold we are moving to we should be more able to evaluate the mobility of these ligands.

Christine A. Caputo addressed Sergii I. Shylin, Peter Brueggeller and Wendy J. Shaw: All of the panelists have utilized some rare and precious metals in their photocatalytic systems. The impetus, of course, is that we move towards systems with more abundant metals for sustainability purposes. Could all of the panelists comment on their choice of metal used in their catalytic system?

Sergii I. Shylin responded: Water oxidation systems consist of several components: an oxygen-evolving catalyst, photosensitizer, electron acceptor, *etc.* Thus, there are so many variables which influence the overall performance of the catalytic system. If we investigate the water oxidation part of the system (indeed it is based on iron, the most abundant transition metal), we need to use the efficient photosensitizer $[Ru(bpy)_3]^{2+}$ with known properties. We could use a noble metal-free light harvester, but our system would be limited by its efficiency. When an efficient catalyst is found and characterized, one can use it in a system to screen potential photosensitizers based on abundant elements. When catalyst and photosensitizer are found, one can vary the third component, and so forth.

Peter Brueggeller answered: Since the main focus of our paper concerning the ligands is on phosphines, it is interesting to note how these ligands change their coordination properties as soon as different metals are involved, *e.g.* using the expensive metal osmium produces chromophores which are stable for more than six weeks (see ref. 1). This is not possible for inexpensive 3d elements like nickel, which is certainly a consequence of the HSAB principle. The same considerations are also valid for WRCs: 3d elements usually show a more complicated coordination chemistry leading to less stable complexes than their more expensive 4d and 5d counterparts.

1 J. Prock, S. Salzl, K. Ehrmann, W. Viertl, R. Pehn, J. Pann, H. Roithmeter, M. Bendig, H. Kopacka, L. Capozzoli, W. Oberhauser, G. Knör and P. Brüggeller, *ChemPhotoChem*, 2018, **2**, 271.

Wendy J. Shaw replied: In our case, we are focused on understanding how the outer coordination sphere works. In order to do this we need a well understood core catalyst and a good attachment point for the outer coordination sphere. The $Rh(PNP)_2$ system provided that. There is a similar system with Co that is also very well understood, unfortunately, it does not have a place to attach a protein to create a scaffold so it would not work to answer the questions we are interested in answering. In principle, the design principles we are developing should be extended to other metals and systems, so we do not see the use of a precious metal as limiting in this respect.

Matthias Beller commented: It is not so important for these examples to move from noble to non-metal, the next step is to move to more integrated devices – it is more important to be using and implementing in real devices. This is interesting for organometallic chemistry but needs to be applied for artificial photosynthesis.

Peter Brueggeller responded: I agree that it is a difficult task to think of real economically possible devices for artificial photosynthesis. However, in my opinion, using environmentally benign and inexpensive metals is an important step towards an affordable device. Of course there are other factors, *e.g.* the TON: a TON of about one million would certainly allow also noble metals, which can be recycled at least in principle.

Another point is of course the sacrificial donor. Nature uses water and this should also be the goal for artificial systems. One has to keep in mind that the hitherto most efficient systems like Nocera's artificial leaf are very expensive, in part also because of the membrane containing palladium. Another highly efficient device is certainly Eisenberg's system using cadmium chalcogenides in the form of quantum dots for the chromophores (ref. 1). Cadmium chalcogenides are so poisonous that they are forbidden within the European Union.

1 Z. Han, F. Qiu, R. Eisenberg, P. L. Holland and T. D. Krauss, *Science*, 2012, **338**, 1321.

Simon T. Clausing then said: Systems for photocatalytic water splitting are often optimized towards their pH value. However, this is usually always the optimization towards a compromise: the sacrificial electron donor (for the water reduction side reaction) functions optimally at a high pH value, and the catalyst might be active at a far lower value, so the full system lies somewhere in between. Ultimately, however, the community wants to leave behind the dependence on sacrificial donors and acceptors. Would it therefore not be sensible to try and find a way to optimize the optimal conditions for only the catalyst, and not a combined catalyst-and-donor system? Have any of you thought of approaches to achieve this goal?

Peter Brueggeller responded: It is true that in sacrificial systems the pH dependence may mainly reflect the pH dependence of the sacrificial donor used. Usually a lower pH is an advantage, since more protons are present. However, it is known in the literature that sacrificial donors like triethylamine show better reduction potential in more basic solutions. Hence, the result is a maximum usually in the range of pH 9–10. So the solution to this problem is of course the independence of sacrificial donors. One possibility is to deliver the electrons electrochemically. The second is to construct so-called combined systems where, as in nature, water is used as the electron donor.

Sergii I. Shylin replied: It is absolutely possible in a photoelectrochemical experiment, with an electric current as the 'pH-independent electron donor'.

David Wakerley asked Sergii I. Shylin: Can you provide evidence on your CV to confirm that the catalytic oxidation wave is O_2 formation and not the oxidation of your ligand?

Sergii I. Shylin replied: We have found using UV–Vis spectroscopy that the complex remains stable after 2 h of bulk electrolysis at 1.4 V *versus* NHE. The spectra of the solution under study taken before and after electrolysis are identical. In addition, in our CV the peak current is linearly dependent on the square root of the scan rate that indicates a homogeneous process in solution, but not on the electrode surface. So the impact of possible decomposition products, such as oxides, on the catalytic current is ruled out.

Han Sen Soo commented: In Fig. 4a of your manuscript (DOI: 10.1039/c8fd00167g), I noticed that the spectra were assigned to $[Ru(bpy)_3]^{2+*}$, but the spectral features are not fully consistent. In the transient absorption spectrum of the Ru* excited state, there is usually both a groundstate bleach as well as the emission spectrum. However, even at very short time scales of 20 ns, the emission component around 600 nm is already absent. I believe that these bands should probably be assigned to the Ru(III) instead.

Joost N. H. Reek enquired: In your conclusion you mention that this catalyst gives the fastest hole scavenging when combined with the ruthenium chromophore, and this is proposed to be important for the stability of the chromophore. Do you understand why this particular complex leads to fast hole scavenging, and could this lead to design criteria for future catalyst development?

Sergii I. Shylin answered: The robust nature of the clathrochelate ligand provides the catalyst with additional stability. Thus the catalyst is able to re-reduce $[Ru(bpy)_3]^{3+}$ repeatedly without degradation. Currently we are working on the modification of the ligand in order to improve its stability.

Joost N. H. Reek continued: So you conclude that it is only a matter of stability of the catalyst?

Sergii I. Shylin responded: It is an important factor, but not the only one. We should also consider thermodynamics, specifically the accessibility of the Ru^{III}/Ru^{II} redox couple by the catalyst.

Christine A. Caputo addressed Peter Brueggeller, Wendy J. Shaw and Sergii I. Shylin: Thinking about structure activity relationships, did any of the panelists use a significant amount of computational chemistry in the design of your molecular systems?

Peter Brueggeller responded: I agree that computationally predicting the best catalysts is still difficult. However, in our case DFT calculations show that during charge transfer and reduction of the WRC the metals seem to become too electron rich in the presence of anionic chlorido ligands. So the calculations predicted that the chlorido ligands are expelled from the coordination sphere as soon as the metals in the precatalysts are reduced. As a consequence we decided to already remove the chlorides from the precatalysts and use solvato complexes. Indeed the catalytic performance of the solvato complexes outperforms that of the corresponding chlorido complexes. This has also practical consequences: the use of salty sea water can be detrimental for the hydrogen evolution reaction.

Wendy J. Shaw responded: We work very closely with computational chemists in all aspects of our work. Unfortunately, computationally predicting the best catalysts is still difficult. There are places they can contribute very well at this stage, for instance, we are using MD to determine the amino acid mutations we should make to control catalysis with a protein-based scaffold. Further understanding how the catalysts work will get us closer to computationally-designed catalysts.

Qian Wang opened a discussion of the paper by Wilson A. Smith: I am curious about the stability of the copper borate oxide layer. Have you checked the sample using XPS after a long-term test?

Wilson A. Smith replied: The stability is based on the operation time and applied potential. If the photocharged sample is held at a constant potential (and under illumination), the current will slowly decrease over time and revert back to the value of the pure $CuWO_4$ electrode. We haven't checked XPS after long term measurements, but we assume the decrease in photocurrent is due to the degradation of the borate layer.

Qian Wang asked: You mentioned that several materials exhibited the same effect. However, different materials have different band structures. I was wondering how you confirm that the formed oxide layer will show suitable band structure matching that of the semiconductor as expected.

Wilson A. Smith responded: It has been shown here that in a borate electrolyte, the $CuWO_4$ photoanode forms a large band gap metal oxide on the surface. Similarly, for the $BiVO_4$ photoanode in a borate electrolyte, a large band gap metal oxide made from borate forms on the surface, and for the BiVO4 photoanode in a phosphate electrolyte, a large band gap metal oxide made from phosphate forms on the surface. It happens that in these cases, metal oxides made with borate and phosphate have larger band gaps than those of $BiVO_4$ and $CuWO_4$ (and many of these materials exhibit glass-like properties). Therefore it happens that borate and phosphate electrolytes form borate- and phosphate-containing metal oxides which have larger band gaps than photoactive metal oxides, leading to beneficial charge separation/catalysis.

Andrew B. Bocarsly asked: You spoke about a chemisorbed species as the top layer of the junction, but you also spoke about it being ~5 nm thick. This is too thick to be a chemisorbed layer. It is either approximately a monolayer or it is a thin film (~5 nm). What picture does your data support?

Wilson A. Smith replied: We believe that the layer really is only chemisorbed, and only mention that the layer is ~5 nm thick because this is the penetration depth of the XPS. Since we still see the signal from the underlying $CuWO_4$ layer we do not think that the layer is complete or as thick as 5 nm.

Virgil Andrei queried: Why are open-circuit conditions required to form the heterojunction? Would the copper borate surface layer form faster when applying a slightly positive or negative potential during irradiation? Under those conditions, no photoelectrochemical oxygen evolution should occur; therefore, any small currents may be traced back to changes in the surface composition.

Wilson A. Smith answered: We have shown that applying a potential close to the photovoltage of the electrode, and for over 20 h in the dark, also promotes an improvement in the photoelectrochemical performance. However, the improvement is much smaller than that during the photo-induced process. This may be because of differences in charge carrier density, diffusion within the bulk of the material, and quasi-fermi level splitting during photon-induced photocharging, which favors passivation of surface states close to the illuminated surface.

Virgil Andrei asked: The paper mentions that copper borates are proven photocatalysts. However, this layer can account for very little light absorption due to its thickness of only 5 nm. In this case, can it really be considered a photo-catalyst, or does it act more as a passivating layer/electrocatalyst?

Wilson A. Smith responded: After finding that we most likely had a copper borate layer on our surface, we looked into the literature and saw that this material has recently been shown to have photocatalytic properties. Since the copper borate layer found on top of our $CuWO_4$ electrode was so thin, we do not believe it contributed to the light absorption and photocatalysis. There-fore, our comment about copper borate being a photocatalyst was from its previous use in the literature and not its activity in our work. For our paper, we believe that it indeed acted as a passivating layer and potentially as an electrocatalyst.

Charles E. Creissen requested: Can you comment on the use of methanol as a hole scavenger and its ability to distinguish between the bulk and surface properties of the semiconductor?

Wilson A. Smith replied: We used methanol as a hole scavenger for reasons of stability and enhanced oxidative performance. In the literature, it is typical to use methanol, hydrogen peroxide, or $NaSO_4$, and here we chose methanol because it worked best with the $CuWO_4$ photoelectrode we use, while we use H_2O_2 for other materials such as $BiVO_4$.

The use of a sacrificial agent should be used to correlate current/voltage characteristics with and without this reactant, which can then provide informa-tion about catalytic and charge separation efficiency. The idea is that a sacrificial agent has a 'maximum' catalytic activity for a given material, so when the same electrode performs water oxidation without it, we can see how close/far from this 'maximum' behavior it is. Likewise, by including the incident photon flux and optoelectronic properties of the material (*i.e.* band gap), we can also roughly estimate the charge separation efficiency.

Charles E. Creissen enquired: How does the degree of surface hydroxylation influence catalysis especially, considering the role of hydroxyl radicals in MeOH oxidation?

Wilson A. Smith answered: We find that surface hydroxylation improves catalysis, and occurs spontaneously in the dark when a (metal oxide) photo-electrode is placed in an electrolyte. After soaking in the electrolyte, and increasing hydroxylation (measured in a separate work by operando X-ray Raman spectroscopy), the PEC performance improves. For the case of MeOH oxidation, this was used as a sacrificial oxidative reaction that should occur more easily than water oxidation, and indeed we see high photocurrents and lower onset potentials for this reaction.

James R. Durrant commented: Your talk gave some very interesting insights. Could you comment on what you mean by the term 'photocharging'? I note that in your figure illustrating photoanode function, your 'photocharged' photoanode is not drawn with more charge.

Wilson A. Smith answered: The term photocharging is indeed somewhat of a misnomer. Many years ago when we first discovered the phenomenon using BiVO$_4$, the PhD student at the time (now Dr Bartek Trzesniewksi), thought the enhanced performance looked like the system had been charged like a battery, and hence called it 'photocharging'. The name does not mean that there is excess charge on the surface (though the surface chemistry does change), so I/we understand that the name is not ideal and may be somewhat misleading if taken literally. We will try to re-name the phenomenon in subsequent work.

James R. Durrant continued: I note that in your electrode energy level diagram the copper borate valence band is deeper than that of copper tungstate. Could you comment on how water oxidation proceeds on this electrode, and particularly where photogenerated holes accumulate to drive this reaction?

Wilson A. Smith answered: The band diagram was drawn for illustrative purposes, and is by no means exact. First, the layer of copper borate is less than 5 nm, and thus may not be thick enough to form an appreciable layer that has a large enough density of states to make a formal or well-defined band structure. In addition, there are 2 structural forms of copper borate (CuB$_2$O$_4$ and Cu$_3$B$_2$O$_6$) with 2 different band gaps (3.1 eV and 2.16 eV). Due to the non-qualitative nature of the XPS results that showed a change of Cu valence, and an increase in a unique O feature and B peak, we cannot definitively say which composition of copper borate we found. As we did not see an appreciable change in the band gap or band edge, we made the implicit assumption for this diagram that the larger band gap copper borate was there, leading to the figure as drawn. We also saw evidence of enhanced charge separation, which is more likely with a hetero-junction with a large band gap material compared to that with a smaller band gap material. Our illustration was not meant to be qualitative, and is only schematic to show the materials properties change we measured (0.4 eV shift in Fermi level), improved catalysis/charge separation, and new Cu/O/B signals from the surface.

Putting this all together, we believe that photogenerated holes in CuWO$_4$ may tunnel through the very thin layer of copper borate, where this layer only provides enhanced charge separation in the space charge region, which may also aid in hole accumulation compared to the bare CuWO$_4$/electrolyte interface.

Chanon Pornrungroj remarked: Could you elaborate more on the comment you made about the day–night cycle (charge/discharge) of the CuWO$_4$ layer and how it would affect the long-term stability of this material?

Wilson A. Smith replied: We observe that the process of the surface hetero-junction forming occurs during illumination, and is removed when the light is turned off. Therefore, in a practical system that turns on and off every day when the sun goes up and down, the surface oxide layer will form, dissolve, and re-form every single day. If the removal of the layer removes Cu or W from the electrode, then each cycle will physically degrade the electrode and over time will remove all of the photoactive material. Therefore, although the photocharging technique brings significant improvement in photoelectrochemical performance, its practical utilization could be hindered by catalyst dissolution in the dark.

Michael Grätzel asked: The hole currents you observed upon polarizing the copper tungstate look like capacitive charging currents and not Faradaic currents. My question is where do the electrons go ?

Wilson A. Smith replied: This is a great question, and it seems this may be different for every material that can photocharge. We see from XPS that the surface valence of Cu changes, but there is a reduction in the Cu^+ peak, which correlates to the oxidation of Cu by anionic adsorption. So far, we cannot account for the photogenerated electrons, but it may relate to bulk defect chemistry which we hope to probe in the coming years.

Andrew B. Bocarsly noted: It is well established that polycrystalline electrodes yield a poorer current–potential and quantum yield response than that observed with single crystal electrodes. This is often attributed to grain boundaries acting as recombination sites. Could grain boundary processes be added to your simulation? Is it possible that the surface layer you are forming is occurring primarily at grain boundaries? If so, how will this impact the expected I–V response?

Wilson A. Smith responded: This is a great question, and one that we share and have already started to address internally. We plan on using *in situ*/operando electrochemical atomic force microscopy (EC-AFM) to observe the surface of our photoelectrodes during operation, and map the current density distribution spatially over the electrode surface. This should allow us to see where certain 'hotspots' are for activity, and if indeed they occur at grain boundaries or along edges/planes/corners.

Michael Grätzel commented: Many researchers are looking for new metal oxide materials. What is promising about your tungsten materials ?

Wilson A. Smith responded: The $CuWO_4$ photoanode we study has a slightly smaller band gap than $BiVO_4$, but after a few years of research it seems that its optoelectronic and catalytic properties cannot be improved like $BiVO_4$ despite efforts of doping, surface passivation and the use of co-catalysts. I think this brings up an important point in the search for 'winning' materials in the field of photoelectrochemistry. Even if a new metal oxide photoelectrode can convert all of the incident sunlight to electrical/chemical energy, I have serious doubts about the scalability of this approach to scale up to meet TW, GW or even MW needs. Efforts should instead focus on understanding mass transport, reactor engineering, and scaling up of this technology, which need to be solved in parallel to materials optimization.

Ravi Shankar added: Following on from Professor Grätzel's comment, what do you think the key limitations are for solar fuel production through photoelectrochemistry? Do you envision some sort of feedback loop?

Wilson A. Smith answered: I think the key limitations for solar fuel production through photoelectrochemistry (PEC) are predominantly engineering: mass transport, reactor engineering, process intensification, process integration, *etc.* The field has been focused on materials properties for four decades, and we are still not in a place close to commercialization or industrial applications. While we do still need to find photo and electrocatalysts to perform key redox reactions (with low overpotential, high stability and high selectivity), we need a massive increase in the areas that look to apply these materials in practical systems. For example, we need to already think: what if we find an 'ideal' material that meets all of the metrics we pose on PEC? Then what? We need to already be thinking about this now, so that when in the next years or decades we do find these materials, we can already put them in a place to upscale.

We cannot wait to find the 'ideal' material, then worry about making the process at the MW, GW or TW scale. In addition, from other fields where large scale electrolysis is performed, it may turn out that the process conditions needed for practical applications are far removed from where a majority of lab-scale testing is done. That means that we may be optimizing the performance and understanding of materials and processes in a regime that is not industrially relevant. Therefore, we absolutely need a positive feedback loop between engineers and materials scientists/chemists to understand what a large scale system may look like in order to give us more realistic conditions to improve materials and systems in a laboratory scale, that can have a faster pathway towards industrial applications.

Sergii I. Shylin enquired: You suggest that formation of a copper borate complex at the electrode surface after the photocharging treatment may contribute to the improved performance of $CuWO_4$. As such, would it be possible to dope the surface of $CuWO_4$ with borate by other methods than light irradiation (*e.g.*, by refluxing electrodes in sodium tetraborate solution)?

Wilson A. Smith responded: This is possible, but we have not tried it. As long as there are borate anionic species in the electrolyte and electronic defects at the surface, there could be spontaneous adsorption, but it may not be energetically favorable to occur at all, or may be very slow (and the photoinduced process we show may simply speed up the reaction.).

Andreas Wagner questioned: Is this effect similar/comparable to work published by Michael J. Rose and co-workers in ref. 1? They show covalent molecular functionalisation of silicon semiconductor surfaces to influence surface recombination and carrier density.

1 D. G. Boucher, J. R. Speller, R. Han, F. E. Osterloh and M. J. Rose, *ACS Appl. Energy Mater.*, 2019, 2, 66.

Wilson A. Smith replied: I think these works are very similar in that anionic adsorption passivates defects at the SLJ. In the case of silicon, dangling bonds or defects in a surface SiO_x layer make preferential adsorption of different molecules, and in our case the anionic borate and phosphate species adsorb to surface defects. In our case the defects are not as clearly characterized, but are most likely related to oxygen vacancies or Cu with non-ideal valence. We are going to probe these surfaces *in situ*/operando with FTIR to see if we can observe which surface species form as a function of potential/electrolyte.

Martijn A. Zwijnenburg asked: This is perhaps a naive question, but would a photoelectrode in a water splitting photoelectrochemical cell routinely be illuminated under open circuit conditions? If not, would photocharging ever occur for such a cell?

Wilson A. Smith replied: That is a great question. For a 'practical' cell, there would need to be at least 2 photoactive materials, as significant modelling work has shown that the best approach to get a high solar-to-hydrogen conversion efficiency is with a tandem absorber configuration. Therefore, if a metal oxide is the top layer, *i.e.* interfacing with the electrolyte, it would always receive a voltage boost from the bottom absorbing material, which would indeed not be true open circuit conditions. However, in the literature it has been shown that photocharging with an applied potential can make the process go even faster, so this may benefit an overall device system.

Ravi Shankar opened a discussion of the paper by Aubrey R. Paris: Aubrey, in Fig. 6 of your paper (DOI: 10.1039/c8fd00177d) you show the different Faradaic efficiencies for your different materials. Could you comment on the influence of the support that you use in your materials as it appears that this seems to have an effect?

Aubrey R. Paris answered: We initially showed that the choice of solid support can influence catalytic activity, specifically for the Ni_3Ga intermetallic, in ref. 1. Following thorough materials characterization, we suggested that the differences in catalytic activity (including product distribution) observed for this species on different carbons could be attributed to changes in surface composition and morphology. In the report prepared for this Faraday Discussion, we decided to probe the carbon support dependence further. The most interesting result of these experiments was the fact that carbon solid supports could not reinstate multi-carbon generation ability for non-intermetallic Ni–Ga films. This result supports our 2018 report, which does not attribute a direct role to the carbon in influencing catalytic behavior (rather, its effect is indirect in dictating the composition and morphology of the surface during catalyst synthesis). However, a consistent point of interest in our 2018 paper and the study presented here is the fact that thin films deposited on glassy carbon and reticulated vitreous carbon (RVC), which are structurally similar materials, generate the same products from CO_2 but in different Faradaic efficiencies. Specifically, RVC-grown films exhibit approximately double the Faradaic efficiency for each carbon-containing product compared to those on glassy carbon. This is unexpected, because the greater surface area of RVC films should not intuitively result in greater Faradaic efficiencies (if anything, a greater current density would be expected). We have yet to understand the chemical reasoning for this observation.

1 A. R. Paris, A. T. Chu, C. B. O'Brien, J. J. Frick, S. A. Francis and A. B. Bocarsly, *J. Electrochem. Soc.*, 2018, **165**, 385.

Ravi Shankar noted: Aubrey, you mentioned in your paper that you presume the CO reduction to be rate-limiting step and that this goes against your previous work. Could you please explain a bit more about this?

Aubrey R. Paris replied: Based on the results of the D_2O electrolyte experiments, it is likely that CO reduction is rate-limiting. However, I should caution that whether or not this is true depends on why the D_2O is affecting the system as it is. For example, if the D_2O is simply slowing down the reaction pathway that is active in H_2O-based electrolyte, then CO reduction should be rate-limiting, but this would not necessarily be the case if the presence of D_2O activates a different catalytic pathway for the system.

In our initial report about the multi-carbon product generation ability of intermetallic Ni_3Al (ref. 1) we used the rate of liquid product formation to hypothesize the rate-determining step of CO_2 electroreduction. In essence, we found that supplying the system with CO feedstock instead of CO_2 resulted in faster product generation, so we suggested that CO_2 reduction to CO might be rate-limiting. Based on the D_2O experiments described in this work, however, that may not actually be the case. A more thorough understanding of the nature of D_2O's effect would likely help discriminate between these two rate-limiting step possibilities.

1 A. R. Paris and A. B. Bocarsly, *ACS Catal.*, 2017, **7**, 6815.

Andreas Wagner asked: Why did you choose to use pH 4.5 sulfate (+bicarbonate) electrolyte, given that pure bicarbonate buffers are much more widely used in CO_2 electrocatalysis? Was the buffer used in the D_2O experiments also prepared in the same manner?

Aubrey R. Paris responded: While bicarbonate buffers are certainly common in this field, they more easily facilitate electrochemical experimentation in the pH 6.8–7.2 range. However, we have shown previously that pH 4.5 is optimal for the parent intermetallic films discussed here, which we presume is due to the high solubility of CO_2 in water at this condition. This actually tends to be the optimal condition for many of the heterogeneous CO_2 reduction catalysts studied in our lab, frequently resulting in our selection of K_2SO_4 as an electrolyte, because it requires a very small amount of $KHCO_3$ to achieve a buffered solution close to pH 4.5. We also often select K_2SO_4 for our electrolyte because we perform pH dependence studies with any new catalyst we develop, and the $K_2SO_4/KHCO_3$ combination allows for examination of a wide pH range (approximately 3.5–7.5). Experiments using D_2O were indeed prepared in the same manner.

Wendy J. Shaw queried: You stated that the role to make multiple carbon chains is indirect. Can you provide any more comments or speculation on this mechanism?

Aubrey R. Paris responded: I would not call the mechanism indirect, but rather the experiments we used to begin probing it were indirect indicators of what it may or may not look like. At this point, we seem to understand that generating multi-carbon chains on these nickel–Group 13 metals is predicated on intermetallic character, suggesting that the role of the two metals is synergistic, or electronic, in nature. This does not rule out the possibility that the two metals on the surface could also provide distinct binding sites to further stabilize CO_2 or the reduction intermediates, but our experiments here do indicate that this "co-stabilization" effect is not the sole reason for multi-carbon product generation. The next steps in further understanding the reduction mechanism should definitely focus on interpreting the H_2O/D_2O effect. Because it is possible to imagine many reasons for this unexpected effect, experiments ruling out certain reasons would be extremely informative. These experiments could include *in situ* studies to examine differences in surface-bound reduction intermediates in the presence of H_2O *versus* D_2O. Importantly, because we have now observed this effect in four different CO_2 reduction systems (*i.e.*, the two discussed here, plus our recent report on a Cr–Ga oxide alloy and another unpublished catalyst from our lab), it seems to be at least a somewhat conserved effect. Understanding it could tell us a lot about CO_2 electroreduction systems as a whole.

Wilson A. Smith commented: The idea of bimetallic or alloy catalysts is interesting from the perspective of changing selectivity according to the composition of active sites for different reactions. However, when you apply the same potential to different metals, not only do they have different selectivities, but they also may have different current densities. What are your thoughts about the activity/current density distribution for bimetallic or alloy catalysts, and how differences in activity on different metallic sites can affect overall selectivity and mechanistic understanding?

Aubrey R. Paris replied: You raise a really important point regarding the design of these electrocatalytic systems. Eliciting two specific reduction events on two distinct surfaces (in the same electrochemical cell) requires compatible electrochemical conditions for the two surfaces, such as an optimized operating potential, pH, and electrolyte salt. This is why I note in the conclusion section of the paper that a single alloy or intermetallic species should theoretically be easier to optimize than coupling two distinct electrodes performing complementary tasks. Nonetheless, this complication does not negate the mechanistic assertions made in the paper, because the product distributions achieved using the "striped" or non-alloyed electrodes seem to be combinations of the product distributions found using single-metal controls.

Sarah Lamaison asked: Some recent articles in the field[1–4] propose to use the binding energies of the reaction intermediates on the individual metals as descriptors to rationalise the selectivity of these alloyed metals. Have you tried to look into that? And if so, are the experimental data consistent with such rationalisation?

1 T. Hatsukade, K. P. Kuhl, E. R. Cave, D. N. Abram, J. T. Feaster, A. L. Jongerius, C. Hahn and T. F. Jaramillo, *Energy Technol.*, 2017, 5, 955.
2 S. Lamaison, D. Wakerley, D. Montero, G. Rousse, D. Taverna, D. Giaume, D. Mercier, J. Blanchard, H. N. Tran, M. Fontecave and V. Mougel, *ChemSusChem*, 2019, 12, 511.
3 A. Bagger, W. Ju, A. S. Varela, P. Strasser and J. Rossmeisl, *ChemPhysChem*, 2017, 18, 3266.
4 Y. C. Li, Z. Wang, T. Tuan, D. –H. Nam, M. Luo, J. Wicks, B. Chen, J. Li, F. Li, F. Pelayo García de Arquer, Y. Wang, C. –T. Dinh, O. Voznyy, D. Sinton and E. H. Sargent, *J. Am. Chem. Soc.*, 2019, 141, 8584.

Aubrey R. Paris responded: It seems likely that intermediate binding energies are at least partially responsible for the results that we see using these intermetallic *versus* distinct metal species. In fact, we are currently finishing up a project which we believe provides experimental evidence for multi-metal catalyst surfaces altering binding energies (*i.e.* compared to the individual constituent metals) for frequently invoked intermediates. So stay tuned for that!

Andrew B. Bocarsly remarked: The two approaches presented in these two papers provide excellent examples of a long debated solar fuels conversion dichotomy. Namely, does the best engineering of a solar fuels system utilize a "single pot" photoelectrochemical cell or is a better system obtained by interfacing a photovoltaic panel to an electrochemical cell containing an optimized electrocatalyst? Please discuss the pros and cons of these two approaches for CO_2 reduction and indicate which approach is preferred.

Aubrey R. Paris replied: Choosing whether an electrochemical or photochemical approach is "preferred" depends largely on one's goal for the research. If the goal is to create a system that could be industrially implemented immediately, electrochemical CO_2 reduction has an upperhand, because much more progress has been made in electrochemical catalytic systems in recent decades. Indeed, many desirable products are already achievable at impressive Faradaic efficiencies using electrocatalysts. This certainly doesn't mean that photoelectrochemical CO_2 reduction research should be abandoned, but it simply doesn't boast the same amount of progress as the electrochemical analog. That said, with more fundamental research, photoelectrochemistry has the potential to help address one of the lasting challenges of electrochemical CO_2 reduction: impractical overpotential requirements.

Wilson A. Smith replied: In the short term (which is the time frame where solutions are needed), the best approach has to be renewable electrolysis (where an electrolyser can be powered by any renewable electricity supply – not just PV) over direct photoelectrochemistry. The technological development of PV panels is too robust, and large electrolysers already exist, but are expensive. However, large CO_2 electrolysers do not exist, so significant efforts should be made towards developing these by translating fundamental lab scale research to an industrial scale. On the other hand, the PEC approach still not only suffers from low efficiencies and material stability issues, but even in the best case scenarios no real upscaling has been done at a practical level. While this avenue should still be interesting to pursue, it may only be viable in a long term timeframe (at best). Therefore the fastest way to engineer a solar fuel system is to use two known and scaled technologies instead of one unproven technology that has never been made at a large scale.

Sergii I. Shylin opened a discussion of the paper by Chia-Yu Lin: Do you get chlorine as a by-product at the anode when electrolyzing seawater?

Chia-Yu Lin replied: We did not notice chlorine generation in this work.

Moritz F. Kuehnel enquired: What is the mechanism by which the electrode corrodes? Is it formation of hypochlorite?

Chia-Yu Lin responded: In the case of the unmodified $CaFeO_x$ electrode, both formation of hypochlorite and low bulk solution pH (<6) after long-term electrolysis are the main reasons for the destabilization of $CaFeO_x$. On the other hand, low bulk solution after long-term electrolysis destabilized $FePO_4$-modified $CaFeO_x$.

Andrew B. Bocarsly asked: You noted that the ocean is buffered around pH 8. This is due to the presence of a borate buffer system. In your synthetic sea water, it appears that borate is not present. Do you expect this to affect your experimental results, given that borate layers have been shown to impact other electrode systems?

Chia-Yu Lin replied: We used phosphate (0.5 M) buffered seawater solution for our experiments. The phosphate concentration is much higher than borate, so its effect would be small. However other ions, such as Mg^{2+}, indeed have a negative effect.

Andrew B. Bocarsly continued: A current density of 10 mA cm^{-2}, is often selected as the operational value for a photoelectrochemical cell. In this specific case, why have you selected this value? Did you rule out the use of focused sunlight?

Chia-Yu Lin replied: We would like to apply this catalyst for photo-electrochemical water splitting in the near future, so we selected a current density of 10 mA cm^{-2} as our applied current density in this work.

Erwin Reisner enquired: You have reported water oxidation to O_2 in the presence of salt water. What is your opinion about the potential economic advantages to oxidising chloride to Cl_2 or other chlorinated products instead? This would allow the solar-driven production of two valuable products – H_2 from aqueous protons and a potentially value-added product from oxidation.

Chia-Yu Lin responded: I think it will be great of interest. Nevertheless, some issues should be addressed. Making chlorine from seawater/salt water should be under acidic conditions, and in this context, a noble metal for chloride oxidization should be used.

Michael Grätzel asked: Are you trying to make a water oxidation catalyst that works best in neutral solution? If so, one compartment of the electrolyzer would be neutral and the other one basic, This is a challenge for the bipolar membrane that would have to be used to separate the two compartments of the electrolyser. As a result there would be additional overvoltage losses. Have you tested your system at pH 0? What pH are you testing? Natural seawater? Not acidified?

Chia-Yu Lin answered: Yes, we tried to develop electrocatalysts for both OER and HER at neutral pH. We did not use a electrolyzer with two compartments with different pHs. In addition, all of the experiments in this work were performed in phosphate (pH 7) buffered electrolytes.

Andrew B. Bocarsly enquired: Given that iron phosphates are reasonably soluble in water, how do you plan to stabilize this surface in seawater? Will your approach require the addition of substantial quantities of phosphate buffer to the electrolyte?

Chia-Yu Lin responded: In this case, sufficient phosphate should be added to the seawater to stabilize the iron phosphate, and of course, this is the limitation of this electrode.

Christine A. Caputo opened a discussion of the paper by Ulf-Peter Apfel: I am curious about your choice of activity comparison – the NiP system with carbon nitride and carbon quantum dots you used for benchmarking the activity of your system. Being an author on both those papers (ref. 1 and 2), I would say that they are not the highest activity systems and so make a weak argument for comparison. Do you think your low activity is just due to the low reduction potential in your system? Is there a way to tune the bandgap of your material?

1 G. A. M. Hutton, B. Reuillard, B. C. M. Martindale, C. A. Caputo, C. W. J. Lockwood, J. N. Butt and E. Reisner, *J. Am. Chem. Soc.*, 2016, **138**, 16722.
2 C. A. Caputo, M. A. Gross, V. M. Lau, C. Cavazza, B. V. Lotsch and E. Reisner, *Angew. Chem. Int. Ed.*, 2014, **53**, 11538.

Ulf-Peter Apfel responded: I personally believe that the low activity is due to the low reduction potential in our system. A possible pathway to further tune the bandgap would be, *e.g.* variation of the metals as well as of the chalcogenides. We are currently investigating such possibilities and hope to find out soon if this is possible.

Christine A. Caputo asked: I have a question about the stability of your systems. Carbon nitride and carbon quantum dots are very stable photosensitisers, in fact we found with those materials that this is quite a challenging experiment to design. It is difficult to find a molecular catalyst that lasts as long as some of the latest photosensitizer materials and will allow for long-term stability studies in hybrid systems. Have you tested your system for long-term stability?

Ulf-Peter Apfel replied: This is a good and very important point. We solely tested the long-term stability under electrochemical conditions. Here, we do not have a problem with long-term stability and the materials operate for more than 3 weeks without decomposition or loss in activity. For the photochemical experiments with the molecular catalysts, we never tested the stability since the shown activity is just too low.

Daniel Antón-García commented: Looking at the Mott–Schottky analysis, there is not enough driving force for H_2 production. You also use NiP, which requires an additional 200 mV of overpotential to produce H_2. Where is this driving force coming from? And why is it that NiP outperforms Pt, which is known to: a) be a more active catalyst and b) require less driving force?

Ulf-Peter Apfel answered: As we discuss in our manuscript, the conduction band minima of all our samples are very close to the hydrogen potential, thus a low hydrogen evolution activity is expected. Interestingly, we found that with NiP decoration hydrogen evolution was observed, however not with Pt decoration. We do not know yet why those two modification behave so differently on our materials, which is one focus of our ongoing work!

David Wakerley said: You present an interesting way to mimic enzyme active sites on a heterogeneous surface. The enzyme active site is quite flexible – will the loss of flexibility on the heterogeneous catalyst present any issues?

Ulf-Peter Apfel responded: The heterogeneous surfaces are still quite flexible. The presented Fe/Ni sulfides for example show an opening of coordination sites at the Fe/Ni centers, concomitant with sulfide loss, at reductive conditions. The catalytic species here is only provided at catalytic conditions – in my mind this is very good compared with an enzyme active site. However, you are absolutely right, the degree of flexibility is reduced compared to that of an enzyme. I personally believe that this is not an issue.

Sylvestre Bonnet asked: Do you have any idea about the type of contact between the solid photosensitising material and the catalyst in solution? Is there any form of adsorption, or is photocatalysis running *via* a form of dynamic quenching?

Ulf-Peter Apfel replied: No not yet. This is something we are currently investigating.

Vivek Badiani commented: What is your opinion on a top-down *vs.* bottom-up approach to material synthesis? What is the flexibility available for the tuning of the materials akin to ligand tuning?

Ulf-Peter Apfel replied: I do not think that there is a big difference between top-down *vs.* bottom-up approaches for material synthesis. Both have specific advantages and disadvantages and none of us currently know the right pathway to gain highly active and robust catalytic materials. I believe that scientists should follow their own ideas without thinking about such things. Concerning the flexibility of the ligand tuning, one always has to ask himself for what purpose this is done. If you aim for an application but you establish a twenty-step synthesis, I hardly can believe that this will go into an industrial application. If you do such things for gaining knowledge on how things can be done, tuning ligands is a wonderful method to understand the underlying principles. There is always a fine line between basic and application driven research. This has to be balanced and communicated openly without any hesitation.

Vivek Badiani commented: You present an interesting take on bioinspiration, are there any other natural active centres you may look to to mimic as catalysts?

Ulf-Peter Apfel replied: Thank you for you comment. We are currently mainly looking for reductive systems such as hydrogenases, CO dehydrogenase and nitrogenases. But I am sure that there will be plenty of interesting systems in the future that we and others have not thought about yet. As an additional example for such a strategy, I want to highlight the work of Philipp Kurz who is using a comparable strategy to ours to mimic the OEC cluster. I am certain that many more catalytic processes can be found where a bioinspired approach might lead to a rationally designed and applicable catalytic system.

Andrew B. Bocarsly enquired: Iron sulfides are characterized as having highly defective crystal structures. This is typically due to the mobility of the metal ions in these structures. Does this play a role in your material's conductivity or doping levels?

Ulf-Peter Apfel responded: This is likewise the case here, but the crystallographic structure of pentlandite is not comparable, *e.g.* with those of FeS2 or NiS. Operando spectroscopy, as well as scanning electrochemical cell microscopy, data indicate that it is the mobility of sulfur/ creating sulfur "holes" instead of the metal mobility that plays a significant role in such materials for their activity. The conductivity, however, seems to be unaffected by such defects.

Andrew B. Bocarsly continued: Given that you have obtained Mott–Schottky plots for your electrodes, you have the data needed to calculate the carrier concentration. Is that value available?

Ulf-Peter Apfel responded: The value is currently not available. But this is a very good point and we will have to determine the carrier concentration now to further receive information on our system.

Michael Grätzel asked: Could your please provide a mechanism that would explain the lower Tafel slopes for H_2 evolution on your Ni/Fe sulfide catalyst compared to that for Pt (30 mV/decade for the Vollmer–Tafel mechanism)?

Ulf-Peter Apfel answered: The Tafel slopes for the Ni/Fe sulfide are not smaller than that for Pt. In addition, the mechanism for the H_2 formation of such sulfides is very different from that of Pt as we could show by operando NRIX measurements. It rather looks like a mechanism that resembles that of [FeNi] hydrogenases rather than that of Pt.

Joost N. H. Reek added: I would like to ask a more philosophical question. In the introduction you use the active site of enzymes such as hydrogenase as inspiration for the preparation of your heterogeneous catalyst materials. The question is whether this is the correct source of inspiration, as the active site only is not a good catalyst. There are thousands of hydrogenase mimics but all require a relatively large overpotential for proton reduction catalysis. Work by Lubitz and Fontecave (ref. 1) demonstrated that an artificial hydrogenase mimic can be put into an apohydrogenase enzyme, restoring full activity. This shows that the activity of the enzyme is for a large part determined by the peptide environment around the active site rather then the active site. If we use just the active site as an inspiration, we may have the wrong starting point, as we do not take into account the environment.

1 G. Berggren, A. Adamska, C. Lambertz, T. R. Simmons, J. Esselborn, M. Atta, S. Gambarelli, J. -M. Mouesca, E. Reijerse, W. Lubitz, T. Happe, V. Artero and M. Fontecave, *Nature*, 2013, **499**, 66.

Ulf-Peter Apfel replied: This is a very interesting question and you are absolutely right that mimics usually operate under different conditions. The example you mentioned is a perfect one. However, if you carefully check the [FeFe]-hydrogenase mimics in solution and when incorporated into HydA, it is obvious that they reveal different structures. The structural change is, as highlighted by you, adjusted by the peptide environment. As such, a different reactivity is expected.

I do, however, believe that bioinspired mimics are still very valuable and that nature gives a lead to come up with new materials. Let me explain this further. Our material contains a cuboidal metal-sulfur structure which can be assumed as a mimic for the [4Fe–4S] clusters that enable electron transport in nature. Likewise to nature, our materials show high conductivity when used in electrochemical experiments. Furthermore, pentlandites reveal short metal–metal interactions that also cover the surface of the material. These structural fragments are comparable to the ones observed in nature and we could recently show that these are important for the material's reactivity. While I agree that this attempt might lead to a wrong starting point, it is one potential inspiration to build new materials and should not be excluded. In the end, we cannot forecast the reactivity of materials and require some kind of inspiration.

Leif Hammarström asked: What about the surface chemistry? Do your band energies shift with pH? Could you get more proton reduction at lower pH values?

Ulf-Peter Apfel answered: We have not tested the behavior under various pH conditions yet. This is certainly a good idea that we will investigate in the near future.

Víctor A. de la Peña O'Shea asked: In Fig. 4 of your paper (DOI: 10.1039/c8fd00173a), you compare the H_2 evolution of $Ni_3Fe_6S_8$ and $NiFe_2O_4$ using different co-catalysts (NiP and Pt). In the case of Pt you used 8 wt%. Do you know the particle size or dispersion of your Pt particles over the semiconductors? Did you consider that this high amount of Pt can lead to a shadow effect that is affecting the interaction of light with the semiconductor? Did you use other amounts of Pt?

Ulf-Peter Apfel replied: No, we have not yet investigated the particle size or dispersion of the Pt particles. The shadow effect is a very interesting hint. We did not think about this and will test this hypothesis by applying different amounts of Pt in the future.

Víctor A. de la Peña O'Shea continued: Did you observe any corrosion of your material? Did you observe changes in iron or nickel oxidation or on their proportion before and after the reactions?

Ulf-Peter Apfel responded: Yes, we did observe corrosion. By operando NRIX measurements we observed that at specific sulfur positions there is sulfur depletion. Subsequently, between two metals there is a void space that we assume is responsible for the high reactivity and can be occupied by protons just as is observed in [FeNi] hydrogenases. Until now, we have no data on the exact oxidation state during the catalysis of Fe or Ni. The proportion of both metals, however, did not change as we could show by SEM-EDX and XPS data.

Andrew B. Bocarsly commented: You indicated that your electrode has some degree of activity as a photocathode for CO_2 reduction, but little activity for H_2 formation. Is there a strategy in this materials class for developing systems that are electrocatalytic for CO_2 reduction?

You saw some CO_2 reduction, usually when looking for a good CO_2 photo-electrode one looks for a lousy hydrogen photoelectrode. Do you have any reasoning for this?

Ulf-Peter Apfel replied: Indeed, we do have a strategy to further alter such materials and to favor the electrocatalytic CO_2 reduction by these materials, *e.g.* by changing the metals and the chalcogenides. However, besides the catalyst itself, the reaction conditions are also of utmost importance and I do believe that you need the proper reactor setup (in addition to the catalyst) to facilitate and to improve a specific reaction. I believe that looking for a lousy hydrogen photo-electrode when you would like to have a good CO_2 photoelectrode is not neces-sarily required. Again, I personally do believe that it is likewise the reaction conditions and the material processing that can be altered to introduce different reactivities. Why not test a good hydrogen electrode under completely different reaction conditions in a different reactor?

Marta C. Hatzell said: With regard to environmental systems, minerals, sands, *etc.*, have these reactions been demonstrated in fields such as photogeochemistry?

Ulf-Peter Apfel responded: Of course. See, for example, deep sea ocean vents. Materials, *e.g.* stable alloys, can be found here that catalyze various reduction processes. Under normal conditions such alloys would normally be surface oxidized and not very reactive. As such, I believe that the environmental condi-tions indeed aid changing redox properties and also stabilize the reactive inter-mediates (or at least do not trigger their decomposition).

Michael Grätzel commented: A few years ago there was lots of excitement in the scientific community about the virtues of MoS_2 as a hydrogen evolution catalyst. How does your Fe/Ni sulfide compare with MoS_2?

Ulf-Peter Apfel replied: While comparing two distinct different sulfides is certainly desired, such a comparison is very difficult, probably misleading and simplifies the problem set too much. The structure of Fe/Ni pentlandite is not at all comparable with MoS_2 and results in different properties of this material. In contrast to MoS_2, the bulk pentlandite is highly conductive due to its unique structure. In addition, pentlandites are very stable under acidic and basic conditions as long as they are kept under reductive conditions. Likewise, the mechanisms of both materials in generating hydrogen seem to be very different. I believe, however, that the major difference (and probably an advantage for application) between pentlandites and MoS_2 is that with the Fe/Ni sulfides we do not require a specific nanostructure to achieve good activity.

Katarzyna Sokol asked: Have you tried or considered other synthesis methods for the particles in order to make them smaller and better control their size? With such a degree of aggregation, and particle sizes in the range of hundreds of nanometers, you will be losing the quantum size effects, and the band gap and conduction band edge should be close to the bulk material. Is there an advantage of using the particles in this case, with so many alternative electrode film deposition methods available? Have you considered making the particles smaller, in the range of 10 nanometers or less, to increase the band gap and possibly shift the conduction band edge to be more negative?

Ulf-Peter Apfel answered: We attempted the synthesis by various methods. The major problem is, however, that the pentlandite has a high temperature phase and low temperatures usually facilitate the formation of other phases. So the formation of defined nanoparticles is a borderline problem. Unfortunately, there are currently no electrode film deposition methods available for this particular material. We are currently working on this issue to make thin films. But at the moment, we have not achieved this goal.

Anna Hankin said: Could you comment on the accuracy of using the Mott–Schottky approach for determining the band edge positions in your materials, especially since the charge carrier densities in the materials are not known? Have you tried to compare Mott–Schottky results with other measurements, such as chopped photocurrent?

Ulf-Peter Apfel answered: My personal impression is that the Mott–Schottky approach is very cumbersome and error-prone. Small variations during the impedance measurements can lead to significant alterations. As such, I do expect a non-negligible error of the presented values. We have not yet measure chopped photocurrents but will certainly perform this measurement.

Andrew B. Bocarsly remarked: From an experimental point of view, the flat band potential is best determined using light intensity dependent open circuit photopotential measurements. This technique avoids the unknown impedances, and frequency dependence of the Mott–Schottky analysis.

Ulf-Peter Apfel responded: You are absolutely right and we plan to do this in the near future.

Flavia Cassiola commented: The material you reported has intriguing properties that should be investigated further as proposed in your paper. During the discussion you commented that the stability of $Fe_3Ni_6S_8$ is not high on your priority list at the moment because the material is considered very cheap and its replacement is consequently not an issue. Thinking ahead, as you develop the potential of the $Fe_3Ni_6S_8$ photocatalytic activity (with or without co-catalysts), one suggestion would be to consider stability and durability of the material in the earliest stages of development. Although the material is cheap and abundant, in a real device in an industrial scale the material should not be replaced in every reaction cycle. One suggestion is to consider stability studies as you proceed with future developments (band gap tuning of the pentlandites, nanoengineering of the particle sizes and further optimization of their elemental compositions as improvements on photocatalytic properties). How would you speculate on the stability of the material with your results and at your current development stage? Does this insight on stability change your current ways of thinking regarding the application of $Fe_3Ni_6S_8$ in a photosynthesis device?

Ulf-Peter Apfel replied: You are absolutely right pointing out that stability is one of the most important parameters for an industrial application. We solely investigated the stability under electrochemical conditions. Herein, stability is not an issue. After talking to some of my industrial partners I am currently not convinced, however, that photocatalytic applications for hydrogen generation will enter industry soon. As such, I did not focus too much on the stability of this material but was rather curious as to whether we can trigger a photocatalytic process with it. It is definitely on my to-do list to also investigate the stability in the near future with a better material and I am certain that, likewise to electrocatalysis, stability is not an issue. But still, even a high stability of our material under photocatalytic conditions would (at the moment) not change my mind regarding its application in a photosynthetic device.

Peter Brueggeller said: At this conference the question of what the best way to find potential photocatalysts is has arisen. One rationale is that the electrocatalytic activity is often only a calculated number, *e.g.* a theoretical proton reduction with a turnover frequency of more than 100 000 per second has been calculated for some DuBois' catalysts in ref. 1. However, the corresponding photocatalytic activity in the form of experimental hydrogen production analysis only leads to a TON of 11.4. Since the compounds in U.-P. Apfel's paper (DOI: 10.1039/c8fd00173a) show an interesting electrocatalytic behaviour with a possible photocatalytic application, I wonder whether the authors of this session could comment on the differences between theoretical electrocatalytic activities and resulting experimental photocatalytic hydrogen determinations.

1 M. L. Helm, M. P. Stewart, R. M. Bullock, M. R. DuBois and D.L. DuBois, *Science*, 2011, **333**, 863.

Ulf-Peter Apfel responded: I am not certain if one should really compare photocatalytic and electrocatalytic conditions. The reaction conditions applied are very different. While an electrocatalyst should be potentially capable of performing this reaction also under photochemical conditions, this is not always the case and usually the photosensitizer and sacrificial electron donor are an important component that do not always receive the right attention when discussing different TON/TOF values. In addition, likewise the product composition can be altered as was shown by the Robert group with their photo- *vs.* electrochemical CO_2 reduction experiments (ref. 1–3).

1 H. Rao, L. Schmidt, J. Bonin and M. Robert, *Nature*, 2017, **548**, 74.
2 H. Rao, C. -H. Lim, J. Bonin, M. Miyake and M. Robert, *J. Am. Chem. Soc.*, 2018, **140**, 17830.
3 H. Takeda, H. Kamiyama, K. Okamoto, M. Irimajiri, T. Mizutani, K. Koike, A. Sekine and O. Ishitani, *J. Am. Chem. Soc.*, 2018, **140**, 17241.

Andrew B. Bocarsly posed a question to Ulf-Peter Apfel and Chia-Yu Lin: Both of you use earth abundant elements in your semiconducting electrodes, but show that these materials introduce all kinds of new complications with respect to electrode stability and response. Do you believe that we are better off using more expensive materials which exhibit fewer electrochemical complications or should our studies continue with less expensive materials and engineer them to be "good" materials?

Ulf-Peter Apfel replied: I personally believe that it is a necessity to proceed with earth-abundant metals/materials. While we certainly do have issues with the stability now, I am convinced that we can overcome these problems. We need materials that are abundant to reduce the current investment costs and to push electro- and photo-chemical devices out of a niche existence into large-scale technology.

Chia-Yu Lin answered: Noble metals do indeed have better performance in most cases, but in some cases, noble metals cannot solve our problems, such as mitigation of chloride oxidation. In addition, some reactions require less expensive materials to activate. Therefore, both kinds of material require further investigation.

Conflicts of interest

There are no conflicts to declare.

Faraday Discussions

PAPER

Sequential catalysis enables enhanced C–C coupling towards multi-carbon alkenes and alcohols in carbon dioxide reduction: a study on bifunctional Cu/Au electrocatalysts

Jing Gao,[a] Dan Ren, [iD]*[a] Xueyi Guo,[b]
Shaik Mohammed Zakeeruddin [iD][a] and Michael Grätzel*[a]

Received 1st December 2018, Accepted 14th January 2019

DOI: 10.1039/c8fd00219c

Electrochemical reduction of carbon dioxide (CO_2) to multi-carbon products such as ethylene, ethanol and n-propanol offers a promising path for utilization of excessive CO_2 and energy storage. Oxide-derived Cu electrodes are among the best electrocatalysts for the selective formation of ethylene and ethanol. However, a large fraction of the faradaic current still goes to hydrogen evolution, even at optimal conditions (electrolyte, potential, etc.). Here we employ the concept of sequential catalysis using judiciously designed CuAu bimetallic catalysts through galvanic exchange between Au^{3+} and Cu_2O nanowires. By controlling the concentration of the Au^{3+} precursor and the exchange time, Au nanoparticles were evenly dispersed onto the Cu_2O nanowires. The optimized oxide-derived CuAu catalyst showed remarkable improvement towards the formation of ethylene, ethanol and n-propanol, in terms of faradaic efficiency and current density. Our analysis of the electrochemical formation of carbon monoxide, ethylene and hydrogen suggests that the presence of Au, an electrocatalyst for CO_2-to-CO conversion, helps enhance *CO-coverage on Cu, thus promoting the production of multi-carbon products and suppressing hydrogen formation on the CuAu catalyst. We propose promising strategies for designing electrochemical systems, which would enable the selective and scalable reduction of CO_2 to ethylene and ethanol.

Introduction

Natural photosynthesis, a process where carbon dioxide (CO_2) and water (H_2O) are converted to carbohydrates in plants using sunlight, provides the ultimate

[a]Laboratory of Photonics and Interfaces, École Polytechnique Fédérale de Lausanne, CH-1015 Lausanne, Switzerland. E-mail: dan.ren@epfl.ch; michael.graetzel@epfl.ch
[b]School of Metallurgy and Environment, Central South University, Changsha, Hunan, 410083, China

energy source for human beings. Mimicking nature, artificial photosynthesis offers a promising technology to store solar energy in chemical bonds and to reduce excessive CO_2 in the atmosphere.[1-3] One efficient approach for this purpose is to integrate photovoltaics with an electrolyzer, where CO_2 is electrochemically reduced to chemical feedstock such as carbon monoxide using solar energy to drive the thermodynamically uphill process.[4] For example, using efficient and selective CuSn based electrocatalysts, our group has recently achieved an energetic solar-to-CO efficiency of 13.4% employing triple-junction III–IV photovoltaics as the electric power source.[5]

One of the most attractive aspects of electrochemical CO_2 reduction is the possibility of directly forming C_1–C_3 (even up to C_5) products.[6-8] Among all metal electrodes, copper (Cu) is a unique electrocatalyst that is able to reduce CO_2 with substantial yields to hydrocarbons and alcohols.[9,10] This is believed to be due to its optimal binding of CO, a key intermediate for the formation of reduction products with $>2e^-$ transfer.[11] Among all hydrocarbons and oxygenates, multi-carbon products such as ethylene (C_2H_4), ethanol (C_2H_5OH) and n-propanol (C_3H_7OH) have much higher commercial value and/or energy density than their C_1 counterparts such as CH_4 and CH_3OH.[12] However, the selectivity of polycrystalline Cu towards the formation of multi-carbon products (C_{2+} products $vs.$ C_1 products) is poor.[6] For example, 30.1% of ethylene and 29.4% of methane, corresponding to a mixture of C_2H_4 and CH_4 were formed on polycrystalline Cu electrode in 0.1 M $KHCO_3$ at an applied current density of -5 mA cm^{-2}.[10]

Extensive research efforts have been undertaken in tuning the selectivity of Cu towards the formation of multi-carbon products. Earlier works by the groups of Hori and Koper established that Cu (100) single crystals with stepped sites showed selectivity preference towards the formation of ethylene.[13-16] Engineering these features into Cu cubic nanoparticles proved effective in improving the faradaic efficiency of ethylene and suppressing the formation of methane.[17-20] Recently, oxide-derived Cu catalysts have been studied intensively for promoting the formation of ethylene, ethanol and even of n-propanol.[8,21-26] Defects such as grain boundaries within oxide-derived catalysts were proposed to be effective in binding reaction intermediates such as CO, thus enabling C–C coupling between two C_1 intermediates.[27,28] Other factors such as a high local pH,[29] Cu^+ species,[23] crystallite size,[30] mass transport[25] and applied potential[31] were investigated to better understand the mechanism of operation for oxide-derived Cu catalysts.

Density functional theory (DFT) has also proved to be helpful in elucidating the reaction pathways towards the formation of multi-carbon products.[32] Using DFT computations, Koper and co-workers proposed that ethylene formation occurs via the dimerization of adsorbed CO, which forms a $*C_2O_2{}^-$ dimer intermediate upon one-electron reduction.[33] This is consistent with the experimental observation that the formation of ethylene is pH-independent on an SHE (standard hydrogen electrode) scale on the Cu (100) surface.[15] Very recently, they managed to probe the protonated dimer $*OCCHO$ using operando IR spectroscopy, delivering the first spectroscopic evidence of a C_2 intermediate during CO_2 reduction.[34] Moreover, Yeo and co-workers studied the formation of ethylene on different Cu single crystals and confirmed that it starts only after sufficient coverage of $*CO$ is achieved.[35] These findings emphasize the importance of achieving a high $*CO$ coverage to boost the formation of multi-carbon products.

The above-mentioned work inspired us to conceive new strategies to improve the selectivity of oxide-derived Cu electrodes towards ethylene formation. By scrutinizing the performance of oxide-derived Cu catalysts reported by different groups, we found that a relatively large fraction of the faradaic current (at least 20%) is still consumed by forming hydrogen *via* the competitive hydrogen evolution reaction. This indicates that the ratio of *CO-coverage *vs.* *H-coverage is not optimal, *i.e.* the *H-coverage is larger than that needed to accomplish the formation of C_2 products on oxide-derived Cu catalysts. We reason that introducing a co-catalyst promoting the reduction of CO_2 to CO would increase the *CO-coverage on the oxide-derived Cu catalysts, thus improving the selectivity and efficiency of forming multi-carbon products.

Based on the above rationale, we first prepared a Cu_2O electrodes covered with Au-nanoparticles *via* a galvanic exchange reactions. A mesoscopic nanowire structure of Cu_2O was chosen to enhance the number of catalytic Cu sites. The oxide-derived bifunctional CuAu catalyst showed an improved selectivity for generating ethylene, ethanol and *n*-propanol, as compared with oxide-derived Cu catalyst. The current density for multi-carbon product generation also increased significantly. The importance of *CO-coverage on CuAu catalysts is elucidated by comparing the CO formation on different CuAu catalysts. We will discuss the important factors affecting the selectivity of C_{2+} products and provide our understanding of how to design bimetallic catalysts for enhanced selectivity of multi-carbon products.

Experimental

General

Deionized water (18.2 MΩ cm, Purelab Ultra, ELGA) was used for washing and the preparation of solutions. All of the electrochemical experiments were carried out at room temperature in an air-conditioned room.

Preparation of Cu_2O nanowires and Au-nanoparticle-covered Cu_2O nanowires

FTO glass was etched using zinc powder (99.9%, Sigma Aldrich) and 10% aqueous hydrochloric acid solution (37%, ACS reagent, Merck) to remove the conductive coating. The resulting substrate was cut into 3 cm × 8 cm, followed by sonication in deionized water for 15 min. These substrates were dried using a compressed air stream. Cu film (99.995%) with a thickness of 1.5 μm was sputtered on the cleaned substrate using a DP 650 sputter coater (Alliance-concept). Then the sputtered film was anodized in 3 M KOH at 8 mA cm^{-2} for 2 min using a potentiostat (Interface 1000, Gamry) in a two-electrode configuration with Cu foil

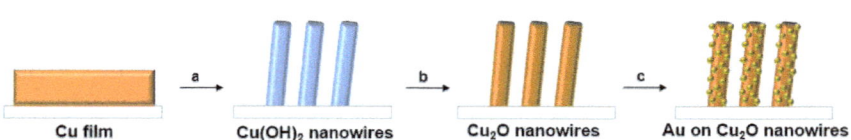

Step a: anodizing in aqueous 3 M KOH at 8 mA cm^{-2} for 2 min;
Step b: annealing at 600 °C for 4 hours in Ar atmosphere;
Step c: immersing in aqueous $HAuCl_4$ solution.

Fig. 1   Schematic preparation of oxide-derived Cu and CuAu catalysts.

(99.99%, Goodfellow) used as the counter electrode (Fig. 1). Afterwards, the $Cu(OH)_2$ film was annealed at 600 °C for 4 h under an Ar atmosphere (99.9999%, Carbagas) in a tube furnace (Lenton).

Aqueous $HAuCl_4$ (99.99%, ABCR) solutions with concentrations of 0.2 mM and 10 mM were prepared. The prepared Cu_2O substrate was cut into units of 3 cm $\times$ 0.5 cm, followed by immersion in the aqueous $HAuCl_4$ solution for galvanic replacement reaction. By varying the reaction time, we prepared Au-nanoparticle-covered Cu_2O samples with different Au loading amount. The resulting film was rinsed with deionized water and dried using compressed air flow.

Materials characterization

The morphology of the catalysts was characterized using a scanning electron microscope (SEM, Zeiss Merlin). The electrochemical surface areas (ECSA) of the Cu_2O derived Cu and CuAu catalysts were estimated from double layer capacitance measurements *via* cyclic voltammetry (Interface 1000, Gamry) at different scan rates of 20 to 150 mV s^{-1} in Ar-saturated 0.1 M $KHCO_3$ (99.99% metal base, Sigma Aldrich).

Electrochemical experiments

The electrochemical CO_2 reduction was performed in an H-type gas-tight electrolytic cell, with the cathodic and anodic compartments separated an anion-exchange membrane (Fumasep FKS-50, Fumatech). A CO_2-saturated aqueous 0.1 M $KHCO_3$ solution (99.99% metal base, Sigma Aldrich) was used as the electrolyte. The volumes of the catholyte and anolyte were 10 and 6 cm^3, respectively. A Ag/AgCl (saturated KCl, Pine) and a Ti foil with electrodeposited IrO_2 film were used as the reference electrode and counter electrode, respectively.[36] The potential of the Ag/AgCl reference electrode was checked regularly against a reversible hydrogen electrode (RHE, Gaskatel). Unless stated otherwise, all of the potentials cited in this work are referenced against an RHE scale.

60 min potentiostatic measurements (Interface 1000, Gamry) at selected potentials were used to evaluate the electrocatalytic performance of the Cu and bifunctional CuAu catalysts. During the measurements, the catholyte was stirred at 1000 rpm and CO_2 was flowed across both compartments (99.999%, Carbagas) at a rate of 10 mL min^{-1} (Bronkhorst High-Tech, F201CV, Bronkhorst). The cathodic gas stream was analyzed using a gas chromatograph (GC, Trace ULTRA, Thermo Scientific) equipped with a pulsed discharge detector (PDD, Vici) and Shincarbon micropacked column (Restek). Helium (99.9999%, Carbagas) was used as the carrier gas and the gas products were injected into the GC every 14 min. The average faradaic efficiency of four individual injections was used as the yield for gaseous products. The liquid phase products were quantified using water suppression ^{1}H Nuclear Magnetic Resonance (NMR, Avance 500 Spectrometer, Bruker). The area ratio of the products to DMSO and Phenol internal standards was compared to the calibration curve to quantify the concentration of different products.[6,37] To assure the reproducibility of each experiment, three individual and fresh samples were tested independently for each potential.

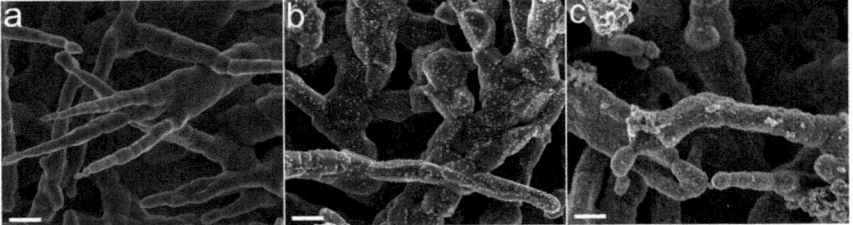

Fig. 2 Morphological characterization of the Cu_2O nanowires and Au-nanoparticle-covered Cu_2O nanowires. SEM images of (a) as-prepared Cu_2O nanowires and Au-nanoparticle-covered Cu_2O nanowires with different loading amounts of Au: (b) galvanic exchange reaction in 0.2 mM $HAuCl_4$ for 1 min and (c) galvanic exchange reaction in 10 mM $HAuCl_4$ for 5 min. Scale bars: 500 nm.

Results and discussions

Materials characterization

The as-prepared Cu_2O and Au-nanoparticle-covered Cu_2O were characterized by SEM (Fig. 2). The Cu_2O film consisted of intertwined nanowires with a length of a few micrometres and a diameter of a few hundred nanometres. These nanowires split into smaller Cu_2O endings that were relatively smooth (Fig. 2a). After immersing Cu_2O in $HAuCl_4$ solution, the simultaneous galvanic exchange of Au^{3+} with Cu^+ results in the formation of Au particles on the nanowire substrate via the following equation:

$$2Au^{3+} + 3Cu_2O + 6H^+ \rightarrow 2Au + 6Cu^{2+} + 3H_2O$$

By changing the concentration of $HAuCl_4$ solution and the reaction time, the loading amount and particle size of Au were systematically tuned. 1 min of galvanic exchange in 0.2 mM $HAuCl_4$ solution led to the formation of isolated Au nanoparticles with diameters < 50 nm. These nanoparticles were evenly dispersed on the surface of the nanowires (Fig. 2b), creating a homogeneous Au-nanoparticle-covered Cu_2O surface. Extending the concentration of the precursor solution and the galvanic exchange time resulted in a higher density of larger Au aggregates on the Cu_2O nanowires (Fig. 2c).

Both the Cu_2O and Au-nanoparticle-covered Cu_2O were electrochemically reduced in the CO_2-saturated 0.1 M $KHCO_3$ solution, thus forming oxide-derived Cu and bifunctional CuAu catalysts. Unless stated otherwise, CuAu catalyst refers to the Au-nanoparticle-covered Cu originating from 1 min of galvanic exchange in 0.2 mM $HAuCl_4$ solution. The electrochemical surface areas of Cu and CuAu were determined by double layer capacitance measurements. We derived the capacitance values of the Cu and CuAu catalysts to be 2.92 and 2.01 mF cm^{-2}, respectively. Thus, the CuAu catalyst has a smaller roughness factor than the Cu catalyst.

Improved C–C coupling on bifunctional CuAu catalysts

The electrocatalytic abilities of the Cu and CuAu catalysts towards CO_2 reduction were assessed using 60 min potentiostatic measurements over a wide range of potentials from −0.60 V to −1.15 V in a custom-built electrochemical cell. Three

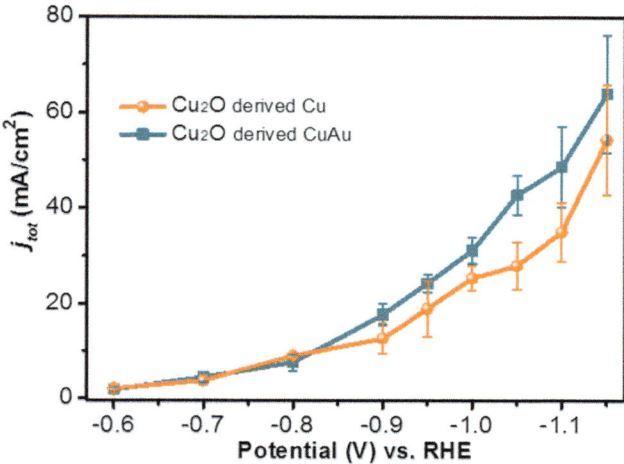

Fig. 3 The total cathodic current densities of CO_2 reduction on the Cu and CuAu (0.2 mM $HAuCl_4$ for 1 min) catalysts as a function of applied potentials. Each data point represents the average of three independent measurements and the error bar represents the standard deviations of these measurements.

independent measurements were carried out for each potential to ensure the reproducibility. The average geometric current density over the course of 60-min chronoamperometry against the applied potential for the Cu and CuAu catalysts is shown in Fig. 3. The total current density on both catalysts correlated positively with the applied overpotential. From −0.6 V to −0.8 V, the total current density observed on the two catalysts was similar. However, at potentials that were more negative than −0.9 V, the total current density on CuAu catalyst became higher than the one on Cu catalyst. For example, the current density observed on CuAu at −1.05 V was 42.9 mA cm^{-2}, which was 1.5× larger than the 28.1 mA cm^{-2} observed on the Cu catalyst at the same applied potential. Considering the lower roughness of the CuAu compared to that of the Cu catalyst, the different J–V

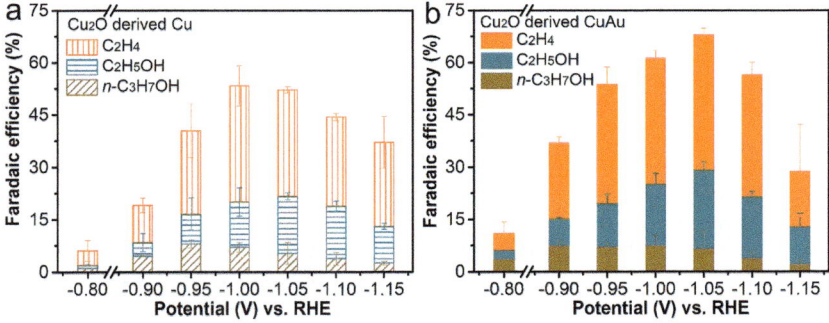

Fig. 4 The faradaic efficiency of ethylene, ethanol and *n*-propanol on (a) Cu and (b) CuAu (0.2 mM $HAuCl_4$ for 1 min) catalysts. Each data point represents the average of three independent measurements and the error bar represents the standard deviations of them.

Table 1 The average faradaic efficiencies of products (%) and total cathodic current density (mA cm^{-2}) obtained from CO_2 reduction on Cu catalyst at different potentials. N.D.: not detectable

Potential (V) vs. RHE	H_2	CO	CH_4	$HCOO^-$	C_2H_4	C_2H_5OH	CH_3COO^-	CH_3COH	C_3H_7OH	j_{tot}
−0.60	59.93	6.83	N.D.	32.97	N.D.	N.D.	N.D.	N.D.	N.D.	2.23
−0.70	44.77	7.07	N.D.	43.77	1.80	N.D.	N.D.	N.D.	N.D.	3.95
−0.80	50.80	5.20	N.D.	34.50	4.17	0.93	N.D.	N.D.	1.20	9.13
−0.90	38.43	3.60	N.D.	33.83	10.80	3.87	0.35	0.65	4.70	12.85
−0.95	39.70	2.23	0.90	15.20	23.83	8.63	0.43	0.53	8.17	19.0
−1.00	31.33	1.13	2.50	9.50	33.20	13.10	0.43	0.93	7.20	25.47
−1.05	34.93	0.53	1.60	2.53	30.47	16.43	0.33	0.75	5.43	28.08
−1.10	41.97	0.50	5.47	1.53	25.47	15.13	0.20	0.57	3.90	35.20
−1.15	50.10	0.60	10.30	0.80	24.07	10.57	0.17	0.43	2.60	54.43

trends of Cu and CuAu correspond to the difference in their catalytic performance.

The cathodic gas stream was periodically analyzed by gas chromatography at 2 min, 16 min, 30 min, 44 min and 58 min of the chronoamperometry. The liquid products were quantified by ^1H nuclear magnetic resonance (^1H NMR) after electrolysis. 4 gaseous products (H_2, CO, CH_4, and C_2H_4) and 5 liquid products ($HCOO^-$, CH_3COO^-, CH_3CHO, C_2H_5OH, and n-C_3H_7OH) were detected (Tables 1 and 2). The faradaic efficiencies for forming alkenes and alcohols on the Cu and CuAu catalysts are shown in Fig. 4.

We observed that the selectivity towards C_{2+} alkene and alcohols on CuAu showed substantial improvement at a lower overpotential. At −0.90 V, the faradaic efficiency of the multi-carbon products nearly doubled from 20% to 38%. The faradaic efficiencies of producing C_{2+} alkene and alcohols on Cu and CuAu catalysts both followed a parabolic trend, peaking at around −1.05 V vs. RHE. For the Cu catalyst, the faradaic efficiencies of ethylene and ethanol were 30–33% and

Table 2 The average faradaic efficiencies of products (%) and total cathodic current density (mA cm^{-2}) obtained from CO_2 reduction on CuAu catalyst at different potentials. N.D.: not detectable

Potential (V) vs. RHE	H_2	CO	CH_4	$HCOO^-$	C_2H_4	C_2H_5OH	CH_3COO^-	CH_3COH	C_3H_7OH	j_{tot}
−0.6	28.87	10.17	N.D.	38.4	N.D.	N.D.	N.D.	N.D.	N.D.	2.00
−0.7	30.33	9.00	N.D.	46.17	1.15	N.D.	N.D.	N.D.	N.D.	4.53
−0.8	30.83	8.10	N.D.	39.83	4.80	2.40	0.50	N.D.	3.90	7.70
−0.9	37.17	4.37	N.D.	18.30	21.70	7.60	0.77	0.50	7.70	17.73
−0.95	36.07	2.40	N.D.	12.30	34.10	12.33	0.60	0.40	7.37	24.32
−1.0	28.30	1.20	N.D.	6.80	36.17	17.50	0.47	0.90	7.73	31.24
−1.05	30.77	1.10	2.20	4.07	38.70	22.60	0.85	1.10	6.83	42.87
−1.1	43.20	0.47	3.15	1.73	34.97	17.50	0.25	0.53	4.13	48.82
−1.15	57.50	0.53	10.8	1.23	15.85	10.70	0.27	0.45	2.40	63.98

13–16%, respectively, with the total faradaic efficiency of multi-carbon alkene and alcohols of 55% at −1.00 V and −1.05 V. At the same time, the faradaic efficiency of CH_4 was suppressed to <3% (Table 1). These observations are consistent with previous reports using other oxide-derived Cu catalysts.[21-23,38] Interestingly, the faradaic efficiencies of ethylene and ethanol increased to 39% and 23%, respectively, on the CuAu catalyst at −1.05 V and the total faradaic efficiency of C_{2+} alkene and alcohols approached ∼70%. Thus, the faradaic efficiency of multi-carbon products improved by nearly 30% on the CuAu catalyst. Considering the 1.5× increase in geometric current density on the CuAu catalyst (Fig. 3), we managed to get a ∼2× increase in the partial current density of multi-carbon products on the CuAu catalyst (∼30 mA cm^{-2}) compared with that of the Cu catalyst (∼15 mA cm^{-2}). This ranks our bifunctional CuAu catalyst among the best catalysts for the formation of multi-carbon alkenes and alcohols.

Selectivity of electrocatalytic ethanol and ethylene formation

We assessed the selectivity of electrocatalytic ethanol (ethanol and n-propanol) and ethylene formation on our Cu and CuAu catalysts by analysing the ratio of the alcohol/ethylene yields as a function of the applied potential (Fig. 5). For Cu, the alcohol to ethylene ratio remained relatively stable, ranging from 0.5 to 0.8. This is consistent with previous studies using various Cu-based catalysts, which showed a preferred C–C coupling towards ethylene instead of alcohol.[39] However, interestingly, with the CuAu the selectivity of alcohol to ethylene was dependent on the applied potential. At −0.80 V, the faradaic yield ratio of alcohol to ethylene increased from 0.5 on Cu to 1.3 on CuAu. However, the yield ratio of alcohol to ethylene on CuAu decreased at potentials that are more negative than −0.90 V (Fig. 5). The CuAu tandem catalyst and Cu_2S–Cu–V (Cu with a vacancy on the surface and sulphur in the subsurface) also exhibited a similar trend, albeit with a much lower faradaic efficiency of multi-carbon products.[40,41] We propose the

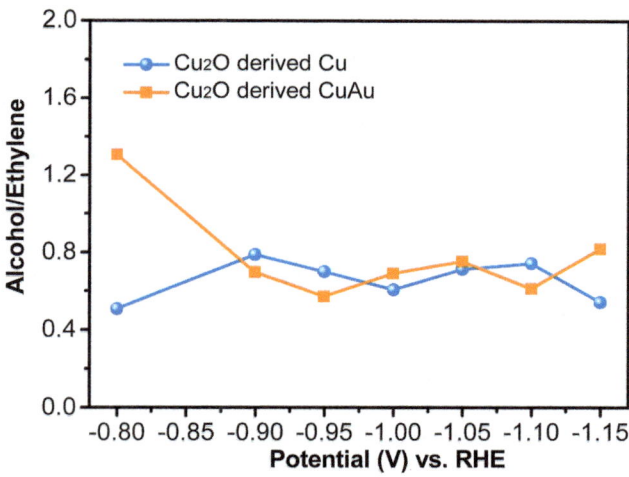

Fig. 5   The faradaic efficiency ratio of alcohol to ethylene as a function of applied potential on the Cu and CuAu catalysts.

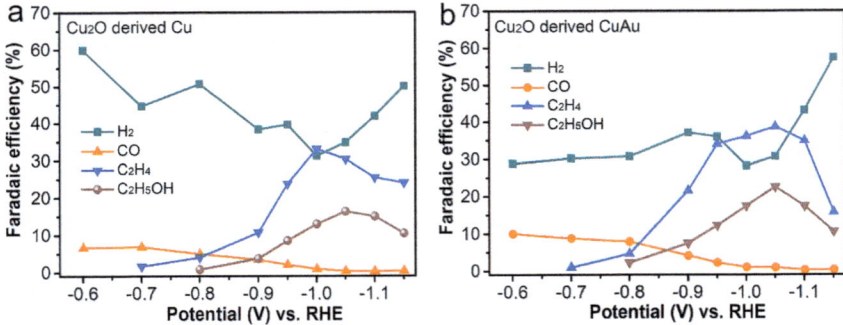

Fig. 6 The faradaic efficiencies of hydrogen, carbon monoxide, ethylene and ethanol as a function of applied potential on (a) Cu and (b) CuAu (0.2 mM $HAuCl_4$ for 1 min) catalysts.

reason behind the potential-dependent alcohol/ethylene ratio on the CuAu catalyst to be a lower coverage of *H at smaller overpotentials.[41]

Reduced *H coverage and enhanced *CO coverage on the CuAu catalyst

The faradaic efficiencies of hydrogen, carbon monoxide, ethylene and ethanol as a function of applied potential are shown in Fig. 6. A clear trend is that the faradaic efficiency of hydrogen decreased significantly on the CuAu catalyst at a lower overpotential. For example, from −0.60 V to −0.80 V, the faradaic efficiency of hydrogen was suppressed to 30% (with the other 70% accounting for carbonaceous products) on the CuAu catalyst, whereas the one on the Cu catalyst ranged from 45% to 60%. The suppressed hydrogen evolution suggests that the presence of Au assists in reducing the *H coverage at low overpotential, which indicates that more active sites are available for binding intermediates for CO_2 reduction, such as CO.

Low *H-coverage has also been found on other Cu-based catalysts. Bell and co-workers have studied different CuAg alloys and proposed that the microstrain caused by the presence of Ag assisted in reducing the coverage of *H.[42] This proposition was corroborated by extensive surface analysis using X-ray photoelectron spectroscopy and cyclic voltammetry. However, we note that the microstrain caused by Ag alloying was at atomic scale whereas our Au nanoparticles were in the scale of 10–50 nm. Thus, the microstrain may not have an important role in determining the low coverage of *H in our CuAu catalyst.

Recent studies on CO reduction using Cu electrodes revealed the low *H coverage in basic electrolyte caused by the low concentration of H^+ at alkaline conditions.[43] Here we assess whether our low *H coverage is due to an increase in local pH.[44] The local pH is correlated with the consumption rate of H^+, which is reflected by the partial current density of non-ionic products (the number of protons transferred equals that of electrons for non-ionic products). By multiplying the total current density and the faradaic efficiency of non-ionic products at low overpotential, we found little difference in the consumption rates of H^+ on the Cu and CuAu. Thus, we believe that the local pH is not the major factor affecting *H-coverage on the CuAu catalysts.

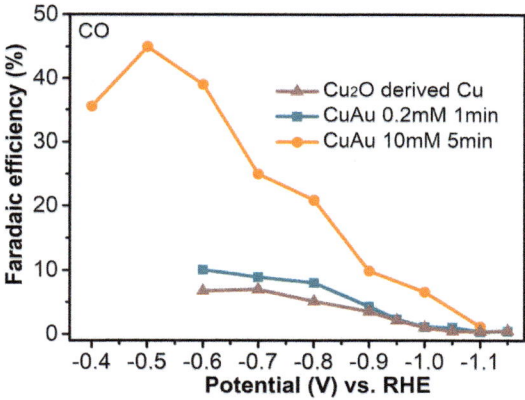

Fig. 7 Trends of CO on Cu, CuAu (0.2 mM HAuCl$_4$ for 1 min) and CuAu (10 mM HAuCl$_4$ for 5 min) as a function of applied potential.

Here we propose that the low *H-coverage is mainly due to the increased coverage of *CO on Cu sites. It is widely believed that hydrogen evolution is a kinetically more facile reaction dominating the reduction of CO_2.[45] Moreover, *CO and *H compete with each other for active sites during CO_2 reduction.[35] On the Cu catalyst at a relatively low overpotential, *i.e.* at −0.60 V, the reduction of CO_2 to CO is limited by insufficient driving potential, as indicated by the high faradaic efficiency of H_2. The coverage of *CO on the Cu catalyst is not comparable with the *H-coverage, the production of which requires less electrochemical driving force.[6] With the introduction of Au nanoparticles in the bifunctional Cu/Au catalyst, the production of CO is accelerated.[46,47] This will increase the *CO coverage on Cu, reducing the number of available sites for *H adsorption.[35] Thus, on the CuAu bifunctional catalyst the *H coverage is reduced, suppressing the formation of hydrogen.

This proposition is further supported by two phenomena: (1) Au catalyst has a better catalytic ability towards the formation of CO compared with that of Cu catalyst, as shown in previous reports.[46,47] Furthermore, the weaker CO adsorption on Au compared to that on Cu facilitates the release of CO.[11] (2) The faradaic efficiency of the CO formation observed on CuAu is slightly higher than that on Cu, indicating that more CO is produced. Note also that the actual increase of CO production should be higher since some CO will be consumed in the formation of multi-carbon products. To further prove our proposition, we analysed CO formation on an Au-enriched CuAu catalyst, prepared by galvanic exchange reaction in 10 mM HAuCl$_4$ solution for 5 min. As shown in Fig. 7, a very high faradaic efficiency of CO was achieved at very low overpotential. For example, at −0.50 V, the faradaic efficiency of CO reached up to 45%, proving that the presence of Au enhances the production of CO.

Sequential catalysis on CuAu bimetallic cathodes

We infer from our study that CO_2 is sequentially reduced to multi-carbon alkene and alcohols on CuAu catalyst as shown in Fig. 8. Au accelerates the reduction of CO_2 to CO, which subsequently adsorbs on the Cu sites, creating a high coverage

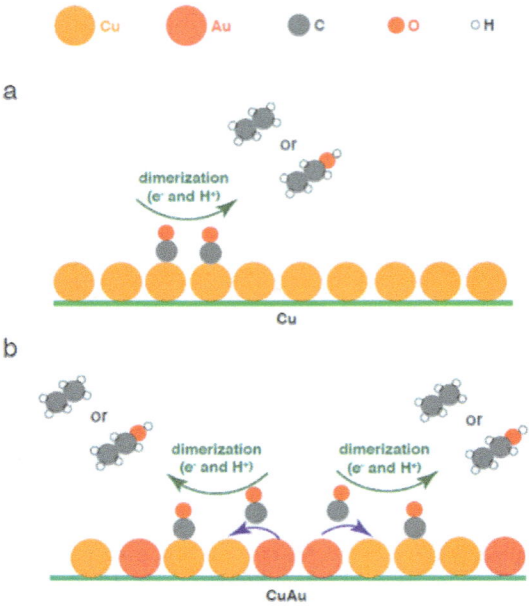

Fig. 8 Proposed mechanism of ethylene and ethanol formation on (a) Cu and (b) CuAu catalysts.

of *CO on Cu. This accelerates the rate of C–C coupling towards the formation of C_{2+} alkene and alcohols. At the same time, the high *CO coverage suppresses the formation of hydrogen, which is the main reaction competing with the reduction of CO_2 to hydrocarbons and alcohols.

With regards to the mechanism for the formation of multi-carbon products, several studies have shown that the formation of ethylene is pH-independent on an SHE reference scale.[29,43,48] The RHE reference scale is better suited to compare pH-dependent reactions (*i.e.* hydrogen evolution reaction) at different pH since it compensates for the pH-dependent equilibrium potential.[49] Thus, if one reaction is pH-dependent on an RHE scale but pH-independent on a fixed scale (*i.e.* SHE), the rate-determining step of the reaction involves no proton transfer. With this in mind, Koper and co-workers have proposed that the dimerization of CO, accompanied by the transfer of one electron, is the key step in the formation of C–C bonds. The adsorbed *CO tends to dimerize once the coverage of *CO reaches a certain level and upon one electron reduction forms the adsorbed *CO dimer anion $*C_2O_2^{-}$. This dimer will be further protonated and reduced to form $*C_2H_4O$.[35] This is the point where the formation of ethylene and ethanol starts to diverge. Since the formation of ethylene has a lower energy barrier than that of ethanol, it is reasonable that ethylene is the preferred C_2 product on the Cu catalyst.[33]

Ideal catalysts for the formation of multi-carbon products should deliver a faradaic efficiency close to 100% for C_{2+} products. This requires suppressed formation of both C_1 products and hydrogen. Previous investigations on oxide-derived Cu catalyst have shown that the suppression of C_1 products can be achieved on these catalysts by selecting appropriate applied potentials.[31,50] However, the suppression of hydrogen evolution remains a great challenge. Here, by accelerating the reduction of CO_2 to CO using Au, we managed to improve the faradaic efficiency and partial current density of C_{2+} alkene and alcohols beyond the limit of pure oxide-derived Cu catalyst. The employment of oxide-derived Cu nanowires as the substrate also helps suppress the formation of CH_4.

Recently, a three-phasic flow cell based on a gas diffusion electrode was developed for CO_2 reduction.[51] With this design, the reaction could take place in alkaline electrolyte at the gas–liquid–electrode interface. High alkaline electrolyte enables the suppression of hydrogen, and gas diffuse layer warrants the achievement of scalable current densities.[52] However, the drawback of this approach, *i.e.* the consumption of KOH, has not been fully addressed yet. Nevertheless, the flow cell infrastructure, together with the rational design of catalysts, is a promising future direction for efficient and scalable systems for artificial photosynthesis.

Another topic of our interest is to study the selectivity between two major C_2 products, ethylene and ethanol, using Cu–X (Au, Ag, Zn) bimetallic systems. Recently, CuAg catalysts were reported to be selective towards the formation of ethanol and a high ratio of ethanol/ethylene was achieved.[53] Interestingly, different CuAg systems showed different performances: Lee *et al.* reported an enhanced selectivity for ethanol when using electrodeposited CuAg, whereas Bell and co-workers found the improved selectivity for oxygenates, especially acetate.[42,53] Elucidating the underlying reasons for the similarities and differences between different Cu–X bimetallic (X = Au, Ag, Zn) will in turn help to design selective catalysts for the formation of different multi-carbon products in CO_2 reduction and achieve tunable selectivity between ethylene and ethanol. Ethylene is by far the preferred product over ethanol. Apart from being a key chemical feedstock its great advantage is that it evolves spontaneously as a gas from the electrolyte while ethanol needs to be separated from solution at high cost and in energy consuming processes.

Conclusions

In this work, we conceived a bifunctional oxide-derived CuAu catalyst operating in sequential fashion to accomplish the selective reduction of CO_2 to multi-carbon alkene and alcohols. This catalyst showed improved selectivity towards the formation of ethylene, ethanol and *n*-propanol, with a faradaic efficiency of 70% for these products. The partial current density of ethylene and alcohol formation nearly doubled compared to that for oxide-derived Cu, one of the best catalysts for the formation of multi-carbon products. The introduction of Au into oxide-derived Cu catalyst enables an improved coverage of *CO, thus enhancing the C–C coupling process towards the formation of ethylene and ethanol while the reduced *H coverage helps suppress the competing hydrogen evolution reaction. We believe that the easily prepared and selective CuAu bimetallic provides a good example of how sequential catalysts enable enhanced selectivity towards the formation of products with high commercial value in CO_2 reduction.

Conflicts of interest

There are no conflicts to declare.

Acknowledgements

This work was funded by the Sino-Swiss Science and Technology Cooperation (SSSTC) 2016 under the project title "All perovskite tandems for solar to CO_2 fixation" from the Swiss National Science Foundation with the grant number IZLCZ2-170294. The authors gratefully thank Aurélien Bornet (NMR, ISIC, EPFL) for the technical support on ^1H NMR. J. G. is financially supported by an overseas exchange scholarship from the China Scholarship Council (No. CSC201706370233).

References

1 A. J. Bard and M. A. Fox, *Acc. Chem. Res.*, 1995, **28**, 141–145.

2 N. S. Lewis and D. G. Nocera, *Proc. Natl. Acad. Sci. U. S. A.*, 2006, **103**, 15729–15735.

3 M. Schreier, L. Curvat, F. Giordano, L. Steier, A. Abate, S. M. Zakeeruddin, J. Luo, M. T. Mayer and M. Grätzel, *Nat. Commun.*, 2015, **6**, 7326.

4 M. R. Singh, E. L. Clark and A. T. Bell, *Proc. Natl. Acad. Sci. U. S. A.*, 2015, **112**, E6111–E6118.

5 M. Schreier, F. Héroguel, L. Steier, S. Ahmad, J. S. Luterbacher, M. T. Mayer, J. Luo and M. Grätzel, *Nat. Energy*, 2017, **2**, 17087.

6 K. P. Kuhl, E. R. Cave, D. N. Abram and T. F. Jaramillo, *Energy Environ. Sci.*, 2012, **5**, 7050–7059.

7 R. Kortlever, I. Peters, C. Balemans, R. Kas, Y. Kwon, G. Mul and M. T. M. Koper, *Chem. Commun.*, 2016, **52**, 10229–10232.

8 S. Lee, D. Kim and J. Lee, *Angew. Chem., Int. Ed.*, 2015, **54**, 14701–14705.

9 Y. Hori, H. Wakebe, T. Tsukamoto and O. Koga, *Electrochim. Acta*, 1994, **39**, 1833–1839.

10 Y. Hori, A. Murata and R. Takahashi, *J. Chem. Soc., Faraday Trans. 1*, 1989, **85**, 2309–2326.

11 K. P. Kuhl, T. Hatsukade, E. R. Cave, D. N. Abram, J. Kibsgaard and T. F. Jaramillo, *J. Am. Chem. Soc.*, 2014, **136**, 14107–14113.

12 J. B. Greenblatt, D. J. Miller, J. W. Ager, F. A. Houle and I. D. Sharp, *Joule*, **2**, 381–420.

13 I. Takahashi, O. Koga, N. Hoshi and Y. Hori, *J. Electroanal. Chem.*, 2002, **533**, 135–143.

14 Y. Hori, I. Takahashi, O. Koga and N. Hoshi, *J. Mol. Catal. A: Chem.*, 2003, **199**, 39–47.

15 K. J. P. Schouten, Z. Qin, E. P. Gallent and M. T. M. Koper, *J. Am. Chem. Soc.*, 2012, **134**, 9864–9867.

16 K. J. P. Schouten, E. Pérez Gallent and M. T. M. Koper, *ACS Catal.*, 2013, **3**, 1292–1295.

17 C. S. Chen, A. D. Handoko, J. H. Wan, L. Ma, D. Ren and B. S. Yeo, *Catal. Sci. Technol.*, 2015, **5**, 161–168.

18 F. S. Roberts, K. P. Kuhl and A. Nilsson, *Angew. Chem., Int. Ed.*, 2015, **54**, 5179–5182.

19 A. Loiudice, P. Lobaccaro, E. A. Kamali, T. Thao, B. H. Huang, J. W. Ager and R. Buonsanti, *Angew. Chem., Int. Ed.*, 2016, **55**, 5789–5792.

20 K. Jiang, R. B. Sandberg, A. J. Akey, X. Liu, D. C. Bell, J. K. Nørskov, K. Chan and H. Wang, *Nat. Catal.*, 2018, **1**, 111–119.

21 R. Kas, R. Kortlever, A. Milbrat, M. T. M. Koper, G. Mul and J. Baltrusaitis, *Phys. Chem.Chem. Phys.*, 2014, **16**, 12194–12201.

22 D. Ren, Y. Deng, A. D. Handoko, C. S. Chen, S. Malkhandi and B. S. Yeo, *ACS Catal.*, 2015, **5**, 2814–2821.

23 H. Mistry, A. S. Varela, C. S. Bonifacio, I. Zegkinoglou, I. Sinev, Y.-W. Choi, K. Kisslinger, E. A. Stach, J. C. Yang, P. Strasser and B. R. Cuenya, *Nat. Commun.*, 2016, **7**, 12123.

24 D. Ren, N. T. Wong, A. D. Handoko, Y. Huang and B. S. Yeo, *J. Phys. Chem. Lett.*, 2016, **6**, 20–24.

25 Y. Lum, B. Yue, P. Lobaccaro, A. T. Bell and J. W. Ager, *J. Phys. Chem. C*, 2017, **121**, 14191–14203.

26 J. Li, F. Che, Y. Pang, C. Zou, J. Y. Howe, T. Burdyny, J. P. Edwards, Y. Wang, F. Li, Z. Wang, P. De Luna, C.-T. Dinh, T.-T. Zhuang, M. I. Saidaminov, S. Cheng, T. Wu, Y. Z. Finfrock, L. Ma, S.-H. Hsieh, Y.-S. Liu, G. A. Botton, W.-F. Pong, X. Du, J. Guo, T.-K. Sham, E. H. Sargent and D. Sinton, *Nat. Commun.*, 2018, **9**, 4614.

27 C. W. Li, J. Ciston and M. W. Kanan, *Nature*, 2014, **508**, 504–507.

28 A. Verdaguer-Casadevall, C. W. Li, T. P. Johansson, S. B. Scott, J. T. McKeown, M. Kumar, I. E. L. Stephens, M. W. Kanan and I. Chorkendorff, *J. Am. Chem. Soc.*, 2015, **137**, 9808–9811.

29 R. Kas, R. Kortlever, H. Yılmaz, M. T. M. Koper and G. Mul, *ChemElectroChem*, 2015, **2**, 354–358.

30 A. D. Handoko, C. W. Ong, Y. Huang, Z. G. Lee, L. Lin, G. B. Panetti and B. S. Yeo, *J. Phys. Chem. C*, 2016, **120**, 20058–20067.

31 D. Ren, J. Fong and B. S. Yeo, *Nat. Commun.*, 2018, **9**, 925.

32 A. A. Peterson, F. Abild-Pedersen, F. Studt, J. Rossmeisl and J. K. Norskov, *Energy Environ. Sci.*, 2010, **3**, 1311–1315.

33 F. Calle-Vallejo and M. T. M. Koper, *Angew. Chem., Int. Ed.*, 2013, **52**, 7282–7285.

34 E. Pérez-Gallent, M. C. Figueiredo, F. Calle-Vallejo and M. T. M. Koper, *Angew. Chem., Int. Ed.*, 2017, **56**, 3621–3624.

35 Y. Huang, A. D. Handoko, P. Hirunsit and B. S. Yeo, *ACS Catal.*, 2017, **7**, 1749–1756.

36 D. Ren, N. W. X. Loo, L. Gong and B. S. Yeo, *ACS Sustainable Chem. Eng.*, 2017, **5**, 9191–9199.

37 J. E. Pander III, D. Ren and B. S. Yeo, *Catal. Sci. Technol.*, 2017, **7**, 5820–5832.

38 D. Kim, S. Lee, J. D. Ocon, B. Jeong, J. K. Lee and J. Lee, *Phys. Chem. Chem. Phys.*, 2015, **17**, 824–830.

39 D. Ren, B. S.-H. Ang and B. S. Yeo, *ACS Catal.*, 2016, **6**, 8239–8247.

40 T.-T. Zhuang, Z.-Q. Liang, A. Seifitokaldani, Y. Li, P. De Luna, T. Burdyny, F. Che, F. Meng, Y. Min, R. Quintero-Bermudez, C. T. Dinh, Y. Pang, M. Zhong, B. Zhang, J. Li, P.-N. Chen, X.-L. Zheng, H. Liang, W.-N. Ge, B.-J. Ye, D. Sinton, S.-H. Yu and E. H. Sargent, *Nat. Catal.*, 2018, **1**, 421–428.

41 C. G. Morales-Guio, E. R. Cave, S. A. Nitopi, J. T. Feaster, L. Wang, K. P. Kuhl, A. Jackson, N. C. Johnson, D. N. Abram, T. Hatsukade, C. Hahn and T. F. Jaramillo, *Nat. Catal.*, 2018, **1**, 764–771.

42 E. L. Clark, C. Hahn, T. F. Jaramillo and A. T. Bell, *J. Am. Chem. Soc.*, 2017, **139**, 15848–15857.

43 L. Wang, S. A. Nitopi, E. Bertheussen, M. Orazov, C. G. Morales-Guio, X. Liu, D. C. Higgins, K. Chan, J. K. Nørskov, C. Hahn and T. F. Jaramillo, *ACS Catal.*, 2018, 7445–7454.

44 K. Schouten, Y. Kwon, C. Van der Ham, Z. Qin and M. Koper, *Chem. Sci.*, 2011, **2**, 1902–1909.

45 A. A. Peterson and J. K. Nørskov, *J. Phys. Chem. Lett.*, 2012, **3**, 251–258.

46 Y. Chen, C. W. Li and M. W. Kanan, *J. Am. Chem. Soc.*, 2012, **134**, 19969–19972.

47 C. W. Li and M. W. Kanan, *J. Am. Chem. Soc.*, 2012, **134**, 7231–7234.

48 Y. Hori, R. Takahashi, Y. Yoshinami and A. Murata, *J. Phys. Chem. B*, 1997, **101**, 7075–7081.

49 M. T. M. Koper, *Chem. Sci.*, 2013, **4**, 2710–2723.

50 J. E. Pander, D. Ren, Y. Huang, N. W. X. Loo, S. H. L. Hong and B. S. Yeo, *ChemElectroChem*, 2018, **5**, 219–237.

51 S. Ma, M. Sadakiyo, R. Luo, M. Heima, M. Yamauchi and P. J. A. Kenis, *J. Power Sources*, 2016, **301**, 219–228.

52 C.-T. Dinh, T. Burdyny, M. G. Kibria, A. Seifitokaldani, C. M. Gabardo, F. P. García de Arquer, A. Kiani, J. P. Edwards, P. De Luna, O. S. Bushuyev, C. Zou, R. Quintero-Bermudez, Y. Pang, D. Sinton and E. H. Sargent, *Science*, 2018, **360**, 783–787.

53 S. Lee, G. Park and J. Lee, *ACS Catal.*, 2017, **7**, 8594–8604.

Faraday Discussions

PAPER

A tandem photoelectrochemical water splitting cell consisting of CuBi₂O₄ and BiVO₄ synthesized from a single Bi₄O₅I₂ nanosheet template†

Yi-Hsuan Lai, [iD] * Kai-Che Lin,‡ Chen-Yang Yen‡ and Bo-Jyun Jiang

Received 16th November 2018, Accepted 21st December 2018

DOI: 10.1039/c8fd00183a

We report on a simple and facile synthetic approach for preparing both a CuBi₂O₄ photocathode and BiVO₄ photoanode from a single Bi₄O₅I₂ nanosheet template. A nanosheet structured Bi₄O₅I₂ (nanoBi₄O₅I₂) template, solvothermally deposited on a conducting glass substrate, can be converted to a textured CuBi₂O₄ photocathode and a BiVO₄ photoanode with coral-like nanostructure by drop-casting a copper ion-containing and a vanadium ion-containing solution, respectively, with a follow-up heat treatment. UV-vis absorption spectra and Mott–Schottky analyses confirm CuBi₂O₄ and BiVO₄ have well-suited band gaps for absorbing a large portion of visible light and complementary band structures for proton reduction and water oxidation, respectively. A photoelectrochemical tandem cell, consisting of a cobalt based catalyst modified CuBi₂O₄ photocathode and a cobalt based catalyst modified BiVO₄ photoanode shows promising photoelectrochemical properties towards water splitting, and can be operated without an external bias. Factors limiting the performance of the tandem cell are also discussed.

1 Introduction

Photoelectrochemical (PEC) water splitting is a promising approach for providing clean and sustainable H₂ fuels from sunlight and water. Among various device configurations, a dual-photoelectrode tandem PEC cell that consists of one photoanode and one photocathode is especially attractive as it has a maximal solar-to-hydrogen (STH) efficiency of 20–30%,[1] outperforming the profitable STH efficiency of 15–20%.[2]

Immense efforts therefore have been devoted to the optimization of the individual performances of photoanodes and/or photocathodes for PEC water

---

Department of Materials and Optoelectronic Science, National Sun Yat-sen University, 70 Lienhai Rd., Kaohsiung 80424, Taiwan. E-mail: yhlai@mail.nsysu.edu.tw

† Electronic supplementary information (ESI) available. See DOI: 10.1039/c8fd00183a

‡ These authors contributed equally to this work.

oxidation and PEC proton reduction, respectively. For example, $BiVO_4$ is a state-of-the-art photoanode, which has a band gap of 2.4–2.5 eV and a suitable valence band edge, capable of using visible light to drive water oxidation.[3] A photocurrent of 6.1 mA cm^{-2} at 1.23 V *vs.* the reversible hydrogen electrode (RHE) under one sun illumination, close to 80% of its theoretical value of 7.6 mA cm^{-2}, was recently achieved by orientation control and facet engineering.[3c] Stability is another important factor for assessing the performance of a photoelectrode. By tuning the electrolyte compositions or modification with a protection layer and/or co-catalyst, the photocorrosion issue of $BiVO_4$ can be efficiently alleviated.[4] A long-term stability of over 1000 h in a neutral pH solution has been successfully demonstrated for $BiVO_4$.[5] Essential progress have also been made for Cu_2O[6] and hematite.[7]

Although the immense progress on optimizing the performance of individual photoelectrodes have been made, there is relatively limited research on an integrated PEC cell for overall solar water splitting and the efficiency is still far below the theoretical value and practical applications.[6a,8] Examples include Cu_2O paired with WO_3,[6a] p-Si paired with $BiVO_4$,[8b] an amorphous Si paired with hematite,[8e] Cu_2O paired with $BiVO_4$,[8a] $CuIn_{0.5}Ga_{0.5}Se_2$ paired with $BiVO_4$,[9] and $CuBi_2O_4$ paired with $BiVO_4$.[10] However, the majority employ arduous and expensive techniques, such as atomic layer deposition and expensive materials and/or catalysts, such as In and Pt. Bottlenecks also arise from complex procedures for preparing photoelectrodes, their incompatibility under device operating conditions, and poor stability of photoelectrodes. Developing simple and facile methods for simultaneously preparing efficient and compatible photoanodes and photocathodes is therefore beneficial for accelerating the progress on dual-photoelectrode tandem PEC cells.

In this discussion, we report on a novel method with simplicity and general applicability for preparing both the photoanode and photocathode from a single template by a solution based conversion method. A nanosheet structured $Bi_4O_5I_2$ (nano$Bi_4O_5I_2$) is first synthesized on a conducting glass substrate by a facile solvothermal method, followed by its conversion into micro-textured $CuBi_2O_4$ (micro$CuBi_2O_4$) and nanocoral structured $BiVO_4$ (nano$BiVO_4$) by drop-casting of a solution containing Cu^{2+} and VO^{2+}, respectively, and subsequent heat treatments. $CuBi_2O_4$ is a recently emerged photocathode material as it has a well-suited conduction band edge to drive proton reduction and a narrow band gap of 1.5–1.8 eV,[8d,11] yielding a theoretical photocurrent of over 17 mA cm^{-2}.[12] A theoretical efficiency of approximately 8% based on the band gaps and minimum reasonable losses is achievable for pairing $CuBi_2O_4$ with $BiVO_4$,[13] which is promising for an all earth-abundant dual-photoelectrode tandem PEC cell. However, research on using $CuBi_2O_4$ as a photocathode in a PEC tandem cell for solar water splitting is still limited.[8d,10] Common $CuBi_2O_4$ electrode preparation methods include drop-casting a precursor solution onto a conductive substrate followed by heat treatments[8d,11a,11d] and electrodeposition of a $CuO–Bi_xO_y$ film followed by conversion to $CuBi_2O_4$ by heating.[11b,11c,11e] $CuBi_2O_4$ also has a very positive valence band edge to provide sufficient internal photovoltage in a tandem PEC cell when paired with a photoanode such as $BiVO_4$.

A tandem PEC cell consisting of a $CuBi_2O_4$ based photocathode and a $BiVO_4$ based photoanode was thereby studied for overall water splitting under solar light irradiation (Scheme 1). Factors affecting the cell performance with respect to the

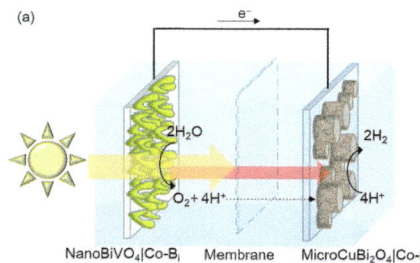 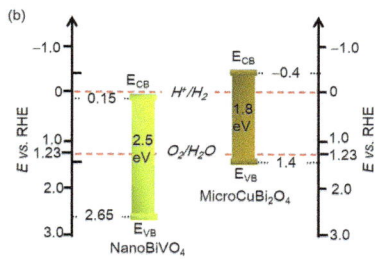

Scheme 1 (a) Schematic representation of the tandem PEC cell for solar water splitting consisting of a CuBi$_2$O$_4$ based photocathode (microCuBi$_2$O$_4$|Co–B$_i$) and a BiVO$_4$ based photoanode (nanoBiVO$_4$|Co–B$_i$) and (b) the band structures of microCuBi$_2$O$_4$ and nanoBiVO$_4$. A near neutral borate buffer solution was used as the electrolyte, whereas a proton exchange membrane was employed between the photocathodes and photo-anodes for suppressing the product crossover.

optical properties, charge carrier separation, and surface catalytic properties of individual photoelectrodes are discussed.

2 Experimental section

Preparation of electrodes

For the growth of nanoBi$_4$O$_5$I$_2$ on FTO, a solution for the solvothermal process was prepared by mixing a diethylene glycol (DEG, Sigma-Aldrich, 99%) solution containing 0.05 M Bi(NO$_3$)$_3$·5H$_2$O (Sigma-Aldrich, 98%) and a methanol (Aencore, 99.9%) solution containing 0.05 M KI (Sigma-Aldrich, ≧99.5%). Subsequently, 16 mL of the precursor solution was put in a 23 mL autoclave. A fluorine doped tin oxide (FTO) coated glass substrate (Pilkington TEC Glass™ 7) was immersed into the precursor solution with FTO facing downward in the autoclave. The autoclave was then sealed and heated at 160 °C for 4 h. Bismuth alkoxide is possibly first formed followed by the gradual formation of BiO$^+$ after the dissolution of Bi(NO$_3$)$_3$·5H$_2$O in the polyol solvent of DEG. BiO$^+$ and I$^-$ ions then nucleated on the FTO substrate followed by subsequent primary particle formation and aggregation and finally Ostwald ripening to from sheet structures in the presence of polyols.[14] Polyols are generally considered as a soft template for directing assembly of sheet structures through the long chain as well as the cross-linking networks by hydrogen bonds between hydroxyl groups. After growth of Bi$_4$O$_5$I$_2$ nanosheets, the resultant pale yellow colored electrodes were rinsed with water and dried in air at room temperature (RT).

The as-prepared nanoBi$_4$O$_5$I$_2$ was subsequently used as the template for the syntheses of microCuBi$_2$O$_4$ and nanoBiVO$_4$ by drop-casting of copper(II) acetate hydrate (0.01 M, based on the molecular mass of anhydrous copper(II) acetate, Sigma-Aldrich, 98%) in methanol and vanadyl acetylacetonate (VO(acac)$_2$, 0.2 M, Sigma-Aldrich, 98%) in dimethyl sulfoxide (DMSO). In the case of microCuBi$_2$O$_4$, the electrodes after drop-casting were placed flat in a vacuum chamber at RT to accelerate the evaporation of the solvent prior to heat treatment at 550 °C for 2 h with a ramping rate of 2 °C min^{-1}. A brown microCuBi$_2$O$_4$ electrode with high uniformity was then obtained (Scheme 2). In the case of nanoBiVO$_4$, the

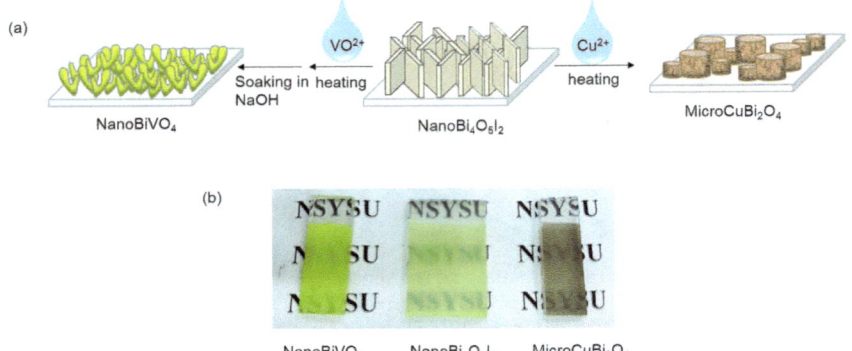

(a)

VO²⁺

Soaking in heating
NaOH

NanoBiVO₄   NanoBi₄O₅I₂

Cu²⁺

heating

MicroCuBi₂O₄

(b)

NanoBiVO₄   NanoBi₄O₅I₂   MicroCuBi₂O₄

Scheme 2 (a) The conversion processes of nanoBiVO₄ and microCuBi₂O₄ photo-electrodes from a single nanoBi₄O₅I₂ template. (b) The photograph image of a nanoBi₄O₅I₂, a nanoBiVO₄, and a microCuBi₂O₄ electrode.

electrodes after drop-casting were heat-treated at 450 °C for 2 h with a ramping rate of 2 °C min⁻¹ followed by soaking in a NaOH solution (1 M) to remove excess V_2O_5 and rinsing with water to obtain a bright yellow nanoBiVO₄ (Scheme 2).

Co–B_i modified photoelectrodes, including microCuBi₂O₄|Co–B_i and nanoBiVO₄|Co–B_i were prepared by a photo-assisted deposition method at 0.3 V *vs.* RHE under solar light irradiation (100 mW cm⁻²) for 150 seconds. The deposition electrolyte is a borate (B_i) buffer solution (pH 9.2) containing 0.5 mM $Co(NO_3)_2 \cdot 6H_2O$ as the Co^{2+} sources.

Physical characterisation

The surface morphologies of the electrodes were characterized by using a JEOL-6330 Field-Emission scanning electron microscope (SEM). The grazing incident X-ray diffraction (GI-XRD) analyses were carried out using a multi-function high power X-ray diffractometer (Bruker D8). UV-vis absorption spectra of the photo-electrodes were recorded by a UV-vis spectrophotometer (Agilent Cary 60) equipped with a diffuse reflectance accessory.

Electrochemical and PEC measurements

All the electrochemical and PEC measurements were recorded with a Metrohm Autolab (PGSTAT 302N) using a two-chamber electrochemical cell separated with a proton exchange membrane (PEM, Nafion 117™). The PEM shows high trans-mittance, up to 90% in the wavelength ranges of 300–800 nm (Fig. S1†). The geometric surface areas of all the photoelectrodes were defined using a polyester tape (0.5 cm²) for electrochemical and PEC analyses, whereas the precise geometric area was determined after the analyses.

For three-electrode measurements, an Ag/AgCl/KCl$_{sat}$ was used as the refer-ence electrode and placed in the same chamber with a working electrode, whereas a Pt foil was used as the counter electrode and placed in another chamber. All the measurements were carried out in a B_i buffer solution (pH 9.2) containing 0.5 M Na_2SO_4 as a supporting electrolyte unless otherwise noted and all the potentials were converted to RHE by using E (V $vs.$ RHE) = E (V $vs.$ Ag/AgCl/KCl$_{sat}$) + 0.197 + 0.059 × pH. Mott–Schottky analyses were carried out in the same three-electrode configuration in the same electrolyte at a frequency of 500 Hz or 1 kHz.

Tandem PEC cell studies were carried out in a two-chamber cell separated with a PEM. A nanoBiVO$_4$|Co–B$_i$ photoanode was used as the front electrode and a microCuBi$_2$O$_4$|Co–B$_i$ photocathode was used as the back electrode in the same light path. The distance between the two photoelectrodes is approximate 8 cm. A solar light simulator (SAN-EI ELECTRIC, XES-40S2-CE, AAA class) equipped with air mass 1.5 global (AM 1.5G) filter was used as the light source and the illumination intensity was calibrated to 100 mW cm^{-2} by a monocrystalline silicon solar cell.

Electrical impedance spectroscopy (EIS) analyses were performed to quantify the increased resistance incurred in moving the working electrode to a tandem (two-electrode) system. For the three-electrode measurements, the solution resistance was typically in the range 20–30 Ω, whereas the solution and membrane resistance in the two-electrode measurement (tandem device) was approximate 80 Ω in a B$_i$ buffer solution (pH 9.2) containing 0.5 M Na$_2$SO$_4$ electrolyte.

3   Results and discussion

NanoBi$_4$O$_5$I$_2$

Bismuth oxyhalides (BiOX, X = Cl, Br, I)[14,15] and bismuth-rich oxyhalides (Bi$_x$O$_y$X$_z$)[16] are promising photocatalysts for pollutant degradation as they have narrow band gaps to utilise visible light in solar spectrum. In particular, their structures are easily controlled to form sheets, flowers, or hollow spheres on the nanoscale due to their layered structure of [Bi–O] slabs placed in between two slabs of halogen ions.[16a] Previously, nanosheet structured BiOI was electro-deposited on a conducting substrate and used as a template for preparing a nanoporous BiVO$_4$ electrode for PEC water oxidation[3a] and a CuO decorated CuBi$_2$O$_4$ electrode for electrocatalytic oxidation of glucose.[17] Beyond the electrodeposition, developing an alternative facile method to prepare nano-structured BiOX or Bi$_x$O$_y$X$_z$ electrodes is desirable for the general applicability of the preparation of photoelectrodes from similar solution based conversion methods.

In this discussion, uniform Bi$_4$O$_5$I$_2$ nanosheets were synthesized on a FTO substrate by a one-step solvothermal method. The precursor solution was prepared by mixing a DEG solution containing Bi(NO$_3$)$_3$·5H$_2$O and a methanol solution containing KI (see Experimental for details). The crystal structure of nanoBi$_4$O$_5$I$_2$ was characterized by GI-XRD (Fig. 1a). Except for the signals rising from the FTO substrate, all signals belong to monoclinic Bi$_4$O$_5$I$_2$.[16b] The SEM images show that the FTO substrate was uniformly covered by nanoBi$_4$O$_5$I$_2$ with a sheet thickness of approximately 25 nm (Fig. 1b and S2a, ESI†). UV-vis spectroscopy shows that nanoBi$_4$O$_5$I$_2$ absorbs light below 550 nm, and the corresponding Tauc plot suggests nanoBi$_4$O$_5$I$_2$ has a band gap of approximately 2.0 eV, close to the reported value of 2.18 eV (Fig. 1c and d).[16b] However, nanoBi$_4$O$_5$I$_2$ shows a negligible photocurrent under simulated solar light irradiation (100 mW cm^{-2}, AM 1.5G, Fig. S2b, ESI†).

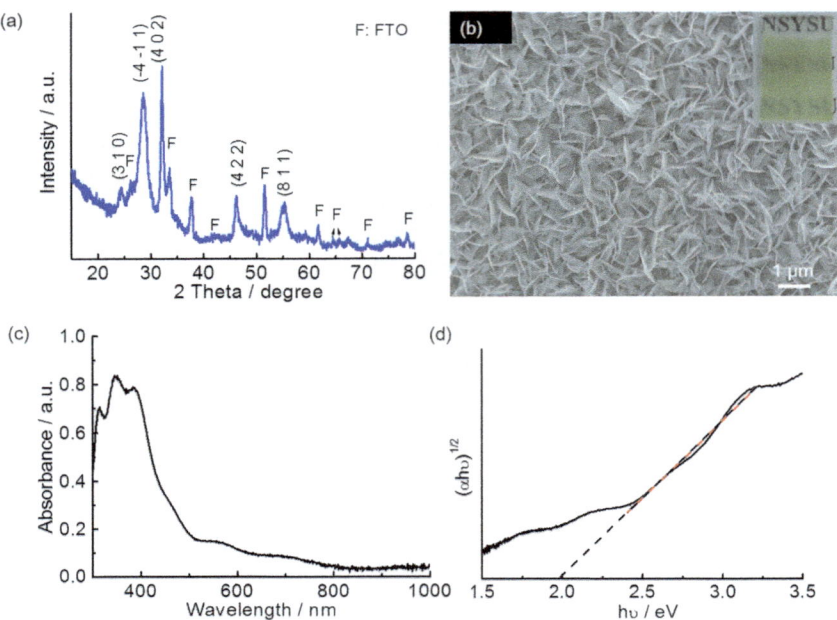

Fig. 1 The (a) GI-XRD, (b) SEM image, (c) UV-vis absorption spectrum, and (d) Tauc plot of nanoBi$_4$O$_5$I$_2$. Inset in (b) shows the photograph image of nanoBi$_4$O$_5$I$_2$.

MicroCuBi$_2$O$_4$

Subsequently, nanoBi$_4$O$_5$I$_2$ was used as a template for the preparation of micro-textured CuBi$_2$O$_4$. The CuBi$_2$O$_4$ photocathode was prepared by drop-casting of a precursor solution containing Cu^{2+} onto nanoBi$_4$O$_5$I$_2$ followed by heat treatments. The drop-casting amount (D) and the heat treatment temperature (T) have significant effects on the compositions and morphologies of prepared CuBi$_2$O$_4$ (Fig. 2). The CuBi$_2$O$_4$ retains the main structure of the nanoBi$_4$O$_5$I$_2$ template and has a porous nanosheet morphology if $T = 450$ °C, whereas the nanosheets start to merge at higher temperature and finally become a textured structure at $T = 550$ °C (Fig. 2a–c). Heat treatment temperature over $T = 600$ °C results in poor reproducibility of the morphologies and performance, and therefore the discussion at this condition is excluded.

GI-XRD analyses were applied to confirm the composition of the CuBi$_2$O$_4$. Strong peaks at 20.88°, 28.01°, and 46.7°, which can be assigned to lattice planes of tetragonal CuBi$_2$O$_4$ (JCPDS no. 48-1886), are observable if the heat treatment temperature is above 450 °C. However, Bi$_2$O$_{2.33}$ (JCPDS no. 27-0051) and CuO (JCPDS no. 48-1548) also exists in samples converted from nanoBi$_4$O$_5$I$_2$ if T is below 550 °C. During the heat treatment, nanoBi$_4$O$_5$I$_2$ is presumably firstly decomposed to I$_2$ and Bi$_2$O$_{2.33}$, and Bi$_2$O$_{2.33}$ reacts with CuO to form CuBi$_2$O$_4$.[11e] PEC characterization (Fig. S3, ESI†) indicates that higher annealing temperature is beneficial in enhancing the photocurrent. The presence of Bi$_2$O$_{2.33}$ and CuO impurity phases might act as recombination centres, resulting in the poor performance of CuBi$_2$O$_4$ samples with T less than 550 °C.

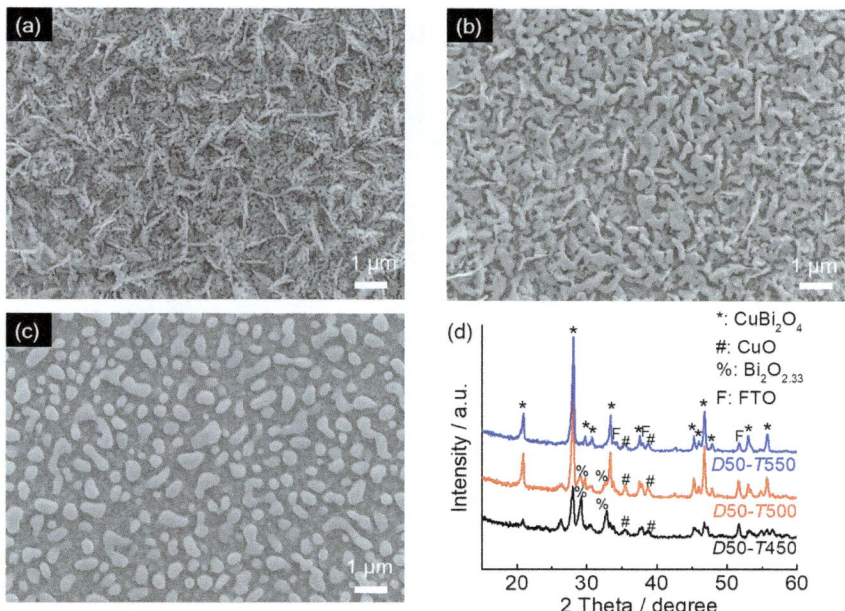

Fig. 2 SEM images of the $CuBi_2O_4$ converted from $nanoBi_4O_5I_2$ with $D = 50$ μL cm^{-2} and heating at (a) $T = 450$ °C, (b) $T = 500$ °C, and (c) $T = 550$ °C. (d) GI-XRD patterns of the $CuBi_2O_4$ converted from $nanoBi_4O_5I_2$ with $D = 50$ μL cm^{-2} at various T.

The second phase of $Bi_2O_{2.33}$ disappears at $T = 550$ °C if a sufficient amount of Cu^{2+} solution is applied to $nanoBi_4O_5I_2$ ($D \geqq 40$ μL cm^{-2}, Fig. S4, ESI†). However, additional peaks belonging to CuO are still observable if D exceeds 40 μL cm^{-2}. The amount of Cu^{2+} solution and heat treatment temperature are therefore optimised to 40 μL cm^{-2} and 550 °C, respectively, to obtain well-crystallized and relatively pure $CuBi_2O_4$ (Fig. 3a). The resultant $CuBi_2O_4$, designated as microCuBi$_2$O$_4$, exhibits a uniformly distributed micro-texture structure on the FTO substrate. UV-vis spectroscopy suggests that microCuBi$_2$O$_4$ absorbs light up to approximately 680 nm and the corresponding Tauc plot indicates that microCuBi$_2$O$_4$ has a direct band gap (E_g) of ~1.8 eV, which is in agreement with the reported value.[11a,11c,11e]

Mott–Schottky analyses were performed for gaining more insights into the electronic structure of microCuBi$_2$O$_4$. The negative slopes in the Mott–Schottky plot confirm the p-type feature of microCuBi$_2$O$_4$. Lines at two different frequencies of 500 Hz and 1 kHz show similar slopes and extrapolated x-axis intercepts, meeting the criteria of the independency on the frequency of Mott–Schottky analyses. The flat band potential therefore can be estimated from the extrapolated x-axis intercepts of approximately 1.4 V $vs.$ RHE, similar to the reported value.[11a,11c] The extremely positive flat band edge of $CuBi_2O_4$ is beneficial for coupling it with a photoanode in a tandem configuration for overall solar water splitting. For a heavily p-doped semiconductor, its valence band is close to its flat band edge,[8d] and the corresponding electronic level positions of microCuBi$_2$O$_4$ are then estimated based on Tauc plot and Mott–Schottky analyses and depicted as shown in the inset of Fig. 3d.

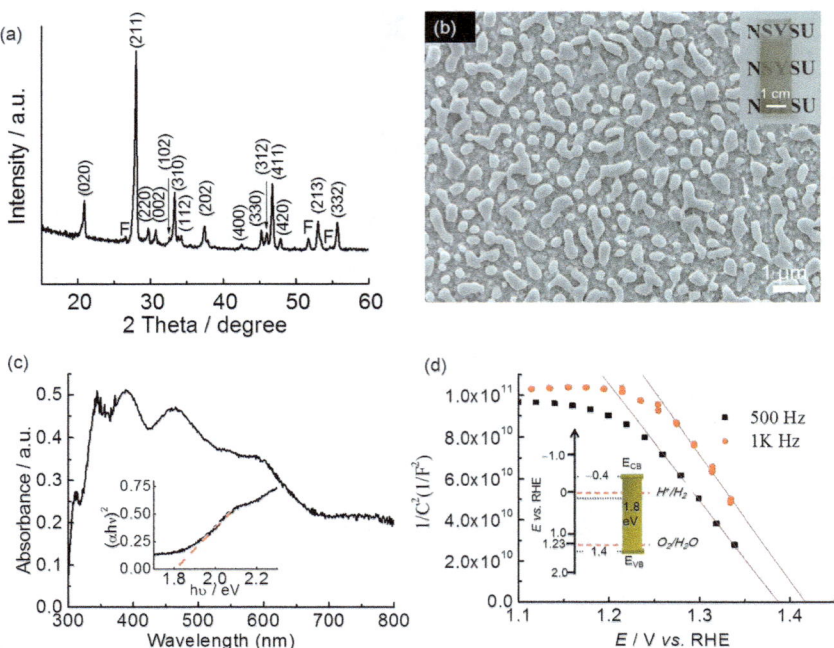

Fig. 3 The (a) GI-XRD pattern, (b) top-view SEM image, (c) UV-vis absorption spectrum, and (d) Mott–Schottky plot of microCuBi$_2$O$_4$. Mott–Schottky analyses were recorded in a 0.1 M B$_i$ buffer solution containing 0.5 M Na$_2$SO$_4$ (pH 9.2) under N$_2$ atmosphere. Insets in (b), (c), (d) are the photograph image, Tauc plot, and schematic band structure of microCuBi$_2$O$_4$, respectively.

The PEC performance of microCuBi$_2$O$_4$ was subsequently investigated in a three-electrode system in a 0.1 M B$_i$ buffer solution containing 0.5 M Na$_2$SO$_4$ supporting electrolyte at pH 9.2. MicroCuBi$_2$O$_4$ exhibits a very positive onset potential at approximately 1.0 V vs. RHE, which is close to its flat band potential of 1.4 V vs. RHE. The photocurrent response of 90 μA cm^{-2} was obtained at an applied potential of 0.3 V vs. RHE ((i) in Fig. 4a). A significant dark current is observable only if the potential is negative than 0.2 V vs. RHE ((i) in Fig. 4a). MicroCuBi$_2$O$_4$ shows a much better photocurrent response in an alkaline solution (Fig. S5, ESI†) with an approximate five-fold increase in photocurrent compared with that in the near neutral solution at 0.4 V vs. RHE. However, compatible operating conditions are necessary for both photocathode and photoanode in a tandem device. BiVO$_4$ likely dissolves in strong alkaline solutions and for this reason, a near-neutral B$_i$ buffer solution was chosen as the suitable electrolyte for the following experiments.

A theoretical absorbing photocurrent current (J_{abs}) of 10.7 mA cm^{-2} is reachable for microCuBi$_2$O$_4$ by assuming 100% absorbed photon to current conversion efficiency for photons with an energy larger than its corresponding E_g of 1.8 eV.[3a,18] However, the practical photocurrent (J_{PEC}) of microCuBi$_2$O$_4$ in (i) of Fig. 4a is much less than its respective theoretical value. In practice, J_{PEC} is determined by (i) the yield of the generated electrons reaching the electrode surface (ψ_{sep}), (ii) yield of the surface-reaching electrons that are transferred into the species in solution (ψ_{ox}) and (iii) J_{abs} by the following equation:

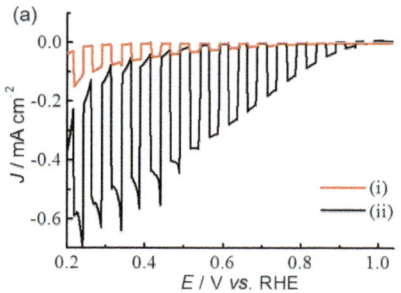
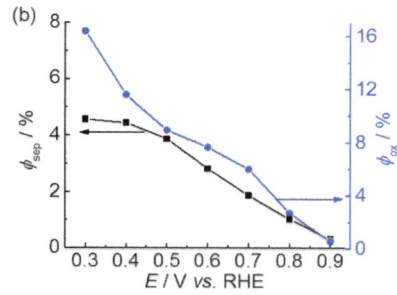

Fig. 4 (a) The current density (J)–potential (E) curve of microCuBi$_2$O$_4$ recorded in an aqueous 0.1 M B$_i$ buffer solution containing 0.5 M Na$_2$SO$_4$ (pH 9.2) under (i) N$_2$ and (ii) O$_2$ atmosphere under solar light irradiation (100 mW cm^{-2}, AM 1.5G) and (b) the corresponding ψ_{sep} and ψ_{ox} of microCuBi$_2$O$_4$.

$$J_{PEC} = J_{abs} \times \psi_{sep} \times \psi_{ox} \qquad (1)$$

To decouple the photon absorption and charge separation from the catalytic ability of the photoelectrodes towards hydrogen evolution, PEC analysis in a O$_2$ saturated solution was also carried out as photo-reduction of O$_2$ is much more kinetically favourable ((ii) in Fig. 4a). The ψ_{sep} of microCuBi$_2$O$_4$ is only approximately 3% at 0.6 V vs. RHE and reaches an optimum of 4.5% at 0.3 V vs. RHE by assuming ψ_{ox} is unity for photo-reduction of O$_2$ (Fig. 4b). On the other hand, the ψ_{ox} of microCuBi$_2$O$_4$ is only 7.6% and 16.4% at 0.6 V and 0.3 V vs. RHE, respectively, derived by dividing the photocurrent obtained in an N$_2$ saturated solution with that obtained in an O$_2$ saturated solution.

A cobalt based catalyst electrodeposited from a buffered solution containing cobalt ions has been identified as a Janus catalyst for hydrogen evolution and oxygen evolution reactions.[19] A layer of cobalt based catalysts was therefore further electrodeposited on microCuBi$_2$O$_4$ (microCuBi$_2$O$_4$|Co–B$_i$) from a B$_i$ solution containing Co^{2+} at 0.3 V vs. RHE under solar light irradiation (100 mW cm^{-2}, AM 1.5G) and a significant enhancement on the photocurrent is observed at potentials positive of 0.7 V vs. RHE (Fig. 5). The ψ_{ox} of microCuBi$_2$O$_4$ is increased from 2.7% to 64% at 0.8 V vs. RHE after modification of Co–B$_i$. However, microCuBi$_2$O$_4$|Co–B$_i$ exhibits a lower photocurrent than that of microCuBi$_2$O$_4$ at more negative potential regions (E < 0.7 V vs. RHE), which is possibly attributed to the pinch-off effect between the semiconductor and metal interface.[20] Platinum is the benchmark hydrogen evolution catalyst and was also deposited on microCuBi$_2$O$_4$ (microCuBi$_2$O$_4$|Pt) for comparison. The photocurrent of microCuBi$_2$O$_4$ is significantly enhanced by the modification with Pt in the whole potential window from 1.0 V vs. RHE to 0.25 V vs. RHE. However, microCuBi$_2$O$_4$|Pt exhibits a slightly lower photocurrent compared with that of microCuBi$_2$O$_4$|Co–B$_i$ at potentials positive of 0.7 V vs. RHE. MicroCuBi$_2$O$_4$|Co–B$_i$ serves as a better candidate in assembling an inexpensive and scalable tandem cell by preventing the use of expensive and scarce Pt. To analyse the PEC performance of microCuBi$_2$O$_4$|Co–B$_i$ in the bottom-absorber position in a tandem PEC cell, its photocurrent was also recorded with incident solar light filtered by a nanoBiVO$_4$|Co–B$_i$ electrode (Fig. S6, ESI†). MicroCuBi$_2$O$_4$|Co–B$_i$ still exhibits an

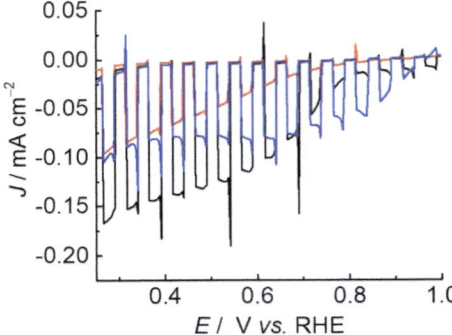

Fig. 5   The J–E curves of microCuBi$_2$O$_4$ (red trace), microCuBi$_2$O$_4$|Co–B$_i$ (blue trace), and microCuBi$_2$O$_4$|Pt (black trace) recorded in a 0.1 M B$_i$ buffer solution containing 0.5 M Na$_2$SO$_4$ (pH 9.2) under chopped solar-light illumination (100 mW cm^{-2}, AM 1.5G) under N$_2$ atmosphere.

onset potential close to 1.0 V *vs.* RHE in the bottom-absorber position in a tandem PEC cell. However, its photocurrent decreased from approximately 90 µA cm^{-2} to 30 µA cm^{-2} at 0.6 V *vs.* RHE.

NanoBiVO$_4$

BiVO$_4$ is a promising photoanode candidate in a dual-photoelectrode tandem PEC cell due to its narrow band gap and a favourable conduction band edge close to the thermodynamic potential of the hydrogen evolution reaction (HER). Since microCuBi$_2$O$_4$ is successfully prepared from the template of nanoBi$_4$O$_5$I$_2$ by a simple solution conversion method (see above), we were interested to see if this similar method can also be applied efficiently to the preparation of BiVO$_4$. NanoBiVO$_4$ was prepared by drop-casting 0.2 M VO(acac)$_2$ in DMSO and dried at 50 °C, followed by a 2 h thermal treatment at 450 °C. The as-prepared nanoBiVO$_4$ was further soaked in NaOH to remove excess V$_2$O$_5$. The amount of VO(acac)$_2$ solution was optimised to $D = 50$ µL cm^{-2} in respect to obtaining the highest PEC performance and reproducibility (Fig. S7, ESI†).

GI-XRD analysis confirms that nanoBiVO$_4$ has a pure monoclinic scheelite structure (JCPDS # 14-0688) without other crystalline impurity phases (Fig. 6a). NanoBiVO$_4$ exhibits a uniform coral-like structure on the FTO substrate and the morphologies are distinct from the sheet structures of the nanoBi$_4$O$_5$I$_2$ template and microCuBi$_2$O$_4$ (Fig. 6b). However, the discrete two-dimensional structure of nanoBi$_4$O$_5$I$_2$ serves as a scaffold to direct the growth of hierarchical three-dimensional nanostructures of BiVO$_4$ with sufficient voids between the crystals. UV-vis absorption spectroscopy and the corresponding Tauc plot suggest that nanoBiVO$_4$ has a bandgap of 2.5 eV and a J_{abs} of 3.7 mA cm^{-2}.

The electronic structure of nanoBiVO$_4$ was also deduced from the UV-vis spectroscopy and the Mott–Schottky analyses (Fig. 6c and d). The positive slopes in the Mott–Schottky plot confirm the n-type feature of nanoBiVO$_4$. NanoBiVO$_4$ shows different slopes at frequencies of 500 Hz and 1 kHz in Mott–Schottky analyses which might be attributed to the existence of surface states.[21] However, it exhibits similar extrapolated x-axis intercepts, and the flat band

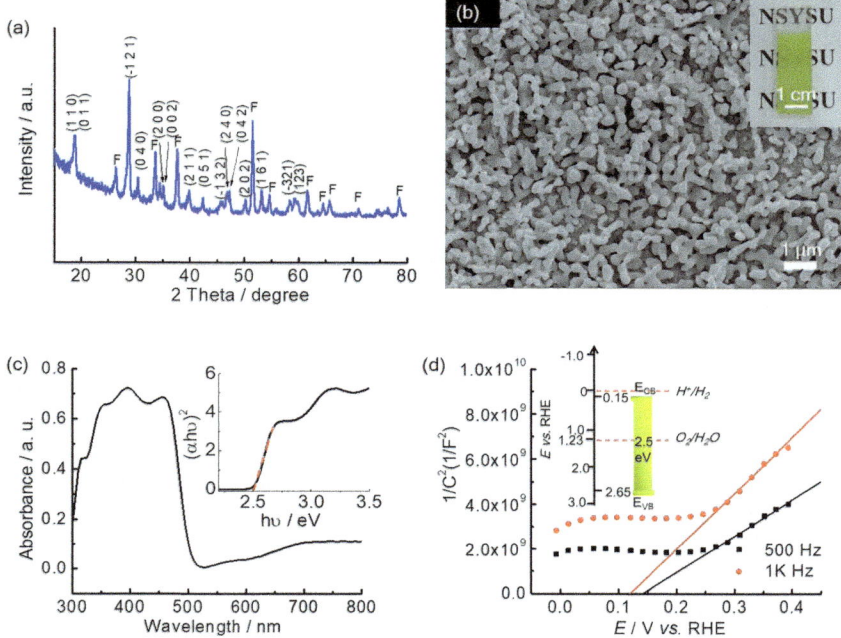

Fig. 6 The (a) GI-XRD pattern, (b) top-view SEM image, (c) UV–vis absorption spectrum, and (d) Mott–Schottky plot of nanoBiVO$_4$. The Mott–Schottky analyses were recorded in an aqueous B$_i$ buffer solution containing 0.5 M Na$_2$SO$_4$ (pH 9.2) under N$_2$ atmosphere. Insets in (b), (c), (d) are the photograph image, Tauc plot, and schematic band structure of nanoBiVO$_4$, respectively.

potential can be estimated to be approximately 0.15 V *vs.* RHE. For an n-type semiconductor, its flat band potential is close to its conduction band edge and therefore the respective valence band edge can be further determined to be approximately 2.65 V *vs.* RHE by considering the band gap of 2.5 eV (Fig. 6).

The PEC performance of nanoBiVO$_4$ was investigated in a three-electrode system in the pH 9.2 B$_i$ solution containing 0.5 M Na$_2$SO$_3$. Na$_2$SO$_3$ serves as a hole-scavenger allowing us to study the performance of nanoBiVO$_4$ in the elimination of surface recombination and to obtain ψ_{sep} by considering that ψ_{ox} is unity for sulfite oxidation. NanoBiVO$_4$ showed an onset photocurrent at approximately 0.2 V *vs.* RHE and reached approximately 1.7 mA cm^{-2} at 1.23 V *vs.* RHE (Fig. 7). The ψ_{sep} of nanoBiVO$_4$ is then estimated to be 46% at 1.23 V *vs.* RHE by eqn (1) with J_{PEC} and J_{abs} equal to 1.7 mA cm^{-2} and 3.7 mA cm^{-2}, respectively. The nanoBiVO$_4$ has a higher ψ_{sep} than that of planar BiVO$_4$, which usually has a ψ_{sep} among 20–30%.[22] This implies that the nanostructure helps the charge separation by decoupling the transport direction of electrons and holes. NanoBiVO$_4$ was further studied for water oxidation in a B$_i$ buffer solution containing 0.5 M Na$_2$SO$_4$ as supporting electrolyte. NanoBiVO$_4$ exhibits a similar onset potential at 0.2 V *vs.* RHE, but a much lower photocurrent of 0.49 mA cm^{-2} at 1.23 V *vs.* RHE than that in sulfite solution, corresponding to a ψ_{ox} of only 29% (Fig. 7) at 1.23 V *vs.* RHE. Co–B$_i$ is an effective water oxidation catalyst in neutral and near-neutral solutions and was electrodeposited on nanoBiVO$_4$ (nanoBiVO$_4$|Co–B$_i$) by the

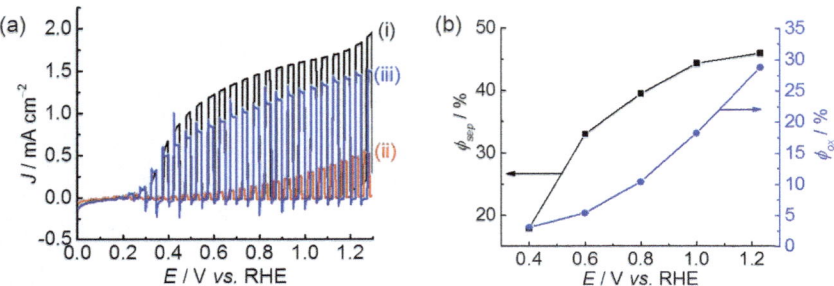

Fig. 7 (a) The J–E curves of nanoBiVO$_4$ for (i) sulfite oxidation and (ii) for water oxidation and (iii) nanoBiVO$_4$|Co–B$_i$ for water oxidation. All measurements were performed in an aqueous pH 9.2 electrolyte solution (0.1 M B$_i$ and 0.5 M Na$_2$SO$_4$ for water oxidation; 0.1 M B$_i$ and 0.5 M Na$_2$SO$_3$ for sulfite oxidation) under chopped irradiation (100 mW cm^{-2}, AM 1.5G). (b) The corresponding ψ_{sep} and ψ_{ox} of nanoBiVO$_4$.

identical method for the preparation of Co based HER catalyst on CuBi$_2$O$_4$. By interfacing with Co–B$_i$, the photocurrent increases to 1.48 mA cm^{-2} at 1.23 V vs. RHE and the corresponding ψ_{ox} increases to approximate 87% (Fig. 7a).

BiVO$_4$–CuBi$_2$O$_4$ tandem cell

A tandem PEC cell was further assembled by pairing nanoBiVO$_4$|Co–B$_i$ and microCuBi$_2$O$_4$|Co–B$_i$ (BiVO$_4$–CuBi$_2$O$_4$) for overall water splitting under solar-light irradiation and their photoelectrochemical performance was investigated in a two-chamber cell separated with a PEM. Solar light was firstly absorbed by nanoBiVO$_4$|Co–B$_i$ as nanoBiVO$_4$ has a larger band gap than that of microCuBi$_2$O$_4$, and the attenuated light was absorbed by the back cell of microCuBi$_2$O$_4$|Co–B$_i$. Both photoanode and photocathode were back-illuminated. The optical properties of nanoBiVO$_4$ did not show significant differences after depositing Co–B$_i$, although a slightly stronger absorbance was observed for Co–B$_i$ modified nanoBiVO$_4$ in the wavelength range of 400–460 nm (Fig. S8†). The opaque effect of Co–B$_i$ on the tandem cell performance can therefore be neglected.

Fig. 8a shows the photocurrent density responses of the tandem PEC cell at various applied biases from 0 to 1.23 V in an aqueous B$_i$ solution (0.1 M, pH 9.2) with Na$_2$SO$_4$ (0.5 M) as supporting electrolyte. A bias-free photocurrent of 36 μA cm^{-2} was observed, which is consistent with the prediction from the half-cell performances of nanoBiVO$_4$|Co–B$_i$ and microCuBi$_2$O$_4$|Co–B$_i$ (Fig. 8). MicroCuBi$_2$O$_4$|Co–B$_i$ has a much lower photocurrent than that of nanoBiVO$_4$|Co–B$_i$, and therefore, the tandem cell performance is limited by the microCuBi$_2$O$_4$|Co–B$_i$ photocathode. A photocurrent of approximately 0.11 mA cm^{-2} was achieved at an applied bias of 0.2 V for the tandem cell, which is in agreement with the half-cell PEC results by considering some losses in polarization due to membrane and solution resistances (Fig. S9, ESI†). EIS analyses were applied to quantify the increased resistance incurred in moving the working electrode to a tandem (two-electrode) system, and the results showed that an additional resistance of 55 ohm was imparted to the ohmic loss in a tandem cell device.

Applied bias solar-to-photocurrent conversion efficiency (ABPE) is a representative indication of the performance of an integrated PEC cell and is defined by the following equation:

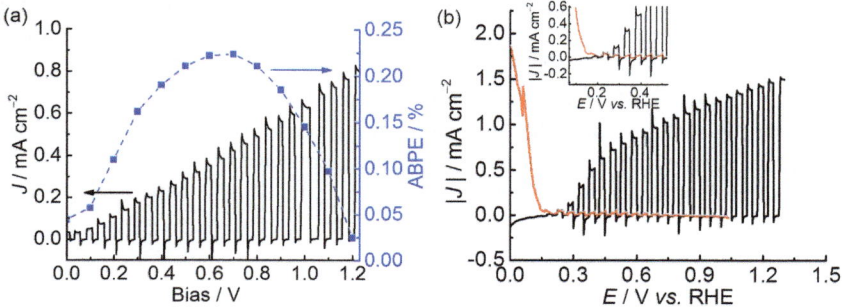

Fig. 8 (a) A linear sweep voltammetry scan and its corresponding ABPE in a two-electrode configuration of a BiVO$_4$–CuBi$_2$O$_4$ tandem cell under chopped solar light irradiation (100 mW cm^{-2}, AM 1.5G). (b) Overlaid $|J|$–E curves in a three electrode configuration of a BiVO$_4$|Co–B$_i$ electrode (black) under chopped solar light irradiation (100 mW cm^{-2}, AM 1.5G) and a microCuBi$_2$O$_4$|Co–B$_i$ electrode placed in the tandem cell position (illumination was filtered by nanoBiVO$_4$|Co–B$_i$, red). The magnified figure is shown in the inset.

$$ABPE = \left[\frac{|J| \times (1.23 - V)}{P_{in}} \right]_{AM\ 1.5G}, \qquad (2)$$

where $|J|$ is the photocurrent density (mA cm^{-2}), V is the applied bias (V) to the tandem cell, and P_{in} is the energy flux of the incident irradiation (mW cm^{-2}). Theoretical water-electrolysis voltage is 1.23 V, and therefore any V greater than 1.23 V is meaningless in a two-electrode PEC system.[23] A maximum ABPE of 0.22% was achieved at a bias of 0.7 V for the tandem PEC cell. However, microCuBi$_2$O$_4$ acts as a dark electrode rather than a photoelectrode in the tandem PEC cell when the external applied bias is higher than 0.2 V (Fig. S9, ESI†). The photocurrent obtained in the regions from 0.2 V to 1.23 V is, therefore, the result of the compromise of the dark current of microCuBi$_2$O$_4$|Co–Bi with the photocurrent of nanoBiVO$_4$|Co–B$_i$. The dark current of microCuBi$_2$O$_4$ is likely the result of the electrode corrosion, and as a result, ABPE obtained in the external applied biases of over 0.2 V is insubstantial for hydrogen generation.

Theoretically, a STH efficiency of approximately 8% is achievable by pairing a back cell having a band gap of 1.8 eV with a front cell having a band gap of 2.5 eV in a stacked configuration.[13] The practical STH efficiency of the BiVO$_4$–CuBi$_2$O$_4$ tandem cell can be calculated by the following equation:

$$STH = \left[\frac{|J_{sc}| \times 1.23 \times FE}{P_{in}} \right]_{AM\ 1.5G} \qquad (3)$$

where $|J_{sc}|$ is the photocurrent density measured at zero applied bias (mA cm^{-2}) and FE is the faradaic efficiency of hydrogen evolution. A STH efficiency of 0.04% is achievable by assuming the FE is unity. However, the STH efficiency of the BiVO$_4$–CuBi$_2$O$_4$ tandem cell is significantly lower than its theoretical STH efficiency. The major limitation is ascribed to the relatively poor performance of the microCuBi$_2$O$_4$|Co–B$_i$ photocathode. Poor ψ_{sep} and ψ_{ox} are the main hurdles in achieving high half-cell performance of microCuBi$_2$O$_4$. Although Co–B$_i$ partially alleviated the poor performance of ψ_{ox} in CuBi$_2$O$_4$, it also results in a pinch-off effect which is deleterious to the photocurrents of CuBi$_2$O$_4$ at more negative potentials $(E < 0.7 V)$. Searching for other inexpensive catalysts that do not level off the photocurrent of $CuBi_2O_4$ will be beneficial for enhancing the operating photocurrent as well as the STH efficiency of the tandem PEC cell. Compared with nanoBiVO$_4$, microCuBi$_2$O$_4$ has more than one order less ψ_{sep} (46% vs. 3%). Tailoring the structure of $CuBi_2O_4$ from microstructure to nanostructure by fine-tuning the preparation conditions might further enhance the ψ_{sep} of $CuBi_2O_4$ by decreasing its bulk recombination.

4   Conclusions

Developing a universal method for facile preparation of both photoanode and photocathode is beneficial for accelerating the research on dual-photoelectrode tandem PEC cells. In this study, a facile and straightforward synthetic approach has been successfully developed for the preparation of micro-textured $CuBi_2O_4$ and nanocoral $BiVO_4$ photoelectrodes from one nanoBi$_4$O$_5$I$_2$ template. The synthesized $CuBi_2O_4$ photocathode and $BiVO_4$ photoanode exhibited an overall photovoltage of approximately 2.0 V, which is beneficial in realizing water splitting without external bias. The interfacial charge transfer efficiency for both photoelectrodes was further improved by surface-modification with a Co–B$_i$ Janus electrocatalyst via photo-assisted deposition. A proof-of-concept tandem PEC cell consisting of a $CuBi_2O_4|$Co–B$_i$ photocathode and a $BiVO_4|$Co–B$_i$ photoanode was finally assembled and generated a bias-free photocurrent of 36 $\mu A\ cm^{-2}$, corresponding to a STH efficiency of 0.04% by assuming the FE is unity. Further improvements on the half-cell performance with respect to the charge separation and charge transfer yields of the $CuBi_2O_4$ photocathode is indispensable for fabricating an efficient $BiVO_4$–$CuBi_2O_4$ tandem PEC cell.

Conflicts of interest

There are no conflicts to declare.

Acknowledgements

Financial support from Ministry of Science and Technology, Taiwan (MOST 105-2218-E-110-004-MY3 and MOST 107-2221-E-110-009) is gratefully acknowledged. We also thank Mr Yu-Ju Lai and Mr Zheng-Lu Liu for their help in GI-XRD and SEM analyses.

References

1 K. T. Fountaine, H. J. Lewerenz and H. A. Atwater, Nat. Commun., 2016, 7, 13706.
2 B. Parkinson and J. Turner, Photoelectrochemical Water Splitting: Materials, Processes and Architectures, The Royal Society of Chemistry, 2013, pp. 1–18.
3 (a) T. W. Kim and K.-S. Choi, Science, 2014, 343, 990–994; (b) B. Lamm, B. J. Trzesniewski, H. Doscher, W. A. Smith and M. Stefik, ACS Energy Lett., 2018, 3, 112–124; (c) H. S. Han, S. Shin, D. H. Kim, I. J. Park, J. S. Kim, P. S. Huang, J. K. Lee, I. S. Cho and X. L. Zheng, Energy Environ. Sci., 2018, 11, 1299–1306.

4 (a) D. K. Lee and K.-S. Choi, *Nat. Energy*, 2018, **3**, 53–60; (b) T. W. Kim and K.-S. Choi, *J. Phys. Chem. Lett.*, 2016, **7**, 447–451.

5 Y. Kuang, Q. Jia, G. Ma, T. Hisatomi, T. Minegishi, H. Nishiyama, M. Nakabayashi, N. Shibata, T. Yamada, A. Kudo and K. Domen, *Nat. Energy*, 2016, **2**, 16191.

6 (a) C.-Y. Lin, Y.-H. Lai, D. Mersch and E. Reisner, *Chem. Sci.*, 2012, **3**, 3482–3487; (b) J. S. Luo, L. Steier, M. K. Son, M. Schreier, M. T. Mayer and M. Grätzel, *Nano Lett.*, 2016, **16**, 1848–1857; (c) W. Z. Niu, T. Moehl, W. Cui, R. Wick-Joliat, L. P. Zhu and S. D. Tilley, *Adv. Energy Mater.*, 2018, **8**, 1702323.

7 (a) A. G. Tamirat, J. Rick, A. A. Dubale, W. N. Su and B. J. Hwang, *Nanoscale Horiz.*, 2016, **1**, 243–267; (b) P. Dias, L. Andrade and A. Mendes, *Nano Energy*, 2017, **38**, 218–231; (c) C. Li, Z. Luo, T. Wang and J. Gong, *Adv. Mater.*, 2018, **30**, 1707502.

8 (a) H. Kim, S. Bae, D. Jeon and J. Ryu, *Green Chem.*, 2018, **20**, 3732–3742; (b) Y.-H. Lai, D. W. Palm and E. Reisner, *Adv. Energy Mater.*, 2015, **5**, 1501668; (c) V. Andrei, R. L. Z. Hoye, M. Crespo-Quesada, M. Bajada, S. Ahmad, M. De Volder, R. Friend and E. Reisner, *Adv. Energy Mater.*, 2018, **8**, 1801403; (d) J. T. Li, M. Griep, Y. S. Choi and D. Chu, *Chem. Commun.*, 2018, **54**, 3331–3334; (e) J.-W. Jang, C. Du, Y. Ye, Y. Lin, X. Yao, J. Thorne, E. Liu, G. McMahon, J. Zhu, A. Javey, J. Guo and D. Wang, *Nat. Commun.*, 2015, **6**, 7447.

9 H. Kobayashi, N. Sato, M. Orita, Y. Kuang, H. Kaneko, T. Minegishi, T. Yamada and K. Domen, *Energy Environ. Sci.*, 2018, **11**, 3003–3009.

10 J. H. Kim, A. Adishev, J. Kim, Y. S. Kim, S. Cho and J. S. Lee, *ACS Appl. Energy Mater.*, 2018, **1**, 6694.

11 (a) S. P. Berglund, F. F. Abdi, P. Bogdanoff, A. Chernseddine, D. Friedrich and R. van de Krol, *Chem. Mater.*, 2016, **28**, 4231–4242; (b) D. W. Cao, N. Nasori, Z. J. Wang, Y. Mi, L. Y. Wen, Y. Yang, S. C. Qu, Z. G. Wang and Y. Lei, *J. Mater. Chem. A*, 2016, **4**, 8995–9001; (c) D. Kang, J. C. Hill, Y. Park and K.-S. Choi, *Chem. Mater.*, 2016, **28**, 4331–4340; (d) H. S. Park, C. Y. Lee and E. Reisner, *Phys. Chem. Chem. Phys.*, 2014, **16**, 22462–22465; (e) N. T. Hahn, V. C. Holmberg, B. A. Korgel and C. B. Mullins, *J. Phys. Chem. C*, 2012, **116**, 6459–6466.

12 Z. B. Chen, T. F. Jaramillo, T. G. Deutsch, A. Kleiman-Shwarsctein, A. J. Forman, N. Gaillard, R. Garland, K. Takanabe, C. Heske, M. Sunkara, E. W. McFarland, K. Domen, E. L. Miller, J. A. Turner and H. N. Dinh, *J. Mater. Res.*, 2010, **25**, 3–16.

13 M. S. Prevot and K. Sivula, *J. Phys. Chem. C*, 2013, **117**, 17879–17893.

14 J. Xiong, G. Cheng, F. Qin, R. Wang, H. Sun and R. Chen, *Chem. Eng. J.*, 2013, **220**, 228.

15 Y. Wang, Y. Long and D. Zhang, *ACS Sustainable Chem. Eng.*, 2017, **5**, 2454–2462.

16 (a) X. L. Jin, L. Q. Ye, H. Q. Xie and G. Chen, *Coord. Chem. Rev.*, 2017, **349**, 84–101; (b) X. Xiao, C. L. Xing, G. P. He, X. X. Zuo, J. M. Nan and L. S. Wang, *Appl. Catal., B*, 2014, **148**, 154–163.

17 C.-H. Wu, E. Onno and C.-Y. Lin, *Electrochim. Acta*, 2017, **229**, 129–140.

18 J. Huang, Y. Zhang and Y. Ding, *ACS Catal.*, 2017, **7**, 1841–1845.

19 S. Cobo, J. Heidkamp, P.-A. Jacques, J. Fize, V. Fourmond, L. Guetaz, B. Jousselme, V. Ivanova, H. Dau, S. Palacin, M. Fontecave and V. Artero, *Nat. Mater.*, 2012, **11**, 802.

20 R. T. Tung, *Appl. Phys. Rev.*, 2014, **1**, 011304.
21 G. V. Govindaraju, G. P. Wheeler, D. Lee and K.-S. Choi, *Chem. Mater.*, 2017, **29**, 355–370.
22 D. K. Zhong, S. Choi and D. R. Gamelin, *J. Am. Chem. Soc.*, 2011, **133**, 18370–18377.
23 (*a*) M. G. Walter, E. L. Warren, J. R. McKone, S. W. Boettcher, Q. X. Mi, E. A. Santori and N. S. Lewis, *Chem. Rev.*, 2010, **110**, 6446–6473; (*b*) T. Hisatomi, J. Kubota and K. Domen, *Chem. Soc. Rev.*, 2014, **43**, 7520–7535.

Faraday Discussions

PAPER

Z-scheme photocatalyst systems employing Rh- and Ir-doped metal oxide materials for water splitting under visible light irradiation†

Akihiko Kudo, [ID] *[ab] Shunya Yoshino,[a] Taichi Tsuchiya,[a]
Yuhei Udagawa,[a] Yukihiro Takahashi,[a] Masaharu Yamaguchi,[a]
Ikue Ogasawara,[a] Hiroe Matsumoto[a] and Akihide Iwase [ID] [ab]

Received 27th November 2018, Accepted 24th January 2019

DOI: 10.1039/c8fd00209f

Various types of Z-scheme systems for water splitting under visible light irradiation were successfully developed by employing Rh- and Ir-doped metal oxide powdered materials with relatively narrow energy gaps (EG): $BaTa_2O_6$:Ir,La (EG: 1.9–2.0 eV), $NaTaO_3$:Ir,La (EG: 2.1–2.3 eV), $SrTiO_3$:Ir (EG: 1.6–1.8 eV), $NaNbO_3$:Rh,Ba (EG: 2.5 eV) and TiO_2:Rh,Sb (EG: 2.1 eV), with conventional $SrTiO_3$:Rh (an H_2-evolving photocatalyst) or $BiVO_4$ (an O_2-evolving photocatalyst), and suitable electron mediators. The Z-scheme systems were classified into three groups depending on the combination of H_2- and O_2-evolving photocatalysts and electron mediator. The Z-scheme systems combining $BaTa_2O_6$:Ir,La with $BiVO_4$, and $NaTaO_3$:Ir,La with $BiVO_4$ were active when a $[Co(bpy)_3]^{3+/2+}$ redox couple was used rather than an $Fe^{3+/2+}$ one. The combination of $SrTiO_3$:Ir with $SrTiO_3$:Rh gave an activity when the $[Co(bpy)_3]^{3+/2+}$ and $Fe^{3+/2+}$ redox couple ionic mediators were used. The Z-scheme systems combining $NaNbO_3$:Rh,Ba and TiO_2:Rh,Sb with $SrTiO_3$:Rh showed activities by using the $[Co(bpy)_3]^{3+/2+}$ and $Fe^{3+/2+}$ redox couples and also *via* interparticle electron transfer by just contact with/without reduced graphene oxide (RGO). These suitable combinations can be explained based on the impurity levels of doped Rh^{3+} and Ir^{3+} toward the redox potentials of the ionic mediators for the Z-scheme systems employing ionic mediators, and p-/n-type and onset potentials of the photocurrent in the photoelectrochemical properties of those photocatalyst materials for the Z-scheme systems working *via* interparticle electron transfer.

[a]Department of Applied Chemistry, Faculty of Science, Tokyo University of Science, 1-3 Kagurazaka, Shinjuku-ku, Tokyo 162-8601, Japan. E-mail: a-kudo@rs.kagu.tus.ac.jp

[b]Photocatalysis International Research Center, Research Institute for Science and Technology, Tokyo University of Science, 2641, Yamazaki, Noda-shi, Chiba-ken 278-8510, Japan

† Electronic supplementary information (ESI) available. See DOI: 10.1039/c8fd00209f

Introduction

Artificial photosynthesis has attracted attention from the view point of solar energy conversion to storable chemical energy. Because solar water splitting is one of the representative reactions, photocatalytic water splitting has extensively been studied.[1-7] Powder-based metal oxide materials are attractive for photocatalytic water splitting because the cost will be low[8-10] and stability will be high compared with other materials such as chalcogenides. Although the efficiency of powder-based photocatalyst systems is behind that of systems of photovoltaic + electrolysis at the present stage, there are some advantages to powder-based photocatalyst systems.

It is crucial to demonstrate a solar water splitting system employing powder-based oxide photocatalysts. The present stage of research of artificial synthesis is along this topic. A reactor system including a gas separation system of evolved hydrogen from oxygen has been studied in addition to the development of photocatalyst materials aimed towards the practical use of solar water splitting.[10] The separation system can be achieved by the use of a suitable separation membrane. Moreover, the safety issue of the co-evolved H_2 and O_2 has also been examined *via* the use of a suitable gas transportation tube. However, even if an excellent reactor system with a gas separation membrane system is established, which photocatalyst is employed for it is still a key issue as high efficient photocatalysts for real solar water splitting into H_2 and O_2 without any sacrificial electron donors and acceptors have not yet been developed. It is essential to develop an efficient photocatalyst for demonstrating a solar water splitting system taking gas separation and safety issues into account.

High efficiency of a photocatalyst can be brought about by a high quantum yield and a response to light with a long wavelength. From this viewpoint, it is vital to develop photocatalysts that can utilize up to 600–700 nm of the solar spectrum. Although the use of noble metals might prevent the photocatalyst materials from practical use, the usage would be allowed if the amount is small and the materials can be recycled.

There are single particulate and Z-scheme systems in powder-based photocatalysts.[2,11] Z-scheme photocatalyst systems have the advantage that photocatalysts active for either H_2 or O_2 evolution can be employed. This means that a Z-scheme system can be constructed with various combinations of H_2- and O_2-evolving photocatalysts.[12-15] $SrTiO_3$:Rh (an H_2-evolving photocatalyst) and $BiVO_4$ (an O_2-evolving photocatalyst) are representative metal oxide materials for the Z-scheme system.[1,2,16] It should be stressed that photocatalyst sheets prepared by a particle transfer method using $SrTiO_3$:Rh,La and $BiVO_4$:Mo powders demonstrate a quantum efficiency of 30% at 420 nm and a solar to hydrogen energy conversion efficiency of 1%.[17-19] This result suggests that the Z-scheme system is a promising photocatalyst system for practical solar water splitting. H_2 evolution separated from O_2 evolution is possible in the Z-scheme system if a suitable reactor is designed.[8] However, this photocatalyst sheet consisting of the $SrTiO_3$:Rh,La and $BiVO_4$:Mo powders responds up to 520 nm because the energy gap (EG) of $SrTiO_3$:Rh,La and the band gap (BG) of $BiVO_4$:Mo are 2.3 eV and 2.4 eV, respectively. So, it is a key issue to develop Z-scheme photocatalyst systems consisting of metal oxide photocatalysts with a response at longer wavelengths than 520 nm.

Transition metal doping is one strategy to make wide band gap photocatalysts responsive to visible light.[1,2] Rh and Ir are effective dopants in this strategy. We have reported that SrTiO$_3$:Rh (EG: 2.3 eV),[20] SrTiO$_3$:Rh,Sb (EG: 2.2–2.4 eV),[21] SrTiO$_3$:Ir (EG: 1.6–1.8 eV),[20,22] BaTa$_2$O$_6$:Ir,La (EG: 1.9–2.0 eV)[23] and NaTaO$_3$:Ir,La (EG: 2.1–2.3 eV)[24] are active for sacrificial H$_2$ evolution in the presence of electron donors such as methanol, while SrTiO$_3$:Rh,Sb,[21] SrTiO$_3$:Ir[20] and TiO$_2$:Rh,Sb (EG: 2.1 eV)[25] are active for sacrificial O$_2$ evolution in the presence of electron acceptors such as Ag$^+$. In these photocatalyst materials, the visible light responses are due to electronic transition from the impurity levels formed by the dopants to the conduction bands of the host materials. These energy and band gaps are close to, or narrower than, those of SrTiO$_3$:Rh (EG: 2.3 eV)[20] and BiVO$_4$ (BG: 2.4 eV).[26,27] So, it is attractive to utilize these Rh- and Ir-doped metal oxide photocatalysts for the construction of Z-scheme systems.

There are several types of Z-scheme system employing different electron mediators such as Fe$^{3+/2+}$ and [Co(bpy)$_3$]$^{3+/2+}$ redox couples,[16,28,29] and reduced graphene oxide (RGO)[30–32] as shown in Fig. 1(a) and (b). Moreover, some Z-scheme systems work even without electron mediators, as shown in Fig. 1(c).[33,34] In this case, electron transfer proceeds through an interface contacted between the particles of the H$_2$- and O$_2$-evolving photocatalysts. So, it is important to examine the electron mediator in the Z-scheme system.

In the present paper, sacrificial H$_2$ and O$_2$ evolutions over Rh- and Ir-doped metal oxide photocatalysts under visible light irradiation were examined first to see the relationship between the photocatalytic properties and their band structures. Then, these H$_2$- and O$_2$-evolving photocatalysts were employed for various types of Z-scheme system for water splitting into H$_2$ and O$_2$ in stoichiometric amounts under visible light irradiation, without any sacrificial reagents. The photocatalytic performances for the sacrificial H$_2$ and O$_2$ evolutions and the Z-schematic water splitting were discussed based on the band structures and

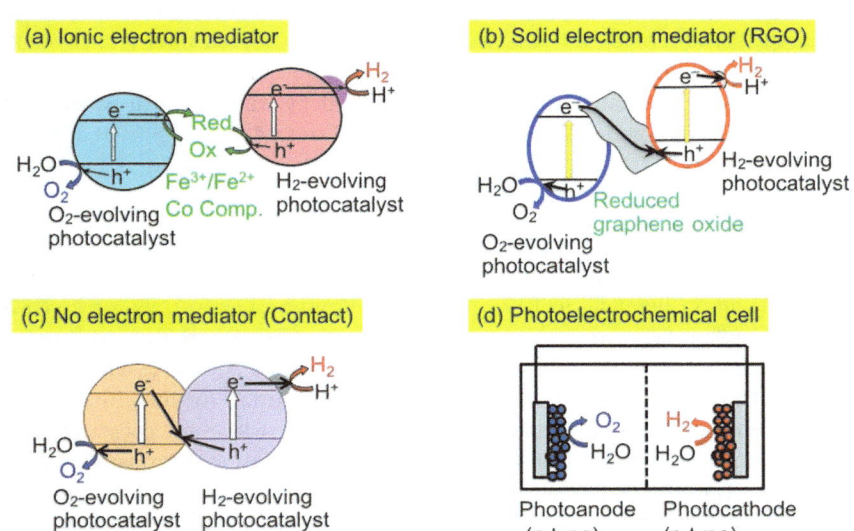

Fig. 1   Various types of powder material-based Z-scheme systems for water splitting.

photoelectrochemical properties of the photocatalyst materials such as p-/n-types and the onset potentials of photocurrents.

Experimental

Preparation of Rh- or Ir-doped metal oxide photocatalysts

$SrTiO_3$:Rh(1%),[29] $SrTiO_3$:Ir(0.2%),[22] TiO_2:Rh(x%),Sb($2x$%) (x = 0.5 or 1.3),[25] $NaTaO_3$:Ir(1%),La(2%),[24] $BaTa_2O_6$:Ir(1%),La(2%)[23] and $BiVO_4$ [26,27] were prepared by a solid-state reaction, a borate-flux method, and a liquid–solid reaction according to previous reports. In addition to them, $NaNbO_3$:Rh(x%),Ba(y%) (x, y) = (1.2, 1.44) or (1.0, 2.0) was newly prepared by a solid-state reaction. The starting materials, Na_2CO_3 (Kanto Chemical; 99.5 or 99.8%), Nb_2O_5 (Kanto Chemical; 99.99% or Kojundo Chemical; 99.99%), Rh_2O_3 (Wako Chemical; 98%), and $BaCO_3$ (Kanto Chemical; 99%), were mixed at a molar ratio of Na/Nb/Rh/Ba = 1.05–1.05y : 1 − x : x : y. An excess of sodium was added in the starting materials to compensate for volatilization. The starting materials mixture was calcined at 1173 K for 1 h, and then 1423–1473 K for 10 h once or twice. The excess sodium was washed out with water after the calcination. The obtained powders had nonspecific shapes with aggregations, judging from the SEM images (Jeol; JSM-6700F) (Fig. S1†). The obtained samples were identified using X-ray diffraction (Rigaku; MiniFlex, Cu Kα). Diffuse reflectance spectra were obtained by a UV-vis-NIR spectrometer (JASCO, V-570) equipped with an integrator sphere and were converted to absorbance measurements *via* the Kubelka–Munk method.

Preparation of an RGO–metal oxide composite

An RGO-incorporated O_2-evolving photocatalyst was prepared by photocatalytic reduction of graphene oxide (GO) on the photocatalyst according to previous reports.[30–32] GO prepared by the Hummers' method[35] and the O_2-evolving photocatalyst were dispersed in an aqueous methanol (Kanto Chemical; 99.8%) solution (50 vol%). The suspension was irradiated with visible light from a 300 W Xe lamp (PerkinElmer; CERMAX PE300BF) with a long pass filter (HOYA; L42) under a N_2 atmosphere with a pressure of 1 atm to obtain the RGO–photocatalyst composite. The methanol was carefully removed by washing with water. The RGO–photocatalyst composite was collected by filtration and was dried at room temperature in air.

Sacrificial H_2 and O_2 evolutions (half reactions of water splitting)

H_2 and O_2 evolutions from aqueous solutions containing the sacrificial reagents CH_3OH (Kanto Chemical; 99.8%) and $AgNO_3$ (Kojima Chemical; 99.9% or Toyo Chemical; 99.9%) that were half reactions of water splitting were examined using a top-irradiation reaction cell with a Pyrex window and a 300 W Xe lamp (PerkinElmer; CERMAX PE300BF). $NaTaO_3$:Ir,La, $BaTa_2O_6$:Ir,La, $NaNbO_3$:Rh,Ba, and TiO_2:Rh,Sb were used as prepared, whereas $SrTiO_3$:Ir without a cocatalyst was reduced under 1 atm of H_2 at 473 K for 2 h as a pretreatment for sacrificial O_2 evolution. The photocatalyst powders (0.1–0.3 g) were suspended in aqueous solutions (120–150 mL) and irradiated with visible light. For the H_2 evolution, Pt (0.3 wt%) cocatalyst, working as an H_2 evolution site, was loaded on photocatalysts by photodeposition from an aqueous methanol solution containing

H_2PtCl_6 (Tanaka Kikinzoku; 37.55% as Pt). The wavelength of the irradiation light was controlled to visible light using long-pass filters (HOYA; L42 and Y44). The amounts of evolved H_2 and O_2 were determined using an online gas chromatograph (Shimadzu; GC-8A, MS-5A column, TCD, Ar carrier).

Z-schematic water splitting

Z-schematic water splitting was conducted using a gas-closed system with a top-irradiation cell with a Pyrex window. H_2-evolving photocatalyst and O_2-evolving photocatalyst powders (0.05 or 0.1 g, respectively) were suspended in 120 mL of water. For the interparticle Z-scheme systems without an electron mediator and with an RGO solid-state electron mediator, water not containing any ionic mediators was used. For the Z-scheme system with ionic mediator, an aqueous solution containing $[Co(bpy)_3]SO_4$ or $FeCl_3$ as a mediator was used. The pH was adjusted with H_2SO_4 in each of the Z-scheme systems with and without electron mediator, if necessary. Ru (0.7 wt%) cocatalyst, functioning as an H_2 evolution site, was loaded on $SrTiO_3$:Rh (an H_2-evolving photocatalyst) by photodeposition from an aqueous methanol solution containing $RuCl_3 \cdot nH_2O$ (Tanaka Kikinzoku; 36% as Ru in $RuCl_3 \cdot nH_2O$). Pt (0.3–1 wt%) was loaded on the $NaTaO_3$:Ir,La and $BaTa_2O_6$:Ir,La (H_2-evolving photocatalysts), and $SrTiO_3$:Ir (an O_2-evolving photocatalyst) by an impregnation method. The photocatalyst powders and an aqueous H_2PtCl_6 solution were placed in a porcelain crucible and dried on a hot plate. The H_2PtCl_6-impregnated $NaTaO_3$:Ir,La and $BaTa_2O_6$:Ir,La powders were calcined at 673 K for 2 h in air, whereas the $SrTiO_3$:Ir was not. The Pt-loaded $NaTaO_3$:Ir,La and Pt-loaded $BaTa_2O_6$:Ir,La were subsequently reduced at 673 K for 1 h under 1 atm of H_2 as a pretreatment, while Pt-loaded $SrTiO_3$:Ir was reduced at 573 K for 1 h. The light source and GC setup were the same as those for the sacrificial H_2 and O_2 evolutions.

Photoelectrochemical measurements

A squeegee method was used to prepare the $SrTiO_3$:Rh(1%) photoelectrode and a drop-casting method was used for the $NaTaO_3$:Ir(1%),La(2%) and $BaTa_2O_6$:-Ir(1%),La(2%) with and without H_2-reduction; $SrTiO_3$:Ir(0.2%) with H_2-reduction; $NaNbO_3$:Rh(1%),Ba(2%) and TiO_2:Rh(0.5%),Sb(1%) without H_2-reduction, and $BiVO_4$ photoelectrodes using powdered photocatalyst materials. For the $SrTiO_3$:-Rh(1%) photoelectrode, a paste consisting of 20 mg of $SrTiO_3$:Rh(1%) photo-catalyst powder, 20 μL of acetylacetone (Kanto Chemical; 99.5%) and 40 μL of distilled water was coated on an indium tin oxide transparent electrode (ITO).[36] For the other photoelectrodes, the photocatalyst powders were dispersed in ethanol (1–2 mg mL^{-1}) by sonication. The suspensions were drop-cast onto a fluorine-doped tin oxide transparent electrode (FTO) to obtain 1–2 mg cm^{-2} of photocatalyst on the FTO. The H_2-reduced $SrTiO_3$:Ir-loaded FTO substrate was not calcined, whereas the other photocatalyst-loaded ITO and FTO substrates were calcined at 573–673 K for 2 h in air. The photoelectrochemical properties were evaluated with a three-electrode system consisting of working, Ag/AgCl reference, and Pt counter electrodes with a potentiostat (Hokuto Denko; HZ-series or HSV-110) using a conventional H-type cell with a Nafion membrane. The electrolyte was 0.1 mol L^{-1} K_2SO_4. 0.025 mol L^{-1} KH_2PO_4 + 0.025 mol L^{-1} Na_2HPO_4 pH buffer was added, if necessary. A 300 W Xe lamp (PerkinElmer; CERMAX PE300BF) with a long-pass filter (HOYA; L42) was employed as a light source.

Results

Photocatalytic activities for sacrificial H_2 and O_2 evolutions over Rh- and Ir-doped metal oxide materials and their band structures

Sacrificial H_2 and O_2 evolutions of half reactions were carried out as test reactions of water splitting over Rh- and Ir-doped metal oxide photocatalysts using a sacrificial electron donor and acceptor to see the ability for photocatalytic H_2 or O_2 evolution, as shown in Table 1, prior to conducting water splitting.

The energy gaps were determined from the diffuse reflectance spectra and wavelength dependence of the photocatalytic activities as shown in Fig. 2 and 3. In general, trivalent and tetravalent Rh and Ir species are doped at the Ti^{4+}, Nb^{5+} and Ta^{5+} sites in metal oxide materials.[20–25,37,38] Among the species, Rh^{3+} and Ir^{3+} contribute to the visible light response for metal oxide photocatalysts.[20,24,37,38] Sb^{5+} was codoped with Rh^{3+} at the Ti^{4+} and Nb^{5+} sites for charge compensation to enhance the formation of Rh^{3+} and Ir^{3+} and suppress the formation of Rh^{4+} and Ir^{4+} as efficient recombination centers between photogenerated electrons and holes, while Ba^{2+} and La^{3+} were replaced at the alkali and alkaline earth metal sites for the same purpose. $SrTiO_3$:Ir was reduced with H_2 at 473 K to form the dopant Ir^{3+}.

In the case of the Ir-doped photocatalysts, $SrTiO_3$:Ir was active for both the sacrificial H_2 and O_2 evolutions,[20,22] whereas $BaTa_2O_6$:Ir,La and $NaTaO_3$:Ir,La were active only for the sacrificial H_2 evolution.[23,24] $BaTa_2O_6$:Ir,La and $NaTaO_3$:Ir,La were not active for sacrificial O_2 evolution even if they were reduced with H_2 at 673 K, as $SrTiO_3$:Ir was, to reduce the Ir^{4+} species. The Rh-doped photocatalysts

Table 1 Sacrificial H_2 and O_2 evolutions under visible light irradiation over Rh- or Ir-doped metal oxide photocatalysts[a]

Photocatalyst	Energy gap/eV	Incident light/nm	Activity/μmol h^{-1} H_2[b]	O_2[c]	Ref.
Pt/NaTaO$_3$:Ir(1%),La(2%)	2.1–2.3	$\lambda > 420$	3.2	—	24
NaTaO$_3$:Ir(1%),La(2%)	2.1–2.3	$\lambda > 420$	—	0	24
Pt/BaTa$_2$O$_6$:Ir(1%),La(2%)	1.9–2.0	$\lambda > 420$	4.6	—	23
BaTa$_2$O$_6$:Ir(1%),La(2%)	1.9–2.0	$\lambda > 420$	—	0	23
Pt/SrTiO$_3$:Ir(0.2%)	1.6–1.8	$\lambda > 440$	8.6	—	20
SrTiO$_3$:Ir(0.2%)[d]	1.6–1.8	$\lambda > 420$	—	4.3	20
Pt/NaNbO$_3$:Rh(1%),Ba(2%)	2.5	$\lambda > 420$	0.2	—	This work
NaNbO$_3$:Rh(1%),Ba(2%)	2.5	$\lambda > 420$	—	5.3	This work
Pt/TiO$_2$:Rh(0.5%),Sb(1%)	2.1	$\lambda > 440$	0	—	This work
TiO$_2$:Rh(0.5%),Sb(1%)	2.1	$\lambda > 440$	—	7.5	25

[a] Photocatalyst: 0.1–0.3 g; light source: 300 W Xe lamp with long-pass filters ($\lambda > 420$ nm or $\lambda > 440$ nm); reaction cell: top-irradiation cell with a Pyrex window. [b] Cocatalyst: Pt (0.3 wt%, photodeposition); reactant solution: 10 vol% aqueous methanol solution. [c] Cocatalyst: none; reactant solution: 0.02–0.05 mol L^{-1} aqueous AgNO$_3$ solution (120–150 mL). [d] Treatment: H_2-reduction at 473 K for 2 h.

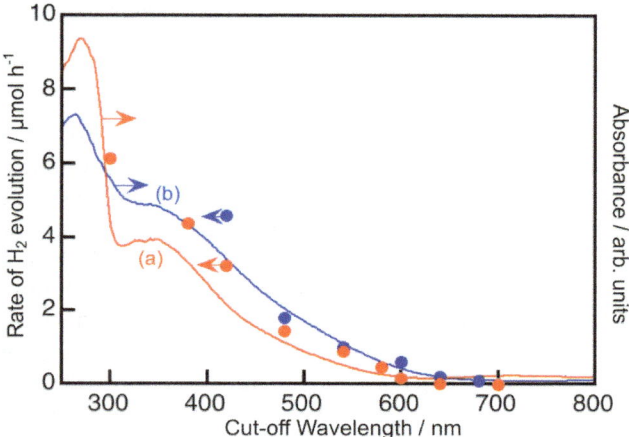

Fig. 2 The wavelength dependence of H_2 evolution from a 10 vol% aqueous methanol solution (closed circles) and diffuse reflectance spectra (solid line) of (a) NaTaO$_3$:-Ir(1%),La(2%) and (b) BaTa$_2$O$_6$:Ir(1%),La(2%). Photocatalyst: 0.1 g, cocatalyst: Pt (photodeposition) for H_2 evolution, reactant solution: 120 mL, light source: 300 W Xe lamp with long-pass filters, and reaction cell: top-irradiation cell with a Pyrex window. Samples of diffuse reflectance spectra were reduced at 473 K.

TiO$_2$:Rh,Sb[25] and NaNbO$_3$:Rh,Ba were active for sacrificial O$_2$ evolution using Ag$^+$ as an electron acceptor. NaNbO$_3$:Rh,Ba showed very low activity for the sacrificial H$_2$ evolution. These properties will be discussed based on the band structure in the Discussion section. These results of the sacrificial H$_2$ and O$_2$ evolutions suggest that NaTaO$_3$:Ir,La and BaTa$_2$O$_6$:Ir,La can be used as H$_2$-evolving

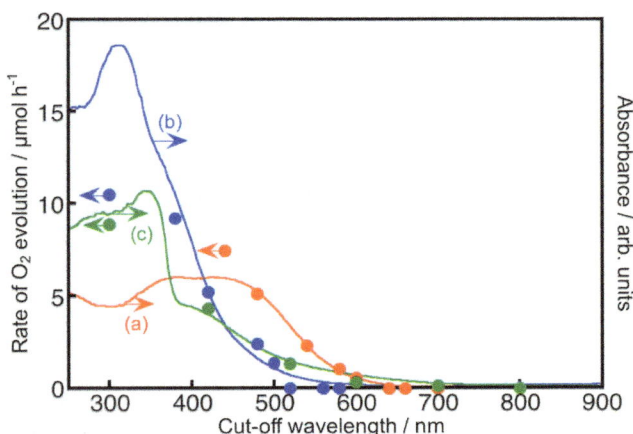

Fig. 3 The wavelength dependence of O$_2$ evolution from a 0.05 mol L^{-1} aqueous silver nitrate solution (closed circles) and diffuse reflectance spectra (solid line) of (a) TiO$_2$:-Rh(1.3%),Sb(2.6%), (b) NaNbO$_3$:Rh(1.2%),Ba(1.44%) and (c) SrTiO$_3$:Ir(0.2%) with H$_2$ reduction at 473 K. Photocatalyst: 0.1–0.3 g, reactant solution: 150 mL, light source: 300 W Xe lamp with long-pass filters, and reaction cell: top-irradiation cell with a Pyrex window.

photocatalysts for the construction of Z-scheme systems, while SrTiO$_3$:Ir, NaNbO$_3$:Rh,Ba and TiO$_2$:Rh,Sb are expected to be employed as O$_2$-evolving photocatalysts.

Fig. 2 and 3 show the diffuse reflectance spectra and wavelength dependence of the photocatalytic H$_2$ and O$_2$ evolutions of the Rh- and Ir-doped metal oxide photocatalysts in the presence of sacrificial reagents. The wavelengths were controlled with long-pass filters. It is vital to see the wavelength dependency of the photocatalytic activity because it is not guaranteed that photocatalysts with visible light absorption bands always give activities under visible light irradiation. The onsets of the wavelength dependence agreed with those of the diffuse reflection spectra. The onset wavelengths for the H$_2$ evolutions were 640 and 600 nm for BaTa$_2$O$_6$:Ir,La and NaTaO$_3$:Ir,La, respectively. These onset wavelengths were longer than the 540 nm of SrTiO$_3$:Rh (a conventional H$_2$-evolving photocatalyst). The onset wavelengths for O$_2$ evolution were 600, 500 and 700 nm for TiO$_2$:Rh,Sb, NaNbO$_3$:Rh,Ba and SrTiO$_3$:Ir, respectively. It is noteworthy that TiO$_2$:Rh,Sb and SrTiO$_3$:Ir responded at longer wavelengths than the BiVO$_4$ (BG: 2.4 eV) (a conventional O$_2$-evolving photocatalyst).

Fig. 4 shows the band structures of the Rh- and Ir-doped metal oxide photo-catalysts. The impurity levels of Rh^{3+} and Ir^{3+} were estimated from the energy gaps determined by diffuse reflection spectra supposing that the valence band consisting of O 2p located at +3 V *vs.* NHE at a pH of 0.[39] The absorption bands in the visible light region shown in Fig. 2 and 3 are due to electronic transition from the impurity levels consisting of Rh^{3+} and Ir^{3+} to the conduction bands of the host materials. The impurity levels formed with electron-filled orbitals of Ir^{3+} were around 1.0–1.2 V for NaTaO$_3$ and BaTa$_2$O$_6$, while Ir^{3+} in SrTiO$_3$ formed an impurity level around 1.4–1.6 V that was deeper than those in the cases of the tantalates. The energy levels formed with electron-filled orbitals of Rh^{3+} located around 2.0–2.1 V that were similar to those of SrTiO$_3$:Rh and SrTiO$_3$:Rh,Sb.[20,21] The reason Rh^{3+} forms a deeper impurity level than Ir^{3+} is due to Ir^{4+} being more stable than Rh^{4+} in metal oxides. Therefore, electronic transition from the Ir^{3+} impurity level to a conduction band is easier than that from Rh^{3+}, resulting in the formation of the shallow impurity level by Ir^{3+}.

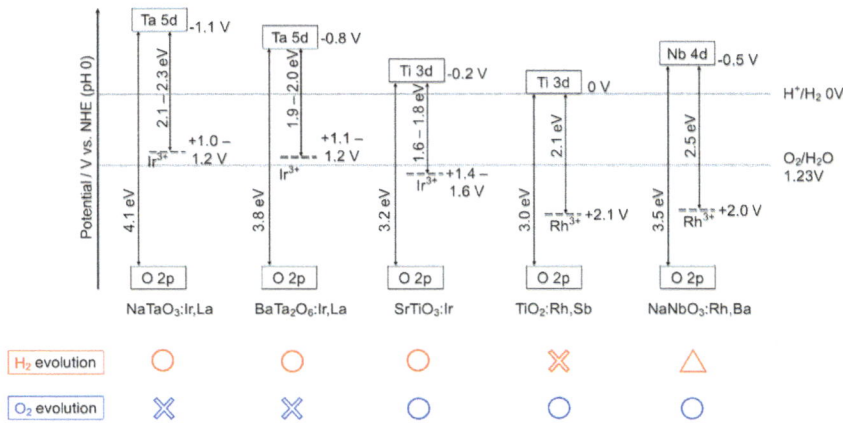

Fig. 4   The band structures of Rh- or Ir-doped metal oxide photocatalysts at pH 0.

Z-schematic systems for photocatalytic water splitting employing Rh- and Ir-doped metal oxide materials

The combination of $SrTiO_3$:Rh and $BiVO_4$ photocatalysts can be a benchmark of a Z-scheme photocatalyst system. The Z-schematic water splitting proceeds using $Fe^{3+/2+}$ and $[Co(bpy)_3]^{3+/2+}$ redox couples as ionic mediators (Fig. 1(a))[16,28,29] and also *via* interparticle electron transfer between $SrTiO_3$:Rh and $BiVO_4$ particles with and without RGO (Fig. 1(b) and (c)).[30,33,34] In general, water splitting *via* Z-schematic interparticle electron transfer by contact between the particles of the H_2- and O_2-evolving photocatalysts with and without RGO can be achieved (Fig. 1(b) and (c)) when the H_2- and O_2-evolving photocatalysts satisfy the following two requirements; (i) H_2- and O_2-evolving photocatalysts possess p- and n-type semiconductor properties, respectively, (ii) there is a certain electrode potential at which the cathodic photocurrent of the p-type semiconductor over-laps with the anodic photocurrent of the n-type semiconductor, being similar to water splitting using a photoelectrochemical cell working with no applied external bias as shown in Fig. 1(d).[31,32] Moreover, H_2- and O_2-evolving photo-catalysts must have contact with each other for interparticle electron transfer. In contrast to this, a Z-scheme system employing an ionic electron mediator could work regardless of the p- or n-type properties of the H_2- or O_2-evolving photo-catalysts, if the photocatalysts have potentials for the reduction or oxidation of redox couple ionic mediators and the adsorption abilities for the redox couples. The $SrTiO_3$:Rh and $BiVO_4$ photocatalysts satisfy these factors resulting in all of the Z-scheme systems showing activities for water splitting into H_2 and O_2 in stoi-chiometric amounts without any sacrificial reagents (Fig. 1(a)–(d)).

Various types of Z-scheme photocatalyst systems employing Ir- and Rh-doped photocatalysts for water splitting into H_2 and O_2 under visible light irradiation are shown in Table 2. Here, the absolute evaluation of the performance of photo-catalyst systems for water splitting is how much H_2 and O_2 is obtained under certain experimental conditions. In this sense, the activities of the different Z-scheme systems in Table 2 are comparable with each other, but not with those of Table 1, because of the almost identical experimental conditions for the water splitting, even if the kinetics would be different among the systems.

For the construction of Z-scheme systems, $BiVO_4$ (an O_2-evolving photo-catalyst) was combined with $BaTa_2O_6$:Ir,La and $NaTaO_3$:Ir,La (H_2-evolving pho-tocatalysts), while $SrTiO_3$:Rh (an H_2-evolving photocatalyst) was combined with TiO_2:Rh,Sb, $NaNbO_3$:Rh,Ba and $SrTiO_3$:Ir (O_2-evolving photocatalysts) as sug-gested by their H_2 and O_2 evolution abilities. In addition to the suspension system, the photoelectrochemical properties of photocatalyst powders immobi-lized on a conducting substrate as shown in Fig. 1(d) were examined using a Pt counter electrode without any sacrificial reagents to see the cathodic or anodic photocurrent and the onset potentials of semiconductor properties, not as a water splitting device, as shown in Fig. 5, in order to consider the potential overlap for Z-schematic water splitting *via* interparticle electron transfer with and without RGO. It has been reported that $SrTiO_3$:Rh[36,40] and $BiVO_4$[32,41–44] function as a photocathode and photoanode, respectively. $NaTaO_3$:Ir,La showed only an anodic photocurrent. Although $BaTa_2O_6$:Ir,La and $SrTiO_3$:Ir showed cathodic and anodic photocurrents, the photocurrents were poor and the onset potentials shifted with the sweeping direction of the CV curves. This result implies that

Table 2 Z-schematic water splitting under visible light irradiation using Rh- or Ir-doped metal oxide photocatalysts[a]

H$_2$-photocat.	O$_2$-photocat.	Mediator	Initial pH	Activity/μmol h^{-1}		Ref.
				H$_2$	O$_2$	
Pt/NaTaO$_3$:Ir(1%),La(2%)[b]	BiVO$_4$	None[e]	4.2	Trace	Trace	This work
Pt/NaTaO$_3$:Ir(1%),La(2%)[b]	BiVO$_4$	RGO[e]	4.2	Trace	Trace	This work
Pt/NaTaO$_3$:Ir(1%),La(2%)[b]	BiVO$_4$	Fe$^{3+/2+}$[f]	2.4	Trace	Trace	This work
Pt/NaTaO$_3$:Ir(1%),La(2%)[b]	BiVO$_4$	[Co(bpy)$_3$]$^{3+/2+}$[g]	4.2	0.8	0.3	This work
Pt/BaTa$_2$O$_6$:Ir(1%),La(2%)[b]	BiVO$_4$	None[e]	4.2	0.9	0.4	This work
Pt/BaTa$_2$O$_6$:Ir(1%),La(2%)[b]	BiVO$_4$	RGO[e]	4.2	0.9	0.4	This work
Pt/BaTa$_2$O$_6$:Ir(1%),La(2%)[b]	BiVO$_4$	Fe$^{3+/2+}$[f]	2.4	0.1	0.4	This work
Pt/BaTa$_2$O$_6$:Ir(1%),La(2%)[b]	BiVO$_4$	[Co(bpy)$_3$]$^{3+/2+}$[g]	4.2	5.9	2.1	22
Ru/SrTiO$_3$:Rh(1%)[c]	Pt/SrTiO$_3$:Ir(0.2%)[d]	None[e]	3.5	0.4	Trace	This work
Ru/SrTiO$_3$:Rh(1%)[c]	Pt/SrTiO$_3$:Ir(0.2%)[d]	RGO[e]	3.5	0.8	Trace	This work
Ru/SrTiO$_3$:Rh(1%)[c]	Pt/SrTiO$_3$:Ir(0.2%)[d]	Fe$^{3+/2+}$[f]	2.4	2.5	1.0	This work
Ru/SrTiO$_3$:Rh(1%)[c]	Pt/SrTiO$_3$:Ir(0.2%)[d]	[Co(bpy)$_3$]$^{3+/2+}$[h]	7.8	5.1	2.4	This work
Ru/SrTiO$_3$:Rh(1%)[c]	NaNbO$_3$:Rh(1%),Ba(2%)	None[e]	2.4	2.1	1.1	This work
Ru/SrTiO$_3$:Rh(1%)[c]	NaNbO$_3$:Rh(1%),Ba(2%)	Fe$^{3+/2+}$[f]	2.4	5.7	2.3	This work
Ru/SrTiO$_3$:Rh(1%)[c]	NaNbO$_3$:Rh(1%),Ba(2%)	[Co(bpy)$_3$]$^{3+/2+}$[h]	3.8	4.1	1.7	This work
Ru/SrTiO$_3$:Rh(1%)[c]	TiO$_2$:Rh(0.5%),Sb(1%)	None[e]	3.5	5.4	2.4	This work
Ru/SrTiO$_3$:Rh(1%)[c]	TiO$_2$:Rh(0.5%),Sb(1%)	Fe$^{3+/2+}$[f]	2.4	12	5.3	This work
Ru/SrTiO$_3$:Rh(1%)[c]	TiO$_2$:Rh(0.5%),Sb(1%)	[Co(bpy)$_3$]$^{3+/2+}$[h]	3.5	28	13	This work

[a] Photocatalyst: 0.05 or 0.1 g each; light source: 300 W Xe lamp with a long-pass filter ($\lambda > 420$ nm); reaction cell: top-irradiation cell with a Pyrex window. [b] Cocatalyst: Pt (0.3 wt%, impregnation at 673 K for 2 h and subsequent H$_2$-reduction at 673 K for 1 h). [c] Cocatalyst: Ru (0.7 wt%, photodeposition). [d] Cocatalyst: Pt (1 wt%, impregnation without calcination and subsequent H$_2$-reduction at 573 K for 1 h). [e] Reactant solution: H$_2$SO$_4$ solution. [f] Reactant solution: 2 mmol L^{-1} FeCl$_3$ solution. [g] Reactant solution: 0.02 mmol L^{-1} [Co(bpy)$_3$]SO$_4$ solution. [h] Reactant solution: 0.5 mmol L^{-1} [Co(bpy)$_3$]$_3$SO$_4$ solution (120 mL).

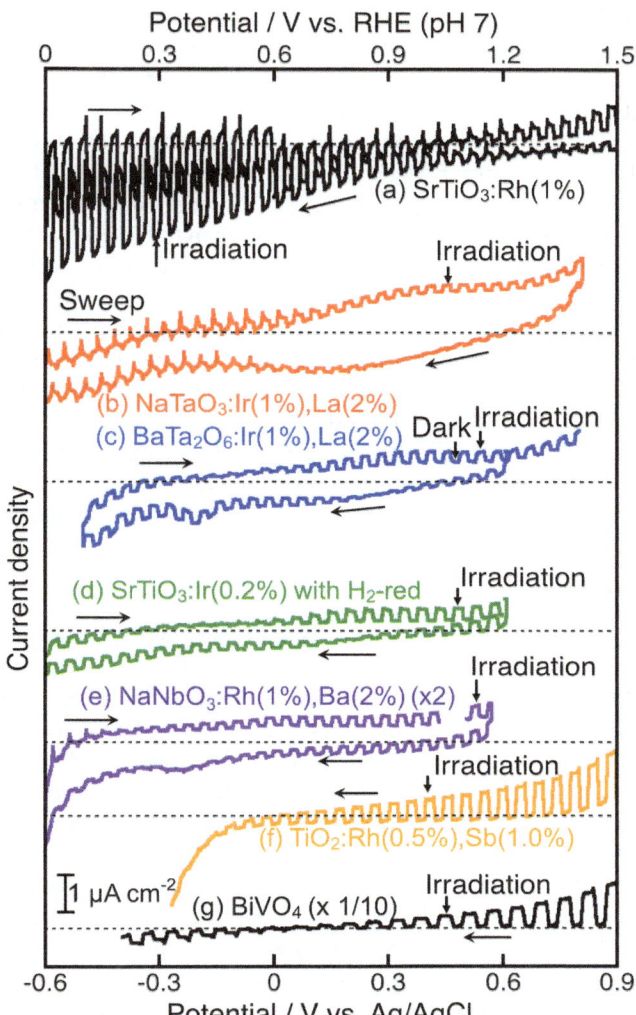

Fig. 5 Current vs. potential curves of (a) SrTiO₃:Rh(1%), (b) NaTaO₃:Ir(1%),La(2%), (c) BaTa₂O₆:Ir(1%),La(2%), (d) SrTiO₃:Ir(0.2%) with H₂ reduction, (e) NaNbO₃:Rh(1%),Ba(2%), (f) TiO₂:Rh(0.5%),Sb(1%), and (g) BiVO₄. Electrolyte: 0.1 mol L⁻¹ K₂SO₄ aqueous solution (pH 7, phosphate buffer was added if necessary), light source: 300 W Xe lamp (λ > 420 nm).

those photocurrents might not be due to H_2 and O_2 evolutions, but possibly due to redox reactions of the doped Ir species. In contrast, TiO_2:Rh,Sb,[45] and $NaNbO_3$:-Rh,Ba gave clear anodic photocurrents indicating an n-type semiconductor character. The onset potential of $NaNbO_3$:Rh,Ba was more negative than that of TiO_2:Rh,Sb, whereas the anodic photocurrent of $NaNbO_3$:Rh,Ba was much smaller than that of TiO_2:Rh,Sb. These anodic photocurrents overlapped with the cathodic photocurrent of $SrTiO_3$:Rh at a certain electrode potential.

The Z-scheme systems were classified into three groups depending on the combination of H_2- and O_2-evolving photocatalysts and an electron mediator;

being active for only a $[Co(bpy)_3]^{3+/2+}$ redox couple, active for $[Co(bpy)_3]^{3+/2+}$ and $Fe^{3+/2+}$ redox couple ionic mediators, active not only for $[Co(bpy)_3]^{3+/2+}$ and $Fe^{3+/2+}$ redox couples but also *via* interparticle electron transfer with and without RGO. $NaTaO_3$:Ir,La + $BiVO_4$ and $BaTa_2O_6$:Ir,La + $BiVO_4$ were active when not $Fe^{3+/2+}$ but a $[Co(bpy)_3]^{3+/2+}$ redox couple was used. Although the $BaTa_2O_6$:Ir,La + $BiVO_4$ showed activities *via* interparticle electron transfer at a pH of 4.2 with and without RGO, the activities were smaller than that with the ionic electron mediator $[Co(bpy)_3]^{3+/2+}$. $SrTiO_3$:Rh + $SrTiO_3$:Ir was active when the $[Co(bpy)_3]^{3+/2+}$ and $Fe^{3+/2+}$ redox couple ionic mediators were used, whereas it was not active *via* interparticle electron transfer with and without RGO at pH 3.5. $SrTiO_3$:Rh + $NaNbO_3$:Rh,Ba and $SrTiO_3$:Rh + TiO_2:Rh,Sb were active for the $[Co(bpy)_3]^{3+/2+}$ and $Fe^{3+/2+}$ redox couples and *via* interparticle electron transfer without RGO. These Z-scheme systems with RGO using $NaNbO_3$:Rh,Ba and TiO_2:Rh,Sb were not examined in detail, because the GOs were not suitably photoreduced on the $NaNbO_3$:Rh,Ba and TiO_2:Rh,Sb due to their poor reducing activities, as expected from their poor H_2 evolution activities as shown in Table 1. $SrTiO_3$:Rh + TiO_2:Rh,Sb showed the best performances among the Z-scheme systems in Table 2.

Discussion

H_2 and O_2 evolution abilities (shown in Table 1) can be considered based on the band structure, as shown in Fig. 4. Photogenerated holes in the Ir^{3+} levels in $NaTaO_3$:Ir,La and $BaTa_2O_6$:Ir,La have no potentials for water oxidation. In contrast, the Ir^{3+} level of $SrTiO_3$:Ir possesses the water oxidation potential, though the driving force is not so large. Conduction bands consisting of Ta 5d orbitals in $BaTa_2O_6$:Ir,La and $NaTaO_3$:Ir,La possess thermodynamically enough potentials for water reduction to form H_2. On the other hand, the photogenerated holes in the Rh^{3+} levels in $NaNbO_3$:Rh,Ba and TiO_2:Rh,Sb have enough potential for water oxidation to form O_2. These energy levels were consistent with the H_2 and O_2 evolution abilities, as shown in Table 1. Of course, other kinetic factors of the active sites would exist for the H_2 and O_2 evolution abilities in addition to the energy levels of the thermodynamic factor.

Let us consider the reason why the Z-scheme systems shown in Table 2 are classified into three groups, based on the band structure and photo-electrochemical properties. Fig. 6 shows energy diagrams for the Z-schemes employing $Fe^{3+/2+}$ and $[Co(bpy)_3]^{3+/2+}$ redox couples at pH 2.4 and 4.0, respectively. It is assumed that the band levels shift with -0.059 V pH^{-1} because those materials are metal oxides.

$NaTaO_3$:Ir,La + $BiVO_4$ and $BaTa_2O_6$:Ir,La + $BiVO_4$ were active with a $[Co(bpy)_3]^{3+/2+}$ redox couple at pH 4.2, but not with $Fe^{3+/2+}$ at pH 2.4. Photogenerated holes in the Ir^{3+} levels in $NaTaO_3$:Ir,La and $BaTa_2O_6$:Ir,La (an H_2-evolving photocatalyst) possess small driving forces for oxidation of Fe^{2+} (an electron mediator) at pH 2.4 as shown in Fig. 6(a), whereas they possess enough potential for oxidation of $[Co(bpy)_3]^{2+}$ at pH 4.2 as shown in Fig. 6(b). The activities from the interparticle electron transfer were very small or negligible because of poor photoresponse and poor overlaps of the potentials giving cathodic photocurrents of $NaTaO_3$:Ir,La and $BaTa_2O_6$:Ir,La and an anodic photocurrent of $BiVO_4$ as shown in Fig. 5, and also may be due to poor contact between $NaTaO_3$:Ir,La or $BaTa_2O_6$:Ir,La (an H_2-evolving photocatalyst) and $BiVO_4$ (an O_2-

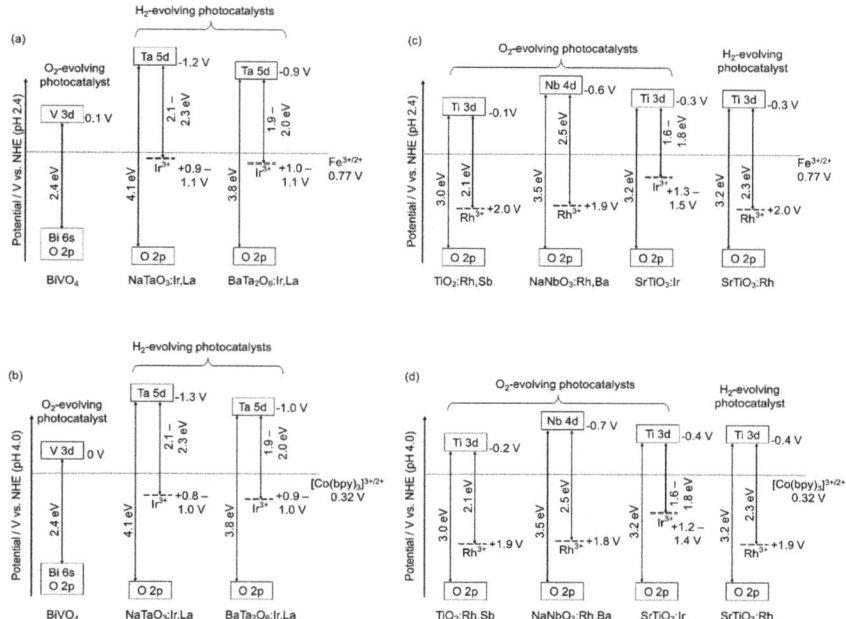

Fig. 6 The band structures of Rh- or Ir-doped metal oxide photocatalysts, and redox potentials of Fe and Co-complex ionic electron mediators at pH 2.4 and 4.0. (a) and (b) represent Z-scheme systems of Ir-doped metal oxides as H_2-evolving photocatalysts combined with $BiVO_4$. (c) and (d) represent Z-scheme systems of Rh-doped metal oxides and $SrTiO_3$:Ir as O_2-evolving photocatalysts combined with $SrTiO_3$:Rh.

evolving photocatalyst) in the suspension at pH 4.2. In contrast to them, $SrTiO_3$:Rh + $SrTiO_3$:Ir was active when the $[Co(bpy)_3]^{3+/2+}$ and $Fe^{3+/2+}$ redox couple ionic mediators were used, because the conduction band of $SrTiO_3$:Ir (an O_2-evolving photocatalyst) has enough potential for the reduction of Fe^{3+} and $[Co(bpy)_3]^{3+}$ (Fig. 6(c) and (d)). However, because the anodic photocurrent of $SrTiO_3$:Ir hardly overlapped with the cathodic photocurrent of $SrTiO_3$:Rh, the activities by interparticle electron transfer with and without RGO were negligible. The conduction bands of TiO_2:Rh,Sb and $NaNbO_3$:Rh,Ba (O_2-evolving photo-catalysts) also possessed enough potential for the reduction of Fe^{3+} and $[Co(bpy)_3]^{3+}$ as well as $SrTiO_3$:Ir (Fig. 6(c) and (d)). Moreover, in these cases, the anodic photocurrents of the TiO_2:Rh,Sb and $NaNbO_3$:Rh,Ba photoelectrodes overlapped enough with the cathodic photocurrent of $SrTiO_3$:Rh. So, it is reasonable that $SrTiO_3$:Rh + $NaNbO_3$:Rh,Ba and $SrTiO_3$:Rh + TiO_2:Rh,Sb were active not only for the $[Co(bpy)_3]^{3+/2+}$ and $Fe^{3+/2+}$ redox couples but also via interparticle electron transfer with and without RGO. The activity of $SrTiO_3$:Rh + TiO_2:Rh,Sb was higher than that of $SrTiO_3$:Rh + $NaNbO_3$:Rh,Ba for all of the types of Z-scheme system. In the cases with the use of ionic electron mediators, it is probably due to the higher activity for O_2 evolution and the narrower energy gaps of TiO_2:Rh,Sb than those of $NaNbO_3$:Rh,Ba, as shown in Table 1. The reason why $SrTiO_3$:Rh + TiO_2:Rh,Sb showed a higher activity than $SrTiO_3$:Rh + $NaNbO_3$:Rh,Ba via interparticle electron transfer is that TiO_2:Rh,Sb gave much larger anodic photocurrents than $NaNbO_3$:Rh,Ba, as shown in Fig. 5.

Conclusions

We have successfully developed Z-scheme photocatalyst systems for water split-ting under visible light irradiation employing Rh- and Ir-doped metal oxide photocatalysts with longer wavelength responses than conventional $SrTiO_3$:Rh and $BiVO_4$. The impurity levels of doped Ir^{3+} and Rh^{3+} that contributed to the visible light responses for the photocatalytic reactions were determined from diffuse reflectance spectra and supposing +3.0 V $vs.$ NHE of a valence band of the oxide photocatalyst materials. The impurity levels of Ir^{3+} in $NaTaO_3$:Ir,La and $BaTa_2O_6$:Ir,La have sufficient potentials for the oxidation of the electron mediator $[Co(bpy)_3]^{2+}$, but insufficient potentials for water oxidation to form O_2 and oxidation of the electron mediator Fe^{2+}. The conduction bands of the $NaTaO_3$:-Ir,La and $BaTa_2O_6$:Ir,La have enough potential for water reduction to form H_2. Therefore, the $NaTaO_3$:Ir,La and $BaTa_2O_6$:Ir,La could be used as H_2-evolving photocatalysts only when a $[Co(bpy)_3]^{3+/2+}$ redox couple was used. The impurity levels of Ir^{3+} in $SrTiO_3$:Ir and Rh^{3+} in TiO_2:Rh,Sb and $NaNbO_3$:Rh,Ba have potentials for water oxidation and their conduction bands possess the potentials for reduction of the electron mediators Fe^{3+} and $[Co(bpy)_3]^{3+}$. This property means that $SrTiO_3$:Ir, TiO_2:Rh,Sb and $NaNbO_3$:Rh,Ba could be employed as O_2-evolving photocatalysts for the construction of Z-scheme systems employing ionic electron mediators. Moreover, photoelectrochemical measurements using photocatalyst powders immobilized on conducting substrate revealed that the n-type characters and relatively negative onset potentials of the TiO_2:Rh,Sb and $NaNbO_3$:Rh,Ba photoanodes enabled the Z-scheme systems to work by interparticle electron transfer. Although the efficiencies of the present Z-scheme systems are low at the present stage, these will be improved by interfacial controls from a kinetic point of view in basic research. These results and discussion will contribute to the design of a highly active photocatalyst system for water splitting into H_2 and O_2, aiming for the demonstration of actual solar water splitting using a suitable reactor.

Conflicts of interest

There are no conflicts to declare.

Acknowledgements

This work was supported by JSPS KAKENHI Grant Numbers 17H06440 and 17H06433 in Scientific Research on Innovative Areas "Innovations for Light-Energy Conversion (I^4LEC)".

References

1 A. Kudo, H. Kato and I. Tsuji, *Chem. Lett.*, 2004, **33**, 1534.

2 A. Kudo and Y. Miseki, *Chem. Soc. Rev.*, 2009, **38**, 253.

3 F. E. Osterloh, *Chem. Mater.*, 2008, **20**, 35.

4 Y. Inoue, *Energy Environ. Sci.*, 2009, **2**, 364.

5 R. Abe, *J. Photochem. Photobiol., C*, 2010, **11**, 179.

6 T. Hisatomi, J. Kubota and K. Domen, *Chem. Soc. Rev.*, 2014, **43**, 7520.

7 K. Maeda and K. Domen, *Bull. Chem. Soc. Jpn.*, 2016, **89**, 627.

8 B. A. Pinaud, J. D. Benck, L. C. Seitz, A. J. Forman, Z. Chen, T. G. Deutsch, B. D. James, K. N. Baum, G. N. Baum, S. Ardo, H. Wang, E. Miller and T. F. Jaramillo, *Energy Environ. Sci.*, 2013, **6**, 1983.

9 D. M. Fabian, S. Hu, N. Singh, F. A. Houle, T. Hisatomi, K. Domen, F. E. Osterloh and S. Ardo, *Energy Environ. Sci.*, 2015, **8**, 2825.

10 T. Yamada and K. Domen, *Chem. Eng.*, 2018, **2**, 36.

11 A. Kudo, in *"Photocatalysis", Contemporary Catalysis; Science, Technology and Applications*, ed. P. Kamer, D. Vogt and J. Thybaut, Royal Society of Chemistry, Cambridge, 2017, ch. 3.7, pp. 326–343.

12 A. J. Bard, *J. Photochem.*, 1979, **10**, 59.

13 A. Kudo, *MRS Bull.*, 2011, **36**, 32.

14 K. Maeda, *ACS Catal.*, 2013, **3**, 1486.

15 Y. Wang, H. Suzuki, J. Xie, O. Tomita, D. J. Martin, M. Higashi, D. Kong, R. Abe and J. Tang, *Chem. Rev.*, 2018, **118**, 5201.

16 H. Kato, M. Hori, R. Konta, Y. Shimodaira and A. Kudo, *Chem. Lett.*, 2004, **33**, 1348.

17 Q. Wang, T. Hisatomi, Q. Jia, H. Tokudome, M. Zhong, C. Wang, Z. Pan, T. Takata, M. Nakabayashi, N. Shibata, Y. Li, I. D. Sharp, A. Kudo, T. Yamada and K. Domen, *Nat. Mater.*, 2016, **15**, 611.

18 Q. Wang, T. Hisatomi, Y. Suzuki, Z. Pan, J. Seo, M. Katayama, T. Minegishi, H. Nishiyama, T. Takata, K. Seki, A. Kudo, T. Yamada and K. Domen, *J. Am. Chem. Soc.*, 2017, **139**, 1675.

19 Q. Wang, T. Hisatomi, M. Katayama, T. Takata, T. Minegishi, A. Kudo, T. Yamada and K. Domen, *Faraday Discuss.*, 2017, **197**, 491.

20 R. Konta, T. Ishii, H. Kato and A. Kudo, *J. Phys. Chem. B*, 2004, **108**, 8992.

21 R. Niishiro, S. Tanaka and A. Kudo, *Appl. Catal., B*, 2014, **150**, 187.

22 S. Suzuki, H. Matsumoto, A. Iwase and A. Kudo, *Chem. Commun.*, 2018, **54**, 10606.

23 A. Iwase and A. Kudo, *Chem. Commun.*, 2017, **53**, 6156.

24 A. Iwase, K. Saito and A. Kudo, *Bull. Chem. Soc. Jpn.*, 2009, **82**, 514.

25 R. Niishiro, R. Konta, H. Kato, W. J. Chun, K. Asakura and A. Kudo, *J. Phys. Chem. C*, 2007, **111**, 17420.

26 A. Kudo, K. Omori and H. Kato, *J. Am. Chem. Soc.*, 1999, **121**, 11459.

27 A. Iwase, H. Kato and A. Kudo, *J. Sol. Energy Eng.*, 2010, **132**, 21106.

28 Y. Sasaki, A. Iwase, H. Kato and A. Kudo, *J. Catal.*, 2008, **259**, 133.

29 Y. Sasaki, H. Kato and A. Kudo, *J. Am. Chem. Soc.*, 2013, **135**, 5441.

30 A. Iwase, Y. H. Ng, Y. Ishiguro, A. Kudo and R. Amal, *J. Am. Chem. Soc.*, 2011, **133**, 11054.

31 K. Iwashina, A. Iwase, Y. H. Ng, R. Amal and A. Kudo, *J. Am. Chem. Soc.*, 2015, **137**, 604.

32 A. Iwase, S. Yoshino, T. Takayama, Y. H. Ng, R. Amal and A. Kudo, *J. Am. Chem. Soc.*, 2016, **138**, 10260.

33 Y. Sasaki, H. Nemoto, K. Saito and A. Kudo, *J. Phys. Chem. C*, 2009, **113**, 17536.

34 Q. Jia, A. Iwase and A. Kudo, *Chem. Sci.*, 2014, **5**, 1513.

35 W. S. Hummers and R. E. Offeman, *J. Am. Chem. Soc.*, 1958, **80**, 1339.

36 K. Iwashina and A. Kudo, *J. Am. Chem. Soc.*, 2011, **133**, 13272.

37 S. Kawasaki, K. Akagi, K. Nakatsuji, S. Yamamoto, I. Matsuda, Y. Harada, J. Yoshinobu, F. Komori, R. Takahashi, M. Lippmaa, C. Sakai, H. Niwa, M. Oshima, K. Iwashina and A. Kudo, *J. Phys. Chem. C*, 2012, **116**, 24445.

38 S. Kawasaki, R. Takahashi, K. Akagi, J. Yoshinobu, F. Komori, K. Horiba, H. Kumigashira, K. Iwashina, A. Kudo and M. Lippmaa, *J. Phys. Chem. C*, 2014, **118**, 20222.

39 D. E. Scaife, *Sol. Energy*, 1980, **25**, 41.

40 S. Kawasaki, K. Nakatsuji, J. Yoshinobu, F. Komori, R. Takahashi, M. Lippmaa, K. Mase and A. Kudo, *Appl. Phys. Lett.*, 2012, **101**, 033910.

41 K. Sayama, A. Nomura, Z. Zou, R. Abe, Y. Abe and H. Arakawa, *Chem. Commun.*, 2003, 2908.

42 K. Sayama, A. Nomura, T. Arai, T. Sugita, R. Abe, M. Yanagida, T. Oi, Y. Iwasaki, Y. Abe and H. Sugihara, *J. Phys. Chem. B*, 2006, **110**, 2908; F. F. Adbi, L. Han, A. H. M. Smets, M. Zeman, B. Dam and R. van de Krol, *Nat. Commun.*, 2013, **4**, 2195.

43 Y. Park, K. J. McDonald and K.-S. Choi, *Chem. Soc. Rev.*, 2013, **42**, 2321.

44 Q. Jia, K. Iwashina and A. Kudo, *Proc. Natl. Acad. Sci. U. S. A.*, 2012, **109**, 11564.

45 M. Yamaguchi, R. Niishiro, Q. Jia, Y. Kuang, K. Kitamura, A. Iwase, T. Minegishi, T. Yamada, K. Domen and A. Kudo, to be submitted.

Faraday Discussions

PAPER

A microfluidic photoelectrochemical cell for solar-driven CO_2 conversion into liquid fuels with CuO-based photocathodes

Evangelos Kalamaras, [ID] *[a] Meltiani Belekoukia, [ID] [a] Jeannie Z. Y. Tan, [ID] [a] Jin Xuan, [ID] *[b] M. Mercedes Maroto-Valer [ID] [a] and John M. Andresen*[a]

Received 19th November 2018, Accepted 15th January 2019

DOI: 10.1039/c8fd00192h

Utilising photoelectrochemical (PEC) devices to produce sustainable fuels from water and CO_2 is a very attractive strategy, in which sunlight is used to convert the greenhouse gas (CO_2) into a usable form of stored chemical energy. While significant progress has been made in the development of efficient photoactive catalysts for PEC reactions, limited efforts have been focused on the reactor design where continuous flow microfluidic PEC reactors are particular promising. In this work, a range of CuO-based thin films were used as photocathodes in a continuous flow microfluidic PEC reactor using CO_2-saturated aqueous $NaHCO_3$ solution under simulated AM 1.5 solar irradiation for up to 12 h. The highest photocurrent density obtained was for the α-Fe_2O_3/CuO photoelectrode yielding -1.0 mA cm^{-2} at 0.3 V $vs.$ RHE and initial results indicated a solar-to-fuel (STF) efficiency of 0.48%. While the CuO, Cu_2O and CuO–Cu_2O photoelectrodes virtually only formed formate, the bilayer α-Fe_2O_3/CuO photocathode produced methanol in addition to formate indicating that combined copper and iron oxides in continuous flow microfluidic PEC cells have great potential of direct solar conversion into useful chemicals.

Introduction

Solar renewable energy could offer a sustainable alternative solution to deal with global warming and fossil fuel depletion. Since sunlight is considered to be the most abundant energy source, especially in converting CO_2 into useful hydro-carbon products,[1] the production of solar fuel from water and CO_2 has attracted much attention.[2,3] Photoelectrochemical (PEC) CO_2 reduction cells using semi-conducting photoelectrodes have been studied as an alternative and promising

[a]Research Centre for Carbon Solutions (RCCS), School of Engineering & Physical Sciences, Heriot-Watt University, Edinburgh, EH14 4AS, UK. E-mail: ek15@hw.ac.uk; j.andresen@hw.ac.uk

[b]Department of Chemical Engineering, Loughborough University, Loughborough, UK. E-mail: j.xuan@lboro.ac.uk

approach to PEC water splitting applications.[4-6] Although PEC CO_2 reduction has been proven feasible, the physicochemical mechanisms involved in the production of carbon-based chemicals at a semiconductor–liquid electrolyte junction are very complicated and challenging.[7]

The PEC CO_2 reduction process is initiated when incident photons are absorbed by a semiconducting material, leading to the creation of electron–hole pairs. Then, the photogenerated charge carriers, driven by the space-charge field, flow to the interface of the semiconductor/liquid electrolyte, in which water oxidation (oxygen evolution reaction – OER) or CO_2 reduction into chemicals, such as CO, HCOOH, CH_3OH, take place. In other words, PEC CO_2 reduction devices involve two half-cell reactions with different kinetics.[4] Despite a large number of efforts being made since Halmann reported a PEC CO_2 reduction cell for the first time in 1978,[8] this technology is still in its early stage of development. A plethora of promising p-type (photocathodes) and n-type (photoanodes) semiconducting materials has been studied, revealing that photocathode-driven devices for CO_2 conversion into carbon-based chemicals can achieve higher efficiency and selectivity.[9] This is mainly related to the fact that in PEC CO_2 reduction cells, p-type semiconductors are coupled with selective and efficient OER catalysts such as Pt, while n-type semiconductors are coupled with electrodes that enable the CO_2 reduction half-reaction such as Cu-based materials showing low selectivity and efficiency due to the competitive water reduction reaction (hydrogen evolution).[10-13]

A p-type semiconducting candidate should absorb a large part of the solar spectrum and simultaneously its conduction band edge should lie at a more negative potential than that of the CO_2 reduction reaction potential.[14] Among these materials, copper oxides seem attractive due to their low cost and abundance, chemical stability and toxicity in comparison with other photocathodic materials. Hence, photocathodes based on cupric oxide (CuO) were synthesized for PEC CO_2 reduction into carbon-based chemicals.

PEC CO_2 reduction systems suffer from low solar-to-fuel (STF) efficiencies that do not exceed 1.2%, the highest yet recorded.[15] The theoretical efficiency limitations of PEC reactions that occur on the electrode/liquid electrolyte interface at a molecular level highly depend on transport phenomena.[16] A promising route to enhance the performance of PEC CO_2 reduction cells is to promote mass transfer of the reactants, intermediates and products at the electrode/liquid electrolyte interface. Therefore, PEC cells should be designed in shape and dimensions that promote transfer of reactants, intermediates and products to and from the active sites of the electrodes. In this study, the performance of a newly-designed continuous microfluidic PEC cell was evaluated using CuO-based photocathodes for PEC CO_2 reduction highlighting the significance of reactor design and igniting a discussion about the benefits of a microscale-based PEC cell approach.

Experimental

Preparation of photocathodes

Synthesis of CuO and Cu_2O thin films. Initially, fluorine-doped tin oxide (FTO) coated glass purchased from Pilkington (8 Ω sq^{-1}) was cut into rectangular pieces and used as substrates. Then, these pieces of FTO glass were ultrasonically cleaned in ethanol and acetone for 30 min, followed by rinsing with deionized water (18.2 MΩ cm at 25 °C) and dried in air. CuO and Cu_2O thin films were synthesized on FTO glass using an electrodeposition method performed on an Autolab PGSTAT 302N electrochemical workstation. As illustrated in Scheme 1a, the electrodeposition of CuO and Cu_2O thin films was carried out using a 3-electrode set-up that consisted of FTO glass as the working electrode, Ag/AgCl (KCl 1 M) as the reference electrode and Pt foil as the counter electrode. Only a certain area (1 cm × 1 cm for linear sweep voltammetry experiments and 3 cm × 1 cm for CO_2 reduction experiments) was exposed to the aqueous electrode-position solution containing 50 mM potassium perchlorate ($KClO_4$ >99%, Sigma-Aldrich) and 4 mM copper(II) nitrate trihydrate ($CuNO_3 \cdot 3H_2O$ >99%, Sigma-Aldrich).[15,17] Working electrodes were held at a constant potential of −0.33 V vs. Ag/AgCl reference electrode. The thickness of the films was controlled by varying the deposition time ranging from 30–180 min. After drying, the electrodeposited thin films on FTO substrates were annealed at 600 °C for 2 h with a heating ramp of 5 °C min^{-1} in the air or Ar atmosphere to fabricate CuO or Cu_2O electrodes, respectively. CuO/Cu_2O photocathodes were also fabricated using the same electrodeposition method. CuO thin films were synthesized on FTO glass followed by Cu_2O deposition on the CuO layer.

Synthesis of Fe_2O_3 thin film. Nanostructured Fe_2O_3 thin films were synthesized on FTO glass using a hydrothermal synthesis method, which was carried out in a Teflon-lined stainless-steel autoclave (Scheme 1b). Briefly, an aqueous hydrothermal solution was prepared containing 0.15 M iron chloride hexahydrate ($FeCl_3 \cdot 6H_2O$ >98%, Sigma-Aldrich) and 1 M sodium nitrate ($NaNO_3$ >99%, Sigma-Aldrich).[18,19] A few drops of hydrochloric acid (HCl 36.5–38%, Sigma-Aldrich) were added to adjust the pH of the solution to 1.5. FTO substrates were placed with the conductive side facing down in autoclave filled with hydrothermal solution. The autoclave was sealed tightly and heated at 100 °C for 3, 6 and 12 h in an oven. After that, the autoclave was naturally cooled down to room temperature and the samples were rinsed with deionized water. Subsequently, a uniform orange thin film (β-FeOOH) deposited on the FTO glass was sintered in air at 600 °C for 3 h with a heating ramp of 5 °C min^{-1}. As a result, a dark red colour thin film was obtained. After that, the electrodeposition method used to synthesize the CuO

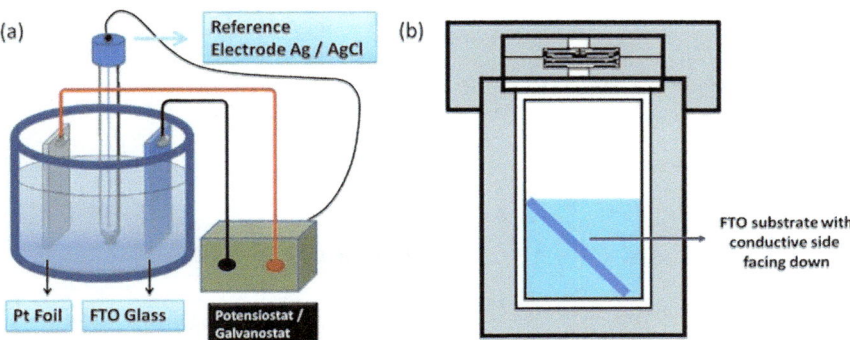

Scheme 1 Schematic illustration of (a) 3-electrode set-up used in the electrodeposition method and (b) hydrothermal synthesis carried out in a stainless-steel autoclave.

thin film was used to fabricate another layer on top of the dark red coloured thin film.

Materials characterization

Optical absorption of all samples was obtained using a UV-Vis spectrometer (PerkinElmer Lambda 950) equipped with a 150 mm integration sphere. Energy band-gaps of the bare semiconducting materials were calculated using the Kubelka–Munk formula:[20,21]

$$F(R) = \frac{(1-R)^2}{2R} \tag{1}$$

where R is the reflectance (absolute values). Crystalline structures of all samples were investigated with a D8 Advance X-ray diffractometer (XRD, Bruker AXS) with Cu Kα radiation source ($\lambda = 1.5418$ Å) and a nickel beta filter ($2\theta = 10$–$80°$). Particle size of all sample electrodes was examined with a Scanning Electron Microscope (SEM, FEI Quanta 3D FEG) at an accelerated voltage of 10 kV. In addition, a FEI Scios dual-beam microscope equipped with an Octane Plus EDS (Energy Dispersive X-ray Analysis, EDAX) detector was used for compositional analysis of electrodes. High resolution transmission electron microscope (HR-TEM, FEI Titan Themis 200) equipped with EDX detector operated at 200 kV was used to examine the nanostructure of the thin film. Prior to TEM experiments, iron oxide particles were diluted in ethanol, sonicated for 5 min and a few drops of this solution were placed on a carbon coated Cu grid.

Fabrication of microfluidic cell

Two different reactors were fabricated for the PEC experiments, as described here. A polymethyl methacrylate (PMMA) homemade batch reactor was fabricated using a Trotec Speedy 300 laser engraver/cutter. The distance between the working and counter electrode was fixed at 10 mm while the total volume of the reactor was 15 mL (50 mm × 30 mm × 10 mm). This reactor was assembled with working electrode, counter electrode and a PMMA chamber. In addition, two stainless steel plates were used to hold together the cell (PEC reactor 1 – PECR1). A

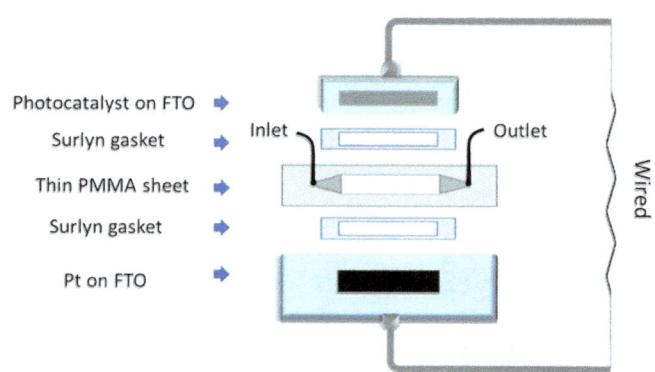

Scheme 2 Schematic representation of wired sandwich-type microfluidic PEC cell (PECR2).

similar reactor with an inlet and outlet of electrolyte was used for the continuous flow PEC measurements. A sandwich-type microfluidic photo-reactor was also designed and fabricated with a working and counter electrode at the bottom and top part of the photo-reactor as illustrated in Scheme 2. A PMMA sheet (100 μm thickness) was cut by Trotec Speedy 300 laser engraver/cutter. Then, the PMMA sheet was placed in between the two electrodes creating a reaction chamber. Finally, two gaskets of a thermoplastic film (Surlyn 30 μm, Dyesol) were cut and placed between the PMMA layer and electrodes in order to seal the reactor (PEC reactor 2 – PECR2). The total volume of the microchamber was ≈50 μL (31 mm × 10 mm × 0.16 mm).

Photoelectrochemical (PEC) measurements

J–V measurements (linear sweep voltammetry) were conducted in the homemade batch reactor as described above. They were performed on an Autolab PGSTAT 302N electrochemical workstation in a 3-electrode configuration consisting of FTO/photocatalyst working electrode (photoelectrode), Ag/AgCl (KCl 1 M) reference electrode and Pt foil as the counter electrode at room temperature. Although Pt is not considered as the best catalyst for water oxidation, it is widely used in PEC CO_2 reduction studies.[9] Among the best catalysts for the water oxidation reaction are IrO_2 and RuO_2.[22] However, in this study, Pt electrodes were selected due to their stability and because they allowed a more fair comparison with the literature. The illuminated surface area of the photoelectrode in direct contact with the electrolyte was 1 cm × 1 cm. The electrolyte for all PEC experiments was 0.2 M $NaHCO_3$ purged with high purity CO_2 (99.99%) or N_2 (99.99%) for 30 min prior to the experiment. $NaHCO_3$ was chosen due to its stability in CO_2 reduction experiments and enhanced CO_2 solubility.[23-25] All PEC experiments were performed under continuous or chopped irradiation using a Newport 92250A solar simulator (AM 1.5 G, 100 mW cm^{-2}) as the light source. During irradiation of the backside (the side that is not contact with the electrolyte) of the photoelectrodes, potentials were swept from −0.6 to 0.4 V $vs.$ Ag/AgCl at a scan rate ranging from 1 to 5 mV s^{-1}. However, the potentials with respect to the Ag/AgCl reference electrode were re-calculated and converted to these with respect to the reversible hydrogen electrode (RHE) via the following equation:

$$V_{RHE} = V_{Ag/AgCl} + 0.222\,V + 0.059\,V \times \text{pH} \tag{2}$$

It should be mentioned that the pH of N_2-purged 0.2 M $NaHCO_3$ solution was 8.4, while the pH for the CO_2-purged solution decreased down to 6.6. In other words, CO_2 purging affected the calculated values for the conversion of $V_{Ag/AgCl}$ to V_{RHE}.

The efficiency of the as-prepared photoelectrodes for PEC CO_2 conversion into carbon-based products was further evaluated with a 2-electrode configuration in both types of reactors. CO_2-purged solution was supplied to PECR1 and PECR2 using a syringe pump (Genie Touch, Kent Scientific Corporation). The liquid flow rate was in the range of 40, 80, 120 and 200 μL min^{-1}. Finally, the liquid solution was collected from the outlet and analyzed. CO_2 conversion into methanol was confirmed and quantified using a Bruker AVIII 300 MHz nuclear magnetic resonance (NMR) spectrometer. ^{1}H NMR allowed the quantification of all liquid phase products identified in these experiments. Standard curves were made using purchased chemicals of interest with internal standards in 0.2 $NaHCO_3$. During the PEC experiments, a fraction of electrolyte was collected (1 mL) from the outlet of the reactor and mixed with deuterium oxide (D_2O, 99.9% Sigma-Aldrich) and dioxane that served as an internal standard. Specifically, 606 μL of sampled electrolyte solution, 24 μL of 5.86 mM aqueous dioxane solution and 70 μL of D_2O were placed in NMR tubes (Norell, 5 mm ultra-precision). Finally, the product peak areas were compared to the internal standard curves to quantify the concentration of CO_2 produced by the reduction reaction enabling the estimation of STF efficiency through the following relationship:[9,26]

$$\sum STF\ efficiency = \frac{r(mmol/s) \times \Delta G^0(kJ\ mol^{-1})}{P_{solar}(mW\ cm^{-2}) \times surface\ area(cm^2)} \times 100\% \qquad (3)$$

where r is the amount of chemical fuel produced (mmol) per second, ΔG^0 is the Gibbs free energy for conversion of gaseous CO_2 into the produced fuel, P_{solar} is the power density of the light source (AM 1.5 G) and surface area is the irradiated photoelectrode area exposed to the electrolyte solution. Overall STF efficiency is calculated by summing the STF efficiencies of every different obtained product.

Results and discussion

Photoelectrode characterization

The films of CuO and Cu_2O on FTO glass in this study were synthesized using an electrodeposition method. FTO glass was utilized as the substrates because it is transparent (allowing the back illumination of the photoelectrode), electrically conductive, resistant to high temperatures and rigid, avoiding cracks of the synthesized photocatalyst due to bending. The crystal structures of the as-prepared semiconducting materials were initially analysed by XRD as illustrated in Fig. 1a. The XRD patterns of electrodeposited and sintered in air CuO samples showed diffraction peaks that matched the monoclinic crystal structure

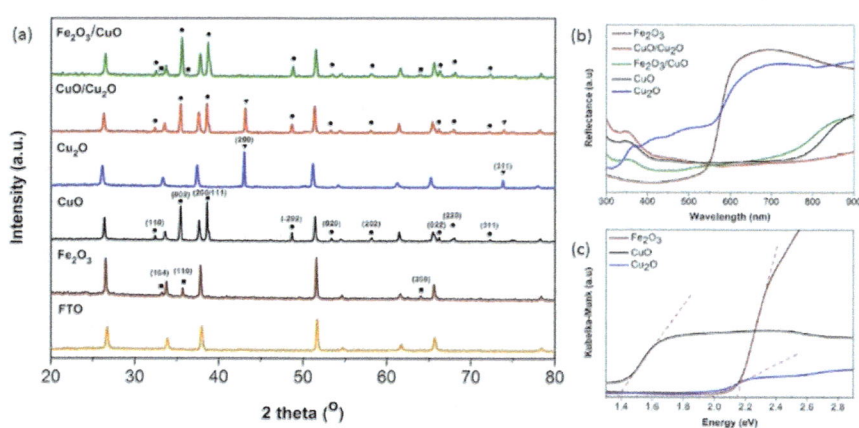

Fig. 1   (a) XRD patterns of all as-prepared photoelectrodes in comparison with bare FTO glass. (b) UV-visible diffuse reflectance spectra of the as-synthesized semiconducting materials on FTO. (c) Plots of the transformed Kubelka–Munk function *versus* the energy of absorbed light for pure CuO, Cu_2O and a-Fe_2O_3 on FTO.

(JCDPS no. 80–1917).[27] Specifically, CuO diffraction peaks were detected at $2\theta =$ 32.6°, 35.6°, 38.8°, 48.9°, 53.7°, 58.5°, 66.4°, 42.6° and 73.9° corresponding to (100), (002), (111, 200), (−202), (020), (202), (022) (220) and (311) crystal planes, respectively. In addition, no characteristic peaks related to Cu_2O, $Cu(OH)_2$ or metallic Cu were detected. The Cu_2O samples that were electrodeposited and annealed in Ar atmosphere presented two main diffraction peaks at $2\theta = 68.2°$ and 72.7° that were assigned to (200) and (311) cubic crystal structure planes of Cu_2O (JCPDS no. 05-0667).[28,29] In this sample, a weak peak ($2\theta = 50.3°$) was detected, which indicates the presence of metallic Cu impurities. XRD spectra of CuO/Cu_2O photoelectrodes showed that both monoclinic CuO and cubic Cu_2O crystal structure planes evidenced the formation of a heterojunction layer on FTO. Fig. 1a also shows the XRD patterns of iron oxide films with diffraction peaks at $2\theta = 33.1°$, 35.6° and 64° that were attributed to (104), (110) and (300) hematite structure planes (JCPDS no. 33-0664). This means that pure rhombohedral phase α-Fe_2O_3 was hydrothermally synthesized on FTO substrates without any peak indicating the presence of other iron oxide states or metallic Fe after annealing at 600 °C for 3 h in air. Both hematite and CuO characteristic diffraction peaks were observed in α-Fe_2O_3/CuO samples indicating the presence of both hematite and CuO on FTO substrate.

In Fig. 1b, the UV-vis diffuse reflection spectra of the as-synthesized Cu_2O, CuO, CuO/Cu_2O, α-Fe_2O_3 and α-Fe_2O_3/CuO films on FTO glass are illustrated. A sharp decrease in reflectance at about 560 nm was observed for sample α-Fe_2O_3. Likewise, an abrupt increase in reflectance at 570 nm was presented for the Cu_2O electrode, while the reflection edge was extended to almost 900 nm for the CuO/Cu_2O electrode. However, a sharp decrease above the wavelength corresponding to the energy band gap of both semiconducting materials is not observed as expected, which is attributed to the presence of authorized energy states within the forbidden energy region. Such states can be associated with oxygen vacancies that can be found in metal oxides synthesized *via* electrodeposition or facile hydrothermal routes. Even after the annealing process at high temperatures, oxygen vacancies can be created due to oxygen that stems from metal oxides.[30] The α-Fe_2O_3/CuO sample showed two shoulders in the reflectance spectrum due to the presence of two semiconducting materials. In general, all CuO-based films showed a broadband absorption in the wavelength region between 400–900 nm due to the low band gap energy of CuO. The band gap of CuO, Cu_2O and Fe_2O_3 was estimated to be 1.4, 2 and 2.1 eV, respectively, that matched well with the reported literature.[15,31–33]

The morphology of the samples was investigated using SEM, TEM and EDX mapping. Fig. 2a shows a top view of micro-coral arrays consisting of cubic-liked CuO particles ranging from 100 to 300 nm. In the top cross-sectional SEM image (Fig. 2b), the as-sythesized CuO photoelectrodes exhibited 2 μm thickness after 1 h of electrodeposition time. It should be mentioned that the thickness of the CuO thin films could be easily controlled by changing the electrodeposition time. The optimised electrodeposition time (1 h) for the PEC CO_2 reduction process was investigated and confirmed by electrochemical measurements of solar irradiated electrodes as shown in the following sub-section. The existence of micro-voids inside the CuO structures or in between the FTO substrate and the CuO layer could be the reason for the high electrochemically active surface area in all CuO-based photoelectrodes (Fig. 2b). For a more detailed analysis of the element

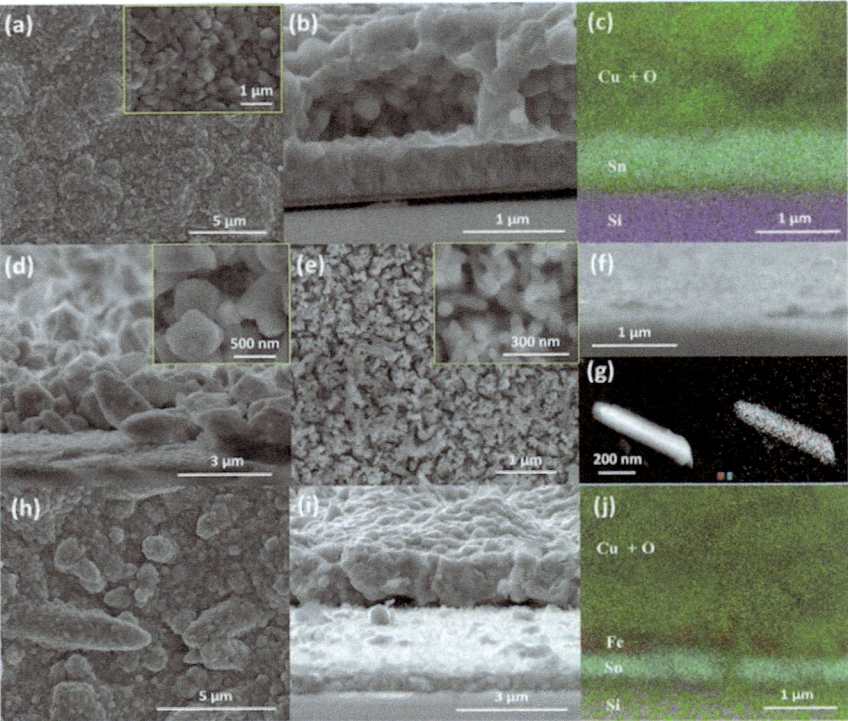

Fig. 2 SEM, TEM and EDAX or EDX characterization of as-synthesized electrodes. (a) Top, (b) cross-sectional SEM images and (c) cross-sectional EDAX element overlay of CuO photoelectrodes. (d) Cross-sectional and top view SEM images of electrodeposited Cu_2O sample. (e) Top, (f) cross-sectional SEM and (g) TEM images combined with elemental mapping of hydrothermally synthesized α-Fe_2O_3 on FTO. (h) Top, (i) cross-sectional SEM images and (j) cross-section EDAX element mapping of α-Fe_2O_3/CuO photoelectrodes.

distribution, EDAX mapping of the bare CuO film on FTO was conducted. EDAX (Fig. 2c) revealed that the glass layer at the bottom, which consisted of silicon and oxygen, was covered with a conductive layer of ~100 nm thickness that consisted of Sn. On top of the Sn layer, the electrodeposited layer consisted of Cu and O, confirming the successive deposition of a uniformly thin CuO film. In addition, large voids (300–500 nm) between the Cu_2O particles were observed (Fig. 2d).

The SEM photo of the hydrothermally synthesized α-Fe_2O_3 sample (Fig. 2e) revealed the formation of well-distributed hematite nanorods oriented upward. A typical hematite nanorod was approximately 500 nm in length and 50 nm in diameter and its elemental composition was confirmed through EDX elemental mapping (Fig. 2g). Most of the hematite nanorods have the as-described dimensions and morphology but it can be seen in the cross-sectional SEM image (Fig. 2f), that a few nanorods had significantly larger length and diameter. This can be attributed to different localized conditions in the autoclave, which is one of the main disadvantages of batch hydrothermal synthesis methods.[34] As illustrated in Fig. 2h, some of these large nanorods were covered with cube-like CuO particles highlighting the full coverage of each substrate *via* this electrodeposition method.

In addition, the surface area of the semiconducting material was significantly increased when CuO particles were electrodeposited on hematite nanorods. The cross-sectional SEM image (Fig. 2i) of the α-Fe$_2$O$_3$/CuO sample illustrated the presence of a bilayer structure consisting of a CuO layer of ~2 μm thickness deposited (1 h deposition time) on a hematite nanorods layer of 200–400 nm (6 h hydrothermal synthesis). Finally, EDAX elemental mapping confirmed that Fe is well-distributed between the conductive substrate of FTO glass and CuO layer (Fig. 2j).

PEC CO$_2$ reduction

PEC activity of the selected as-synthesized photoelectrodes was investigated in CO$_2$-purged 0.2 M NaHCO$_3$ aqueous solution. Fig. 3 shows the current density as a function of potential under solar light illumination (AM 1.5 G; 100 mW cm^{-2}). CuO and Cu$_2$O photoelectrodes showed a reductive photocurrent corresponding to PEC CO$_2$ reduction.[15,17,35,36] More specifically, a positive onset potential of ~0.8 V vs. RHE was achieved for bare CuO electrodes (Fig. 3a). The effect of CuO layer thickness was scrutinized because this property has a great impact on light absorption properties and resistance that play a crucial role in PEC activity of photoelectrodes.[37] Although CuO thickness did not affect the onset potential, the

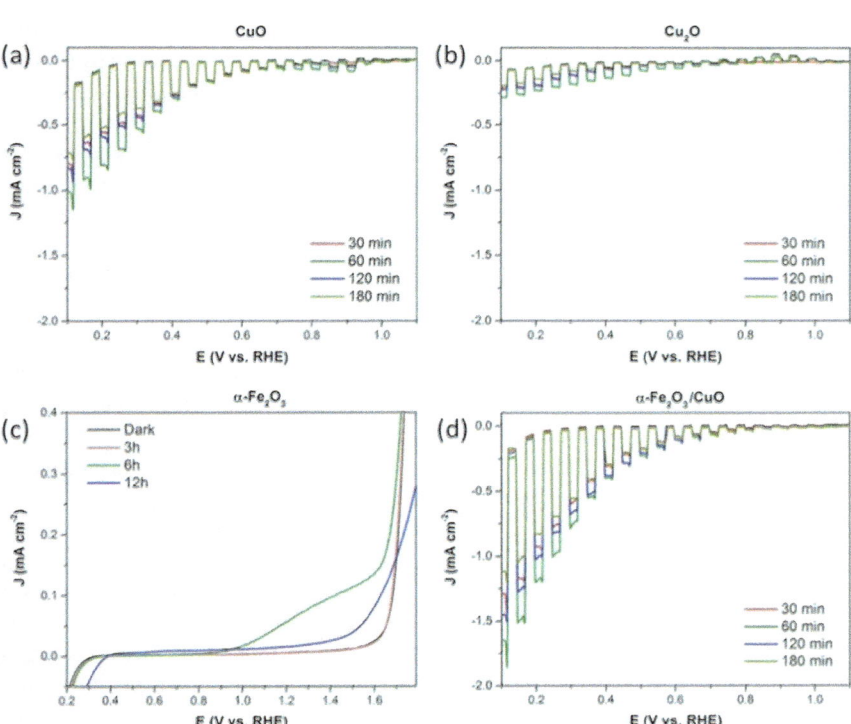

Fig. 3 Current density vs. potential (J–V) curves of (a) CuO, (b) Cu$_2$O, (c) α-Fe$_2$O$_3$ and (d) α-Fe$_2$O$_3$/CuO photoelectrodes in 0.2 M NaHCO$_3$ showing the effect of electrodeposition or hydrothermal synthesis time. All the J–V curves were obtained under continuous or chopped illumination of solar light (AM 1.5 G; 100 mW cm^{-2}) in PECR1.

photoresponse was notably enhanced leading to a photocurrent density of -0.80 from -0.55 mA cm^{-2} at 0.25 V $vs.$ RHE when CuO electrodeposition time increased from 30 to 60 min. However, a further increase in electrodeposition time resulted in a decreased photoresponse. In general, the photoabsorber's thickness can influence the photoelectrochemical activity because there is a trade-off relation between charge collection and light absorption.[37] For instance, a thin film enables efficient collection of holes at the back contact of the photoabsorber and electrons at the photoabsorber–electrolyte interface to drive the CO_2 reduction reaction but achieves low light absorption. On the other hand, although thick photoabsorber layers exhibit enhanced light absorption, electron and hole recombination limits the electron transfer process at the semiconductor interface and thus hinders subsequent oxidation or reduction reactions. Therefore, it is crucial to optimise these two key factors in order to synthesize the optimal CuO film on FTO substrate. The same trend of photoabsorber thickness on photo-response of photoelectrodes was observed in bare Cu_2O samples. Generally, Cu_2O showed significantly lower PEC performance than CuO films and this was attributed to lower absorbance of solar light (larger band gap) and the presence of empty spaces on the surface of FTO glass after electrodeposition and calcination steps. In addition, for both CuO and Cu_2O photoelectrodes, anomalies were observed in the J–V curves at 0.75 and 0.90 V $vs.$ RHE, which are attributed to Cu^{2+}, Cu^+ or Cu_2O reduction, respectively, under illumination and applied bias potential.[38]

PEC performance of hematite nanorod arrays on FTO was also studied showing a water-oxidizing photoanode behaviour. The optimum hydrothermal synthesis time was determined based on the PEC activity of the photoelectrodes. In hydrothermal synthesis, increasing the growth time will result in longer nanorods causing two opposite phenomena: (i) enhanced light absorption that increases PEC activity and (ii) broadening of the length that photogenerated electrons must diffuse in order to be collected at the FTO back contact that decreases the PEC activity. However, hematite photocatalyst has a very short lifetime of its photogenerated charge carriers (<10 ps), and thus, only very thin hematite films can be used in PEC applications.[39]

Hematite photoelectrodes with hydrothermal growth time of 6 h presented the highest photoresponse (Fig. 3c). In particular, a photocurrent density of 0.1 mA cm^{-2} at 1.20 V $vs.$ RHE and an onset potential of 0.9 V were recorded. The best hematite nanorod thin films deposited on FTO glass were used as substrates for CuO electrodeposition. Therefore, a p–n bilayer photoelectrode was formed as can be seen from SEM-EDAX images (Fig. 1a and c) and J–V curves (Fig. 3a and c). The PEC activity of these samples is presented in Fig. 3d as a function of electrode-position time of CuO. Thus, the effect of CuO thickness over hematite nanorod arrays on the overall PEC performance of photoelectrodes was explored. However, a similar behavior to the CuO and Cu_2O samples was obtained showing that the optimal electrodeposition time was 1 h. The α-Fe_2O_3/CuO samples prepared under optimal synthesis conditions (6 h hydrothermal synthesis time of hematite thin film and 1 h electrodeposition time of the CuO photoelectrodes) reproducibly achieved photocurrent densities of -1.20 mA cm^{-2} at 0.3 V $vs.$ RHE and onset potentials of 0.7 V $vs.$ RHE.

The enhanced photoelectrochemical performance of CuO layer loaded on Fe_2O_3 nanorods compared to bare CuO (Fig. 4a) can be attributed to the increased

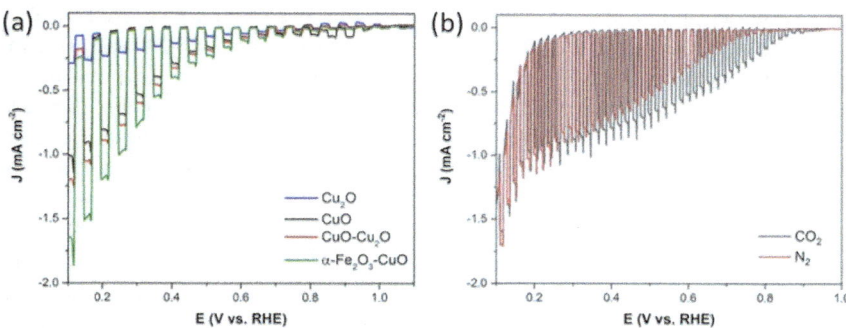

Fig. 4 (a) Comparison of current density *vs.* potential (*J–V*) curves of CuO (black), Cu_2O (blue), $CuO–Cu_2O$ (red) and α-Fe_2O_3/CuO (black) in 0.2 M $NaHCO_3$. (b) Linear sweep voltammograms of α-Fe_2O_3/CuO electrodes in 0.2 M $NaHCO_3$ electrolyte solutions purged with N_2 (red) or CO_2 (black). Both *J–V* curves were obtained under chopped illumination of solar light (AM 1.5 G; 100 mW cm^{-2}) (PECR1).

surface area as observed in SEM images, and formation of a p–n heterojunction at the interface between the two photocatalysts. Specifically, p–n junctions can provide effective separation of the photogenerated charge carriers and thereby enhance the overall photoelectrochemical activity of the α-Fe_2O_3/CuO photoelectrode. Another factor that can lead to enhanced current densities of mixed copper and iron oxide photoelectrodes for CO_2 reduction is a synergistic effect between Cu and Fe particles.[17,40,41]

To determine if charge transfer is selective to CO_2 reduction compared to hydrogen production through water splitting, linear sweep voltammetry measurements were conducted for both N_2 and CO_2 pre-purged electrolyte. As presented in Fig. 4b, the onset potential for N_2-purged electrolyte was approximately 0.75 V *vs.* RHE, while for CO_2-purged electrolyte the onset potential was anodically shifted to 0.90 V *vs.* RHE indicating that the α-Fe_2O_3/CuO heterojunction could be an efficient photoelectrode for CO_2 conversion into fuels. In addition, enhanced photocurrent density in CO_2-purged $NaHCO_3$ electrolyte can be attributed to the fact that both water and dissolved CO_2 can serve as electron acceptors in PEC cells. Finally, it should also be mentioned that sharp spikes observed in each chop cycle of all chop-lighted linear sweep voltammograms corresponded to capacity currents.[42] This effect is more intense in p–n junctions because of the formation of a depletion layer between CuO and α-Fe_2O_3 layer that results in additional capacitance in the photoelectrode.[43]

The majority of PEC reduction of CO_2 studies has focused on the development of materials with high PEC activity utilising laboratory-scale batch reactors for their evaluation. Generally, only a few studies have explored the effect of continuous flow reactors in photocatalytic or PEC systems for CO_2 reduction.[44–46] Among them, Homayoni *et al.* utilized a continuous flow PEC reactor with microchannels showing an interesting shift of products from methanol to longer chain carbon-based products. Therefore, this has already highlighted the importance of reactor design to the complex PEC CO_2 conversion process, which has a lot of similarities with both photocatalytic and electrocatalytic CO_2 conversion routes. In this study, a sandwich-type planar microfluidic PEC reactor was designed and fabricated consisting of a photoelectrode, Pt coated on FTO (counter electrode), laser cut PMMA parts and Surlyn gaskets (PECR2) without any separator between the electrodes. Although ion-conducting separators or membranes are critical components in electrochemical and PEC cells, they add substantial cost, increase the complexity of reactor design, increase the ionic resistance and are vulnerable to degradation effects.[47] Therefore, both PECR1 and PECR2 were designed and fabricated without an ion-conducting membrane.

Fig. 5a shows the total photogenerated products as a function of photo-electrosynthesis time using as-prepared photocathodes of 3 cm^2 in PECR1. CuO, Cu_2O and composite CuO/Cu_2O photoelectrodes converted CO_2 to formate and according to eqn (2) their STF efficiency after 12 h was 0.10, 0.04 and 0.14%, respectively. However, in the case of the bilayer α-Fe_2O_3/CuO photoelectrode, methanol was also obtained in addition to formate, boosting the STF efficiency to

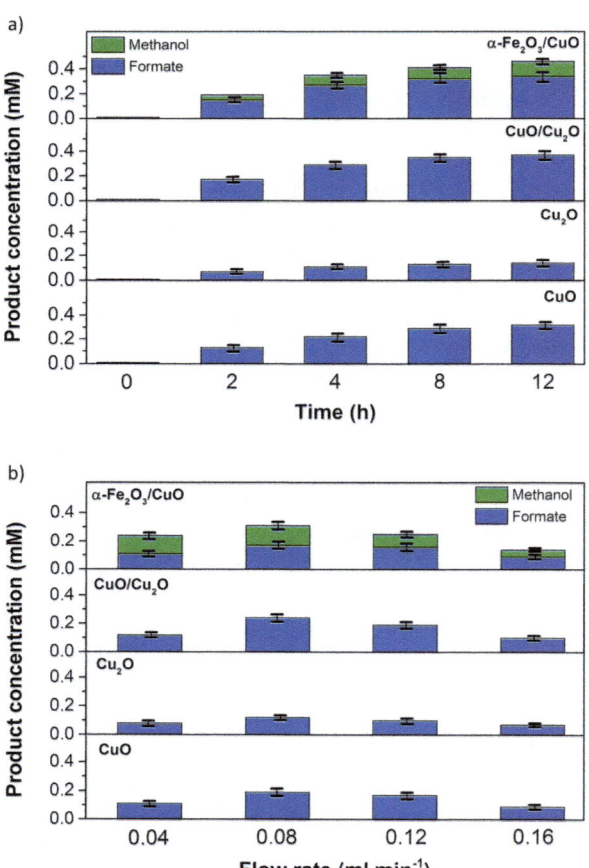

Fig. 5 (a) Product concentration in 0.2 M $NaHCO_3$ electrolyte as function of photo-electrosynthesis time for CuO, Cu_2O, CuO/Cu_2O and α-Fe_2O_3/CuO photoelectrodes used in PECR1, and (b) product concentration in 0.2 M $NaHCO_3$ electrolyte as a function of flow rate in PECR2 for CuO, Cu_2O, CuO/Cu_2O and α-Fe_2O_3/CuO photoelectrodes after 1 h of PEC CO_2 reduction experiment. In all cases photoelectrodes were directly wired with counter electrodes while not biased.

0.2%. Nonetheless, the presence of iron oxide can generate a synergistic effect between Cu and Fe atoms that can shift CO_2 reduction products to higher energy density chemicals. In addition, all curves did not illustrate a linear increase *vs.* photoelectrosynthesis time because back reactions took place simultaneously, such as methanol and formate oxidation on the dark anode.

The absence of a separator between the two electrodes allowed product crossover. In addition to liquid crossover (formate and methanol), gas crossover was observed because oxygen evolved out of the liquid electrolyte due to the water oxidation reaction that took place on the dark anode electrodes. The formation of oxygen bubbles is a deleterious effect as it reduces the active electrochemical area of the electrodes and enhances product mixing.[48] One way to minimize this effect was the utilization of a continuous flow reactor instead of a batch reactor. Membraneless continuous flow cells can reduce the ionic resistance due to the greater ion-mobility in liquid electrolyte solutions.[48] Flow reactors can also significantly reduce the cost and improve the stability of PEC systems bringing them a step closer to an economically viable industrial application. For these reasons, the performance of the as-fabricated microfluidic reactor (PECR2) was also evaluated.

As illustrated in Fig. 5b, in all cases a flow rate of 0.08 mL min^{-1} was optimal. Initially, an increase in flow rate led to an enhancement of CO_2 conversion into methanol and formate but when the flow rate increased higher than 0.08 ml min^{-1}, the efficiency of the system decreased. This effect was associated with the efficient reactant supply and residence time. A longer residence time means that dissolved CO_2 in the electrolyte remains in contact with the photoelectrode for sufficient time to complete the desired reaction but simultaneously the products in contact with the illuminated semiconducting material or Pt can be re-oxidized. Another plausible limiting factor in the efficiency of PEC CO_2 reduction at low flow rates is the formation of oxygen bubbles in the microchamber of PECR2. Oxygen bubbles that significantly decrease the efficiency of the cell were observed in flow rates lower than 0.01 mL min^{-1} due to small fluidic drag force while such bubbles were not observed at flow rates higher than 0.02 mL min^{-1} showing stable operation. In other words, as the flow rate increased the drag force became larger decreasing the size of the bubbles. On the other hand, at high flow rates the diffusion of the products across the electrodes is ruling the CO_2 reduction process. Another difference between PECR1 and PECR2 is that in batch reactors the large inter-electrode distance results in large current gradients, while in continuous flow microreactors with a small distance between electrodes this disadvantageous effect can be minimized.[49] In addition, high surface-area-to-volume ratio of both photocatalysts and electrocatalysts improved the mass transfer and decreased the reaction time.[46] Therefore, the optimal α-Fe_2O_3/CuO photoelectrode achieved the highest overall STF efficiency of 0.48% after 1 h with 0.08 mL min^{-1} continuous flow rate in microfluidic reactor with close spacing of 160 μm. Although the highest STF efficiency of 0.48% in this study is low, it is among the highest recorded in such PEC CO_2 reduction systems[9] while the best STF efficiency reported so far is 1%.[50] Overall, the low STF efficiency that was observed in all cases can also be attributed to low solubility of CO_2 in water-based electrolytes (0.034 M) at standard pressures.[45,51] The limitation of CO_2 solubility in water sets a maximum limit of catalytic current density and therefore STF efficiency for PEC CO_2 reduction.

Fig. 5b illustrated that utilization of PECR2 led to production of higher concentrations of formate and methanol and enhanced percentage of methanol formation in the case of α-Fe$_2$O$_3$/CuO photoelectrodes. Thus, the continuous flow microfluidic PEC reactor promoted the formation of methanol which has significantly higher energy density than formate. A possible chemical route for CO$_2$ conversion into methanol is that methanol formation commenced after formate is already formed. An indication for this scenario is that the product distribution changed as a function of irradiation time. As illustrated in Fig. 5a, the product distribution shifted away from formate toward methanol for photo-electrosynthesis times longer than 2 h. Methanol formation is enhanced in PECR2 (Fig. 5b) due to the high surface-area-to-volume ratio of both photocatalyst and electrocatalyst and the small inter-electrode distance that increase the interaction of the generated products.

Conclusions

In this work, an electrodeposition method was employed to synthesize CuO-based photoelectrodes for PEC CO$_2$ conversion into liquid fuels. Hydrothermal deposition of a hematite thin layer prior to CuO electrodeposition significantly increased the PEC activity of the electrodes due to a higher surface area and formation of a p–n junction between the two layers. Therefore, coupling α-Fe$_2$O$_3$ nanorods with CuO considerably enhanced the PEC CO$_2$ reduction efficiency and it seems a promising heterojunction for artificial solar fuel generation. In addition, a continuous flow microfluidic PEC cell was developed for PEC CO$_2$ reduction experiments showing the ability to convert CO$_2$ into formate and methanol. Both products were measured to evaluate the feasibility and performance of the system. It was found that the optimal flow rate was 0.08 ml min^{-1} and the α-Fe$_2$O$_3$/CuO under optimal conditions achieved an STF efficiency of 0.48%. Finally, it was proven that fabrication of a continuous flow reactor from low-cost materials can offer the ability to investigate the potential benefits of microfluidics in PEC CO$_2$ reduction using existing technology of electrochemical and PEC cells.

Conflicts of interest

There are no conflicts to declare.

Acknowledgements

The authors would like to acknowledge the Engineering and Physical Sciences Research Council (EPSRC) for financial support through the projects EP/K021796/1, EP/N009924/1, and EP/R012164/2. The electron microscopy facility in the School of Chemistry, University of St. Andrews, which is supported by the EPSRC Capital for Great Technologies Grant EP/L017008/1, is acknowledged.

References

1 S.-M. Park, A. Razzaq, Y. H. Park, S. Sorcar, Y. Park, C. A. Grimes and S.-I. In, *ACS Omega*, 2016, **1**, 868–875.
2 L. Hammarström, *Faraday Discuss.*, 2017, **198**, 549–560.

3 O. Ola and M. M. Maroto-Valer, *J. Photochem. Photobiol., C*, 2015, **24**, 16–42.

4 J. L. White, M. F. Baruch, J. E. Pander Iii, Y. Hu, I. C. Fortmeyer, J. E. Park, T. Zhang, K. Liao, J. Gu, Y. Yan, T. W. Shaw, E. Abelev and A. B. Bocarsly, *Chem. Rev.*, 2015, **115**, 12888–12935.

5 J. Ronge, T. Bosserez, D. Martel, C. Nervi, L. Boarino, F. Taulelle, G. Decher, S. Bordiga and J. A. Martens, *Chem. Soc. Rev.*, 2014, **43**, 7963–7981.

6 J. Zhao, X. Wang, Z. C. Xu and J. S. C. Loo, *J. Mater. Chem. A*, 2014, **2**, 15228–15233.

7 M. Mikkelsen, M. Jorgensen and F. C. Krebs, *Energy Environ. Sci.*, 2010, **3**, 43–81.

8 M. Halmann, *Nature*, 1978, **275**, 115–116.

9 E. Kalamaras, M. M. Maroto-Valer, M. Shao, J. Xuan and H. Wang, *Catal. Today*, 2018, **317**, 56–75.

10 J. Cheng, M. Zhang, J. Liu, J. Zhou and K. Cen, *J. Mater. Chem. A*, 2015, **3**, 12947–12957.

11 J. H. Kim, G. Magesh, H. J. Kang, M. Banu, J. H. Kim, J. Lee and J. S. Lee, *Nano Energy*, 2015, **15**, 153–163.

12 G. Magesh, E. S. Kim, H. J. Kang, M. Banu, J. Y. Kim, J. H. Kim and J. S. Lee, *J. Mater. Chem. A*, 2014, **2**, 2044–2049.

13 M. Deguchi, S. Yotsuhashi, H. Hashiba, Y. Yamada and K. Ohkawa, *Jpn. J. Appl. Phys.*, 2013, **52**, 08JF07.

14 A. Kudo and Y. Miseki, *Chem. Soc. Rev.*, 2009, **38**, 253–278.

15 U. Kang, S. K. Choi, D. J. Ham, S. M. Ji, W. Choi, D. S. Han, A. Abdel-Wahabe and H. Park, *Energy Environ. Sci.*, 2015, **8**, 2638–2643.

16 *Photoelectrochemical Hydrogen Production*, ed. R. van de Krol and M. Grätzel, Springer US, New York, 1st edn, 2012, DOI: 10.1007/978-1-4614-1380-6.

17 U. Kang and H. Park, *J. Mater. Chem. A*, 2017, **5**, 2123–2131.

18 A. Annamalai, P. S. Shinde, A. Subramanian, J. Y. Kim, J. H. Kim, S. H. Choi, J. S. Lee and J. S. Jang, *J. Mater. Chem. A*, 2015, **3**, 5007–5013.

19 L. Vayssieres, N. Beermann, S.-E. Lindquist and A. Hagfeldt, *Chem. Mater.*, 2001, **13**, 233–235.

20 A. B. Murphy, *Sol. Energy Mater. Sol. Cells*, 2007, **91**, 1326–1337.

21 E. L. Simmons, *Appl. Opt.*, 1975, **14**, 1380–1386.

22 Y. Lee, J. Suntivich, K. J. May, E. E. Perry and Y. Shao-Horn, *J. Phys. Chem. Lett.*, 2012, **3**, 399–404.

23 H. Zhong, K. Fujii and Y. Nakano, *J. Electrochem. Soc.*, 2017, **164**, F923–F927.

24 H. Zhong, K. Fujii, Y. Nakano and F. Jin, *J. Phys. Chem. C*, 2015, **119**, 55–61.

25 M. D. Salazar-Villalpando, *ECS Trans.*, 2011, **33**, 77–88.

26 M. M. May, D. Lackner, J. Ohlmann, F. Dimroth, R. van de Krol, T. Hannappel and K. Schwarzburg, *Sustainable Energy & Fuels*, 2017, **1**, 492–503.

27 C. V. Niveditha, M. J. J. Fatima and S. Sindhu, *J. Electrochem. Soc.*, 2016, **163**, H426–H433.

28 A. K. Sasmal, S. Dutta and T. Pal, *Dalton Trans.*, 2016, **45**, 3139–3150.

29 S. M. Pawar, J. Kim, A. I. Inamdar, H. Woo, Y. Jo, B. S. Pawar, S. Cho, H. Kim and H. Im, *Sci. Rep.*, 2016, **6**, 21310.

30 M. M. El-Nahass, H. S. Soliman and A. El-Denglawey, *Appl. Phys. A*, 2016, **122**, 775.

31 K. Rajeshwar, N. R. de Tacconi, G. Ghadimkhani, W. Chanmanee and C. Janaky, *ChemPhysChem*, 2013, **14**, 2251–2259.

32 J. Luo, L. Steier, M.-K. Son, M. Schreier, M. T. Mayer and M. Grätzel, *Nano Lett.*, 2016, **16**, 1848–1857.

33 E. Kalamaras, V. Dracopoulos, L. Sygellou and P. Lianos, *Chem. Eng. J.*, 2016, **295**, 288–294.

34 A. Ali, H. Zafar, M. Zia, I. Ul Haq, A. R. Phull, J. S. Ali and A. Hussain, *Nanotechnol., Sci. Appl.*, 2016, **9**, 49–67.

35 G. Ghadimkhani, N. R. de Tacconi, W. Chanmanee, C. Janaky and K. Rajeshwar, *Chem. Commun.*, 2013, **49**, 1297–1299.

36 P. Wang, Y. H. Ng and R. Amal, *Nanoscale*, 2013, **5**, 2952–2958.

37 H. Tada and M. Tanaka, *Langmuir*, 1997, **13**, 360–364.

38 C. Li, T. Hisatomi, O. Watanabe, M. Nakabayashi, N. Shibata, K. Domen and J.-J. Delaunay, *Energy Environ. Sci.*, 2015, **8**, 1493–1500.

39 Gurudayal, S. Y. Chiam, M. H. Kumar, P. S. Bassi, H. L. Seng, J. Barber and L. H. Wong, *ACS Appl. Mater. Interfaces*, 2014, **6**, 5852–5859.

40 P. Q. Li, H. Jing, J. F. Xu, C. X. Wu, H. Peng, J. Lu and F. S. Lu, *Nanoscale*, 2014, **6**, 11380–11386.

41 X. Yang, E. A. Fugate, Y. Mueanngern and L. R. Baker, *ACS Catal.*, 2017, **7**, 177–180.

42 A. Martínez-García, V. K. Vendra, S. Sunkara, P. Haldankar, J. Jasinski and M. K. Sunkara, *J. Mater. Chem. A*, 2013, **1**, 15235–15241.

43 T. Jiang, T. Xie, W. Yang, L. Chen, H. Fan and D. Wang, *J. Phys. Chem. C*, 2013, **117**, 4619–4624.

44 K. Li, X. An, K. H. Park, M. Khraisheh and J. Tang, *Catal. Today*, 2014, **224**, 3–12.

45 H. Homayoni, W. Chanmanee, N. R. de Tacconi, B. H. Dennis and K. Rajeshwar, *J. Electrochem. Soc.*, 2015, **162**, E115–E122.

46 X. Cheng, R. Chen, X. Zhu, Q. Liao, L. An, D. Ye, X. He, S. Li and L. Li, *Energy*, 2017, **120**, 276–282.

47 G. D. O'Neil, C. D. Christian, D. E. Brown and D. V. Esposito, *J. Electrochem. Soc.*, 2016, **163**, F3012–F3019.

48 S. M. H. Hashemi, M. A. Modestino and D. Psaltis, *Energy Environ. Sci.*, 2015, **8**, 2003–2009.

49 A. A. Folgueiras-Amador, K. Philipps, S. Guilbaud, J. Poelakker and T. Wirth, *Angew. Chem., Int. Ed.*, 2017, **56**, 15446–15450.

50 U. Kang, S. K. Choi, D. J. Ham, S. M. Ji, W. Choi, D. S. Han, A. Abdel-Wahab and H. Park, *Energy Environ. Sci.*, 2015, **8**, 2638–2643.

51 C. S. Wong, P. Y. Tishchenko and W. K. Johnson, *J. Chem. Eng. Data*, 2005, **50**, 817–821.

Faraday Discussions

DISCUSSIONS

Demonstrator devices for artificial photosynthesis: general discussion

Ryu Abe, [iD] Catherine M. Aitchison, [iD] Virgil Andrei, [iD]
Matthias Beller, [iD] Daniel Cheung, [iD] Charles E. Creissen, Víctor A. de la
Peña O'Shea, [iD] James R. Durrant, [iD] Michael Grätzel,
Leif Hammarström, [iD] Sophia Haussener, [iD] Su-Il In,
Evangelos Kalamaras, [iD] Akihiko Kudo, [iD] Moritz F. Kuehnel, [iD]
Pramod Patil Kunturu, Yi-Hsuan Lai, [iD] Chong-Yong Lee, [iD]
Marcelino Maneiro, [iD] Esther Edwardes Moore,
Huu Chuong Nguyen, [iD] Aubrey R. Paris, [iD] Chanon Pornrungroj, [iD]
Joost N. H. Reek, [iD] Erwin Reisner, [iD] Murielle Schreck, [iD]
Wilson A. Smith, [iD] Han Sen Soo, [iD] Reiner Sebastian Sprick,
Anirudh Venugopal, [iD] Qian Wang, Dominik Wielend [iD]
and Martijn A. Zwijnenburg [iD]

DOI: 10.1039/C9FD90023C

Wilson Smith opened discussion of the paper by Michael Grätzel: For the Cu–Au electrodes, the proposed mechanism is that CO_2 is converted to CO on the Au site, then diffuses on the surface to a Cu site where it hydrogenates to form higher order products. Given that the system is highly porous and the local pH may be higher than the bulk solution, can you comment on the likelihood for the proposed mechanism (surface diffusion of CO), or if CO is formed in the gas phase, then re-adsorbs on the Cu surface from the solution? The two pathways have large implications on mechanistic understandings, and different approaches would be needed to optimise these steps in different ways.

Michael Grätzel replied: We propose indeed that the CO directly spills over directly from the Au nanoparticles to the Cu support. Nevertheless your point is well taken that the CO adsorption on Au is reversible and hence that there is an equilibrium between the adsorbed and free CO. So the path of CO transfer from Au to Cu *via* the gas phase is a viable alternative to the direct spillover. It is favored by the fact that desorption of CO can occur from any Au surface site while in order for the spillover to occur the Au site must have adjacent Cu sites. More analysis is required to determine the predominant mechanism.

Huu Chuong Nguyen queried: What would be the reduction mechanism with the zinc system?

Michael Grätzel replied: We are still examining the details of this mechanism. However it is well known that the Topsoe catalyst promoting the formation of methanol from CO/CO_2 also uses a bimetallic Cu/Zn formulation.

James Durrant commented: Michael, you showed impressive results on the electrochemical reduction of CO_2. Could you comment on the issue of CO_2 solubility in aqueous solutions, and at what current densities gas diffusion electrodes are needed?

Michael Grätzel answered: We find that for our H-cells the current densities for CO_2 reduction in aqueous electrolyte are limited to ca. 50 mA cm^{-2}. Going above this limit requires the use of gas diffusion electrodes for which the current densities can attain several hundred mA cm^{-2}.

Qian Wang opened the discussion of the paper by Yi-Hsuan Lai: What is the real Faradic efficiency of your tandem cells? Have you checked the gas evolution? If the hydrogen produced by your system is undetectable, how do you confirm the observed photocathode current was due to the water reduction?

Yi-Hsuan Lai answered: We have tried to detect the hydrogen gas by using gas chromatography. However, at this stage, the photocurrent in our system is far below the detection limit of gas chromatography. The performance of our system must be further improved in order to allow the products to be detected.

Esther Edwardes Moore asked: Looking at Fig. 4a of the paper (DOI: 10.1039/ c8fd00183a), at potentials less than 0.5 V *vs.* RHE, the dark current increases on the introduction of O_2 into the system. Could you comment on the cause of this effect?

Yi-Hsuan Lai answered: The increased dark current on the introduction of O_2 at potentials less than 0.5 V *vs.* RHE possibly results from the reduction of O_2 to peroxide ion *via* the two-electron pathway.[1]

1 X. Ge, A. Sumboja, D. Wuu, T. An, B. Li, F. W. T. Goh, T. S. A. Hor, Y. Zong and Z. Liu, *ACS Catal.*, 2015, 5, 4643–4667.

Ryu Abe commented: It is a nice idea that you prepared two different semiconductor materials, p-type $CuBi_2O_4$ and n-type $BiVO_4$, from a single precursor. Have you ever tried to prepare these two materials on one side of the FTO substrate? In the present study, you eventually prepared these two materials separately, probably due to different conditions required to obtain optimum performance. I understand the difficulty, but if you can prepare such an electrode on which n-type and p-type semiconductors contact directly, as well as *via* the FTO substrate, some interesting properties might be obtained.

Yi-Hsuan Lai responded: Many thanks for the fruitful comment. Yes, we have considered preparing such an electrode on which n-type and p-type semiconductors contact directly and it is indeed our final goal. However, since the performance of $CuBi_2O_4$ is much lower than that of $BiVO_4$, enhancing the performance of $CuBi_2O_4$ is our top priority at the current stage.

Sophia Haussener remarked: You mention that hydrogen and oxygen evolution in one location or chamber is possible. Is this however implementable in a practical and scalable reactor?

Yi-Hsuan Lai replied: I am sorry that I do not remember I have mentioned hydrogen and oxygen evolution in one location or chamber is possible. However, I believe a suitable membrane is necessary for a practical and scalable reactor to prevent product crossover.

Sophia Haussener asked: What are the optical properties of your membrane?

Yi-Hsuan Lai replied: The UV–Vis transmittance spectrum suggests the membrane has high transmittance of up to 90% in the wavelength range from 300 to 800 nm.

Chanon Pornrungroj commented: According to the manuscript, there seems to be a similarity in the fabrication process of the $CuBi_2O_4$ photocathode and $BiVO_4$ photoanode. Have you ever tried to engineer both types of electrodes on a single FTO at the same time? It could be interesting to use the inkjet printer to engineer and print different patterns of Cu^{2+} and VO^{2+} precursors on the nano$Bi_4O_5I_2$ to achieve an effective overall water splitting.

Yi-Hsuan Lai answered: Many thanks for the suggestion. We have not tried to engineer both types of electrodes on a single FTO, although we have indeed considered that. Enhancing the performance of $CuBi_2O_4$ by engineering its synthetic conditions is our top priority at this stage. The two types of semi-conductor materials will then be engineered on a single FTO if the performance of the $CuBi_2O_4$ photocathode can be comparable with that of the $BiVO_4$ photoanode.

Sophia Haussener asked: Can you comment on the stability of your electrodes and predict long term performance of your photoanode and photocathode?

Yi-Hsuan Lai responded: The photocurrent of the photoanode gradually decreased over time and had a half-life time of approximately one hour at a potential of 1.23 V *vs.* RHE. On the other hand, the photocurrent of the photocathode quickly dropped within the first five minutes and remained roughly at only 15% of its initial photocurrent at a potential of 0.5 V *vs.* RHE.

Charles Creissen asked: With respect to the photodegradation of $CuBi_2O_4$, is this a process that can be limited through efficient extraction of electrons, for example through immobilisation of a catalyst?

Yi-Hsuan Lai replied: Yes, since the photodegradation of $CuBi_2O_4$ might result from the reduction of copper ions, we believe immobilising an efficient catalyst on the surface of the $CuBi_2O_4$ electrode can enhance the stability of $CuBi_2O_4$.

Murielle Schreck opened the discussion of the paper by Akihiko Kudo: You measure the activity of your systems in μmol h^{-1} and you use different amounts of photocatalyst depending on the material. Why do you divide only by the time and not by the mass too and what effect does the use of different masses of photocatalysts have on their activity? Can you also make a comment on quantum yield or quantum efficiency?

Akihiko Kudo responded: The optimum amount of photocatalyst depends on the photocatalyst, photocatalytic reaction, light source and setup for the measurement. The optimum amount of photocatalyst should be determined with your experimental setup as well as the reaction condition, such as pH. So, I just show the activity in μmol h^{-1} and indicating the experimental setup and condition without any normalization. The normalization of a photocatalytic activity by mass of photocatalyst is incorrect, because a photocatalytic activity is not always proportional to the mass.[1] A photocatalytic activity is fundamentally limited not by the mass but a photon flux. An activity should be indicated using an optimum amount of photocatalyst for each experimental setup. A quantum yield and a solar energy conversion efficiency are values to evaluate absolutely the performance of a photocatalytic reaction. These values should not depend on an experimental condition, if measurement of a photocatalytic reaction is suitably conducted. The solar energy conversion efficiency (STH: solar to hydrogen conversion efficiency) will be the absolute indication of performance of artificial photosynthesis. For example, the amount of H_2 obtained using a solar simulator or actual sunlight should be measured.

1 A. Kudo and Y. Miseki, Heterogeneous photocatalyst materials for water splitting, *Chem. Soc. Rev.*, 2009, **38**, 253–278.

Catherine Aitchison commented: It is particularly interesting that some of your systems work without redox mediators (when catalysts are just in suspension). Could the activities of these Z-schemes be improved by synthesising heterostructures or Janus-like particles of the two photocatalysts? Do you think this is a worthwhile approach or will systems with redox mediators always be more viable?

Akihiko Kudo answered: Synthesizing heterostructures or Janus-like particles of the two photocatalysts will be an attractive strategy in order to improve a Z-schematic photocatalyst system, because the contact becomes better and separations of charge and reaction sites may be enhanced by the asymmetric structure. We reported such a composite photocatalyst of $SrTiO_3$:Rh and $BiVO_4$ (ref. 1) though the morphology is not beautiful compared with the ideal Janus-like particles. It is an advantage for the composite system that no additives such as an electron mediator are required. If the morphology and the structure are suitably designed, the activity will be much enhanced. However, there are other issues for the heterostructure. Undesirable back electron transfer, for example electron transfer from a conduction band of a H_2-evolving photocatalyst not to water molecules but a conduction band of an O_2-evolving photocatalyst, might also be enhanced due to firm contact resulting in a decrease in a photocatalytic activity. So, I can not say at the present stage which is better, with an electron mediator or just by contact.

1 Q. Jia, A. Iwase and A. Kudo, BiVO$_4$–Ru/SrTiO$_3$:Rh composite of Z-scheme photocatalyst for solar water splitting, *Chem. Sci.*, 2014, **5**, 1513–1519.

Virgil Andrei asked: Some of the photocathodes present a large hysteresis between the forward and backward scans of the cyclic voltammograms in Fig. 5 of the paper (DOI: 10.1039/c8fd00209f). What causes this hysteresis, and why is it not observed in the case of the photoanodes? Could that indicate a capacitive charging of your electrodes, or is there another reason for the dark currents?

Akihiko Kudo responded: It is not clear at the present stage. The hysteresis would mainly be due to the redox behaviour of doped Rh and a capacitance property of powdered materials. In general, such a hysteresis is observed for a material with a capacitance property. We used a powdered material with a rough surface for the photoelectrodes. Such behavior could be suppressed when the quality of a photoelectrode is high. Actually, such a hysteresis is not observed for the photoelectrodes of single crystalline thin films prepared by pulse laser deposition.[1,2] TiO$_2$:Rh and Sb does not show the hysteresis, because a TiO$_2$-based photoanode is an efficient photoanode as well as BiVO$_4$.

1 S. Kawasaki, K. Akagi, K. Nakatsuji, S. Yamamoto, I. Matsuda, Y. Harada, J. Yoshinobu, F. Komori, R. Takahashi, M. Lippmaa, C. Sakai, H. Niwa, M. Oshima, K. Iwashina and A. Kudo, Elucidation of Rh-Induced In-Gap States of Rh:SrTiO$_3$ Visible-Light-Driven Photocatalyst by Soft X-Ray Spectroscopy and First-Principles Calculations, *J. Phys. Chem. C*, 2012, **116**, 24445–24448.
2 S. Kawasaki, R. Takahashi, T. Yamamoto, M. Kobayashi, H. Kumigashira, J. Yoshinobu, F. Komori, A. Kudo and M. Lippmaa, Photoelectrochemical water splitting enhanced by self-assembled metal nanopillars embedded in an oxide semiconductor photoelectrode, *Nat. Commun.*, 2016, **7**, 11818.

Reiner Sebastian Sprick asked: When going to large scale applications, it would probably be desirable to use saline water. I would expect that the ionic species used as mediators would be affected by salt content. Have you studied this and how does it effect the performance?

Akihiko Kudo replied: Yes, chloride ions in a salt affect the performance. When saline water was used for water splitting using a Z-scheme photocatalyst employing an Fe$^{3+/2+}$ redox couple as a mediator, the activity was about half compared with that in pure water. In other words, it worked. Oxidation products of chloride ions such as Cl$_2$ and HClO might have a negative effect on water splitting activity by the redox reactions of those species. If chloride ions coordinate to metal complexes of an electron mediator, some properties of the metal complex such as a redox potential will change. I am not sure that such a coordination effect is significant to the activity. River water is also a candidate, unless it contains heavy metal cations. It will be much better than saline water. If it contains some organic compounds working as an electron donor, more hydrogen could be obtained.

Reiner Sebastian Sprick queried: Would you be able to comment on how to scale up these systems as agitation of suspension on a large scale would be an additional challenge?

Akihiko Kudo responded: We suggest not using an agitation system but instead the photocatalyst sheet system proposed by Prof. Domen for large-scale practical use.[1,2] It will be easy to handle it.

1 Q. Wang, T. Hisatomi, M. Katayama, T. Takata, T. Minegishi, A. Kudo, T. Yamada and K. Domen, Particulate photocatalyst sheets for Z-scheme water splitting: advantages over powder suspension and photoelectrochemical systems and future challenges, *Faraday Discuss.*, 2017, **197**, 491–504.
2 T. Yamada and K. Domen, *ChemEngineering*, 2018, **2**, 36.

Moritz F. Kuehnel asked: Could you comment on the prospects and challenges of scaling up one-pot water splitting? The formation of H_2 and O_2 at the same time in the same vessel is supposedly going to be not only a safety problem, but also a cost factor. How much will implementing a separation of H_2 mechanism (*e.g.* membrane) lower the overall efficiency of the process, in particular considering that photocatalytic water splitting has a relatively low efficiency even without separation?

Akihiko Kudo replied: Yes, safety and cost issues are important. Please see the literature for reply to your comments.[1,2]

The most important issue is still to develop photocatalyst materials showing high solar energy conversion efficiency by utilizing a wide solar spectrum after overcoming the safety and cost issues. The cost issue may rely mainly on the system, such as the reactor, not the photocatalyst.

1 T. Setoyama, T. Takewaki, K. Domen and T. Tatsumi, *Faraday Discuss.*, 2017, **198**, 509–527.
2 T. Yamada and K. Domen, *ChemEngineering*, 2018, **2**, 36.

Han Sen Soo said: In your Z-scheme systems, Rh and Ir were used as dopants. However, the Ir and Rh are not used as catalysts but mainly as light absorbers. In this case, to reduce the costs, have you considered using first row transition metals, Mo, or W as the dopants instead?

Akihiko Kudo responded: Yes, we tested various dopants including first row transition metals. Ir and Rh are the best dopants at the present stage, though they are noble metals. Mn and Ru are also effective dopants to give a visible light response.[1] It is reasonable judging from their suitable redox properties. I do not think Mo and W can be used instead of Ir and Rh, because those dopants lower the conduction band of $SrTiO_3$. It results in loss of H_2-evolution ability.

1 R. Konta, T. Ishii, H. Kato and A. Kudo, Photocatalytic Activities of Noble Metal Ion Doped $SrTiO_3$ Under Visible Light Irradiation, *J. Phys. Chem. B.*, 2004, **108**(26), 8992–8995.

Martijn Zwijnenburg asked: Knowledge of the band structure of materials is, based on your work, clearly important when trying to understand which combination of materials, or which combination of materials and a given redox mediator, will form a successful Z-scheme for overall water splitting. Could you please tell us more about how you obtain such information, especially the relative positions of the conduction band minima and valence band maxima *vs.* NHE as shown in Fig. 6 of the paper (DOI: 10.1039/c8fd00209f).

Akihiko Kudo responded: It would be nice if we could experimentally determine the absolute positions of conduction and valence bands and an impurity level, but it is not easy, especially for powdered samples. You can use UPS for the determination of Fermi energy if you have high quality thin film materials. You may see the valence band position by looking at the valence band region in XPS. Here, we have estimated the band positions as follows.[1] In many metal oxide semiconductor photocatalysts consisting of MO_6 polyhedra, the valence bands are formed with O2p orbitals. The valence band position is around 3.0 V *vs.* NHE at pH 0. The photocatalysts in Fig. 6 are supposed to be on the line. The positions of a conduction band and an impurity level can be determined from the valence band level and band gaps derived from absorption spectra, usually diffuse reflectance spectra for solid materials. These processes give the band structure of Fig. 4 and 6.

1 A. Kudo and Y. Miseki, Heterogeneous photocatalyst materials for water splitting, *Chem. Soc. Rev.*, 2009, **38**, 253–278.

Wilson Smith opened the discussion of the paper by Evangelos Kalamaras: It was interesting to see that as a function of flow rates, the formate and methanol selectivity was very high across the range tested (especially at low flow rates), but in a batch reactor, only formate was formed. Can you comment on how you think the effects of mass transport improve methanol synthesis? Does this imply formate is a precursor to methanol?

Evangelos Kalamaras responded: Fig. 5b in the paper (DOI: 10.1039/c8fd00192h) showed that utilization of a continuous flow PEC reactor led to production of higher concentrations of formate and methanol and increased the percentage of methanol formation in the case of α-Fe_2O_3/CuO photoelectrodes. We believe that this is highly related to the improved mass transport of the flow reactor because the percentage of methanol production is significantly lower in a batch reactor. In addition, crossover of products due to the absence of a separator was significantly higher in the batch reactor leading to re-oxidation of products. A plausible scenario for CO_2 reduction into methanol is that its formation is initiated after some amount of formate is already formed. Rajeshwar and co-workers observed a similar behaviour of a continuous PEC flow system for CO_2 reduction into ethanol. In order to better understand the proposed mechanism, experiments with electrolytes containing formate should be conducted.

1 H. Homayoni, W. Chanmanee, N. R. de Tacconi, B. H. Dennis and K. Rajeshwar, *J. Electrochem. Soc.*, 2015, **162**, E115-E122.

Dominik Wielend asked: I have two questions concerning the issue you mentioned in your Faraday Discussions paper about the "non-linearity of the product formation over time" in your batch-reactor experiment:

Did you actually test if your catalyst is stable after this 12 h reaction by some surface analysis technique?

Or did you also investigate the flow-reactor experiment for this longer time of 12 h instead of the reported 1 h to check if the catalyst still has the same activity after that reaction time?

Evangelos Kalamaras responded: Yes, the photoelectrodes were stable for more than 12 h. Photoelectrodes consisting of similar semiconducting materials have shown remarkably stability for more than one week as Kang and co-workers reported.[1] However, in our case we have not performed any surface analysis technique apart from SEM after the reaction. The non-linearity of the product formation in a PEC batch reactor was attributed to back reactions due to the absence of a membrane. For this reason, we believe that the continuous flow PEC reactor showed more linear hourly production. The experiments in the continuous flow system were conducted for 6 to 12 h depending on the flow rate. We could not do experiments for more than 12 h in the continuous-flow system due to limitations of the syringe pump equipment. Generally, the stability of the photoelectrodes was tested by using each photoelectrode for 5 different PEC measurements showing similar performance.

1 U. Kang, S. K. Choi, D. J. Ham, S. M. Ji, W. Choi, D. S. Han, A. Abdel-Wahab and H. Park, *Energy Environ. Sci.*, 2015, **8**, 2638–2643.

Han Sen Soo asked: Since you proposed that methanol may be formed from formate, did you examine if formate could be used as a substrate to produce methanol? Are there any other intermediates that can also generate methanol? This can provide some insights about the mechanism of the reduction.

Evangelos Kalamaras replied: We have not conducted experiments with electrolyte containing formate in order to have more evidence about the proposed mechanism. As far as I'm aware, there are no other experimental research works that studied the mechanism of PEC CO_2 reduction into methanol in order to provide more insights. Therefore, I believe that further investigation of the CO_2 reduction mechanism using *in situ* and operando characterization techniques is required to unveil the intermediate steps.

Sophia Haussener queried: In order to tailor your products, could it make sense to use a sequential reactor that includes not just one electrode but a second, tailored one?

Evangelos Kalamaras replied: A proposed scenario for CO_2 reduction into methanol is that its formation is initiated after some amount of formate is already formed. However, further experiments are required to study the mechanism of PEC CO_2 reduction. The existence of intermediate products will definitely favour the utilization of such a reactor with subsequent photoelectrodes allowing the CO_2 reduction reaction to take place in a series of multi-electron steps on different photocatalysts. In this way, STF efficiency of the system will be maximized.

Sophia Haussener asked: How would you scale up this microfluidic system? What are the perfect operating conditions, in terms of velocity, *etc.*?

Evangelos Kalamaras replied: In our system the optimum flow rate was 0.08 ml min^{-1}.

One cost-effective way that microfluidic PEC devices can be scaled-up is the fabrication and utilization of large planar electrodes separated by an electrolyte

(1-D scaling-up). Another way to scale-up microfluidic PEC cells is by numbering-up, which requires the incorporation of parallel arrangements of single micro-fluidic cells (2-D parallelization scaling-up). The main advantage of this approach is that electrochemical processes observed in lab-scale cells will be maintained in each unit at any level of scaling out. Finally, it should be mentioned that new developments in fabrication techniques such as high-resolution additive manufacturing are needed in order to overcome the existing challenges in such numbering-up strategies with complex fluidic interconnections.

Sophia Haussener asked: Why do you use microfluidics? Do we need such small scales or would millimeter-scale also work?

Evangelos Kalamaras replied: Utilization of microfluidic reactors with inter-electrode distance up to 1 mm minimizes the Ohmic drop avoiding short-circuiting the cell, offers high-surface-area-to-volume ratio of photocatalytic and/or electrocatalytic materials and improves mass transfer from the bulk electrolyte solution to the surface of the electrodes. A millimetre-scale reactor would achieve lower efficiency than the proposed reactor since lower separation distance can lead to large fluidic resistances, while larger separation distance can result in large losses from ionic transport.[1]

1 M. A. Modestino, D. Fernandez Rivas, S. M. H. Hashemi, J. G. E. Gardeniers and D. Psaltis, *Energy Environ. Sci.*, 2016, 9, 3381–3391.

Sophia Haussener commented: Without a membrane you have to go to very low velocities in order to suppress mixing. Is this useful for a practical reactor where you want large production rates?

Evangelos Kalamaras responded: For membrane-less operation in the PEC flow reactor, we were limited to flow rates higher than 0.04 ml min^{-1} to avoid crossover and lower than 0.2 ml min^{-1} to give sufficient time to complete the desired reaction (Fig 5b in the paper (DOI: 10.1039/c8fd00192h)). Although membranes are very important components for PEC applications minimizing anolyte and catholyte crossover, they suffer from degradation and significantly increase the total cost of a PEC cell. Therefore, we believe a promising approach to achieve high solar fuel production rates is the utilization of parallel continuous flow microfluidic PEC systems.

Reiner Sebastian Sprick said: I was wondering what your view on scale-up of continuous flow microfluidic PEC reactors is. These systems seem to have potential, but it seems challenging to do this on a very large scale.

Evangelos Kalamaras replied: Scaling-up of conventional reactors requires increased dimensions of the reactor to enhance the production leading to numerous problems. In contrast, the significant advantage of microfluidic reactors stems from the fact that a parallel network with multiple microreactors of the same size can be used for scaling-up. Another significant advantage of this approach is that by using multiple microreactors, the electrochemical processes taking place in each one remains the same at any level of scaling-out. However, scale-up strategies for microfluidic reactors have not yet been comprehensively evaluated.

Aubrey Paris returned to the discussion of the paper by Michael Grätzel: Much of the CO_2 reduction literature on improving the selectivity of copper electrodes either focuses on methane or two-carbon product selectivity. In your case, you can tune for a particular two-carbon product over the other (*i.e.*, ethylene over ethanol, or *vice versa*). However, your electrode system – and other copper-based systems – can also generate propanol. I have not seen reports in the literature that have been able to selectively make propanol (or other C_{3+} products). Do you have any ideas regarding strategies to selectively generate propanol or other higher-order products from CO_2? While C_2 products like ethylene may be commercially attractive, it seems beneficial to expand the product scope of CO_2 electroreduction to give this technology the greatest probability of succeeding in real-world applications.

Michael Grätzel answered: So far we have not specifically targeted C_3 or higher molecular weight products. This is a challenging task which is certainly worth pursuing. For example, Siemens has launched last year a joint venture to produce butanol from sunlight, CO_2 and water. The process is divided up in two steps, *i.e.* the light reaction where CO is first generated from CO_2 followed up a dark reaction where microorganisms convert the CO to butanol in a bioreactor.

Wilson Smith asked: Regardless of the mechanism that is found for the electrodes tested in an aqueous H-cell, do you think the mechanism and selectivity will transfer to operation with gas diffusion electrodes with high current density and high local pH?

Michael Grätzel replied: It remains to be seen how the increase in current density and pH on a gas diffusion electrode will affect the product distribution. For our Zn-doped Cu electrodes the ratio of ethanol over ethylene remained around 4 over a wide range of electrode potentials while for the Au/Cu electrocatalyst the product distribution was more sensitive to the applied voltage. Hence we would expect that the use of a gas diffusion electrode will not change significantly the product selectivity for the Zn/Cu catalyst.

Virgil Andrei remarked: The Cu/Au nanostructured electrocatalysts present a very nice product distribution as a function of the applied potential, with the catholyte stirred at 1000 rpm. Have you looked at whether the product distribution is changed at different stirring speeds or under no stirring? The stirring velocity may influence the mass transport and local pH gradient in the vicinity of the electrode, and therefore also the product distribution.

Michael Grätzel answered: Thank you for raising this important point. So far we have always applied the same stirring speed of 1000 rpm but we plan to extend our studies to examine the effect of stirring speed on the product distribution.

Anirudh Venugopal opened a general discussion of the papers by Michael Grätzel, Yi-Hsuan Lai, Akihiko Kudo and Evangelos Kalamaras: This is a general question (to everyone in the room) regarding the future of photoelectrochemical water splitting. If we look at some of the opinion/perspective articles from the last year or so, there seems to be general scepticism about the future of PEC water splitting. One article in particular (ref. 1) comes to my mind. Their arguments are that the PV+electrolyzer route has advanced so much in the recent past that the PEC route cannot really compete even in an ideal/best case scenario for PEC water splitting. Additionally, even though the concept of PEC looks simple and elegant, working with it has made people realize that it is tremendously difficult to optimize each aspect in a PEC system like the right bandgap, band edge positions, stability, catalytic activity *etc.* without effecting each other. In that sense a decoupled system like a PV+electrolyzer makes more sense for easier optimization. What is the opinion of everyone in the PEC community present here regarding the statements made in such articles? Do you also have similar thoughts?

1 T. Jesper Jacobsson, Photoelectrochemical water splitting: an idea heading towards obsolescence?, *Energy Environ. Sci.*, 2018, **11**, 1977–1979.

Akihiko Kudo replied: There are mainly PV+electrolyzer, PEC and powdered photocatalyst systems for solar water splitting. The PV+electrolyzer actually gives the highest efficiency among those systems at the present stage. However, criteria of actual artificial photosynthesis systems are not only the efficiency but also the cost, scalability, recyclability, etc, as discussed in the literature.[1,2] Actually, PV+electrolyzer gives a high solar energy conversion efficiency (STH), though it is still not at the stage of practical use implying that there are several problems for such a system, for example the cost of an electrolyzer. Therefore, mainly Japanese research groups believe that the powder based photocatalyst material will be the final and ideal system for the business by practical artificial photosynthesis.[1–3] But, anyway, we should develop several kinds of technologies to achieve the practical artificial photosynthesis, because they have advantages and disadvantages. We should not focus on just one technology at the present stage. We should search any possibilities.

1 T. Setoyama, T. Takewaki, K. Domen and T. Tatsumi, *Faraday Discuss.*, 2017, **198**, 509–527.
2 T. Yamada and K. Domen, *ChemEngineering*, 2018, **2**, 36.
3 B. A. Pinaud, J. D. Benck, L. C. Seitz, A. J. Forman, Z. Chen, T. G. Deutsch, B. D. James, K. N. Baum, G. N. Baum, S. Ardo, H. Wang, E. Miller and T. F. Jaramillo, Technical and economic feasibility of centralized facilities for solar hydrogen production *via* photocatalysis and photoelectrochemistry, *Energy Environ. Sci.*, 2013, **6**, 1983–2002.

Michael Grätzel responded: It is true that the PV-electrolyzer can operate at higher current densities than PECs depending on whether solar light is concentrated or normal sunlight is used, however this comes at the price of higher overvoltage losses for the Faradaic electrode reactions at the electrodes and for transport of ions between the electrodes. Furthermore, acid electrolyzers require an expensive NAFION membrane (PEM) and noble metal catalysts (Pt and Ir) for the hydrogen and oxygen evolution. Hence industry uses mainly the much cheaper alkaline electrolyzers, which, however, cannot sustain the discontinuous operation imposed for PV applications *via* the day and night cycle. The PEC devices have the advantage over PV-electrolyzers that the charges photo-generated in the PEC material can be immediately converted to fuels. There is no need for electricity collection, which requires expensive wiring and interconnection of solar panels. Finally we should keep in mind that natural photosynthesis is nothing else but a photo-electrochemical converter which has worked beautifully over more than 2 billion years. Hence we have every reason to mimic this crucial life-sustaining process on earth by artificial photosynthetic systems such as PECs.

Anirudh Venugopal addressed Michael Grätzel and Akihiko Kudo: Continuing on the discussion, a PV + electrolyzer can operate at high current densities, which tremendously decreases the size of the electrolyzer and its cost. It also means that the amount of precious catalysts that we use are also smaller. New cheaper catalytic materials like MoS_2 to replace platinum and $NiOOH/FeOOH$ have also been shown to work well with a normal electrolyzer. Does this take out any economic cost argument in favor of PEC?

Akihiko Kudo answered: Development of efficient electrocatalysts consisting of non-noble metals will decrease the cost of an electrolyzer. But I am not sure if those electrocatalysts make the breakthrough toward the cost issue of the total system of a PV + electrolyzer.

Michael Grätzel replied: It is true that the PV-electrolyzer can operate at high current densities but this comes at the price of higher overvoltage losses for the Faradaic electrode reactions at the electrodes and for transport of ions between the electrodes. Furthermore acid electrolyzers require an expensive NAFION membrane (PEM) and noble metal catalysts (Pt and Ir) for the hydrogen and oxygen evolution. Hence industry uses mainly the much cheaper alkaline electrolyzers, which, however, cannot sustain the discontinuous operation imposed for PV applications by the day and night cycle. Finally the PEC devices have the advantage over PV-electrolyzers that the charges photo-generated in the PEC material can be immediately converted to fuels. There is no need for electricity collection which requires expensive wiring and interconnection of panels.

Leif Hammarström commented: The paper referred to in the earlier question from Anirudh Venugopal expresses a personal opinion, but lacks scientific and technical background for this opinion. It ignores the research and analysis that has already been published on the topic of PV+electrolysis *vs.* direct solar-to-fuel conversion.

Akihiko Kudo replied: I agree with this opinion.

Víctor A. de la Peña O'Shea continued the discussion of the paper by Akihiko Kudo: In your reactions you are using two different semiconductors, a molecular mediator and graphene, and during the reactions all of these materials are separate.
How is the electron transfer produced at the same timescale in these different materials? Do you know the maximum distance between the materials to have an efficient charge transfer?

Akihiko Kudo replied: What you pointed out is important. But it is hard to say the distance you mention. Lifetimes of photogenerated carriers in a H_2-evolving photocatalyst would be different from those in an O_2-evolving photocatalyst.

Collision is an indispensable process for electron transfer in the case of a reduced graphene oxide, because RGO/O_2-evolving photocatalyst composite particles are suspended with H_2-evolving photocatalyst particles. It means that the frequency of the collision between the different particles is important rather than the distance. It is clarified that p-n semiconductor character is a key issue for electron transfer during the collision. The distance depends on the concentration of photocatalysts and mediators in the case of a molecular mediator. We roughly optimize the concentration as listed in the footnotes of the Tables and Figures in the paper (DOI: 10.1039/c8fd00209f). If redox reactions on an O_2-evolving photocatalyst is smoother than that on a H_2-evolving photocatalyst, electrons would be accumulated in a molecular mediator as its reduced form or RGO as an electron pool. In the case of the reverse, holes are accumulated in a mediator as its oxidized form.

Su-Il In addressed Akihiko Kudo, Michael Grätzel, Evangelos Kalamaras and Yi-Hsuan Lai: I'm asking this question to all of you. Prof. Domen looked at thousands of materials for water splitting. Water splitting materials are complicated. How do you select materials? What is your strategy? Trial and error? Serendipity? Band gap alignment of the materials?

Akihiko Kudo answered: Suitable band gaps and positions of conduction and valence bands are required for water splitting photocatalyst materials. Those are basically determined by the elements and crystal structures. We can design new photocatalyst materials with the idea based on the band and crystal engineering involving the strategies of doping and substitution of other elements.[1,2] Charge separation and migration are also important processes as well as band gap excitation. The band structure calculation by DFT can prove the strategy and further develop other photocatalysts with a suitable band structure as well as giving information on carrier mobility. However, surface redox reactions on the surface of photocatalyst are very important: that can hardly be solved by calculation at the present stage. Electrocatalysts give some hints to the surface properties for the redox reactions including for cocatalysts. AI might be useful for finding new photocatalyst materials in future.

1 A. Kudo, H. Kato and I. Tsuji, Strategies for the Development of Visible-light-driven Photocatalysts for Water Splitting, *Chem. Lett.*, 2004, **33**(12), 1534–1539.
2 A. Kudo and Y. Miseki, Heterogeneous photocatalyst materials for water splitting, *Chem. Soc. Rev.*, 2009, **38**, 253–278.

Evangelos Kalamaras responded: In our case, photoelectrochemical CO_2 reduction, we conducted a theoretical study based on modelling work of Vesborg and co-workers.[1] In this study, we found that a combination of two semiconductors with band gap ranging from 1.4 eV to 2.2 eV achieved the best theoretical Solar-to-Fuel efficiency. This is one of the main reasons that we selected semiconducting materials such as hematite and copper oxides.

1 P. C. K. Vesborg and B. Seger, *Chem. Mater.*, 2016, **28**, 8844–8850.

Michael Grätzel responded: Yes, water-splitting materials are complicated. But we learned a great deal over the last 3 decades. p-type oxides would be my current preference. For example, Cu_2O is a good start for a photoelectrochemical material. It has a band gap of about 2 eV, shows a luminescence and if used with an overlayer of Ga_2O_3 it develops a photovoltage of over 1 volt. Photocurrents in AM 1.5 sunlight reach over 10 mA cm^{-2}. One needs to protect it from direct water contact since there is a risk of photocorrosion. Starting from Cu_2O binary oxides such as perovskites and delafossites should be tested.

Yi-Hsuan Lai replied: In our case, we first select a state-of-the-art photoanode material, $BiVO_4$, followed by finding a photocathode material that can provide sufficient photovoltage for pairing with $BiVO_4$.

Daniel Cheung returned to the discussion of the paper by Akihiko Kudo: My question is directed primarily at Professor Akihiko Kudo but if other delegates would like to contribute that would be useful.

I am also working on Z-scheme photocatalysts based on CdS and WO_3 semiconductor nanoparticles. I have synthesised the materials and have characterised them using SEM and TEM. SEM images show that the CdS and WO_3 are present in the final material, however, attempts to acquire high-resolution (HR-TEM) images of the materials have been unsuccessful. HR-TEM images are vital for my research because they would provide direct evidence for the formation of a heterojunction between the two semiconductor materials.

I have seen similar heterojunctions in the literature that have been imaged successfully using HR-TEM; the micrographs show clearly that heterojunction formation is possible for these materials. Do you have any advice on how to tackle my problem? Thank-you.

Akihiko Kudo answered: HR-TEM is a powerful technique to see the actual structure. However, CdS is often unstable for electron microscopes because CdS volatilizes during the measurement depending on the acceleration voltage, even for SEM. So you should be advised on the measurement condition by a specialist of HR-TEM. But the important point is not only the geometric structure observed by electron microscopes but the energy structure of the heterojunction. There are two charge separation processes in the composite of CdS and WO_3; (i) Z-schematic charge transfer and (ii) electron transfer from a conduction band of CdS to that of WO_3 and simultaneous hole transfer from a valence band of WO_3 to that of CdS according to the potential difference. If you obtain the activity for water splitting into H_2 and O_2 at a stoichiometric ratio, not for sacrificial H_2 evolution, you can say that the Z-schematic charge transfer proceeds. I do not think that making such a composite for sacrificial H_2 evolution is a proper direction for the design of photocatalysts because CdS itself gives a quite high quantum yield close to 90% in the presence of sacrificial electron donors.[1] In process (ii), although the charge separation would be enhanced, abilities of the migrated electrons and holes for redox reactions thermodynamically become low losing water splitting ability.

1 J. Yang, H. Yan, X. Wang , F. Wen, Z. Wang, D. Fan, J. Shi and C. Li, Roles of cocatalysts in Pt-PdS/CdS with exceptionally high quantum efficiency for photocatalytic hydrogen production, *J. Catal.*, 2012, **290**, 151–157.

Sophia Haussener asked: The mediator is very important in these particle-based devices. Could you share some advice on the thermodynamics and kinetics of the mediator? Should we develop catalysts for specific mediators?

Akihiko Kudo replied: From the thermodynamics, it is required that the redox potential is between the conduction band level of an O_2-evolving photocatalyst and the valence band level of a H_2-evolving photocatalyst at a certain pH. From a viewpoint of kinetics, an adsorption property on the surface of the photocatalyst is important as well as the electron transfer for redox reactions between a photocatalyst and a mediator. But we need to suppress backward reactions by an oxidized form of the mediator on the surface of a H_2-evolving photocatalyst and those of a reduced form on an O_2-evolving photocatalyst, because those backward reactions compete with water reduction and oxidation, respectively.[1]

It is also a key issue for a Z-schematic photocatalyst to find efficient mediators as you suggest.

1 H. Kato, Y. Sasaki, A. Iwase and A. Kudo, Role of Iron Ion Electron Mediator on Photocatalytic Overall Water Splitting under Visible Light Irradiation using Z-Scheme Systems, *Bull. Chem. Soc. Jpn.*, 2007, **80**(12), 2457–2464.

Marcelino Maneiro commented: I would like to discuss further about the role of electron mediators in the work presented here by Kudo *et al.* Different photocatalytic activities were found depending on the mediator. With the $[Co(bpy)_3]^{3+/2+}$ mediator, the activity more than doubles for the same metal oxide material. Have you tried to use other cobalt complexes with different polypyridine ligands? Do you think that other mediators of this type may enhance the water splitting activity of these metal oxide photocatalysts?

Akihiko Kudo responded: We have tried to use $[Co(phen)_3]^{3+/2+}$.[1] It worked as well as $[Co(bpy)_3]^{3+/2+}$ for several Z-scheme photocatalysts for water splitting. In contrast to them, Fe complexes were not effective for our systems. But, anyway, it is important to find other mediators as you pointed out. There is a factor of affinity between a mediator and a photocatalyst.

1 Y. Sasaki, H. Kato and A. Kudo, $[Co(bpy)_3]^{3+/2+}$ and $[Co(phen)_3]^{3+/2+}$ Electron Mediators for Overall Water Splitting under Sunlight Irradiation Using Z-scheme Photocatalyst System, *J. Am. Chem. Soc.*, 2013, **135**, 5441–5449.

Chong-Yong Lee addressed Akihiko Kudo, Michael Grätzel and Yi-Hsuan Lai: From the talks in this session, obviously nanostructuring plays a key role in enhancing performances of electrocatalytic, photocatalytic, as well as photoelectrocatalytic systems. Looking forward to the future, especially for commercial applications in large-scale device fabrication, do you think nanostructuring is the way to go? Due to the cost and being readily available in large amounts, the current industrial systems often involve microscale catalysts rather than nanostructured. Would there be issues with the cost, structural homogeneity *etc.* when upscaling nanostructured materials? Furthermore, for the electrocatalytic system, such as gas diffusion electrodes that commonly involving mixing with binders and/or support materials *etc.*, is nanostructuring still relevant in such a system?

Akihiko Kudo replied: Nanostructure will be fine if a high efficiency is obtained due to it. Of course, the cost depends on the synthetic process. The cost of preparation of nanostructured materials will be not so serious a problem, if it is a bottom-up process that can make the mass production possible. For the electrocatalytic system, a gas diffusion electrode is OK. It is actually commercially used for soda industries.

Yi-Hsuan Lai responded: I believe nanostructuring photoelectrode materials is necessary to improve their efficiency, especially for those materials having charge carrier diffusion lengths shorter than their respective light absorption length. In our study, we also found nanostructuring indeed plays an important role in the enhancement of our photoelectrode performances.

Michael Grätzel answered: It is true that some types of nanostructures require expensive chemical precursors and sophisticated equipment, such as the high vacuum chambers employed by molecular beam epitaxy (MBE) for their realization. However other types of nanostructures, such as the mesoscopic oxides we employ in photovoltaics, photo-electrochemistry and catalysis use cheap precursors and low cost methods for their production. For example the Cu_2O nanowires that we use as cathodes for the photo-generation of hydrogen and reduction of CO_2 are readily and cheaply prepared by anodic oxidation of Cu films. Up-scaling this preparation method is straightforward. Gas diffusion electrodes (GDEs) would not be required for the photo-electrochemical fuel producing systems since the photo-currents generated in full sunlight would remain in the 10–30 mA cm^{-2} range. They are used for CO_2 reduction in electrochemical reactors. They are normally planar but we are presently testing Cu_2O nanowires as well for this electrocatalytic application.

Pramod Patil Kunturu addressed Akihiko Kudo: How should we choose new semiconductor photocatalysts for an efficient PEC water splitting reaction? Are there any software screening techniques available like the combinatorial synthesis technique used by JCAP?

Akihiko Kudo responded: Combinatorial synthesis would be a powerful technique if you use it for surveying some parameters, for example changing a ratio of components and an amount of cocatalyst, around a known photocatalyst. However, it is not easy to discover completely new materials with the combinatorial synthesis. In future, AI may be the technique for it.

Sophia Haussener asked: Following up on your response on selection of materials: do we need to explore a whole range of possibilities or should we stay very focused on one/two materials and mediators?

Akihiko Kudo answered: I am sure we should go to both ways: to find new material systems and to improve the present material systems. So, material design is still a key issue.

Leif Hammarström addressed Michael Grätzel: High throughput screening sounds attractive, but you already highlighted the importance of morphology and nano-structuring. Moreover, a device will most likely have different materials for light harvesting and charge separation on the one hand, and catalysis for the other, and screening junctions is much harder than screening single materials. How do you see the role of high throughput screening in the field?

Michael Grätzel replied: It is correct that high throughput screening has a limited potential given the high risk of misjudging potential of the material. One useful metric that can be employed in the rapid screening process is luminescence. The quantum yield of emission directly gives the maximum photovoltage that can be obtained with a given light harvesting material.

Joost Reek returned to the general discussion of the papers by Michael Grätzel, Yi-Hsuan Lai, Akihiko Kudo and Evangelos Kalamaras: There is a lot of progress in making solar fuel devices, and for most set-ups a membrane is required. Do we have sufficient choice of membranes for future applications or should there be more development in this area?

Akihiko Kudo answered: I do not think that it is already sufficient. To develop membranes is also a key issue for devising artificial photosysnthesis systems. For example, it is necessary for separation of hydrogen from oxygen in the mixed gas. We can use a membrane as a separator and supporter for self-standing Z-scheme photocatalyst devises.

Yi-Hsuan Lai answered: I believe there should be more membranes developed in this area. For example, in our experiments, we use a slightly alkaline solution as the electrolyte and Nafion 117™ as the membrane. However, Nafion is a proton exchange membrane and is more suitable for being applied in acidic solutions. Therefore, developing other membranes such as bipolar and/or anion exchange membranes should be helpful for the progress of solar fuel devices.

Michael Grätzel replied: There is clearly a need for developing better membranes. For example it would be advantageous to use PEMs that transport hydroxide anions instead of protons.

Wilson Smith addressed Evangelos Kalamaras and Yi-Hsuan Lai: Membrane development will play a big role in the coming years for optimising electrochemical systems. For example, the bipolar membrane has already been used for doing electrolysis in two different pH compartments, but can still be further optimised and is understood to improve selectivity/stability. For gas diffusion electrodes, integrating catalysts and membranes in a membrane-electrode-assembly (MEA) will be very interesting to lower overpotentials and improve overall performance. Here, finding the right combinations of membrane/polymers and catalysts, and ways to fabricate and coat them will be very fruitful for the field.

Evangelos Kalamaras replied: Although ion-conducting separators or membranes are critical components in electrochemical and PEC cells, they are vulnerable to degradation effects, add substantial cost, increase the complexity of reactor design and increase the ionic resistance. For all the reasons above, we developed a membrane-less continuous flow PEC cell for CO_2 reduction. However, for membrane-less operation in the PEC flow reactor, we were limited to flow rates higher than 0.04 ml min^{-1} to avoid significant product crossover.

Yi-Hsuan Lai replied: Yes, I agree that the research of membranes is extremely beneficial for the progress of artificial photosynthesis.

Víctor A. de la Peña O'Shea addressed Evangelos Kalamaras: Taking into account that you are not using membranes between both electrodes on your device and some or your experiments are in the batch regime, are you sure that your products are not re-oxidized during the reactions? Did you perform any surface study to determine changes in your photocatalyst after the reactions?

Evangelos Kalamaras replied: The photoelectrodes were characterized before and after PEC measurements using Scanning Electron Microscopy (SEM). The obtained SEM images showed no significant alterations on the surface morphology of the photoelectrodes. We believe that re-oxidation of the products took place due to the absence of a membrane which decreased the STF efficiency of the PEC system. However, in the future we need to perform XPS analysis for the photoelectrodes after PEC measurements to further confirm the back reactions.

Erwin Reisner addressed everyone: We have been discussing solar-to-fuel conversion using 1 Sun irradiation so far. What would be the potential benefits and disadvantages of using concentrated solar light for solar fuel synthesis?

Sophia Haussener replied: Concentrated radiation can make PEC approaches and devices more interesting as it intensifies the input and allows for larger energy and power density devices (usually connected to economic and sustainability advantages). Concentrated irradiation also allows the utilization of more precious materials, as their mass and volume can be reduced roughly by the factor of the irradiation concentration. Furthermore, concentrated radiation will result in large energy inputs, requiring thermal management in order to sustain this input (which is challenging). At the same time, thermal management can be used to enhance the performance of PEC devices (for example: larger temperatures reduce kinetic overpotentials). Additionally, the matching of energy and mass flows in a PEC device utilizing liquid water as the reactant could require concentrated irradiation in order to optimally use this "concentrated" stream of reactant (in contrast, the dilute 1 Sun irradiation will always have low reactant conversion as there are too few charge carriers compared to the available reactant). A large challenge is long term operation: devices operating under such intense irradiation conditions and generally at larger current densities could potentially suffer from stronger degradation (to be investigated).

Michael Grätzel responded: Some advantages of using concentrated sunlight for hydrogen generation are (i) increase of photovoltage provided with light intensity and (ii) higher temperature would decrease the voltage require for water splitting and smaller photo-electrolyser size reducing cost. On the other hand, caveats for this technology are (i) the need to have an absolutely clear, dust-free sky, (ii) the high cost of the concentrators including their maintenance and (iii) increased overvoltage losses due to higher photocurrents.

Conflicts of interest

There are no conflicts to declare.

PAPER

Utilising excited state organic anions for photoredox catalysis: activation of (hetero) aryl chlorides by visible light-absorbing 9-anthrolate anions†

Matthias Schmalzbauer, [ID] Indrajit Ghosh [ID] and Burkhard König [ID] *

Received 14th November 2018, Accepted 5th February 2019

DOI: 10.1039/c8fd00176f

The tricyclic aromatic ketone 9-anthrone and its derivatives are under basic conditions in equilibrium with their corresponding anionic forms. Unlike the neutral species, the 9-anthrolate anions can be excited by blue LED light and thus, are able to initiate a photoinduced electron transfer (PET) reaction. To demonstrate the synthetic applicability of the catalytic system, various (hetero)aryl chlorides were converted in C–C and C–Het bond-forming reactions affording the corresponding arylation products in moderate to excellent yields. The reactions proceed under very mild conditions without the need for a sacrificial electron donor. Besides 9-anthrone, other closely related derivatives were synthesised and investigated concerning their ability to catalyse demanding reductive transformations. Based on spectroscopic findings and radical trapping experiments a conceivable mechanism is proposed.

Introduction

Within the past few years, photoredox catalysis evolved into a versatile tool in organic synthesis. In the course of this, photoreduction of aryl halides to generate highly reactive aryl radicals has emerged as a valuable transition-metal-free alternative to the well-established Pd- and Rh-catalysed C–H activation chemistry.[1–8] The formation of complex molecules starting from cheap, readily available and stable starting materials has always been an important request in synthetic organic chemistry. Among (hetero)aryl halides, (hetero)aryl chlorides in particular can be considered as such reagents and thus, approaches to activate the $C(sp^2)$–Cl bond are of great interest. However, converting (hetero)aryl chlorides to forge new bonds upon mesolytic cleavage is still challenging, as the bond-dissociation energy (*e.g.*, Ar–Cl; BDE *ca.* 407 kJ mol^{-1}) is high and reduction potentials are usually difficult to overcome using commonly employed photocatalysts.[2,9,10] To

Institut für Organische Chemie, Universität Regensburg, Universitätsstraße 31, Regensburg, 93053, Germany. E-mail: Burkhard.Koenig@chemie.uni-regensburg.de; Fax: +49-941-943-1717; Tel: +49-941-943-4575

† Electronic supplementary information (ESI) available. See DOI: 10.1039/c8fd00176f

expand the accessible substrate scope in photoredox chemistry towards less activated molecules, the development of photocatalytic systems with stronger reducing power is required. In this context, several synthetically useful, mild and transition-metal-free strategies have recently been reported and among others, two are discussed here.

(I) Using a photocatalyst with large S1 ← S0 band gap: the group of de Alaniz showed that activated aryl chlorides can be reduced by using 10-phenyl-phenothiazine (PTH) which possesses a very large S1 ← S0 band gap and is therefore able to convert high energy photons into redox equivalents ($\lambda_{ex} = 380$ nm, $\lambda_{em} = 445$ nm; $E^*_{1/2} = -2.1$ V *vs.* SCE).[11]

Very recently a modified carbazole scaffold was reported to act as a strong reductant ($E^*_{1/2} \approx -2.75$ V *vs.* SCE),[2] which converted both unactivated aryl and alkyl chlorides (using an UVA-LED) to their corresponding dechlorinated products in good yields. However, C–C bond forming reactions using demanding aryl chlorides and electron-rich pyrroles (or benzene) were achieved only in low yields. Altering the reaction conditions to elevated temperature (90 °C), high excess of trapping reagent and by using an inorganic base instead of a typical amine electron donor, DIPEA, led to slightly improved yields of C–H arylated products.

(II) Consecutive photoinduced electron transfer (conPET) process: recently, our group demonstrated that the reduction power of a photocatalyst can be easily enhanced by using the energy of two photons in one catalytic cycle. Perylene diimide (PDI) was used as a photocatalyst and the excitation of the ground state radical anion (PDI˙⁻, generated *in situ via* a photoinduced electron transfer process, Scheme 1) led to the formation of a highly reducing doublet excited state (*i.e.*, PDI˙⁻*) capable of activating acceptor substituted aryl halides for C–H

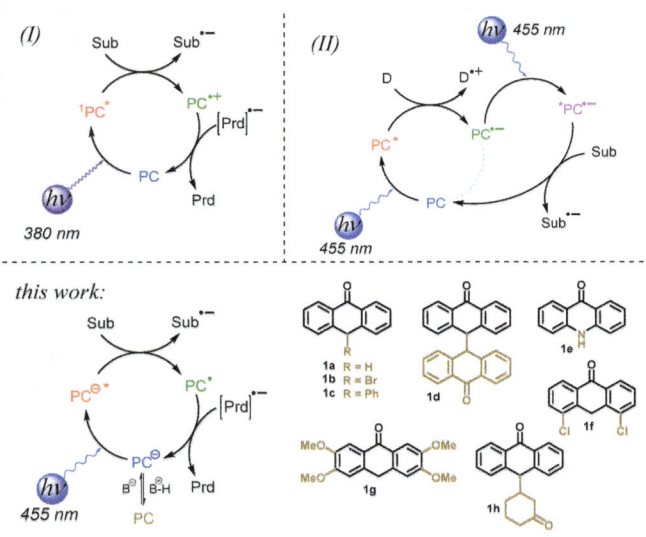

Scheme 1 Overview of discussed activation strategies: (I) using a photocatalyst with large S1 ← S0 band gap; (II) consecutive photoinduced electron transfer (conPET); (bottom, this work) concept for excitation of visible light-absorbing organic anions and the herein examined photocatalysts.

arylation reactions.[12] In this effort, the König group also demonstrated that the use of rhodamine 6G under very similar reaction conditions allows for activating electron-rich aryl bromides in C–C bond forming reactions.[13,14] However, only a few examples of aryl chlorides are discussed and electron-rich pyrrole derivatives or isocyanides were mainly used as trapping reagents.

In this work we report a series of visible-light absorbing anionic ground state photocatalysts **1** (Scheme 1) that act as strong reducing agents from their photoexcited states. We got inspired by literature reports in which the excited states of cationic organic dyes, such as 9-mesityl-10-methylacridinium perchlorate or triphenylpyrylium tetrafluoroborate, have been utilised frequently as powerful single electron-accepting photocatalysts.[15–17] We envisioned that photoexcited anions of organic molecules accordingly may also act as very strong electron donors. Additionally, intermediates of a PET between excited anionic donor and neutral ground state acceptor are free of electrostatic attraction resulting in an accelerated separation of the radical pair and hence, a decrease in back electron transfer (BET) rate.[18,19] To the best of our knowledge, organic anions are a barely explored class of photoredox catalysts[20–22] and thus, we investigated 9-anthrolate and some derivatives as visible light-absorbing anionic photocatalysts which enable the activation of several aryl and heteroaryl chlorides *via* a single electron transfer process for C(sp^2)–C and C(sp^2)–heteroatom bond-forming reactions.

Results and discussion

In solution, 9-anthrone (**1a**) is present in a tautomeric equilibrium and the keto–enol ratio strongly depends on the hydrogen bonding ability of the solvent. A value of $K_{enol} = 1.6 \pm 1$ in DMSO is reported for **1a**, *i.e.* both forms are coexistent in considerable amount.[23] Hence, a gain in energy due to aromatic stabilisation by forming the enol 9-anthrole is less pronounced compared to phenol. Furthermore, resonance stabilisation of 9-anthrolate **2a** is pivotal for the distinct acidity of **1a**.[24–26] Under basic conditions 9-anthrone is known to get deprotonated[27] (*cf.*, ^{1}H-NMR spectra, Fig. 1) and the resulting anionic species possesses an altered absorption spectrum with a prominent absorption maximum around 400 nm and a new broad absorption band in the visible region ranging from 410 to 560 nm (Fig. 1).[22] Thus, working in alkaline solution allows for the photoexcitation of 9-anthrolate (**2a**) with blue LED (455 nm) light. To our delight, the examined enolate **2a** turned out to be emissive, which allowed us to record its emission spectrum and to determine the luminescence lifetime ($\tau = 18.7$ ns, see ESI†) of the chromophore in a time-dependent experiment. Additional data regarding photophysical and electrochemical properties of other investigated anthrone derivatives, including compound **1b**, which has been used exclusively for synthetic reactions, are provided in the ESI.†

Screening of catalysts

With these spectroscopic data in hand, we began our synthetic investigations using methyl 4-chlorobenzoate **3a** (0.1 mmol) as the model substrate, N-methylpyrrole **4a** (1.5 mmol) as the trapping agent, which has recently been reported to trap aryl radicals efficiently, and catalytic amounts of **1a–h** (10 mol% w.r.t. **3a**). To

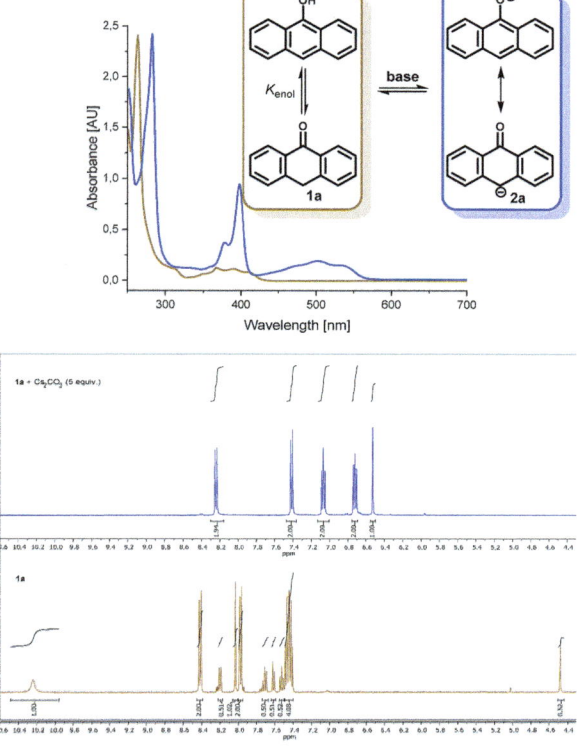

Fig. 1 (top) UV-Vis spectra of 9-anthrone (**1a**, 50 µM) in absence (brown) and presence (blue) of Cs_2CO_3. Spectra were recorded in a gastight quartz cuvette (10 × 10 mm) under N_2 atmosphere in dry degassed DMSO; (bottom) stacked ^{1}H-NMR spectra recorded in deuterated DMSO in the presence (blue) and absence (brown) of base. Integration over signals suggests the formation of the anion in the presence of base.

activate the chromophore, caesium carbonate was added as the base and reactions were carried out under a nitrogen atmosphere as **1a** and its anion are prone to oxidation in the presence of air.[28] The yield of coupling product **5a** was determined by calibrated GC-FID analysis with biphenyl as internal standard. All anthrone derivatives (**1a–1h**, see Scheme 1 for the respective chemical structures) investigated herein, were effective in catalysing the model reaction to afford the desired product in moderate to good yields (Table 1, entry 1–8).

When **1a** was used as a photocatalyst, the desired C–H arylated product was obtained in 66% yield. Other derivatives such as **1c** and **1h** gave very similar yields, and product formation dropped slightly when **1f** was used as a photocatalyst. It is worth mentioning here that the intention behind synthesising compounds **1c** and **1h** was to introduce steric hindrance into the 10-position of the molecule bearing a bulky phenyl or cyclohexanone residue. Nicewicz *et al.* reported that Mes-Acr-Me$^+$ and its reduced radical species are shielded by the sterically demanding mesityl group against nucleophilic and radical addition and thus diminish the bleaching of the dye.[17] However, no enhancement in catalytic activity could be determined compared to the unsubstituted 9-anthrone. Efforts

Table 1 Catalyst screening and control reactions[a]

Entry	PC	Conv.[b] [%]	Yield[b] [%]
1	**1a**	85	66
2	**1b**	91	77
3	**1c**	78	64
4	**1d**	100	79
5	**1e**	100	81
6	**1f**	69	54
7	**1g**	100	84
8	**1h**	85	65
9[c]	**1b**	15	0
10[d]	**1b**	6	6
11[e]	—	47	22

[a] Reaction conditions: a mixture containing of substrate **3a** (0.1 mmol), catalyst **1** (10 mol% w.r.t. **3a**) and Cs_2CO_3 (2.0 mmol) was dissolved in dry, degassed DMSO (1.0 mL) in a sealed snap vial equipped with stirring bar and under N_2 atmosphere. **4a** (1.5 mmol) was added via syringe needle. While stirring, the reaction was exposed to LED light (455 ± 15 nm) for 18 hours; [b] Yield and conversion were determined by calibrated GC-FID analysis with internal standard method. [c] Reaction was examined in the dark. [d] Reaction in absence of Cs_2CO_3. [e] Reaction without catalyst.

to synthesise 10-mesityl-9-anthrone were unsuccessful. The use of 10-bromo-9-anthrone (**1b**) as a photocatalyst increased the product yield to 77% and showed an enhanced selectivity towards the desired coupling product (cf. entry 1–2). Other employed anthrone derivatives, for example the dimer of anthrone (**1d**) or the nitrogen analogue **1e**, gave very similar yields with respect to **1b**. The highest yield of 84% was obtained using the electron-rich tetramethoxy anthrone **1g** (entry 7) however, it is not commercially available and is typically prepared in a three-step synthesis. Hence, we continued using **1b**, which is easily synthesised from the commercially available 9-anthrone via bromination, for further reaction optimisation and synthetic investigations. The control reactions in the dark (entry 9) or in the absence of base (entry 10) gave none, or only traces, of the desired product. Interestingly, a detectable amount of product (22%, entry 11) was formed when the reaction mixture was irradiated only in the presence of Cs_2CO_3 (2.0 equiv.). Very recently, Rossi et al. proposed the formation of a dimsyl anion as a key intermediate in the initiation step of photoinduced base-promoted homolytic aromatic substitution reactions.[29] According to their report, strong bases such as KO^tBu or NaH can facilitate electron transfer reactions in DMSO under visible-light irradiation. However, the authors have also reported that the use of Cs_2CO_3 or K_2CO_3 did not lead to any substrate conversion and hence may not be able to generate dimsyl anions. To get more insights, the reaction progress was monitored in the absence and presence of **1b** by GC-FID (Table S1, ESI†). The obtained results indicate an induction period with almost no detectable amount of product and a little substrate conversion after 2 hours of irradiation. After 4 hours, the product was formed, however, the reaction rate in the presence of **1b** was significantly faster accompanied by higher selectivity towards **5a**.[30]

Optimisation of the reaction conditions

The polarity of the solvent seems to play a crucial role as the product formation was observed in aprotic polar solvents such as DMSO, DMF or MeCN, whereas no substrate conversion was detected in less polar methylene chloride. Among the aprotic polar solvents, DMSO was found to be the most suitable one under our reaction conditions. Therefore, the optimisation reactions were carried out in DMSO under nitrogen atmosphere using **1b** as a photocatalyst,[31] **3a** as the model substrate and **4a** as the trapping agent and the reactions were examined in the presence of a broad range of inorganic or organic bases and additive. The results are summarised in Table 2. Interestingly, the use of Li_2CO_3 did not lead to any

Table 2 Optimisation of the reaction conditions[a]

Entry	1b [mol%]	Base [equiv.]	Additive [equiv.]	Time [h]	Yield[b] [%]
1	10	2 (Cs_2CO_3)	—	18	77 (91)
2	10	1 (Cs_2CO_3)	—	18	47 (79)
3	**10**	**1.2 (Cs_2CO_3)**	**0.6 (18c6)**	**18**	95[c]
4	5	1.2 (Cs_2CO_3)	0.6 (18c6)	18	92 (98)
5	20	1.2 (Cs_2CO_3)	0.6 (18c6)	18	88 (94)
6	10	2 (Li_2CO_3)	—	18	0 (0)
7	10	2 (Na_2CO_3)	—	18	4 (13)
8	10	2 (K_2CO_3)	—	18	47 (53)
9	10	2 (K_3PO_4)	—	18	46 (70)
10	10	2 (NaAsc)	—	18	2 (2)
11	10	1.2 (TMG)	—	18	42 (63)
12	10	1.2 (DBU)	—	18	35 (65)
13	10	1.2 (DABCO)	—	18	0 (0)
14	10	1.2 (Cs_2CO_3)	0.6 (18c6)	10	79 (88)
15[d]	10	1.2 (Cs_2CO_3)	0.6 (18c6)	18	8 (31)
16[e]	10	1.2 (Cs_2CO_3)	0.6 (18c6)	18	30 (51)
17[f]	10	1.2 (Cs_2CO_3)	0.6 (18c6)	18	34 (59)
18[f,g]	10	1.2 (Cs_2CO_3)	0.6 (18c6)	18	83 (100)

[a] Reaction conditions: a mixture containing substrate **3a** (0.1 mmol), catalyst **1b**, base and additive was dissolved in dry, degassed DMSO (1.0 mL) in a sealed snap vial equipped with stirring bar and under N_2 atmosphere. **4a** (1.5 mmol) was added *via* syringe needle. While stirring the reaction was exposed to LED light (455 ± 15 nm). [b] Yields were determined by calibrated GC-FID analysis with internal standard method, numbers in parenthesis refer to the conversion of **3a**. [c] Isolated yield. [d] 1 equiv. of **4a** was used. [e] Green LED (535 nm) was used as the irradiation source, however radiant flux is lowered by a factor of 8 compared to blue LED (see ESI). [f] Reactions were run in air. [g] Catalyst **1e** was used.

product formation (entry 6), however, detectable amounts of products were obtained when Na_2CO_3 or sodium ascorbate (NaAsc) were used as bases (entry 7 and 10). The product yield increased by a reasonable amount in the presence of K_2CO_3 or K_3PO_4 (entry 8–9). Under similar reaction conditions, the use of Cs_2CO_3 led to 77% product formation. Employing strong non-nucleophilic organic bases, such as tetramethylguanidine (TMG) or 1,8-diazabicyclo[5.4.0]undec-7-ene did not increase the yield of the desired product, although they are completely soluble in DMSO. The weaker amine base triethylenediamine (DABCO) resulted in no product formation (entries 11–13).

With respect to the amount of base: it is not only necessary to activate the catalyst, but furthermore to neutralise the HCl formed in the course of the reaction. Hence, lowering the amount of Cs_2CO_3 to only one equivalent with respect to **3a** caused a decrease in product yield to 47% (entry 2). Carbonates are poorly soluble in DMSO. However, adding crown ether (18-crown-6, Table 3) increases both solubility and reactivity of the carbonate base in DMSO and an increased amount of product (95%) could be isolated. Simultaneously, it allowed the amount of base to be reduced to 1.2 equivalents (Table 2, entry 3). Under these reaction conditions, a doubling of the catalyst loading to 20 mol% had little influence (*cf.*, entry 3 and 5). The product yield and substrate conversion dropped significantly however, when **4a** was employed in a stoichiometric amount.[32] From the recorded UV-vis spectrum of **1b** in the presence of base (Fig. S3†) it is evident that excitation should also be possible with less-energetic and more economical light sources. To provide evidence, we run a reaction using a weak green light LED (535 nm, entry 16) which also resulted in product formation. Anthrolate anions **2** are known to get oxidised in the presence of air. However, acridanone **1e**, the oxidised acridine, is less reactive towards oxygen in alkaline media. Reactions in air revealed a significant decrease in product yield using catalyst **1b**, whereas a remarkable catalytic activity was found for catalyst **1e** even under a non-inert atmosphere (entry 17–18).

Scope of chlorinated substrates for C–H arylation

Under the optimised reaction condition (Table 2, entry 3), a range of aryl or biologically important heteroaryl chlorides could be effectively converted in coupling reactions. The isolated yields are summarised in Table 3. When **4a** is used as a trapping reagent, the direct C–H arylated products using methyl chlorobenzoates (**5a** and **5b**) were obtained in 95% and 51% isolated yield. 2-Chlorobenzonitrile was readily converted to give the desired biaryl product **5f** in excellent 92% yield. Interestingly, 2,6-dichlorobenzonitrile (**3g**) gave the twofold arylated product in 84% isolated yield with excellent selectivity towards the difunctionalised product.[33]

Our developed reaction conditions allowed for the activation of 2-chlorobenzotrifluoride (**3e**), which was found to be challenging (for CV measurements see ESI†). Nevertheless, the C–H arylated product **5e** could be obtained in 52% isolated yield. Notably, when employing catalyst **1a** the isolated yield could be increased to 71% (Table 3, entry 5e). 4-Chloroacetophenone (**3c**) and 4-chlorobenzaldehyde (**3d**) possess very similar reduction potentials with respect to methyl benzoate **3a**,[34] but react less well and give the corresponding coupling products in low 36% and 15% yields.[35] Arylated heteroarenes constitute an

Table 3 Scope of (hetero)aryl chlorides[a]

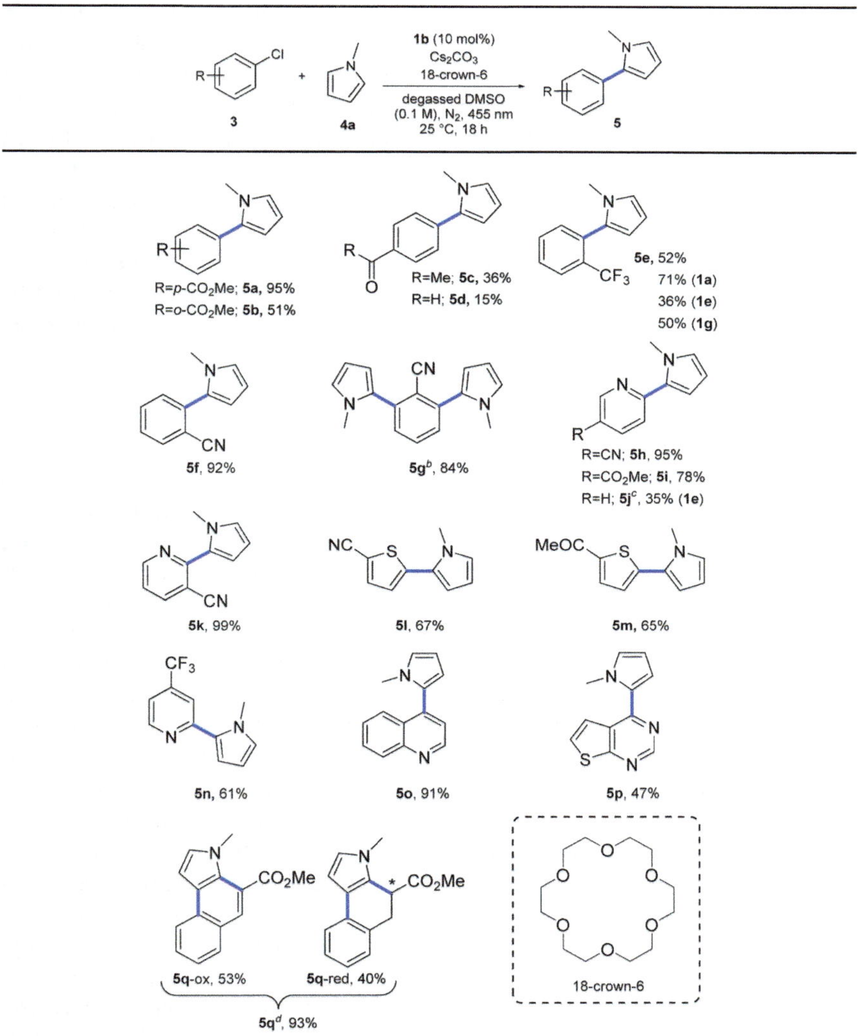

1b (10 mol%)
Cs2CO3
18-crown-6
degassed DMSO
(0.1 M), N2, 455 nm
25 °C, 18 h
3
4a
5
R=p-CO2Me; 5a, 95%
R=o-CO2Me; 5b, 51%
R=Me; 5c, 36%
R=H; 5d, 15%
5e, 52%
71% (1a)
36% (1e)
50% (1g)
5f, 92%
5g[b], 84%
R=CN; 5h, 95%
R=CO2Me; 5i, 78%
R=H; 5j[c], 35% (1e)
5k, 99%
5l, 67%
5m, 65%
5n, 61%
5o, 91%
5p, 47%
5q-ox, 53%
5q-red, 40%
5q[d], 93%
18-crown-6

[a] Reaction conditions: a mixture containing substrate **3** (0.1 mmol), catalyst **1b** (10 mol% w.r.t. **3**), crown ether (0.06 mmol) and Cs_2CO_3 (1.2 mmol) in a sealed snap vial equipped with stirring bar and under N_2 atmosphere, was dissolved in dry, degassed DMSO (1.0 mL) and **4a** (1.5 mmol) was added *via* syringe needle. While stirring, the reaction was exposed to LED light (455 ± 15 nm) for 18 hours from the plane bottom side of the vial. [b] Only the disubstituted product was obtained. [c] 2.0 mmol **4a**, 0.2 mmol Cs_2CO_3, 0.1 mmol crown ether and 20 mol% **1e** were used. Reaction was irradiated for 24 h. [d] Methyl (*Z*)-α-chlorocinnamate was used as the substrate and products **5q**-ox and **5q**-red could be isolated separately.

important structural motif in material and pharmaceutical sciences due to their optoelectronic and biological properties.[36,37] Hence, we were pleased to see that arylated derivatives of nicotinic acid nitriles are formed in excellent 95% and 99% yields (**5h, 5k**). The presence of an electron withdrawing group attached to the pyridine ring facilitates the product formation; under slightly altered reaction conditions (see Table 3 and **5j**) we were also able to transform unsubstituted 2-chloropyridine ($E_{red} = -2.40$ V *vs.* SCE, ESI†). Chlorinated thiophenes and tri-fluoromethylpyridine gave the desired products in 65–67% and 61% isolated yields, respectively. Chloroquinoline **3o** gave the desired arylated product in excellent 91% yield. The chlorinated thienopyrimidine is also useful as a substrate providing **5p** in 47% yield. Notably the heteroarene functionalised thienopyrimidines have been shown to possess a variety of interesting biological properties.[38] Our photocatalytic protocol also allows cascade bond forming processes providing interesting tricyclic compounds, for example **5q**, in almost quantitative yield. Notably, such cascade reactions were previously reported by Reiser and coworkers,[39,40] using vinyl halides as the precursors of relatively stable vinyl radicals.

Scope of tolerated trapping agents

The arylation reactions with **4a** led to product formation in moderate to excellent yields. From previous work it is known that electron deficient aryl radicals react readily with electron rich acceptor molecules.[6,12,41] Furthermore, isocyanides are reported to react with aryl radicals.[14,42] In contrast to activation strategy (II) mentioned in the introductory part, no sacrificial electron donor is necessary under the reaction conditions discussed in this work. Hence, excluding electron sources like tertiary amines, which are efficient H-atom donors, throttle the predominant side reaction causing the dehalogenated, reduced substrate. We envisioned that this enables the use of less reactive reagents as coupling partners, *i.e.*, furans, thiophens or even functionalised benzenes. C–H arylation reactions of furan and thiophene using aryl sulfonyl chlorides[43] or aryl diazonium salts[44] are well known and the desired hetero-cyclic products could be obtained in good yields. However, it has been observed that such heteroarenes are not efficient aryl radical trapping reagents under reductive reaction conditions (typically in the presence of a sacrificial electron donor) using (hetero)aryl chlorides as substrates. Consequently, photoredox catalytic protocols to functionalise furan or thiophene using aryl halides are barely reported.[6] Interestingly, this includes reactions with the amine donor used in sub-stoichiometric amount.[45] To our delight, the developed catalytic protocol allows for a variety of heteroarene functionalisations (Table 4): for example, when furan was used as trapping reagent the C–H arylated products were isolated in 49% and 62% yields, respectively (**6a**, **6b**).[46] Similarly, thiophene (**6d**) or alkoxy-substituted heteroarenes (**6c**, **6e**) gave the desired products in moderate yields. Compound **6d** was also obtained when other derivatives of **1** were employed. In the case of 1-phenylpyrrole, the excess amount of trapping agent was reduced (10 equiv. w.r.t. **3**). The corresponding coupling products (**6f**, **6g**) could nonetheless be isolated in reasonable yields. Aromatic hydrocarbons, such as benzene, mesitylene, or *p*-xylene gave the corresponding C–H arylated products (example **6h**), however a low substrate conversion was observed. Recently, efficient trapping and cascade cyclisation of aryl radicals with isocyanobiphenyls and 2-isocyanoaryl thioethers were reported.[14,42] Hence, isocyanides **4′i** and **4′k** were synthesised and reactions with several aryl chlorides were examined.[47] The corresponding

Table 4   Scope of suitable trapping agents[a]

a Reaction conditions: a mixture containing substrate 3′ (0.1 mmol), catalyst 1b (10 mol% w.r.t. 3′), crown ether (0.06 mmol) and Cs$_2$CO$_3$ (1.2 mmol) in a sealed snap vial equipped with stirring bar and under N$_2$ atmosphere, was dissolved in dry, degassed DMSO (1.0 mL). Trapping agent 4′ (1.5 mmol) was added via syringe needle (liquids) or before sealing the snap vial (solids). While stirring, the reaction was exposed to LED light (455 ± 15 nm) unless otherwise stated for 18 hours from the plane bottom side of the vial. b 2.0 mmol of furan was used. c Cs$_2$CO$_3$ (0.3 mmol), perylene (0.01 mmol) and DIPEA (50 mol%) were used. d 4′f (1.0 mmol), Cs$_2$CO$_3$ (0.3 mmol) and DIPEA (25 mol%) were used. e Isocyanide (0.5 mmol) was used. f Isocyanide (0.3 mmol) was used. g Compound was not isolated, yield determined by crude NMR with internal standard method.

coupling products 6i–6m could be isolated in moderate to good yields. Furthermore, aryl phosphonate 6n was obtained in good yield when triethyl phosphite was used as the trapping reagent and in addition, the pinacol ester of the aryl boronic acid 6o was formed in the presence of bis(pinacolato) diboron.

Mechanistic investigations

Excited state interactions (*e.g.*, single electron transfer) between the fluorescent catalyst molecule and a quencher cause a decrease in luminescence intensity and excited state lifetime. Therefore, a steady state and time-resolved fluorescence titration was performed to elucidate the mechanism of the discussed reaction.

The spectroscopic investigations were performed using **1a** in the presence of a base as fluorophore and 2-chlorobenzonitrile (**3f**), that possesses a reduction potential of −2.01 V *vs.* SCE, as an emission quencher. Notably, compound **3f** was easily activated in the presence of **1b** to afford the desired coupling product with **4a** in 92% isolated yield (example **5f**). Hence, it is also expected to quench the excited state of anion **2a** *via* an electron transfer process. The steady-state luminescence titration experiment revealed a decrease of emission intensity upon addition of the substrate **3f** (Fig. 2, top). Moreover, in the time-resolved study a decline in the excited state lifetime was observed. A Stern–Volmer analysis revealed a linear correlation indicating a dynamic quenching of the excited **2a** by **3f**. In a control experiment, no quenching, neither in the steady-state nor in the time-resolved experiment, was found when 2-chloroanisole (Fig. S5, ESI†) was used. Chloroanisole exhibits a very negative reduction potential of −2.83 V *vs.* SCE and remains inert both in C–H arylation reactions and in the quenching experiment.

Next, we identified radical intermediates in the reaction mixture by trapping experiments (Scheme 2). The persistent radical TEMPO is known to trap other

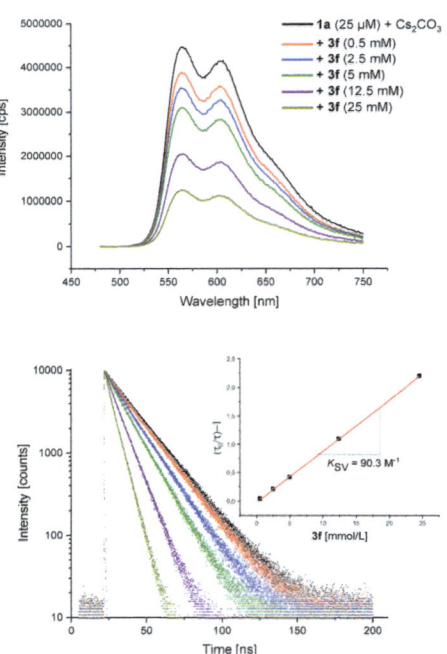

Fig. 2 (top) Steady-state luminescence quenching of **1a** in the presence of base. Superimposed emission spectra at different concentrations of **3f** (bottom) time-resolved luminescence quenching of **1a** in the presence of base. Stacked luminescence decays at different concentrations of **3f**; inset: Stern–Volmer plot of the time-resolved experiment.

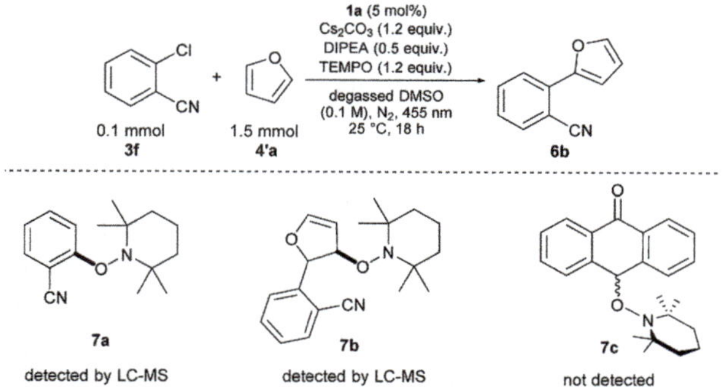

Scheme 2 (top) Reaction conditions for the TEMPO radical trapping experiment; (bottom) structural proposals for detected and expected adducts.

radicals efficiently to form stable adducts, which can be detected by mass analysis.

Adding TEMPO (1.2 equiv.) to the reaction mixture of 2-chlorobenzonitrile (**3f**), furan (**4′a**) and catalyst **1a** led, besides the coupling product **6b**, to detectable amounts of TEMPO adducts, which could be assigned to structures **7a** and **7b** by LCMS analysis. These findings support the hypothesis of an initially formed aryl radical (**7a**) and the subsequent radical reaction with the present (hetero)arene (**7b**). In a control experiment examined in the dark, no TEMPO adducts were detected, supporting the fact that irradiation of the reaction mixture is of importance for the generation of radical intermediates.

Based on the results obtained in spectroscopic measurements and radical trapping experiments we propose the following mechanism (Scheme 3) for the photocatalysed coupling reaction of (hetero)aryl chlorides with hetero(arenes). The visible light-absorbing catalyst **2** is in an acid–base equilibrium with the

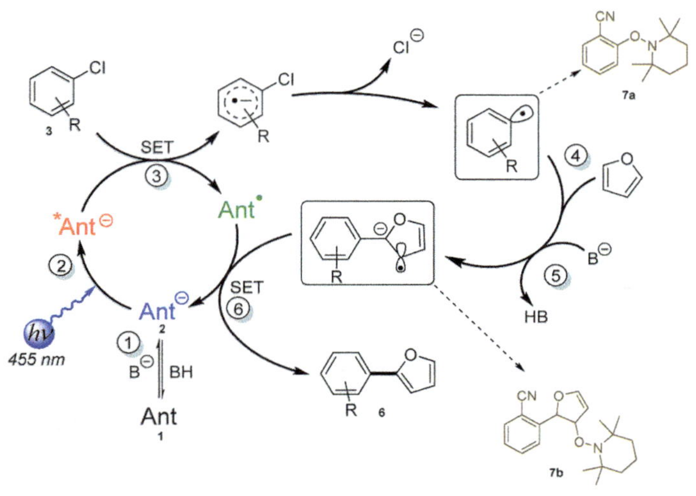

Scheme 3 Proposed photocatalytic mechanism based on experimental findings.

neutral non-absorbing pre-catalyst **1** (1). UV-vis absorption and luminescence spectroscopy revealed that **2** gets photoexcited by visible light (2). Emission quenching studies showed that the excited state of **2a** interacts with suitable substrates.

We propose a single electron transfer from the photoexcited anion **2a*** to the (hetero)aryl chloride (3) to form a radical anion of the substrate **3**. Upon mesolytic bond-cleavage, the generated radical anion fragments to release the reactive aryl radical intermediate, which is trapped by the present (hetero)arene (4) to give a bicyclic radical intermediate. Walton recently showed that radicals adjacent to certain functional groups cause a remarkable increase in acidity. Moreover, cyclohexadienyl type radicals formed after addition of a radical to an (hetero) aromatic ring exhibit an enhanced acidity caused by the neighbouring radical.[48,49] Hence, such compounds are easily deprotonated in the presence of base (5) to form a radical anion intermediate. Regeneration of the oxidized catalyst by single electron transfer (SET) from the radical anion intermediate provides the desired product **6** and closes the catalytic cycle (6). The formation of the aryl radical as well as the bicyclic radical intermediate was confirmed by TEMPO trapping experiments (Scheme 2).

Conclusions

A novel photocatalytic concept based on photoexcitation of an organic anionic ground state catalyst is reported. Instead of using high-energy light sources or accumulating the energy of two photons (conPET), we show that an initial chemical activation (*i.e.*, deprotonation) of a ground state molecule followed by subsequent photoexcitation generates a strong anionic reducing agent, featuring a remarkably long excited state lifetime. The reaction methodology was used to activate a variety of aryl chlorides and the corresponding C–H arylation or coupling products were obtained in moderate to excellent yield. Furthermore, the reaction operates under very mild reaction conditions and avoids the use of sacrificial electron donors, which are typically required in conPET processes. Thus, the reported metal-free protocol allows efficient $C(sp^2)$–$C(sp^2)$ (het)arylation and many applications in organic synthesis can be envisaged.

Conflicts of interest

There are no conflicts to declare.

Acknowledgements

We would like to thank Dr Rudolf Vasold (University of Regensburg) for his technical support regarding GC-FID and GC-MS measurements, Regina Hoheisel (University of Regensburg) for her assistance in recording cyclic voltammetry and spectroelectrochemical measurements and Julia Zach (University of Regensburg) for her kind support regarding spectroscopic matters. Financial support from the German Science Foundation (DFG) (GRK 1626, Chemical Photocatalysis and KO 1537/18-1) is acknowledged.

Notes and references

1 I. Ghosh, L. Marzo, A. Das, R. Shaikh and B. König, *Acc. Chem. Res.*, 2016, **49**, 1566–1577.

2 R. Matsubara, T. Yabuta, U. Md Idros, M. Hayashi, F. Ema, Y. Kobori and K. Sakata, *J. Org. Chem.*, 2018, **83**, 9381–9390.

3 D. Liu, M.-J. Jiao, Z.-T. Feng, X.-Z. Wang, G.-Q. Xu and P.-F. Xu, *Org. Lett.*, 2018, **20**, 5700–5704.

4 I. Ghosh, R. S. Shaikh and B. König, *Angew. Chem., Int. Ed.*, 2017, **56**, 8544–8549.

5 M. Neumeier, D. Sampedro, M. Májek, V. A. de la Peña O'Shea, A. Jacobi von Wangelin and R. Pérez-Ruiz, *Chem.–Eur. J.*, 2018, **24**, 105–108.

6 J. I. Bardagi, I. Ghosh, M. Schmalzbauer, T. Ghosh and B. König, *Eur. J. Org. Chem.*, 2018, **2018**, 34–40.

7 T. Fukuyama, Y. Fujita, H. Miyoshi, I. Ryu, S.-C. Kao and Y.-K. Wu, *Chem. Commun.*, 2018, **54**, 5582–5585.

8 E. T. Nadres, A. Lazareva and O. Daugulis, *J. Org. Chem.*, 2011, **76**, 471–483.

9 C. Costentin, M. Robert and J.-M. Savéant, *J. Am. Chem. Soc.*, 2004, **126**, 16051–16057.

10 M. Montalti, A. Credi, L. Prodi and M. T. Gandolfi, *Handbook of Photochemistry*, CRC Press, Boca Raton, 3rd edn, 2006.

11 E. H. Discekici, N. J. Treat, S. O. Poelma, K. M. Mattson, Z. M. Hudson, Y. Luo, C. J. Hawker and J. R. de Alaniz, *Chem. Commun.*, 2015, **51**, 11705–11708.

12 I. Ghosh, T. Ghosh, J. I. Bardagi and B. König, *Science*, 2014, **346**, 725–728.

13 I. Ghosh and B. König, *Angew. Chem., Int. Ed.*, 2016, **55**, 7676–7679.

14 X. Li, D. Liang, W. Huang, H. Sun, L. Wang, M. Ren, B. Wang and Y. Ma, *Tetrahedron*, 2017, **73**, 7094–7099.

15 S. Fukuzumi and K. Ohkubo, *Org. Biomol. Chem.*, 2014, **12**, 6059–6071.

16 A. Joshi-Pangu, F. Lévesque, H. G. Roth, S. F. Oliver, L.-C. Campeau, D. Nicewicz and D. A. DiRocco, *J. Org. Chem.*, 2016, **81**, 7244–7249.

17 N. A. Romero and D. A. Nicewicz, *Chem. Rev.*, 2016, **116**, 10075–10166.

18 W. P. Todd, J. P. Dinnocenzo, S. Farid, J. L. Goodman and I. R. Gould, *J. Am. Chem. Soc.*, 1991, **113**, 3601–3602.

19 B. Legros, P. Vandereecken and J. P. Soumillion, *J. Phys. Chem.*, 1991, **95**, 4752–4761.

20 E. Hasegawa, K. Mori, S. Tsuji, K. Nemoto, T. Ohta and H. Iwamoto, *Aust. J. Chem.*, 2015, **68**, 1648–1652.

21 E. Hasegawa, Y. Nagakura, N. Izumiya, K. Matsumoto, T. Tanaka, T. Miura, T. Ikoma, H. Iwamoto and K. Wakamatsu, *J. Org. Chem.*, 2018, **83**, 10813–10825.

22 C. Kerzig and M. Goez, *Phys. Chem. Chem. Phys.*, 2015, **17**, 13829–13836.

23 S. G. Mills and P. Beak, *J. Org. Chem.*, 1985, **50**, 1216–1224.

24 G. M. McCann, C. M. McDonnell, L. Magris and R. A. More O'Ferrall, *J. Chem. Soc., Perkin Trans. 2*, 2002, 784–795.

25 B. Freiermuth, B. Hellrung, S. Peterli, M.-F. Schultz, D. Wintgens and J. Wirz, *Helv. Chim. Acta*, 2001, **84**, 3796–3809.

26 F. G. Bordwell, R. J. McCallum and W. N. Olmstead, *J. Org. Chem.*, 1984, **49**, 1424–1427.

27 T. Fujii, S. Mishima, N. Tanaka, O. Kawauchi, K. Kodaira, H. Nishikiori and Y. Kawai, *Res. Chem. Intermed.*, 1997, **23**, 829–839.

28 Y. Ogata, Y. Kosugi and K. Nate, *Tetrahedron*, 1971, **27**, 2705–2711.

29 M. E. Budén, J. I. Bardagí, M. Puiatti and R. A. Rossi, *J. Org. Chem.*, 2017, **82**, 8325–8333.

30 The species responsible for the background reaction is not well understood at present and requires further investigations.

31 During the photoreaction the photocatalyst may bleach, and the degraded or photo-dimerised products (wherever applicable) may also participate in the chemical transformation affording the desired coupling products.

32 Efforts to generate twofold substituted pyrroles by using an excess amount of **3a** (Table S2, ESI†) remained unsuccessful with low conversion of the starting material. We speculate that under our reaction conditions a fast and efficient conversion of the aryl radical intermediate is important for an effective catalytic transformation.

33 The monosubstituted product was obtained only in traces.

34 Determined reduction potentials using CV measurements: 4-chloroacetophenone (E_{red} = −1.82 V *vs.* SCE), 4-chlorobenzaldehyde (E_{red} = −1.72 V *vs.* SCE) and methyl 4-chlorobenzoate (E_{red} = −1.95 V *vs.* SCE). For further information see ESI.†

35 Although the reduction potential values suggest a very similar thermodynamically favourable electron transfer from the excited catalyst to these compounds compared to **3a**, a competing side reaction might be the reduction of the carbonyl group (E_{red} = −1.93 V *vs.* SCE for 4-chlorobenzaldehyde) giving the ketyl radical anion.

36 C. Wang, H. Dong, W. Hu, Y. Liu and D. Zhu, *Chem. Rev.*, 2012, **112**, 2208–2267.

37 S. Suzuki and J. Yamaguchi, *Chem. Commun.*, 2017, **53**, 1568–1582.

38 S. I. Panchamukhi, A. K. Mohammed Iqbal, A. Y. Khan, M. B. Kalashetti and I. M. Khazi, *Pharm. Chem. J.*, 2011, **44**, 694–696.

39 S. Paria and O. Reiser, *Adv. Synth. Catal.*, 2014, **356**, 557–562.

40 T. Föll, J. Rehbein and O. Reiser, *Org. Lett.*, 2018, **20**, 5794–5798.

41 I. Ghosh and B. König, *Angew. Chem., Int. Ed.*, 2016, **55**, 7676–7679.

42 W. C. Yang, K. Wei, X. Sun, J. Zhu and L. Wu, *Org. Lett.*, 2018, **20**, 3144–3147.

43 P. Natarajan, A. Bala, S. K. Mehta and K. K. Bhasin, *Tetrahedron*, 2016, **72**, 2521–2526.

44 D. P. Hari, P. Schroll and B. König, *J. Am. Chem. Soc.*, 2012, **134**, 2958–2961.

45 A. Arora and J. D. Weaver, *Org. Lett.*, 2016, **18**, 3996–3999.

46 We found that adding perylene (10.0 mol%) to the reaction mixture led to a further increase in the product yield. Using **3f** as the starting material, the product was obtained in 88% GC-yield (Table S3, ESI†).

47 These trapping reagents are deeply coloured oils. To allow efficient light penetration through the reaction solution, the addition of isocyanide was kept between 0.3 and 0.5 mmol.

48 J. C. Walton, *J. Phys. Chem. A*, 2018, **122**, 1422–1431.

49 A. Studer and D. P. Curran, *Nat. Chem.*, 2014, **6**, 765–773.

Faraday Discussions

PAPER

Influence of carbonaceous species on aqueous photo-catalytic nitrogen fixation by titania†

Yu-Hsuan Liu,[a] Manh Hiep Vu,[b] JeongHoon Lim,[c] Trong-On Do [ID][b] and Marta C. Hatzell [ID] *[c]

Received 19th November 2018, Accepted 5th February 2019

DOI: 10.1039/c8fd00191j

For decades, reports have suggested that photo-catalytic nitrogen fixation by titania in an aqueous environment is possible. Yet a consensus does not exist regarding how the reaction proceeds. Furthermore, the presence of an aqueous protonated solvent and the similarity between the redox potential for nitrogen and proton reduction suggest that ammonia production is unlikely. Here, we re-investigate photo-catalytic nitrogen fixation by titania in an aqueous environment through a series of photo-catalytic and electrocatalytic experiments. Photo-catalytic testing reveals that mineral phase and metal dopants play a marginal role in promoting nitrogen photofixation, with ammonia production increasing when the majority phase is rutile and with iron dopants. However, the presence of a trace amount of adsorbed carbonaceous species increased the rate of ammonia production by two times that observed without adsorbed carbon based species. This suggests that carbon species play a potential larger role in mediating the nitrogen fixation process over mineral phase and metal dopants. We also demonstrate an experimental approach aimed to detect low-level ammonia production from photo-catalysts using rotating ring disk electrode experiments conducted with and without illumination. Consistent with the photocatalysis, ammonia is only discernible at the ring with rutile phase titania, but not with mixed-phase titania. Rotating ring disk electrode experiments may also provide a new avenue to attain a higher degree of precision in detecting ammonia at low levels.

1 Introduction

Reports of photocatalytic nitrogen fixation on titania have surfaced for nearly eight decades.[1] Demonstrations have occurred in both the gas and aqueous

[a]School of Civil and Environmental Engineering, Georgia Institute of Technology, Atlanta, Georgia 30313, USA
[b]Department of Chemical Engineering, Laval University, Rue de l'Université, Québec, QC G1V 0A6, Canada
[c]George W. Woodruff School of Mechanical Engineering, Georgia Institute of Technology, Atlanta, Georgia 30313, USA. E-mail: marta.hatzell@me.gatech.edu

† Electronic supplementary information (ESI) available. See DOI: 10.1039/c8fd00191j

phase, and with various environmental conditions (relative humidity, temperature, pressure). Nearly all experimental observations suggest that abiotic nitrogen photofixation may be possible at ambient temperature and pressure.[2,3] This is impactful as it promotes the possibility for nutrient based fertilizer production from environmentally abundant materials (minerals) using only the sun as a source of energy.[33] Considering the state-of-the-art thermochemical approach to fertilizer production is responsible for consuming 1% of global energy and emitting 3% of global CO_2, photocatalytic nitrogen fixation could have a significant influence on improving the environmental impact associated with ammonia production.[2]

Photocatalytic nitrogen fixation also has environmental significance, as nitrogen fixation is a critical catalytic process in the nitrogen cycle. Although biological fixation is the most notable natural entry point for dinitrogen in soils, additional entry points driven through photon based reactions on earth abundant minerals are not outside the realm of possibility. A prominent Indian soil scientist (N. Dhar) in the 1940s suggested that the generation of fixed nitrogen in soils was possible through light based interactions.[1] Schrauzer and Guth explored this hypothesis further in the late 1970s through conducting a series of experiments aimed at discerning the photocatalytic nitrogen fixation activity on environmental catalysts (global sands).[3-5] This work concluded that the presence of titania in sands was responsible for nitrogen photofixation. Despite these intriguing findings, the potential impacts of photocatalytic nitrogen fixation on the nitrogen cycle is still limited to only back-of-the envelope based calculations.[6]

This slow progress is due to a lack of understanding regarding the catalytic processes that enable nitrogen fixation to occur on titania based photocatalysts. The wide band gap of the mineral coupled with the location of the band edges suggest that nitrogen reduction is unlikely. Furthermore, in environments were hydrogen evolution is possible (aqueous conditions), selectivity toward nitrogen reduction would require a low overpotential. Furthermore, the low measurable yields, high propensity for contamination and numerous sources of measurement error have all contributed to the significant debate regarding the probability of photocatalytic nitrogen fixation occurring on titanium dioxide.[7-11] Most prior studies have emphasized the role iron dopants and oxygen vacancies play in activating dinitrogen on titania.[12-14] Significant evidence supports the hypothesis that iron dopants aid in stabilizing oxygen vacancies, and that the presence of vacancies weakens the dinitrogen triple bond, enabling dinitrogen reduction. While possible, a more recent hypothesis suggests that carbon may play a critical role in nitrogen photofixation on titania.[15] Specifically, gas phase AP-XPS based experiments demonstrated that the presence of adventitious carbon was important for promoting interactions between titanium dioxide and nitrogen. The researchers hypothesized that a photocatalyzed carbon radical in concert with defects may promote nitrogen photofixation to ammonia. These gas phase experiments promote the need to understand if photocatalyzed carbon radicals also play a similar role in an aqueous photocatalytic environment.

Here, we aim to expand upon work completed in the gas phase; to explore the potential role carbonaceous species may play in promoting photo-catalytic nitrogen fixation in aqueous based experiments. While adventitious carbon and carbon dioxide may be the dominant carbonaceous species present in gas phase photocatalysis, here organic hole scavengers (methanol, ethanol and formic acid) are the most prominent sources of carbon. We begin by probing the photo-catalytic activity of rutile, mixed phase and iron-doped titania in the presence and absence of carbonaceous species. Rotating ring disk electrode experiments con-ducted on the catalyst also aim to determine if ammonia is detectable under illumination. Finally, the surface based active sites (oxygen vacancies and lattice iron) are mapped using bulk and surface based characterization.

2 Materials and methods

2.1 Catalysts preparation

A commercial rutile titanium dioxide and mixed phase titanium dioxide were purchased from US Research Nanomaterials Inc and Alfar Aesar (Haverhill, MA). Iron-doped titanium dioxide was prepared by standard sol–gel methods.[16] Tita-nium dioxide sols were prepared through the addition of 5 mL of titanium tet-raisopropoxide (TTIP) solution into 50 mL of absolute ethanol. Then, the TTIP/ethanol solution was added (dropwise) into 50 mL of distilled water which was adjusted to pH 1.5 with chloric acid under vigorous stirring at room temperature. After continuously stirring for 24 h, the resulting transparent solution was evaporated using a rotary heater at 50 °C and dried in the oven at 70 °C overnight. The resulting particles were calcined at 500 °C for 1 h under air, and stored in air tight vessels. Catalyst cleaning prior to use occurred through a calcination process at 400 °C for 4 h. This cleaning procedure prevented the detection of any ammonia during controls, which were conducted in the absence of nitrogen (argon gas and light) and in the absence of light (nitrogen gas and dark).

2.2 Photocatalysis experiments

The photocatalytic nitrogen fixation experiments were conducted in a 500 mL reaction vessel at ambient temperature and atmospheric pressure. The temper-ature was fixed at 28 °C during the tests and was monitored to ensure a constant temperature existed under illumination. A 300 W Xenon–Mercury lamp (Newport Corporation – Irvine, CA) was used which contained a ultraviolet (UV) cutoff filter ($\lambda > 320$ nm) and an infrared (IR) cutoff filter. In all tests, 300 mg of photocatalyst was suspended in 300 mL of aqueous electrolyte. The aqueous electrolyte con-sisted of deionized water, or a mixture of deionized water with a hole scavenger. The hole scavenger concentration remained fixed at five volume percent. Meth-anol, ethanol and formic acid were chosen as representative hole scavengers. The photocatalytic reactor was deaerated using nitrogen gas (purity $\geq 99.999\%$) which was bubbled through this solution for thirty minutes in the dark. Various times were investigated to ensure that the solution was completely saturated with nitrogen and that all residual dissolved oxygen was removed. After saturation was reached, the reactor was irradiated with ultra-violet light ($\lambda > 320$ nm) from the top of the reactor. The reactor was continuously fed nitrogen gas during the entire twenty-four hour test period. Controls were conducted to probe the presence of contamination. In the first control, nitrogen gas was replaced by argon gas (purity $\geq 99.999\%$). All other conditions remained constant, including the degree of illumination. A second control was conducted in the dark with nitrogen gas. During all tests (including controls), approximately 5 mL of the suspension was removed from the reaction vessel at given intervals to test for ammonium.

2.3   Ammonium measurement

Two milliliters of dilute acid (0.005 M dilute H_2SO_4) was added to the catalyst suspension removed from the photocatalytic reactor. The dilute acid was used to promote ammonia desorption from the catalyst surface.[4] The particle suspension was then filtered by a 0.22 μm precision-glide syringe needle (Sigma-Aldrich, St. Louis, Mo) and syringe filters (Hach Co). The desorbed ammonium concentration was analyzed by ion chromatography using a cation exchange column (Aquion Dionex, Thermo Fisher). Ammonia was easily distinguished from other cations and had a retention time of 5.24 minutes. Standard solutions of ammonium chloride were used to create a calibration curve for the ion chromatograph. While many standard methods rely on optical measurements (UV-vis) whereby an organic compound reacts with ammonia (indophenol blue), we found these methods were susceptible to erroneous data, and a lack of reproducibility, consistent with prior works.[17]

2.4   Electrochemical characterization

Electrochemical testing was conducted using a rotating ring disk electrode experimental set up (Pine Research, Durham, NC). The catalyst was dispersed onto the disk, and the ring was poised to sense product (ammonia) formation. The catalyst was dispersed onto the disk using standard methods. A stock solution of 20% isopropanol and 0.5% Nafion solution was prepared by mixing 20 mL of isopropanol with 70 mL of distilled water and 10 mL of 5 wt% Nafion solution in a 100 mL volumetric flask. Then 25 mg of catalyst was added into the Nafion/IPA solution and sonicated for one hour. After the suspension was well dispersed without aggregation or sedimentation, 10 μL of the ink was drop cast onto the glassy carbon disk (6.2 mm diameter) and rotated at 1000 rpm for one hour using the air dry method.[18,19] A lamp (Asahi Spectra, Torrance, CA) was situated below the disk outside of the glass reactor. The lamp was used to irradiate the disk during experiments. A platinum ring (inner and outer diameter, 7 and 8.4 mm) served as the working electrode. A silver–silver chloride reference electrode (saturated potassium chloride) and a platinum wire counter electrode were also used. Cyclic voltammetry was performed while rotating the shaft of the working electrode at 500 rpm with a scan rate of 5 mV s^{-1}. Simultaneously, the ammonia oxidation reaction was investigated at the ring. In all testing the supporting electrolyte consisted of sodium hydroxide. Alkaline conditions were deemed necessary in order to promote the ammonia oxidation reaction.

2.5   Material surface characterization

The morphology of the titanium dioxide materials was investigated by scanning electron microscopy (SEM) coupled with an energy dispersive X-ray spectrometer (EDS). EDS coupled with XPS analysis confirmed that the iron-doped samples exhibited 0.69 atomic percent iron in the titania lattice. The as-prepared materials were identified through X-ray powder diffraction (XRD) using a Pananalytical Xpert Pro Alpha-1 XRD system for crystalline identification with Cu Kα radiation ($\lambda = 1.54051$). Step scanning was used with 2θ intervals from $10°$ to $90°$ with a residence time of 1 s. The chemical states of the surface elements were recorded using an X-ray photon spectrometer (XPS) using a Thermo K-Alpa XPS spectrometer equipped with monochromatic Al Kα radiation as the X-ray source. The Ti2p, O1s, C1s, and Fe2p peaks were analyzed to access the surface properties (oxygen vacancies and dopants) of each catalyst.

3 Results and discussion

3.1 Photocatalytic testing with carbonaceous species

Photocatalysis in an aqueous phase reactor was conducted with the three catalysts. Two off the shelf titanium dioxide catalysts were chosen. The first off-the-shelf catalyst had a mineral composition that was primarily rutile phase. The second off-the-shelf catalyst had a mineral composition that was a mixture of anatase and rutile phases. A third catalyst was synthesized in the lab and contained iron dopants at 0.69 atomic percent, as confirmed through energy dispersive X-ray spectroscopy and XPS. Iron was chosen as the representative metal dopant, due to the demonstrated high activity reported in prior photocatalytic nitrogen fixation investigations.[4,14]

Photocatalytic experiments were conducted in three phases. In phase one, suspended particles were deaerated with either argon or nitrogen gas prior to illumination to remove all dissolved oxygen. The reactor was then illuminated for twenty-four hours, with samples removed at timed intervals. For all tests with argon, no ammonia was detected over the entire twenty four hour test. Dark controls in the presence of nitrogen also yielded no detectable ammonia. This confirmed that little nitrogen based contamination was present on the catalyst or photo-catalytic reactor prior to and during testing. Thus all ammonia detected was a result of a light-based reaction.

During the phase one experiments, the rate of ammonia production did not change regardless of the catalyst (rutile, mixed phase, and iron-doped titanium dioxide). The rate of ammonia production ranged from $0.25–0.38 \pm 0.03$ μM h^{-1} g^{-1}, indicating that the phase (rutile or anatase) and metal dopants (iron) did not play a significant role in improving the rate of nitrogen photofixation (Fig. 1 – phase 1). Prior investigations have highlighted that the phase (rutile)[4] and concentration of oxygen vacancies are critical material properties needed in order to increase the rate of ammonia production.[13] Furthermore, rutile phase catalysts have been shown both experimentally and theoretically to be active for both photo and electrocatalytic nitrogen fixation.[4,12]

Oxygen vacancies or point defects in the titanium dioxide lattice are also well known to exist in non-stoichiometric samples.[20] However, in the presence of oxygen or water, as is the case with aqueous-based photocatalytic experiments, vacancies can heal resulting in a near stoichiometric crystal. Thus, while oxygen vacancies may be active sites, the lifetime and concentration of these active sites are uncertain in a aqueous environment. We also note that the initial rates observed are in line with prior investigations, which contained catalysts with a low concentration of oxygen vacancies.[13]

After the initial phase, the minerals were separated and recycled back to the reactor for phase two. In phase two, a hole scavenger was introduced into the system. Traditional hole scavengers are organic compounds (e.g. EDTA, methanol and ethanol) which can be easily oxidized at the valance band by photogenerated holes (Fig. 1 – phase 2). Here, the aim in introducing a hole scavenger was not to minimize the rate of charge carrier recombination, but rather to probe the

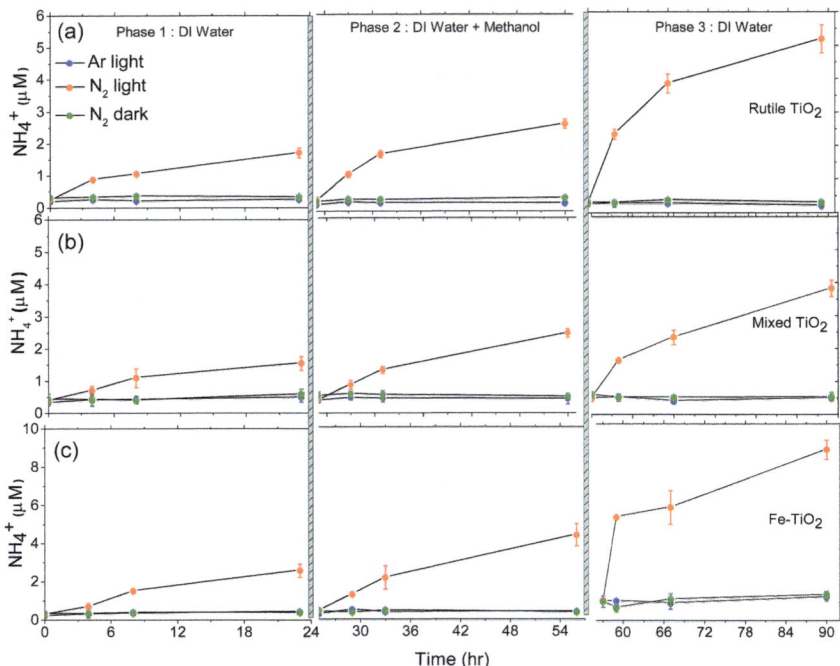

Fig. 1 Ammonia yield with time during photo-catalytic testing under a 300 W lamp. Experiments were conducted in three phases (labeled phase 1, 2 and 3 in figure), and with three catalyst (a) rutile titania, (b) mixed phase titania and (c) iron-doped titania.

hypothesis that nitrogen photofixation occurs due to an interaction with a photogenerated carbon radical. Methanol was chosen as a representative hole scavenger, as methanol is well studied with titania based photocatalytic systems, has been used as a hole scavenger in prior nitrogen fixation investigations, and has a well-documented reaction pathway.[21] Methanol oxidation is suspected to occur predominately through a dissociative pathway resulting in adsorbed methoxy species and hydroxyl groups.[22,23]

$$CH_3OH_{(a)} + O_{(a)} \rightarrow CH_3O_{(a)} + OH_{(a)} \tag{1}$$

The methoxy species have furthermore been shown to generate carbon-based radicals that have been detected through both infrared and electronic paramagnetic resonance based spectroscopy.[22] With this goal in mind, phase two proceeded with methanol present in the deionized water. Methanol was supplied to the deionized water at 5 vol%, which is consistent with the concentrations traditionally supplied to photocatalytic systems, whereby methanol acts as a hole scavengers. We will also note that in addition to the carbon radical generation due to the presence of methanol, a degree of adventitious carbon is also well known to exist on all titania based photocatalysts.

For each catalyst, the rate of ammonia production increased on average by 60% when compared to testing conducted without an organic hole scavenger, with the rates increasing to $0.4–0.65 \pm 0.04\ \mu M\ h^{-1}\ g^{-1}$. This increase in the rate of ammonia produced is traditionally ascribed to the system no longer being limited by the hole driven reaction (water splitting). Instead of oxidizing water, methanol is oxidized by the large oxidative potential of the holes generated in titanium dioxide, preventing charge carrier recombination. In phase two, the performance of rutile, mixed phase and iron-doped titania was similar, indicating that phase and metal dopants again did not alter performance significantly. In addition, controls (nitrogen dark and argon light) resulted in no appreciable ammonia production.

In an effort to probe the role of the carbon radical more directly, the catalysts were recycled a third time back in deionized water which contained no methanol. Here, by recycling the catalyst exposed to methanol oxidation, the aim was to remove the methanol, while maintaining a degree of adsorbed methoxy groups (or other oxidized carbon species) on the catalyst. Prior investigations have shown that methanol desorbs easily from the catalyst and solution at 308–373 K, while the methoxy groups remain adsorbed ($T_{des} = 460$ K).[23,24] Thus we hypothesize that in phase three, carbon radical generation may be possible in the absence of methanol due to these adsorbed species. In phase three, each catalyst experienced an increase in the rate of ammonia production (Fig. 1 – phase 3) over testing with methanol (phase two), and more importantly over results observed with just deionized water (phase one). In addition, the rate of ammonia production differed between the three individual catalysts. The rate of ammonia produced on the iron-doped titania, rutile titania and mixed phase titania were 1.3 ± 0.07 μM h^{-1} g^{-1}, 0.76 ± 0.06 μM h^{-1} g^{-1} and 0.56 ± 0.03 μM h^{-1} g^{-1}. Thus the order of photocatalytic activity for the three catalysts was iron-doped > rutile > mixed phase (Fig. 1). We also conducted control experiments (rutile titania) in which the catalysts were recycled three times without an organic scavenger. In these tests, that the ammonia yield remained at 0.2 ± 0.008 μM h^{-1} g^{-1} (consistent with the results obtained from phase 1), indicating that the increased rates are not a result of catalyst recycling and handling.

Phase three photocatalytic testing clearly emphasizes that the adsorbed carbon radicals most likely play a role in mediating the nitrogen fixation process on titania, as the performance increase cannot be ascribed to the presence of the hole scavenger. Additional testing also conducted with ethanol and formic acid resulted in similar trends (Fig. SI3†), with slight improvements observed with formic acid. Prior theoretical insight suggests that the carbon radicals have free energy (-1.89 eV) for nitrogen, which ultimately can aid in adsorbing and reacting dinitrogen.[12] Thus, the observed photocatalytic activity demonstrated here in an aqueous environment aligns with the prior gas phase experiments. With a range of reported values observed with photocatalytic nitrogen fixation based experiments on titania based photocatalysts,[4,7,13] here we highlight that in addition to material properties (bulk and surface), the carbon species prominently used within the photocatalytic reactor may also influence the nitrogen fixation reaction mechanism and catalytic activity.

3.2   Electrocatalytic characterization of titania

In addition to monitoring the ammonia production through bulk phase photocatalytic testing, we also probed the formation of ammonia through a series of rotating ring disk electrode (RRDE) experiments. The three catalysts (rutile, mixed phase and iron-doped titanium dioxide) were investigated through dispersing the catalyst on the disk, while the ring was then poised at the ammonia oxidation potential. The aim of the ring is to act as a sensor to detect products formed as a result of the photocatalyzed reactions (hole and electron driven) on the titanium dioxide. The disk therefore was alternatively excited and not excited by light that was situated below the catalyst. The experiments were carried out with nitrogen and argon saturated electrolytes in the dark and light. A key difference between the photocatalytic experiments and the electrocatalytic experiments is the choice of electrolyte. While ideally, DI water would have been investigated, there were two chief challenges. The first being that ammonia oxidation in a highly resistive cell would produce very little current. With the goal being to correlate photo-catalyzed ammonia production to ring current, we needed to limit resistance through the use of a supporting electrolyte. Another challenge was that ammonia oxidation occurs through the following proposed mechanism on Pt:

$$NH_3 + OH^- \rightarrow NH_{2,ad} + H_2O + e^- \tag{2}$$

$$NH_{2,ad} + OH^- \rightarrow NH_{ad} + H_2O + e^- \tag{3}$$

$$NH_{x,ad} + NH_{y,ad} \rightarrow N_2H_{x+y,ad} \tag{4}$$

$$N_2H_{x+y,ad} + OH^- \rightarrow N_2 + x + yH_2O + x + ye^- \tag{5}$$

Since the ring utilized here was Pt, in order to promote the ammonia oxidation reaction, OH^- is needed in the supporting electrolyte at significant concentrations. For this reason an alkaline electrolyte was chosen, as prior investigations have shown that ammonia oxidation on Pt cannot be detected in electrolytes with a pH lower than eight, and it is more preferable to operate with electrolytes that have a pH greater then ten.[25] It should be noted that titanium dioxide was not operated as an electrocatalyst, as titania was not electrochemically biased during the experiments.

In the control experiments, 0.1 M NH_4Cl was added to the supporting electrolyte and the ring voltage was swept from 0.2 vs. RHE to 1.2 vs. RHE. A noticeable peak was observed in all tests at $\approx$ 0.7 vs. RHE. This peak was attributed to ammonia oxidation and was used as a reference for future testing conducted with titanium dioxide (Fig. 2a).[26] Initial testing conducted with rutile titanium dioxide resulted in no peak during initial tests without light (Fig. 2b – 0 h). However, with the addition of light, the ammonia oxidation peak was observed. The catalyst was tested over a 10 hour period while illuminated, and the peak position did not vary significantly, while the peak current increased slightly. The extended experiment suggests that the rate of ammonia produced and oxidized remained constant with time (Fig. 2b – 5, 10 h). The other peaks besides ammonia oxidation are Pt–O formation (potential region 0.8–1 V), Pt–O reduction (potential region 1.2–0.7 V vs. RHE), Pt–H reduction (ca. 0.3 V vs. RHE) and Pt–H oxidation (ca. 0.35 V vs. RHE).

The three catalysts were first evaluated under similar conditions tested in the photocatalytic reactor. Namely, in an electrolyte saturated with nitrogen and argon gas, and under dark and light conditions. With the rutile phase titanium dioxide photocatalyst, the ammonia oxidation peak (ca. 0.7 V vs. RHE) was observed with nitrogen gas and light (Fig. 3a). No ammonia oxidation peak was present when

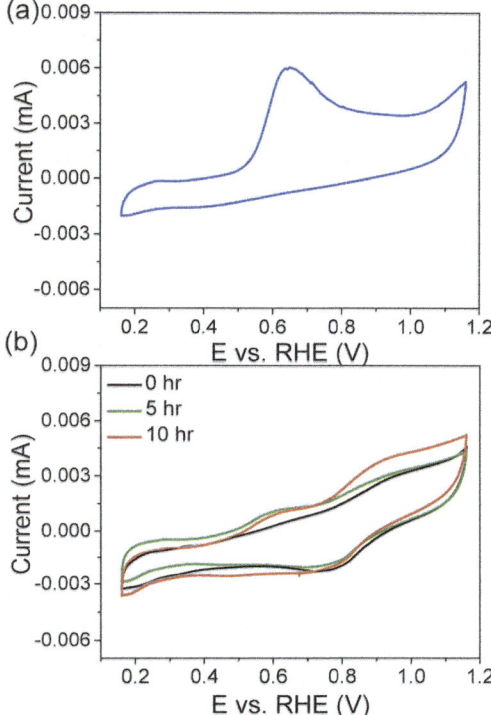

Fig. 2   (a) Ammonia oxidation peak observed on the Pt ring with 0.1 M NH$_4$Cl in the alkaline supporting electrolyte. (b) Ammonia oxidation of rutile TiO$_2$ under light and nitrogen gas.

nitrogen was replaced by argon, or when the sample was not illuminated (dark testing with nitrogen gas). In the potential region from 0.8–1.0 V *vs.* RHE, a Pt–O peak was observed. This is due to the presence of oxygen, being formed under oxidative potentials on Pt. Furthermore, when scanning negatively in the potential region from 1.2–0.7 V *vs.* RHE, Pt–O reduction occurred. A Pt–H desorption peak was observed in the potential region around 0.3 V *vs.* RHE under the negative scan and its redox couple was present around 0.35 V *vs.* RHE due to Pt–H oxidation.[18] These peaks were consistent for all catalysts, which is to be expected as the Pt based reactions are independent of the disk catalyst.

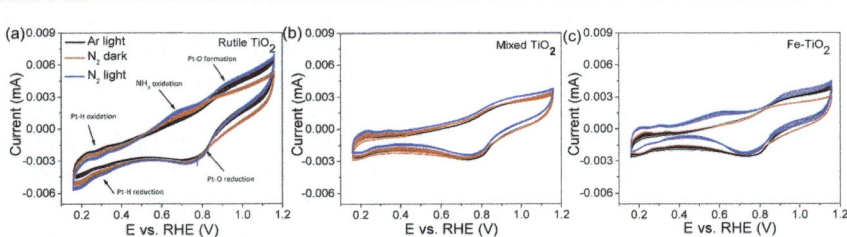

Fig. 3   Rotating ring disk electrode experiments with and without light in argon and nitrogen atmospheres with (a) rutile (b) mixed phase and (c) iron-doped titania.

When the mixed phase titanium dioxide was used as the catalyst, no ammonia oxidation peak was visible under nitrogen (light and dark) and argon (light). Prior investigations have suggested that a mixed phase photocatalyst is non-ideal for nitrogen fixation, as rutile may be the most active phase. Furthermore, mixed phase photocatalysts are most active for hydrogen evolution. Since hydrogen evolution competes with nitrogen reduction it is not surprising that ammonia was not detected with the mixed phased photocatalyst (Fig. 3b). With the iron-doped titania sample a more obvious ammonia oxidation peak was observed under nitrogen in the light. Again, controls with argon and nitrogen (dark) did not result in a peak (Fig. 3c). The observed peak currents (ammonia oxidation peaks) measured in the ring electrode were in-line with the photocatalytic testing activity, which suggests that rutile phase and iron-doped titania are active for nitrogen fixation, whereas the mixed phase photocatalyst is the least active catalyst. The current and the charge from the ammonia oxidation area can be used to esti-mate the rate of nitrogen photofixation. The observed rates for the rutile and iron-doped titanium dioxide were on the order of 10^{-6} mol per second, whereas the mixed phase catalyst was on the order of 10^{-8} mol per second. Furthermore, the cyclic voltammetry of the Pt disc electrode under nitrogen and argon gas flows, with and without light are shown in ESI Fig. SI1.† As shown, the Pt–O formation/reduction and Pt–H oxidation/reduction remained, but no ammonia oxidation was found with these three conditions indicating that Pt does not participate in ammonia formation with light illumination. Also, the most active iron-doped catalyst has been selected to perform the same phases 1–3 used during photocatalytic testing. The results were shown in Fig. SI2.† In phase one, iron-doped TiO_2 was examined under nitrogen flow and light. After phase one, the disk is recycled back in an electrolyte which contains methanol. In phase two methanol is oxidized when the disk is illuminated (presumably generating carbon radicals). The methanol oxidation peak was observed at 0.8 V *vs.* RHE. In phase 3, the electrolyte is replaced (no methanol), while adsorbed methoxy groups remain. The result showed that the residual adsor-bed surface groups again increased the ammonia formation, observable through the increase in the peak area (*ca.* 0.7 V *vs.* RHE). The observed rate of ammonia oxidation in phase three is 2 times higher than that in phase one. Also, the CV showed that no adsorbed methanol remained on the surface of the catalyst since there is no methanol oxidation presented in the phase three experiment.

3.3   Catalyst surface and bulk characterization

Both the photocatalytic testing and rotating ring disk electrode experiments provide evidence that the rate of nitrogen photofixation is greatest on the iron-doped titanium dioxide, followed by the rutile phase catalyst and the mixed phase catalyst. Material characterization was obtained by X-ray photoelectron spectroscopy (XPS) and X-ray powder diffraction (XRD). XPS provided insight into the degree of oxygen vacancies that were present in the catalyst, and the degree of iron dopant at the surface of the catalyst. It should be noted that the *ex situ* analysis provides indirect evidence regarding the likelihood that a given catalyst has oxygen vacancies. It is not capable of determining how stable the vacancies are when immersed in the aqueous solution.

For the XPS, the Ti2p, O1s, C1s, and Fe2p regions were all analyzed. Determination of the degree of oxygen vacancies can be observed through monitoring the differences in the O1s and Ti2p regions (Fig. 4). The O1s spectra typically contains two sub-peaks, one which represents the lattice oxygen and a second which details the surface Ti_2O_3 groups. The Ti2p is typically fit with two peaks, $Ti2p_{3/2}$ and $Ti2p_{1/2}$ (Fig. 4b). In a well coordinated sample with no oxygen vacancies these peaks are sharp and are assigned to Ti^{4+} atoms. As the $Ti2p_{3/2}$ and $Ti2p_{1/2}$ broaden, two additional peaks can be introduced which are assigned to Ti^{3+} sites. Thus the two signatures of oxygen vacancies is the increase in Ti_2O_3 groups (Fig. 4a) and the increase in the concentration of Ti^{3+} sites. For the photocatalysts tested, we see that the most active samples have a higher degree of these two signatures, indicating that in the dry state, the amount of oxygen vacancies is indeed higher.[27,28]

According to the XPS analysis of Ti2p, we observed that the area of Ti^{4+} is 80% and Ti^{3+} is 20% in rutile TiO_2. Meanwhile, the area of Ti^{4+} is 86% and Ti^{3+} is 14% for iron-doped TiO_2. However, mixed TiO_2 consisted of 93% Ti^{4+} and 7% Ti^{3+}. Similarly in the O1s spectra, TiO_2 revealed 90% lattice oxygen and 10% Ti_2O_3. The iron-doped TiO_2 consisted of 91% lattice oxygen and 9% Ti_2O_3. The mixed TiO_2 has 97% lattice oxygen and 3% Ti_2O_3. The results reveal that both iron-doped and rutile TiO_2 have comparable oxygen vacancies or defects and mixed TiO_2 has the least. The iron dopant on the surface of TiO_2 plays a role as an electron carrier which will help create more Ti^{3+} and aid nitrogen adsorption onto the surface of

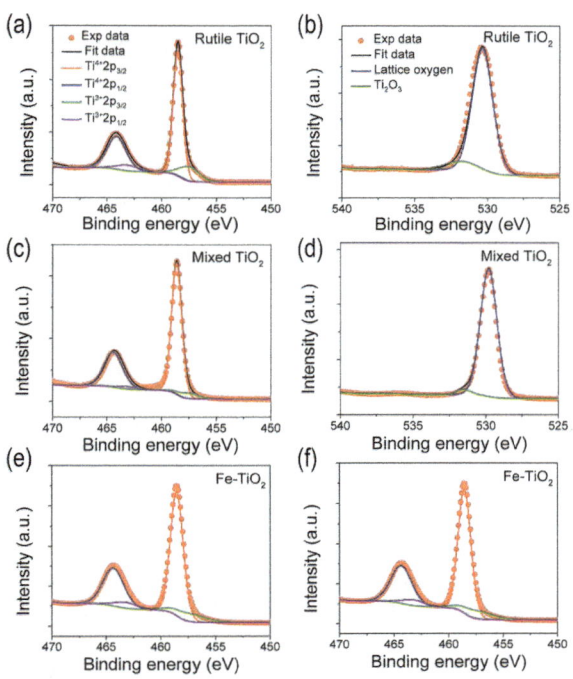

Fig. 4 XPS of rutile, mixed phase and iron-doped titanium dioxide samples. (a, c and e) Ti2p region and (b, d and f) O1s region of the XPS spectra for the three samples.

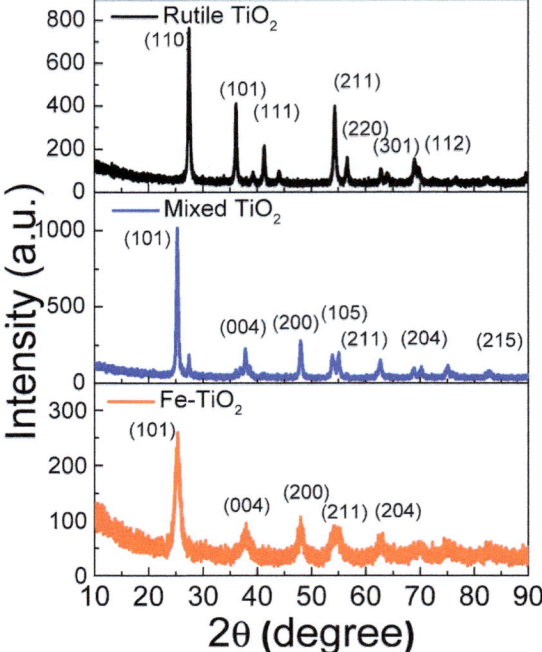

Fig. 5 XRD analysis of the rutile, mixed phase and Fe-doped titanium dioxide based photocatalysts.

TiO_2.[29,30] Therefore the results are consistent with the photocatalytic testing with iron-doped TiO_2 >rutile TiO_2 > mixed TiO_2.

In addition to the surface properties, XRD aided in identifying the structure of the three catalysts. All diffraction lines are relatively strong indicating a high crystallinity for all the samples. Furthermore, the peak positions and relative intensities of the diffraction lines match well with the standard diffraction data for different titanium dioxide phases (*i.e.* rutile and anatase). For rutile titanium oxide, the major peaks observed at 2θ values of 27.4°, 36.1°, 41.3°, 54.4°, 56.3°, 69.3° and 70.1° could be indexed to (110), (101), (111), (211), (220), (301) and (112) (Fig. 5). The pattern is consistent with the standard XRD data (JCPDS, no. 21-1276). The pattern of mixed titanium dioxide is consistent with the XRD standard data (JCPDS, no. 21-1272) which showed a strong anatase crystalline phase.[31,32] The XRD spectra of the synthesized iron-doped titanium dioxide revealed that the material was more likely to be anatase than rutile.

4   Conclusions

Photocatalytic nitrogen fixation on titanium dioxide has been heavily investigated for nearly eighty years due to the potential relevance toward achieving the long standing goal of clean ammonia synthesis. Nitrogen photofixation is also an intriguing entry point to the abiotic nitrogen cycle, yet has not been thoroughly investigated due to a lack of understanding regarding how the catalytic process occurs on this wide-band gap material. Here, we aimed to test an emerging hypothesis which emphasizes that carbon species may play an integral role in catalyzing the dinitrogen adsorption and reduction to ammonia. Through a controlled experimental procedure whereby methanol is oxidized to produce adsorbed carbon radical based species, we show that the rate of ammonia produced can be increased by as much as two times that observed through testing without adsorbed carbonaceous species. This work extends previous work conducted in the gas phase, which also highlighted the potential role adventitious carbon plays in promoting gas-phase nitrogen fixation. We also investigate a method to detect ammonia at low levels using a rotating ring disk electrode experimental set up. The results show that a noticeable ammonia oxidation peak is observed on the rutile and iron-doped titanium dioxide samples, yet is not detected with the mixed phase titania. With many questions regarding the robust nature of low level ammonia measurements, the potential to use current rather than optical or chromatography based methods could improve the resolution of micromolar measurements which is important for both photo and electro-chemical nitrogen fixation based studies.

Conflicts of interest

There are no conflicts to declare.

Acknowledgements

The researchers thank Georgia Tech start up funding for supporting this work, as well as the Georgia Tech TI:GER program. This material is based upon work supported by the National Science Foundation under Grant No. 1846611.

References

1 N. Dhar, E. Seshacharyulu and N. Biswas, *Proc. Natl. Acad. Sci., India*, 1941, **7**, 115–131.
2 A. J. Medford and M. C. Hatzell, *ACS Catal.*, 2017, 2624–2643.
3 G. N. Schrauzer, in *Photoreduction of Nitrogen on TiO$_2$ and TiO$_2$-Containing Minerals*, ed. L. Zang, Springer, London, 2011, pp. 601–623.
4 G. Schrauzer and T. Guth, *J. Am. Chem. Soc.*, 1977, **99**, 7189–7193.
5 G. N. Schrauzer, N. Strampach, L. N. Hui, M. R. Palmer and J. Salehi, *Proc. Natl. Acad. Sci. U. S. A.*, 1983, **80**, 3873–3876.
6 T. A. Doane, *ACS Earth Space Chem.*, 2017, **7**, 411–421.
7 J. G. Edwards, J. A. Davies, D. L. Boucher and A. Mennad, *Angew. Chem., Int. Ed.*, 1992, **31**, 480–482.
8 D. L. Boucher, J. A. Davies, J. G. Edwards and A. Mennad, *J. Photochem. Photobiol., A*, 1995, **88**, 53–64.
9 G. Richardson, J. Davies and J. Edwards, *Fresenius' J. Anal. Chem.*, 1991, **340**, 392–394.
10 L. F. Greenlee, J. N. Renner and S. L. Foster, *The Use of Controls for Consistent and Accurate Measurements of Electrocatalytic Ammonia Synthesis from Dinitrogen*, 2018.
11 Y. Song, D. Johnson, R. Peng, D. K. Hensley, P. V. Bonnesen, L. Liang, J. Huang, F. Yang, F. Zhang, R. Qiao, *et al.*, *Sci. Adv.*, 2018, **4**, e1700336.

12 B. M. Comer and A. J. Medford, *ACS Sustainable Chem. Eng.*, 2018, **6**, 4648–4660.

13 H. Hirakawa, M. Hashimoto, Y. Shiraishi and T. Hirai, *J. Am. Chem. Soc.*, 2017, **139**, 10929–10936.

14 J. Soria, J. C. Conesa, V. Augugliaro, L. Palmisano, M. Schiavello and A. Sclafani, *J. Phys. Chem.*, 1991, **95**, 274–282.

15 B. M. Comer, Y.-H. Liu, M. B. Dixit, K. Hatzell, Y. Ye, E. J. Crumlin, M. C. Hatzell and A. J. Medford, *J. Am. Chem. Soc.*, 2018, **45**, 15157–15160.

16 J. Choi, H. Park and M. R. Hoffmann, *J. Phys. Chem. C*, 2009, **114**, 783–792.

17 X. Gao, Y. Wen, D. Qu, L. An, S. Luan, W. Jiang, X. Zong, X. Liu and Z. Sun, *ACS Sustainable Chem. Eng.*, 2018, **6**, 5342–5348.

18 Y. Garsany, O. A. Baturina, K. E. Swider-Lyons and S. S. Kocha, *Experimental methods for quantifying the activity of platinum electrocatalysts for the oxygen reduction reaction*, 2010.

19 B. Su, Y. Ma, Y. Du and C. Wang, *Electrochem. Commun.*, 2009, **11**, 1154–1157.

20 S. Wendt, R. Schaub, J. Matthiesen, E. K. Vestergaard, E. Wahlström, M. D. Rasmussen, P. Thostrup, L. Molina, E. Lægsgaard, I. Stensgaard, *et al.*, *Surf. Sci.*, 2005, **598**, 226–245.

21 D. A. Panayotov, S. P. Burrows and J. R. Morris, *J. Phys. Chem. C*, 2012, **116**, 6623–6635.

22 D. C. Hurum, A. G. Agrios, K. A. Gray, T. Rajh and M. C. Thurnauer, *J. Phys. Chem. B*, 2003, **107**, 4545–4549.

23 M. Shen and M. A. Henderson, *J. Phys. Chem. Lett.*, 2011, **2**, 2707–2710.

24 Q. Yuan, Z. Wu, Y. Jin, L. Xu, F. Xiong, Y. Ma and W. Huang, *J. Am. Chem. Soc.*, 2013, **135**, 5212–5219.

25 A. Kapałka, S. Fierro, Z. Frontistis, A. Katsaounis, S. Neodo, O. Frey, N. De Rooij, K. M. Udert and C. Comninellis, *Electrochim. Acta*, 2011, **56**, 1361–1365.

26 K. Endo, Y. Katayama and T. Miura, *Electrochim. Acta*, 2005, **50**, 2181–2185.

27 B. Bharti, S. Kumar, H.-N. Lee and R. Kumar, *Sci. Rep.*, 2016, **6**, 32355.

28 M. J. Jackman, A. G. Thomas and C. Muryn, *J. Phys. Chem. C*, 2015, **119**, 13682–13690.

29 R. Grau-Crespo and U. Schwingenschlögl, *J. Phys.: Condens. Matter*, 2011, **23**, 334216.

30 X. Pan, M.-Q. Yang, X. Fu, N. Zhang and Y.-J. Xu, *Nanoscale*, 2013, **5**, 3601–3614.

31 Y. Wang, L. Li, X. Huang, Q. Li and G. Li, *RSC Adv.*, 2015, **5**, 34302–34313.

32 J. Wang, J. Yu, X. Zhu and X. Z. Kong, *Nanoscale Res. Lett.*, 2012, **7**, 646.

33 B. M. Comer, P. Fuentes, C. O. Dimkpa, Y. Liu, C. A. Fernandez, P. Arora, M. Realff, U. Singh, M. C. Hatzell and A. J. Medford, *Joule*, 2019, **3**, 1–28.

Faraday Discussions

PAPER

p-Type dye-sensitized solar cells based on pseudorotaxane mediated charge-transfer†

Tessel Bouwens, Simon Mathew and Joost N. H. Reek ⓘD *

Received 7th November 2018, Accepted 23rd January 2019

DOI: 10.1039/c8fd00169c

The efficiency of p-type dye-sensitized solar cells (DSSCs) remains low compared to that of n-type congeners due to charge recombination events. We report a supramolecular approach to reduce recombination at the NiO–dye interface, realized by using the cyclophane cyclobis(paraquat-p-phenylene) ring ($\mathbf{RING^{4+}}$/$\mathbf{RING^{3\cdot+}}$) as a redox mediator and a dye ($\mathbf{P_N}$) functionalized with a 1,5-dioxynaphthalene (DNP) recognition site, promoting the supramolecular formation of a pseudorotaxane capable of directing charge transfer away from the NiO–dye interface. The binding affinity of $\mathbf{RING^{4+}}$ to $\mathbf{P_N}$ is high ($K_{ass} = 3.4 \times 10^4$ M^{-1}), with quenching of the photoexcited dye ($\mathbf{P_N}$*) ascribed to reduction of $\mathbf{RING^{4+}}$ to $\mathbf{RING^{3\cdot+}}$. The reduced $\mathbf{RING^{3\cdot+}}$ exhibits a lower binding affinity to $\mathbf{P_N}$, facilitating exchange with the excess $\mathbf{RING^{4+}}$ present in solution. This supramolecular phenomenon was implemented into p-type DSSCs by anchoring the $\mathbf{P_N}$ dye on a NiO photocathode in conjunction with the $\mathbf{RING^{4+}}$/$\mathbf{RING^{3\cdot+}}$ redox couple, yielding a 10 fold enhancement in the short-circuit photocurrent (J_{SC}) compared to control devices utilizing $\mathbf{P1}$ dye or the methylviologen ($\mathbf{MV^{2+}}$/$\mathbf{MV^{\cdot+}}$) redox couple that cannot form pseudorotaxanes.

Introduction

Dye-sensitized solar cells (DSSCs), invented by O'Regan & Grätzel in 1991,[1] have complementary advantages to traditional silicon solar cells including low-cost manufacturing,[2] low-light performance,[3] and tunable colors, making them useful for indoor applications and building integrated photovoltaics (BIPV).[4] High-efficiency n-type DSSCs employ sterically demanding dyes and/or surface passivation to shield the TiO$_2$–electrolyte interface and suppress the recombination of photoinjected electrons.[5,6] This design principle has enabled the use of

Homogeneous Supramolecular and Bio-inspired Catalysis, Van't Hoff Institute for Molecular Sciences (HIMS), University of Amsterdam (UvA), Science Park 904, 1098 XH Amsterdam, The Netherlands. E-mail: j.n.h.reek@uva.nl

† Electronic supplementary information (ESI) available. See DOI: 10.1039/c8fd00169c

alternative redox mediators yielding enhancements of open-circuit voltages (V_{OC}) that translate to power conversion efficiencies (PCEs) beyond 14%.[7,8]

Complementary p-type DSSCs open avenues to tandem solar cells to attain greater PCEs while demonstrating utility for both photovoltaics and solar-driven fuel generation, but they are relatively unexplored compared to their n-type analogue.[9-13] Intrinsic to the operation of p-type DSSCs is electron transfer from the Valence Band (VB) of the p-type semiconductor NiO to the excited state dye (D*).[14,15] However, currently the highest efficiencies of p-type cells are 10 fold lower compared to those of the n-type analogues due to recombination phenomena.[16,17] Electron propagation in p-type DSSCs and recombination phenomena are outlined in Fig. 1. Ideally the absorption of light to generate an excited state dye (D*) will proceed *via* an electron transfer from the VB of NiO (Process 2, Fig. 1) to afford the radical dye anion (D⁻). Regeneration of the ground-state dye proceeds *via* a second electron transfer to the redox mediator (Process 3, Fig. 1) allowing the reduced species of the redox couple to diffuse to the counter electrode, regenerate and complete the circuit (Process 4).

Unfavorable recombination phenomena can lower the PCE by creating competition for the forward electron transfer to the redox couple.[18,19] Ultra-fast recombination between the radical dye anion and the hole in the VB of NiO (Process 6, Fig. 1) is ameliorated by the molecular engineering of donor–π–acceptor dyes. Spatial separation of the donor near the semiconductor surface and the acceptor away from the NiO surface promotes vectorial electron transfer away from the surface upon light absorption, increasing the lifetime of the photoreduced excited state.[20-22] The second recombination process involves the electron transfer from the reduced redox couple to the VB of the NiO (Process 5, Fig. 1). As charge migration is relatively slow in NiO (4×10^{-8} cm² s⁻¹), this process also significantly contributes to the low efficiencies.[23]

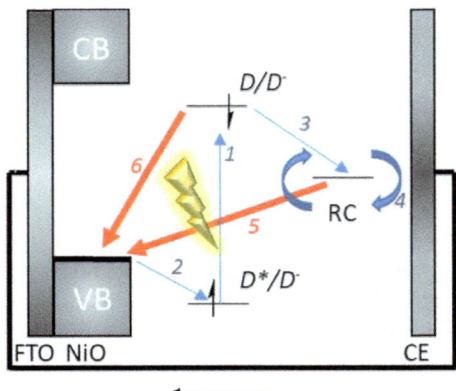

Photocathodic current

Fig. 1 A schematic representation of a p-type DSSC. The blue arrows represent the forward electron propagation processes and the red arrows show the unfavorable recombination phenomena. FTO = Fluorine Tin Oxide. RC = Redox Couple. CB = Conduction Band. VB = Valence Band. D = Dye. D* = excited state dye. D⁻ = radical dye anion.

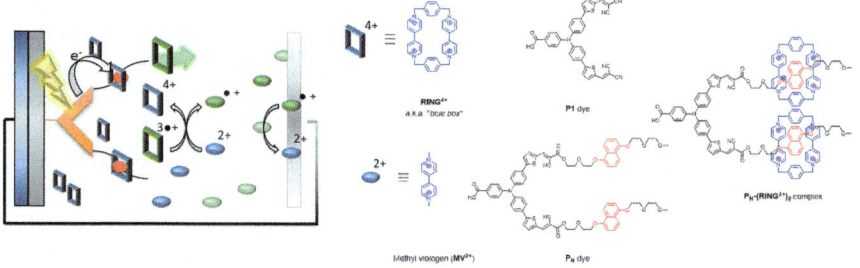

Fig. 2 The proposed operational principle of the pseudorotaxane-based p-type DSSC. The ring in the oxidized form (**RING^{4+}**, blue) has a high affinity for the binding site (DNP, in red) in contrast to the reduced ring (**RING$^{3\cdot+}$**, green), leading to a high local concentration of **RING^{4+}** close to the dye. The molecular structures of **P$_N$**, **P1**, the ring **RING^{4+}**, **MV^{2+}** and the **P$_N$–(RING^{4+})$_2$** complex (right).

With these challenges in mind, we propose a novel approach to the construction of a p-type DSSC in which we functionalize the dye with a 1,5-dioxynaphthalene (DNP) recognition site for the ring to form a pseudorotaxane with the cyclophane cyclobis(paraquat-*p*-phenylene) ring, **RING^{4+}**, being the oxidized form of the redox couple.[24] Electron transfer from D$^-$ reduces **RING^{4+}**, yielding **RING$^{3\cdot+}$** which loses affinity for the recognition site and is driven away from the dye–semiconductor interface, facilitating charge rectification within the device (Fig. 2).

This self-assembly strategy is unknown for p-type cells based on the I$^-$/I$_3^-$ redox couple.[25] Pre-organization of the redox couple should result in improving the competitiveness of forward electron propagation in the device compared to charge recombination. This is in contrast to the operation of current liquid DSSCs where the electron transfer between the dye and redox mediator is dictated by diffusion (collisional) phenomena.

In this paper we report the synthesis of dyes with recognition sites that facilitate pseudorotaxane formation and implementation into a p-type DSSC where the ring acts as a redox mediator. Initial experiments confirm that this supramolecular approach results in a ~10 fold improvement in photocurrent compared to that of control cells without supramolecular motifs for pre-organization.

Results and discussion

Design and description of the components

Creating a switchable pseudorotaxane-based DSSC requires the components to meet specific criteria. Firstly, the oxidized form of the ring component in the pseudorotaxane must have a high affinity (high binding constant, K_{ass}) for the recognition site, accompanied by a decrease in K_{ass} upon reduction of the ring, facilitating ring exchange and charge transport. Secondly, the redox potentials of all of the components must promote forward electron propagation by an electron cascade triggered by the absorption of light to reduce the ring as the redox mediator. With these requirements in mind, we considered pseudorotaxanes developed by Stoddart and coworkers as suitable.[24,26] The electron-deficient ring

structure **RING⁴⁺**, or the famous "blue box", exhibits a strong electrostatic interaction with the electron rich DNP (Fig. 2, in red).[27] Upon reduction of **RING⁴⁺** to
RING³⁺⁺ *via* chemical,[28] electrochemical[29] or photochemical stimuli,[30,31] the
increase in electron density causes a loss of affinity for the DNP recognition
site.[26,32]

We decided to modify the **P1** dye system[20,33] as it can accommodate the
incorporation of the DNP recognition sites at the acceptor part with minimal
perturbation of the dye electronic structure (Fig. 2) so that the energy levels of the
NiO VB, the dye and the redox couple match to allow for forward electron propagation within the system (Fig. 3). Initial experiments showed that high
concentrations of the ring caused detachment of the **P_N** dye, a problem that was
solved by using low concentrations of **RING⁴⁺** and **MV²⁺** as the main redox
mediator as the former is essentially two connected **MV²⁺** molecules.

Synthesis of the building blocks

The **P1** dye[20] and the [**RING·4PF₆**][34] were prepared according to literature procedures. The **P_N** dye functionalized with the DNP recognition site was prepared
using a modified procedure, outlined in Scheme 1. The bisaldehyde **4** was reacted
with malononitrile to obtain **P1**.[20] The last step in the preparation of **P_N** involves
a Knoevenagel condensation of **4** with the cyanoacrylate-functionalized DNP unit
1. After purification by column chromatography the **P_N** dye was obtained as a waxy
red solid in 11% yield. The compound was comprehensively characterized by ¹H
NMR, ¹³C NMR, FD-MS, ATR-IR, CV, and UV-vis.

Optical and electrochemical properties

The absorption maxima and electrochemical properties of **P1** and **P_N** are presented in Fig. 4 and summarized in Table 1. A strong absorption between 400–
680 nm was ascribed to intramolecular charge transfer, with the maximum for **P_N**

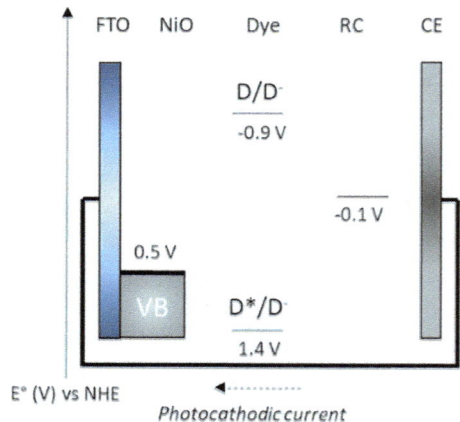

Fig. 3 A schematic energy diagram for the p-type DSSC based on the **P_N** pseudorotaxane
dye. Energy levels are represented in *V versus* NHE. This effectively creates a local high
concentration of oxidized redox mediator in close proximity to the dye, facilitating the
productive transfer processes compared to recombination pathways.

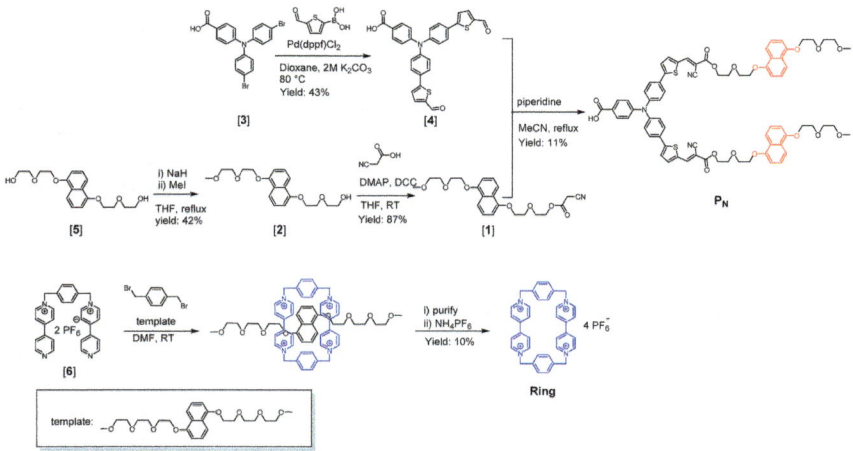

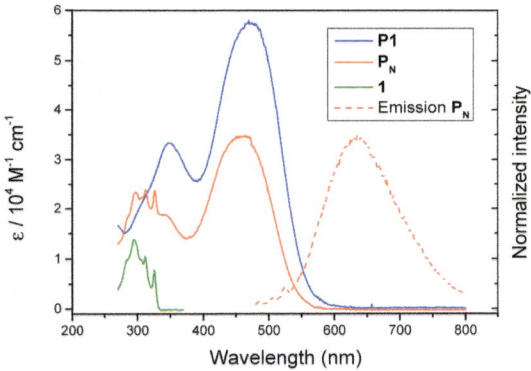

Scheme 1 Synthesis of **P$_N$** dye with DNP recognition sites and synthesis of the ring.

Fig. 4 UV-vis spectra of **P1** (solid blue line), **P$_N$** (solid red line) and compound **1** (solid green line) together with the normalized emission spectrum of **P$_N$** upon excitation at 450 nm (dashed red line).

(450 nm) slightly blue-shifted in comparison to that for **P1** (475 nm). This blue shift arises from the slight difference in electron withdrawing capability of the acceptor due to the substitution of one cyano group with an ester. The absorption of the DNP unit is a clearly visible feature between 270–340 nm in the UV-vis spectrum of the **P$_N$** dye, and the absorption spectrum of **P$_N$** indeed approximates a superposition of the UV-vis spectra of both compound **1** and **P1** (Fig. 4). Furthermore, the DNP unit in **P$_N$** exhibits a structured fluorescence band with a maximum at 344 nm upon excitation at 310 nm, while the dye part of **P$_N$** shows a broad fluorescence feature at 630 nm upon excitation at 450 nm as is shown in Fig. 4. There is no electronic communication between the DNP group and the dye part of **P$_N$** as these fluorescence features are separated from each other and only appear upon excitation at 310 nm or 450 nm, respectively.

Table 1 Optical and electrochemical properties of **P1** and **P$_N$** (0.5 mM) in MeCN. The redox potentials were determined in 0.1 M TBAPF$_6$ with a glassy carbon working electrode, a leakless Ag/AgCl reference electrode and a Pt wire as the counter electrode

Dye	λ_{max} (nm)	ε (L mol^{-1} cm^{-1})	E_{0-0} (eV)	$E_{(D^+/D)}$ (V vs. NHE)	$E_{(D/D^-)}$ (V vs. NHE)
P1	485	5.8×10^4	2.27	1.30	-0.67
P$_N$	450	3.5×10^4	2.30	1.32	-0.98

NMR investigation of binding phenomena

The binding of **RING^{4+}** to the **P$_N$** dye is qualitatively confirmed by a 5 point ^{1}H NMR titration (Fig. S13†). Mixing of the components resulted in an upfield shift for the resonances originating from **RING^{4+}** ($\delta = 0.02$ ppm) and the DNP ($\delta = 0.20$ ppm) moiety with concomitant peak broadening observed in the ^{1}H NMR spectra. DOSY NMR (Fig. S14†) established that both the **P$_N$** dye and **RING^{4+}** components in the **P$_N$**–**RING^{4+}** complex demonstrate identical diffusion constants (5.50×10^{-10} m^2 s^{-1}) that are slightly smaller than that of the **P$_N$** dye (1.02×10^{-9} m^2 s^{-1}) and significantly smaller than that of **RING^{4+}** (1.26×10^{-10} m^2 s^{-1}). This observation is consistent with the **P$_N$**–**RING^{4+}** complex exhibiting rigidification upon complex formation at the DNP recognition site, manifesting in an increase in the hydrodynamic radius. Quantification of the binding constant (K_{ass}), along with the photoinduced charge transfer in the **P$_N$**–(**RING^{4+}**)$_2$ complexes, was investigated by fluorescence quenching studies.

Fluorescence quenching studies

The K_{ass} of **RING^{4+}** to the **P$_N$** dye was measured by titration, monitoring the quenching of emission from the DNP moiety in **P$_N$** ($l_{em} = 344$ nm) upon the addition of aliquots of **RING^{4+}**.[35] The K_{ass} obtained by fitting the titration curve using a Host–Guest–Guest (HGG) model, as introduced by Hunter[36] showed that binding of **RING^{4+}** to the two binding sites exhibited almost the same association constant ($K_{ass} = 3.4 \times 10^4$ M^{-1}). The cooperativity factor α is determined to be 0.7, suggesting a minor negative cooperativity, i.e. the first binding event for the formation of **P$_N$**–(**RING^{4+}**) is slightly stronger than the second binding event to form **P$_N$**–(**RING^{4+}**)$_2$. Overall the association constant is of the same order of magnitude as was previously reported for a related system.[37]

Next, it was investigated whether the ring is able to quench the excited state of the dye when bound to the DNP recognition site. Upon formation of the **P$_N$**–(**RING^{4+}**)$_2$ complex, 83% of the fluorescence of the dye at 630 nm was quenched. No other emission was observed, which is in line with a photoinduced electron transfer process.

Photovoltaic properties

The NiO working electrodes were functionalized with sensitizers by soaking the electrodes in dye solution (**P1** or **P$_N$** in MeCN) for 16 hours. Adsorption of the dye was confirmed by the loss of the $\nu_{C=O}$ bands of the acidic group of the dye in the ATR-IR spectra (Fig. S8†). Dye desorption experiments show that the dye coverage is 1.6×10^{-8} mol cm^{-2} for **P$_N$**. To prove the effect of pseudorotaxane-mediated pre-organization, MV^{2+} on its own as the electrolyte was employed as the control for $RING^{4+}$. They both demonstrate similar reduction potentials (Table S3†), differing in their ability to participate in pseudorotaxane formation, *i.e.* MV^{2+} is not able to form a pseudorotaxane whereas $RING^{4+}$ is.[38,39] DSSCs employing four different combinations of dye (either $P1$ or P_N) and redox couple (either 25 mM $MV^{2+/\cdot+}$ or 24.925 mM $MV^{2+/\cdot+}$ with 75 µM (0.3 mol%) $RING^{4+/\cdot 3+}$) were fabricated. This relatively low concentration of $RING^{4+}$ is used in the cell as initial experiments revealed that the use of 25 mM (*i.e.* 100%) $RING^{4+/3\cdot+}$ as the electrolyte results in leaching of the P_N dye from the surface, observed as the sensitized orange color of the photocathode disappearing upon introduction of the electrolyte (for further explanation see ESI 8†). The redox couple was added as a mixture of the reduced and non-reduced species (*i.e.* $MV^{\cdot+} : MV^{2+} = 9 : 1$) by combining a pre-reduced 25 mM $MV^{\cdot+}$ solution with a non-reduced 25 mM MV^{2+} solution. As the reduced electrolyte containing $MV^{\cdot+}$ and $RING^{3\cdot+}$ is sensitive to oxygen, all of the DSSCs were constructed under inert atmosphere inside a nitrogen filled glovebox. The properties of the DSSCs were measured directly after fabrication in a two-electrode system, connected to a potentiostat to measure the Short-Circuit photocurrent density J_{SC} (*i.e.* at zero bias) under AM1.5 G illumination (100 mW cm^{-2}) in 30 second cycles (Fig. 5), with the data summarized in Table 2.

The DSSCs based on the $P1$ and the P_N dye with $MV^{2+/\cdot+}$ as the electrolyte (Entry 1 and 2) give identical photocurrents, indicating that the new P_N dye functions similarly to the known $P1$ dye in this set-up. Entry 2 shows that, in the presence of $RING^{4+/\cdot 3+}$ and in the absence of the recognition site, similar currents to those in Entry 1 and 3 are observed. Only DSSCs utilizing the P_N dye and the $RING^{4+/\cdot 3+}$ redox mediator elements, conducive to pseudorotaxane formation, exhibit a dramatic increase in photocurrent (Entry 4). In addition the photocurrent profile features an initial spike at -17 µA cm^{-2} which decays down to a constant

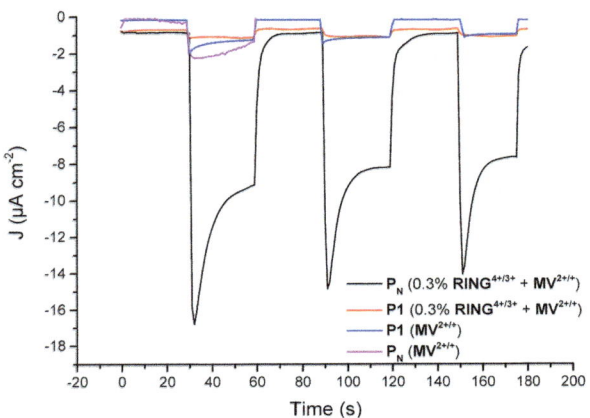

Fig. 5 The photocurrents obtained by switching the solar simulator on and off in cycles of 30 seconds connected to a potentiostat to measure the photocurrent at zero bias under AM1.5 G illumination (100 mW cm^{-2}). The black line (P_N with 75 µM 0.3 mol% $RING^{4+}$ and 24.925 mM $MV^{2+/\cdot+}$) indicates 10 times more photocurrent than the pink line (P_N with 25 mM $MV^{2+/\cdot+}$), the red line ($P1$ 75 µM 0.3 mol% $RING^{4+}$ and 24.925 mM $MV^{2+/\cdot+}$) and the blue line ($P1$ with 25 mM $MV^{2+/\cdot+}$).

Table 2 Obtained J_{SC} values for the four different constructed DSSCs

Entry	Dye	Redox couple	J_{SC} (μA cm^{-2})
1	P1	$MV^{2+/\cdot+}$ only	−1.31
2	P1	0.3% $RING^{4+}$ + $MV^{2+/\cdot+}$	−1.02
3	P_N	$MV^{2+/\cdot+}$ only	−1.41
4	P_N	0.3% $RING^{4+}$ + $MV^{2+/\cdot+}$	−9.56

value of −9.56 μA cm^{-2}, indicative of a mass transport limitation as the diffusion of $MV^{2+/\cdot+}$ to the counter electrode is slower than the I^-/I_3^- redox couple due to its larger molecular size. This is a well-known phenomenon in n-type DSSC with redox couples based on larger molecules than the I^-/I_3^- redox couple.[40,41]

Although the photocurrents are low for the control DSSC, in these experiments the typical spike phenomenon is also observed (Fig. 5). These initial experiments demonstrate that the presence of a recognition site, enabling pre-organization of an electron acceptor by pseudorotaxane formation that also acts as a redox couple, leads to DSSCs exhibiting higher photocurrents. This suggests that the pseudorotaxane approach may indeed be a useful strategy to reduce charge recombination in p-type DSSCs.

Conclusions

We proposed here a new type of p-type DSSC where the redox acceptor is pre-organized to the dye, directing charge transfer away from the dye–electrolyte interface. We constructed a new dye, P_N, based on the previously reported P1 dye, functionalized with DNP moieties that can form pseudorotaxanes with the electron acceptor $RING^{4+}$, providing competition with recombination by directing charge transfer away from the dye. $RING^{4+}$ shows a high binding affinity (3.4 × 10^4 M^{-1}) for the DNP recognition sites installed on the P_N dye and the affinity for the reduced ring $RING^{3\cdot+}$ is known to be low. Implementation into p-type DSSCs showed that pseudorotaxane formation afforded photocurrents almost one order of magnitude higher than those of the control experiments. Future research will be focused on time-resolved characterization of the system to provide quantification of reductions in charge recombination. Also, as the eventual efficiencies of the fabricated DSSCs are still rather low, we aim to further improve this using the concept introduced in this paper.

Material and instruments

All of the reagents and solvents were obtained from Sigma-Aldrich, Fluorochem and VWR, and were used without purification. Compound **3** was obtained following a literature procedure.[20] MeCN and THF were dried in a solvent purification system. Column chromatography was performed using silica gel (Sili-Cycle, SiliaFlash P_{60}, 40–63 μm, 230–400 mesh) while fractions were analyzed using TLC (TLC silica gel 60 F_{254} from Merck KGaA) visualized with 254/350 nm light. The products were analyzed using MS, NMR and IR. ^{1}H-NMR and ^{13}C NMR spectra were recorded on a Bruker AV300 and AV400 spectrometer and are reported in ppm using a solvent residual signal as an internal standard (7.26 ppm for $CDCl_3$ and 2.05 ppm for $(CD_3)_2CO$). 2D ^1H-DOSY spectral data were acquired with temperature and magnetic gradient calibration prior to the measurements, and the temperature was kept at 298 K during the measurements. IR spectra were recorded on a Bruker Alpha FTIR machine. The exact mass of the compounds was obtained on an AccuTOF GC v 4g, JMS-T100GCV Mass spectrometer (JEOL, Japan) equipped with an FD Emitter, Carbotec or Linden (Germany), FD 13 μm. The current rate was 51.2 mA min^{-1} over 1.2 min *via* a machine using field desorption (FD) as the ionization method. UV-vis measurements were performed on a single beam Hewlett Packard 8453 spectrometer in a 10 mm path length quartz cuvette using MeCN as the background. (Photo)electrochemical experiments were performed with a g-stat potentiostat from Autolab and an Oriel LCS-100 solar simulator. The light intensity was set to 100 mW cm^{-2} *via* a calibrated DSSC Newport, 91150-2000. Fluorescence spectra were recorded on a Fluorolog Jobin Yvon-SPEX together with their corresponding UV-vis spectra (Shimadzu UV-2700 Spectrophotometer).

Dye-sensitized solar cell fabrication and characterization

The NiO was bought as a paste (Ni-Nanoxide N/SP from Solaronix) and screen printed on FTO glass obtained from Sigma Aldrich (2.2 mm, 7 Ω/sq) as 0.5 cm diameter circles. Sandwich cells were constructed with carbon based counter electrodes separated from the working electrode using Meltonix polymer 1170-60 (Solaronix, Switzerland) with a thickness of 60 μm. The preparation of electrolyte and the DSSCs was performed under inert atmosphere (N_2) in a glovebox since the reduced species $MV^{\cdot+}$ and $RING^{3\cdot+}$ are reactive towards oxygen. The reduced $MV^{\cdot+}$ was prepared by the addition of Zn dust (2 equiv.) to a 25 mM solution of MV^{2+} in 1 M LiTFSI as the supporting electrolyte in N_2 degassed MeCN. Reduction is clearly visible as the color changes from colorless to dark blue. After stirring for 1 hour, the final electrolyte solution was prepared by adding 9 parts filtered reduced solution with 1 part oxidized MV^{2+}, both 25 mM (either with 25 mM MV^{2+} or 24.925 mM MV^{2+} with 75 μM (0.3 mol%) $RING^{4+}$). Filtration of the reduced solution is required to remove excess Zn dust and precipitated Zn^{2+}. Then the electrolyte was introduced inside the cells *via* a pre-drilled hole. After cleaning and wiping the outside of the cell, the hole was sealed with the Meltonix polymer and optical glass plate by melting the polymer using a 350 °C heating source. The DSSCs were measured directly after fabrication using a g-stat potentiostat and solar simulator. The cells were masked to make sure 0.126 cm^2 of the NiO surface was irradiated, to stay consistent throughout all of the measurements.

Electrochemistry

Cyclic voltammograms and differential pulse voltammograms were recorded on a potentiostat PGSTAT302N from Autolab with a glassy carbon working electrode (MetrOhm, diameter 3 mm), a leakless Ag/AgCl reference electrode (eDAQ, ET069) and a Pt wire as the counter electrode. The analyte solution was prepared by degassing a 100 mM $TBAPF_6$ solution with N_2 and adding this to the analyte to obtain a 0.5 mM final analyte solution.

Binding studies

In order to determine the K_{ass} of the ring to the dye, a solution of $\mathbf{P_N}$ was kept at a constant concentration (5 μM) and titrated with an increasing concentration of ring (2 mM and 10 mM) inside a fluorescence cuvette with a path length of 10 mm. After each addition of the ring the fluorescence at 344 nm was recorded by irradiation with 310 nm light and the UV-vis spectrum was measured to correct for the inner-filter effect following the formula described by Turner and coworkers:[42]

$$F_{corr} = F_{obs} \times 10^{\left(\frac{A_{exc}+A_{emm}}{2}\right)}$$

The corrected data was fitted to a $1 : 2$ host–guest model using a Matlab script.[43]

Excited state quenching studies

The excited state quenching studies were performed using a similar approach to that of the determination of K_{ass}. The $\mathbf{P_N}$ was kept at a constant concentration (10 μM) and was titrated with an increasing concentration of ring (4 mM and 20 mM). The decrease in the fluorescence at 630 nm upon excitation with 450 nm light was directly used to determine the amount of quenching and no correction was needed as there was no self-absorption of the sample at 630 nm.

Synthesis

Compound 2. To a solution of compound 5 (2.00 g; 5.94 mmol; 1 equiv.) in dry THF (15 mL), NaH (242.6 g; 6.06 mmol; 1.02 equiv.) in mineral oil (65%) was added portion wise. Upon addition, a precipitate was formed and a gas was released. After the mixture was heated at reflux for 1 hour, MeI (614 mg; 268 μL; 7.13 mmol; 1.2 equiv.) was added all at once. After heating at reflux for 4 more hours the reaction mixture was quenched by the dropwise addition of water. The water layer was extracted with EtOAc ($3\times$). The combined organic layers were dried with Na_2SO_4. After filtering of the Na_2SO_4 the organic solvents were evaporated. The product was purified by column chromatography (silica, DCM/EtOAc, $1 : 1$, the second fraction). Compound 2 was obtained as a yellow oil. Yield: 875 mg (42%). The ^{1}H-NMR was consistent with the literature.[39]

Compound 1. Compound 2 (825 mg; 2.35 mmol; 1 equiv.), cyanoacetic acid (210.3 mg; 2.47 mmol; 1.05 equiv.) and a pinch of DMAP were dissolved in 10 mL of dry DCM. DCC (510 mg; 2.47 mmol; 1.05 equiv.) was added and direct precipitation of DCU was observed upon addition. After stirring overnight at RT, DCU was filtered of and DCM was removed and the crude was purified by column chromatography (silica, DCM/EtOAc, 4 : 1). Compound 1 was obtained as an off-white waxy solid. Yield: 854 mg (87%). ^{1}H NMR (400 MHz, CDCl$_3$) δ 7.85 (dd, $J = 19.5, 8.5$ Hz, 2H), 7.35 (td, $J = 8.0, 4.1$ Hz, 2H), 6.83 (dd, $J = 7.6, 4.4$ Hz, 2H), 4.39–4.33 (m, 2H), 4.28 (dt, $J = 9.8, 4.8$ Hz, 4H), 3.97 (dt, $J = 11.7, 4.7$ Hz, 4H), 3.87–3.74 (m, 4H), 3.62–3.55 (m, 2H), 3.39 (d, $J = 4.4$ Hz, 5H). ^{13}C NMR (75 MHz, CDCl$_3$) δ 162.99, 154.40, 154.18, 126.80, 126.71, 125.24, 125.06, 114.84, 114.38, 112.85, 105.76, 105.74, 77.23, 72.03, 70.89, 69.86, 69.46, 68.84, 67.93, 65.82, 63.64, 59.10, 24.61. IR ν_{max} (cm^{-1}) 2924 (b, $-CH_2-$ glycol), 2226 (w, CN),

1749 (s, C=O). FD-MS (in MeCN) m/z: calculated for $C_{22}H_{27}N_1O_7$ $[M]^+$: 417.1788. Found 417.1808.

Compound 4. To a mixture of compound 3 (140 mg; 0.31 mmol; 1 equiv.) and boronic acid (195 mg; 1.25 mmol; 4 equiv.) in dioxane (28 mL), 3.5 mL of 2 M K_2CO_3 in water was added. The mixture was degassed with N_2 for one hour before the Pd(dppf)Cl$_2$ catalyst (42 mg 0.05 mmol; 15 mol%) was added. The mixture was stirred at 80 °C overnight. The dioxane was removed and water was added. The water layer was acidified to approximately pH 3 after which the water layer was extracted with EtOAc. The organic layers were combined and dried with Na$_2$SO$_4$. After filtration the EtOAc was evaporated. The crude mixture was purified by column chromatography (silica, DCM/MeOH/acetic acid, 98% : 2% : 0.002%). Yield: 67.9 mg (43%). The ^{1}H NMR spectrum corresponds to the literature.[20]

P_N dye. Compound 1 (189 mg; 0.45 mmol; 3.5 equiv.), compound 4 (66 mg; 0.13 mmol; 1 equiv.) and 4 drops of piperidine were dissolved in MeCN (22 mL) and heated at reflux for 4 hours. The MeCN was evaporated and acidified water was added (pH 3.5). The organic compounds were extracted with DCM. After drying the organic layers with Na$_2$SO$_4$ and filtration, the product was purified by column chromatography (silica, DCM/MeOH/acetic acid, 98% : 2% : 0.002%). The red fractions containing product were checked for their purity by MS and pure fractions were combined to yield the product P_N as a waxy, red solid. Yield: 18.7 mg (11%). ^{1}H NMR (300 MHz, (CD$_3$)$_2$CO) δ 8.21 (s, 2H) 8.01 (d, J = 8.4 Hz, 2H), 7.85 (t, J = 8.7 Hz, 4H), 7.63 (t, J = 6.8 Hz, 6H), 7.34 (dd, J = 9.4, 6.3 Hz, 6H), 7.17 (dd, J = 14.1, 8.4 Hz, 6H), 6.83 (dd, J = 19.8, 7.7 Hz, 4H), 4.50 (dd, J = 5.9, 3.6 Hz, 4H), 4.29 (dt, J = 16.6, 4.8 Hz, 9H), 4.05 (t, J = 4.7 Hz, 4H), 3.97 (q, J = 4.8 Hz, 8H), 3.78 (dd, J = 5.6, 3.6 Hz, 4H), 3.59 (dd, J = 5.6, 3.6 Hz, 4H), 3.39 (s, 6H). ^{13}C NMR (75 MHz, CDCl3) δ (ppm): 170.49, 163.07, 154.41, 154.39, 153.93, 151.29, 147.43, 146.76, 139.55, 134.78, 131.97, 128.93, 127.97, 126.87, 125.54, 125.31, 125.19, 124.16, 123.68, 122.48, 116.04, 114.83, 114.70, 105.88, 105.78, 97.58, 72.58, 70.99, 70.09, 69.96, 69.29, 68.12, 68.00, 65.57, 59.21. IR: ν_{max}/cm^{-1} 3061 (b, OH conj. acid) 2924 (b, –CH$_2$– glycol), 2217 (w, CN), 1713 (s, C=O ester) 1680 (m, C=O conj. acid). FD-MS (in MeCN) m/z: calculated for $C_{37}H_{69}N_3O_{16}S_2$ $[M]^+$: 1307.4119. Found 1307.4095.

The ring [RING·4PF$_6$]

The ring was synthesized based on a protocol described in the literature with minor modifications.[34]

Compound 6 was combined with bromoxylene and the template in dry DMF and stirred for 30 days. After one night, a purple precipitate was observed. After 30 days the product was precipitated by the addition of ether. The purple solid was dissolved in 1 M NH$_4$Cl (50 mL) and to this, 50 mL of DCM was added. This mixture was heated at reflux under gentle stirring until the purple color disappeared. The DCM layer was separated from the water layer, after which the water layer was heated again at reflux with DCM. This was repeated two times until the purple color had disappeared completely. Upon addition of NH$_4$PF$_6$ to this water solution, a white precipitate was obtained. The solid was filtered and washed thoroughly with Milli-Q and ether. The product was obtained from recrystallization from acetonitrile. The ^{1}H NMR corresponds to the literature.[34]

Salt exchange [MV·2(PF$_6$)]. Methyl viologen dichloride (**MV·Cl$_2$**) (1.00 g; 7.78 mmol) was dissolved in Milli-Q water. To this, NH$_4$PF$_6$ was added as a solid until no more precipitation was observed. The white solid was filtered and washed thoroughly with Milli-Q and ether after which it was dried in a vacuum oven. The white solid was used without further purification.

Conflicts of interest

There are no conflicts to declare.

Acknowledgements

This study was supported by the Holland Research School for Molecular Sciences (HRSMC) and the University of Amsterdam. We would like to acknowledge Dr. Remko Detz and Lidy van der Burg for synthesizing the **P1** dye; Sandra Nurttila and René Becker for their help with the binding studies; and Prof. Dr S. Woutersen for fruitful discussions.

References

1 B. O'Regan and M. Grätzel, *Nature*, 1991, **353**, 737–740.
2 M. Freitag, J. Teuscher, Y. Saygili, X. Zhang, F. Giordano, P. Liska, J. Hua, S. M. Zakeeruddin, J. E. Moser, M. Grätzel and A. Hagfeldt, *Nat. Photonics*, 2017, **11**, 372–378.
3 B. P. Lechêne, M. Cowell, A. Pierre, J. W. Evans, P. K. Wright and A. C. Arias, *Nano Energy*, 2016, **26**, 631–640.
4 Y. Cui, C. Yang, H. Yao, J. Zhu, Y. Wang, G. Jia, F. Gao and J. Hou, *Adv. Mater.*, 2017, **29**, 1–7.
5 A. Hagfeldt, G. Boschloo, L. Sun, L. Kloo and H. Pettersson, *Chem. Rev.*, 2010, **110**, 6595–6663.
6 H. N. Tsao, C. Yi, T. Moehl, J. H. Yum, S. M. Zakeeruddin, M. K. Nazeeruddin and M. Grätzel, *ChemSusChem*, 2011, **4**, 591–594.
7 S. Mathew, A. Yella, P. Gao, R. Humphry-Baker, B. F. E. Curchod, N. Ashari-Astani, I. Tavernelli, U. Rothlisberger, M. K. Nazeeruddin and M. Grätzel, *Nat. Chem.*, 2014, **6**, 242–247.
8 K. Kakiage, Y. Aoyama, T. Yano, K. Oya, J. I. Fujisawa and M. Hanaya, *Chem. Commun.*, 2015, **51**, 15894–15897.
9 T. E. Rosser, M. A. Gross, Y. H. Lai and E. Reisner, *Chem. Sci.*, 2016, 7, 4024–4035.
10 J. J. Leung, J. Warnan, D. H. Nam, J. Z. Zhang, J. Willkomm and E. Reisner, *Chem. Sci.*, 2017, **8**, 5172–5180.
11 F. Li, K. Fan, B. Xu, E. Gabrielsson, Q. Daniel, L. Li and L. Sun, *J. Am. Chem. Soc.*, 2015, **137**, 9153–9159.
12 J. He, H. Lindström, A. Hagfeldt and S. E. Lindquist, *Sol. Energy Mater. Sol. Cells*, 2000, **62**, 265–273.
13 N. Põldme, L. O'Reilly, I. Fletcher, J. Portoles, I. V. Sazanovich, M. Towrie, C. Long, J. G. Vos, M. T. Pryce and E. A. Gibson, *Chem. Sci.*, 2018, 20–22.
14 J. He, H. Lindström, A. Hagfeldt and S.-E. Lindquist, *J. Phys. Chem. B*, 1999, **103**, 8940–8943.

15 Z. Huang, G. Natu, Z. Ji, P. Hasin and Y. Wu, *J. Phys. Chem. C*, 2011, **115**, 25109–25114.
16 F. Odobel and Y. Pellegrin, *J. Phys. Chem. Lett.*, 2013, **4**, 2551–2564.
17 J. C. Freys, J. M. Gardner, L. D'Amario, A. M. Brown and L. Hammarström, *Dalton Trans.*, 2012, **41**, 13105–13111.
18 A. Morandeira, G. Boschloo, A. Hagfeldt and L. Hammarström, *J. Phys. Chem. B*, 2005, **109**, 19403–19410.
19 W. B. Swords, S. J. C. Simon, F. G. L. Parlane, R. K. Dean, C. W. Kellett, K. Hu, G. J. Meyer and C. P. Berlinguette, *Angew. Chem., Int. Ed.*, 2016, **55**, 5956–5960.
20 P. Qin, H. J. Zhu, T. Edvinsson, G. Boschloo, A. Hagfeldt and L. C. Sun, *J. Am. Chem. Soc.*, 2008, **130**, 17629.
21 A. Nattestad, A. J. Mozer, M. K. R. Fischer, Y. B. Cheng, A. Mishra, P. Bäuerle and U. Bach, *Nat. Mater.*, 2010, **9**, 31–35.
22 J. Preat, A. Hagfeldt and E. A. Perpète, *Energy Environ. Sci.*, 2011, **4**, 4537–4549.
23 S. Mori, S. Fukuda, S. Sumikura, Y. Takeda, Y. Tamaki, E. Suzuki and T. Abe, *J. Phys. Chem. C*, 2008, **112**, 16134–16139.
24 P. L. Anelli, P. R. Ashton, D. Philp, M. Pietraszkiewicz, M. V. Reddington, N. Spencer, J. F. Stoddart, C. Vicent, R. Ballardini, V. Balzani, M. T. Gandolfi, L. Prodi, M. Delgado, T. T. Goodnow, A. E. Kaifer, A. M. Z. Slawin and D. J. Williams, *J. Am. Chem. Soc.*, 1992, **114**, 193–218.
25 C. J. Wood, C. A. McGregor and E. A. Gibson, *ChemElectroChem*, 2016, **3**, 1827–1836.
26 P. R. Ashton, V. Balzani, O. Kocian, L. Prodi, N. Spencer and J. F. Stoddart, *J. Am. Chem. Soc.*, 1998, **120**, 11190–11191.
27 F. M. Raymo, K. N. Houk and J. F. Stoddart, *J. Org. Chem.*, 1998, **63**, 6523.
28 R. A. Bissell, E. Córdova, A. E. Kaifer and J. F. Stoddart, *Nat. Chem.*, 1994, **369**, 133–137.
29 J. W. Choi, A. H. Flood, D. W. Steuerman, S. Nygaard, A. B. Braunschweig, N. N. P. Moonen, B. W. Laursen, Y. Luo, E. Delonno, A. J. Peters, J. O. Jeppesen, K. Xu, J. F. Stoddart and J. R. Heath, *Chem.–Eur. J.*, 2006, **12**, 261–279.
30 H. Li, C. Cheng, P. R. McGonigal, A. C. Fahrenbach, M. Frasconi, W. G. Liu, Z. Zhu, Y. Zhao, C. Ke, J. Lei, R. M. Young, S. M. Dyar, D. T. Co, Y. W. Yang, Y. Y. Botros, W. A. Goddard, M. R. Wasielewski, R. D. Astumian and J. F. Stoddart, *J. Am. Chem. Soc.*, 2013, **135**, 18609–18620.
31 H. Li, A. C. Fahrenbach, A. Coskun, Z. Zhu, G. Barin, Y. L. Zhao, Y. Y. Botros, J. P. Sauvage and J. F. Stoddart, *Angew. Chem., Int. Ed.*, 2011, **50**, 6782–6788.
32 P. R. Ashton, R. Ballardini, V. Balzani, A. Credi, K. R. Dress, E. Ishow, C. J. Kleverlaan, O. Kocian, J. A. Preece, N. Spencer, J. F. Stoddart, M. Venturi and S. Wenger, *Chem.–Eur. J.*, 2000, **6**, 3558–3574.
33 P. Qin, M. Linder, T. Brinck, G. Boschloo, A. Hagfeldt and L. Sun, *Adv. Mater.*, 2009, **21**, 2993–2996.
34 M. Asakawa, W. Dehaen, G. L'abbé, S. Menzer, J. Nouwen, F. M. Raymo, J. F. Stoddart and D. J. Williams, *J. Org. Chem.*, 1996, **61**, 9591–9595.
35 M. Venturi, S. Dumas, V. Balzani, J. Cao and J. F. Stoddart, *New J. Chem.*, 2004, **28**, 1032–1037.
36 C. A. Hunter and H. L. Anderson, *Angew. Chem., Int. Ed.*, 2009, **48**, 7488–7499.
37 R. Castro, K. R. Nixon, J. D. Evanseck and A. E. Kaifer, *J. Org. Chem.*, 1996, **61**, 7298–7303.

38 M. Jonsson, A. Houmam, G. Jocys and D. D. M. Wayner, *J. Chem. Soc., Perkin Trans. 2*, 1999, **117**, 425–429.

39 P. R. Ashton, L. Pérez-García, J. F. Stoddart, R. Ballardini, V. Balzani, A. Credi, M. T. Gandolfi, L. Prodi, M. Venturi, S. Menzer, A. J. P. White and D. J. Williams, *J. Am. Chem. Soc.*, 1995, **117**, 11171–11197.

40 A. Yella, S. Mathew, S. Aghazada, P. Comte, M. Grätzel and M. K. Nazeeruddin, *J. Mater. Chem. C*, 2017, **5**, 2833–2843.

41 J. J. Nelson, T. J. Amick and C. M. Elliott, *J. Phys. Chem. C*, 2008, **112**, 18255–18263.

42 T. Larsson, M. Wedborg and D. Turner, *Anal. Chim. Acta*, 2007, **583**, 357–363.

43 X. Wang, S. S. Nurttila, W. I. Dzik, R. Becker, J. Rodgers and J. N. H. Reek, *Chem.–Eur. J.*, 2017, **23**, 14769–14777.

Faraday Discussions

Photo-generation of cyclic carbonates using hyper-branched Ru–TiO$_2$

Stelios Gavrielides, [ID] Jeannie Z. Y. Tan, [ID] Eva Sanchez Fernandez and M. Mercedes Maroto-Valer*

Received 16th November 2018, Accepted 5th February 2019

DOI: 10.1039/c8fd00181b

Anthropogenic CO$_2$ is the main contributor to the increased concentration of greenhouse gases in the atmosphere, and thus utilising waste CO$_2$ for the production of valuable chemicals is a very appealing strategy for reducing CO$_2$ emissions. The catalytic fixation of CO$_2$ with epoxides for the production of cyclic carbonates has gained increasing attention from the research community in search of an alternative to the homogeneous catalytic routes, which are currently being used in industry. A novel photocatalytic heterogeneous approach to generate cyclic carbonates is demonstrated in this work. Hyper-branched microstructured Ru modified TiO$_2$ nanorods decorated with RuO$_2$ nanoparticles, supported on fluorine-doped tin oxide (FTO) glass were fabricated for the first time and were used to catalyse the photo-generation of propylene carbonates from propylene oxides. Propylene carbonate was used as a reference for cyclic carbonates. The photo-generation of cyclic carbonates from epoxides and CO$_2$ was carried out at a maximum temperature of 55 °C at 200 kPa in a stainless steel photoreactor with a quartz window, under solar irradiation for 6 h. The best performing photocatalyst exhibited an estimated selectivity of 83% towards propylene carbonates under the irradiation of a solar simulator.

Introduction

Humankind relies heavily on fossil fuels as the primary energy source, producing CO$_2$ as a by-product which is released into the atmosphere. The increased level of anthropogenic CO$_2$ emissions is one of the most significant contributors to climate change. In this regard, the research community has been trying to address this challenge by developing alternative energy sources and increasing their efficiency, and researching carbon capture, utilisation and storage (CCUS) techniques.[1] Utilising CO$_2$ in the production of commercially valuable chemicals and fuels has recently attracted much attention.[2] CO$_2$ is often a good source of C-building blocks for various organic syntheses, as it is abundant and low-cost.[3] Using waste CO$_2$ as a feedstock for chemical reactions not only offers a more

Research Centre of Carbon Solutions (RCCS), School of Engineering & Physical Sciences, Heriot-Watt University, Edinburgh, UK. E-mail: M.Maroto-Valer@hw.ac.uk

sustainable route to generate value-added products, but also contributes to CO_2 fixation.[1,4]

The synthesis of cyclic carbonates (CCs) through the coupling of CO_2 and epoxides has been commercially available for over 60 years and is one of the most promising CO_2 utilisation industrial applications.[1] CCs are valuable as mono-mers, small molecule[5] and polymer intermediates,[6] pharmaceuticals,[7] and fine chemical intermediates.[3] In addition, the most important and rapidly growing application of CCs lies in Li-ion batteries, in which CCs are used as electrolytes.[8] Nowadays, Li-ion batteries power most portable electronic devices and their application in electronic vehicles is becoming very popular.[8] Due to the projected future demands for Li-ion batteries, the current industrial production of Li-ion is rapidly increasing.

The current commercial process for the synthesis of CCs uses homogeneous catalysts, such as quaternary ammonium[9] or phosphonium halides. Although these catalysts are inexpensive, they suffer from low efficiencies and require high temperature and pressure.[10] More specifically, the reaction conditions have been reported to be 100–200 °C, and 50–100 bar.[11] Additionally, due to their corrosive nature, special reactor materials are necessary, which further increases the production cost. Hence, in recent decades, new approaches have been developed for the production of CCs.[12–15] Many researchers have used homogeneous cata-lysts, such as (salen)Cr(III) complexes,[3] (salen)Co(III),[16] (salen)Mn(III),[17] and supramolecular metal complexes[18] to catalyse the production of CCs. However, several detected drawbacks have been reported for these developed homogeneous catalysts, including low catalyst stability, reactivity and air sensitivity.[3,8] Further-more, the recovery of these homogeneous catalysts, and the purification of the product is challenging, expensive and time-consuming.[1]

Recently, heterogeneous catalysts for the synthesis of CCs have been developed because of the easy separation of fluids from the solid catalyst, and their conve-nient handling and catalyst regeneration.[1,19] For example, Zn–ZIF-67–MOFs,[20] Fe(III)[21] and bimetallic complexes[22] have been developed to catalyse the produc-tion of CCs from CO_2 and epoxide.

Photocatalysis has received much attention from the research community in recent years due to its sustainability aspect and its high performance in a variety of applications, including photo-generation of H_2,[23–26] photoreduction of CO_2 for the production of fine chemicals and solar fuels,[27–29] and air and water purifica-tion.[30–32] TiO_2 is the most widely used photo-catalyst for these applications because it is non-toxic, thermally and chemically very stable, abundantly available and has a good photo-response under UV light irradiation.[33,34]

Very recently, Prajapati *et al.* demonstrated the first photocatalytic synthesis of CCs using cobalt phtalocyanine grafted on TiO_2.[34] The synthesized photocatalyst revealed a conversion of 94.2% after 24 h of irradiation at 25 °C and 1 atm. Additionally, the photocatalysis route has shown promising performance and provides a safer route with milder reaction conditions compared to those of other heterogeneous catalysts, which have been reported to require temperatures of up to 80–150 °C and pressures of 10–30 bar.[19]

To further explore the photogeneration of CCs using TiO_2-based heteroge-neous photocatalysts, the use of a thin film photocatalyst is proposed here to eliminate the need for a separation step to recycle the catalyst, as well as to ease the handling of the photocatalyst. To the best of the authors' knowledge, the photo-generation of CCs using a thin film photocatalyst has not been reported in the literature. TiO_2 is a wide band gap semiconductor, and thus, it is photo-active predominantly in the UV range. To address this issue, the addition of foreign elements that are active in the visible light region is employed. Decorating TiO_2 with RuO_2 is known to increase the overall absorbance of the RuO_2/TiO_2 material in the visible light region.[35] Therefore, the combination of Ru^{4+} doping and RuO_2 decoration on the TiO_2 can greatly increase the absorbance in the visible light region, and in turn make the use of solar light viable.[36] Furthermore, RuO_2 has high chemical stability, and electrical conductivity. The combination of RuO_2 with TiO_2 has been reported to improve charge separation, hindering the recombination process which should enhance the photocatalytic activity of the material.[35] This effect is attributed to the position of the RuO_2 valence and conduction bands relative to those of TiO_2. Therefore, in this work, a Ru_x–TiO_2 thin film supported on fluorine-doped tin oxide (FTO) glass is proposed for the photogeneration of CCs.

Moreover, in order to provide a large surface area with more exposed active sites and superior light scattering capability, fast electron transport and efficient charge collection, hyper-branched nanorods (HBN) are synthesized. These HBN have been reported to exhibit improved optical and photoelectrical properties which resulted in improved photo-conversion reactions.[37] Hence, the HBN were modified with Ru to be used as a novel Ru_x–TiO_2 photocatalyst for the photo-generation of CCs.

Experimental

Materials

Fluorine-doped tin oxide (FTO) glass (TEC-15) was purchased from Ossila (dimensions of 2.5 cm × 2.5 cm, roughness of 12.5 nm, FTO layer thickness of 200 nm, 83.5% transmission and a resistivity of 12–14 Ω cm^{-1}). Potassium titanium oxide oxalate dihydrate (PTO, $\geq$98.0%), diethylene glycol (DEG, 99.0%), bis(cyclopentadienyl)ruthenium (($C_5H_5)_2$Ru, 98.0%), n-hexane (C_6H_{14}, 95.0%), propylene oxide (PO, 99.0%), dichloromethane (DCM, CH_2Cl_2, 99.8%), 4-(dimethylamino)pyridine (DMAP, 99.0%), propylene carbonate (PC, 99.7% anhydrous) and polyethylene glove AtmosBags were purchased from Sigma-Aldrich. Isopropanol (IPA, 99.5%), acetone (>95.0%) and ethanol (99.0%) were procured from Fisher Scientific. All of the chemicals were used without any further purification. All of the aqueous solutions were prepared using Milli-Q ultrapure type 1 water (18.2 $M\Omega$ cm) collected from a Millipore system.

Synthesis

The FTO glass was cleaned prior to use with a solution of H_2O and IPA, and acetone in a ratio of 1 : 1 : 1, for 1 h in the sonicator and air dried at 75 °C for 30 min.

Hyper-branched nanorods (HBNs) of TiO_2 were fabricated using a hydro-thermal approach. PTO was dissolved in a mixture of H_2O and DEG with a ratio of 1 : 7. The concentration of PTO was 0.05 M. After 30 minutes of vigorous stirring, the precursor solution was transferred to a 100 mL Teflon-lined autoclave along with the FTO glass. The FTO glass was positioned resting against the Teflon-lined walls with the conductive side facing down, at approximately 60°. The hydrothermal synthesis was carried out at 180 °C for 9 h. After the reaction time was complete, the autoclave was allowed to cool down to room temperature. The TiO_2 nanorods were rinsed several times with Milli-Q type 1 water and ethanol, and then calcined at atmospheric conditions at 550 °C for 1 h.

The Ru loaded TiO_2 (Ru_x-TiO_2) was synthesized under a dry nitrogen atmosphere, in a polyethylene Atmos glovebag. A known amount of bis(cyclopentadienyl)ruthenium $(C_5H_5)_2Ru$, was dissolved in n-hexane (C_6H_{14}) and stirred vigorously at 40 °C until a clear solution was obtained. Three different concentrations of Ru precursor were synthesised, namely 0.05 M, 0.01 M, and 0.005 M, that are denoted as $Ru_{05}-TiO_2$, $Ru_{01}-TiO_2$, and $Ru_{005}-TiO_2$, respectively. The TiO_2 loaded FTO glass was heated to 150 °C for 1 h to remove any adsorbed water and then allowed to cool to room temperature in dry air. The TiO_2 loaded FTO glass was then placed into the Teflon liner resting against its walls at roughly 60° with the coated surface facing down. The ruthenium precursor liquid was added to the Teflon liner to cover the entire FTO surface (25 mL). The Teflon liner was then transferred into the autoclave and placed in the oven at 180 °C for 30 h. The Ru_x-TiO_2 FTO glass was then rinsed with n-hexane in dry nitrogen. It was then calcined to 400 °C for 10 h under a dry air atmosphere with a ramp rate of 10 °C min^{-1}.

Characterisation

The morphology of the synthesized products was examined by a field emission scanning electron microscope (FE-SEM, FEI Quanta 200 F). Further investigation of the morphology and the elemental composition of the samples was carried out using a transmission electron microscope (TEM) and high resolution HR-TEM (FEI Titan Themis 200) equipped with an energy dispersive X-ray spectroscopy (EDX) detector operated at 200 kV. The samples were sonicated in ethanol for 5 min and then a few drops of the solution were placed on a carbon-coated copper TEM grid. Crystallinity and phase identification of the synthesized products were performed *via* powder X-ray diffraction (XRD), using a Bruker D8 Advanced Diffractometer equipped with Cu Kα radiation ($\lambda = 1.5418$ Å). Raman spectra were collected using a Renishaw inVia Raman Microscope with a 785 nm excitation source. The diffuse reflectance was measured using a Perkin Elmer Lambda 950 UV-vis equipped with an integrating sphere (150 mm) and the band gap energy was estimated using the Kubelka Munk function. X-ray photoelectron spectroscopy (XPS) was performed on a Scienta 300 XPS machine with a rotating Al Kα X-ray source operating at 13 kV × 333 mA (4.33 kW). Electron analysis was performed using a 300 cm radius hemispherical analyser and lens system. The electron counting system consisted of a multichannel plate, phosphorescent screen and CCD camera. All multichannel detection counting was done using proprietary Scienta software. The elements present were determined *via* a wide energy range survey scan (200 mW step, 20 ms dwell time, 150 eV pass energy and summed over 3 scans). The high resolution scans were performed at a similar pass energy (150 eV), but with a step size of 20 mV. A dwell time of 533 ms was used and accumulated over 3 scans. The instrument operated at a base pressure of 1×10^{-9} mbar; the energy scale was calibrated using the Au 4f, Ag 3d and Cu 2p emission lines, where the half width of the Au 4f$_7$ emission line is approximately 1.0 eV. All of the data analysis and peak fitting were performed using the CaseXPS software.

Photocatalysis

The reaction solution was prepared using DCM and PO in a 1 : 8 ratio, adding 1.022 mg of DMAP per mL of solution, and was kept in the dark at room temperature. The solution was then transferred to the stainless steel Teflon-lined pressurized photo-reactor (Scheme 1) along with the photocatalyst. Two control experiments were performed, one in the absence of photocatalyst and light, and one in the absence of light, keeping the rest of the reaction conditions the same. The temperature throughout the experiment was measured using a pyrometer from the outside of the reactor that ranged between 40–55 °C. For the experiments incorporating photocatalysts, four pieces of Ru_x–TiO_2 coated FTO glass were supported on a custom-made photocatalyst stage, as shown in Scheme 1. The stage was positioned directly below the quartz glass to be irradiated with the AM1.5G solar simulator directly (1 sun equivalent, 100 Wm^{-2}, 92250 A Newport, USA) while being submerged in the reaction solution. The reaction solution was bubbled with CO_2 and pressurised in a CO_2 atmosphere, throughout the experiment. The 500 mL photo-reactor was supplied with continuous CO_2 and the pressure was maintained at 200 kPa using a bubbler. The reaction was conducted under the solar simulator for 6 h. After the reaction time was completed, the unreacted substrate and solvent were removed *in vacuo* and the products were then isolated using a rotary evaporator. The products were then identified using an FTIR spectrometer (Perkin Elmer Frontier).

Results and discussion

The as-prepared TiO_2 and the fabricated Ru_x–TiO_2 thin film samples, synthesized *via* solvothermal treatment followed by calcination, exhibited homogeneous coating on the FTO glass. All of the prepared samples were examined *via* XRD. In addition to the XRD pattern of the FTO coating on the glass, the crystallinity of anatase TiO_2 was also observed in all of the samples (Fig. 1). The peaks positioned at 25.4 and 48.1° were assigned to the (101) and (200) planes of anatase. No ruthenium related peak was observed in the XRD patterns of the fabricated Ru_x–

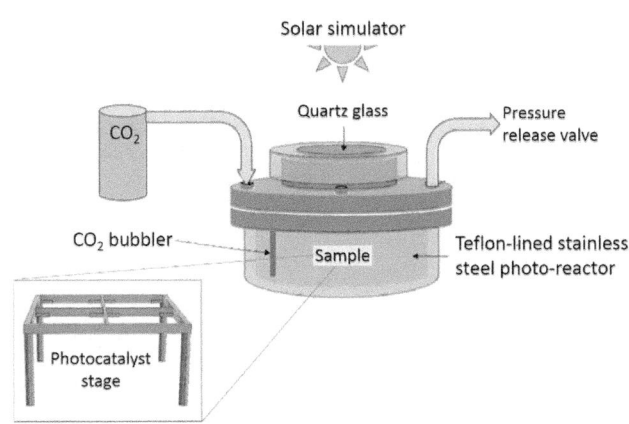

Scheme 1 A schematic diagram of the photocatalysis setup and photocatalytic stage.

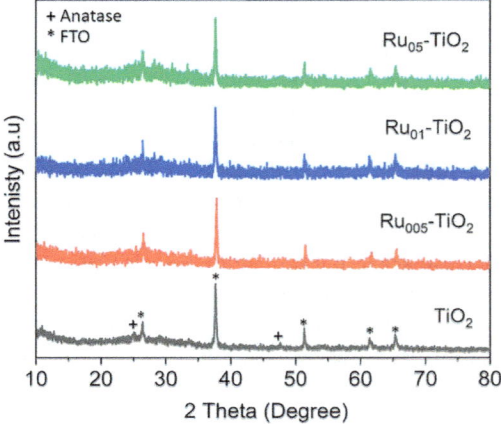

Fig. 1 XRD patterns of the as prepared TiO_2 and Ru_x–TiO_2 samples.

TiO_2 samples. This was probably due to the low concentration of Ru deposited on the glass. However, it is observed that with a higher concentration of Ru loading the crystallinity of the sample steadily decreased, suggesting that the Ru particles have been doped into the crystal lattice of the TiO_2 HBNs.

Raman spectroscopy is more sensitive and capable of characterising the crystal structure,[38] therefore, it was used to further investigate the crystal phases present on the TiO_2-based thin film samples. The Raman patterns of the as-prepared TiO_2 and Ru_x–TiO_2 samples are shown in Fig. 2. The Raman features of the anatase phase, positioned at 143, 395, 517, and 638 cm^{-1}, which are associated with the E_g, B_{1g}, A_{1g}, and E_g vibrations, respectively, were present in all of the fabricated samples.[39-43] Under close inspection, the main anatase peak of the Ru_{05}–TiO_2 sample positioned at 143 cm^{-1} was shifted to 149 cm^{-1}, whereas the peak positioned at 638 cm^{-1} was shifted to 621 cm^{-1}. When the concentration of Ru was lowered to 0.01 and 0.005 M, these shifts were reduced. These shifts are

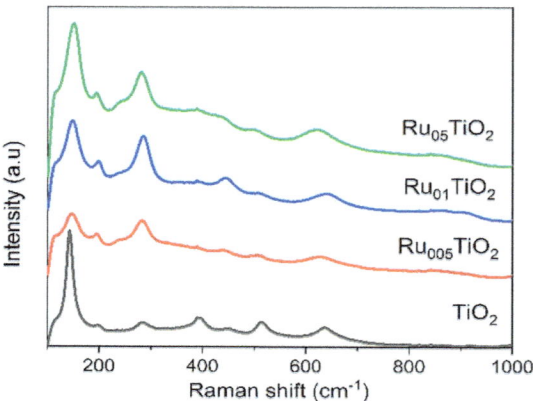

Fig. 2 Raman patterns of the as-prepared TiO_2 and Ru_x–TiO_2 samples.

speculated to be attributed to lattice substitution of Ru into TiO_2.[44] Additionally, Raman features of potassium titanate (PT), positioned at 285 and 444 cm^{-1}, were also observed, indicating potassium contamination from the titania precursor, PTO (Fig. 2).[45] The PT and anatase peak intensities seem to have an inverse relationship, where one increases while the other one decreases. However, a general trend can be observed, where the PT peaks are found to increase in the Ru_x–TiO_2 samples. This corresponds to the loss of crystallinity observed in the XRD patterns (Fig. 1) due to lattice substitution of Ti with Ru, which lowers the intensity of the anatase phase peaks. No ruthenium related peak was found on the Raman spectra of the Ru_x–TiO_2 samples. This is probably due to the small particle size of RuO_2 (discussed in the next section) and low concentration, along with the low intensity Raman excitation source (785 nm).

The morphology of the HBN structure of the titania was examined using SEM (Fig. 3). The microstructured as-prepared TiO_2 was evenly coated on the FTO (Fig. 3a). Each nanorod exhibited a vertically oriented spine with a highly branched nanostructure (inset of Fig. 3a). When different concentrations of Ru were loaded onto the TiO_2 HBN *via* solvothermal treatment, no significant alteration was observed in the Ru_x–TiO_2 samples (Fig. 3b–d). However, there were noticeable nanoparticles deposited on the spine and the nano-branches of the TiO_2 HBNs, which could be attributed to agglomeration of the RuO_2 nanoparticles (inset of Fig. 3b–d).

The cross-section view of the as-prepared TiO_2 coating reveals the morphology of the HBNs attached to the FTO glass (Fig. 4a). The growth of nano-branches of TiO_2 HBNs was perpendicular to the FTO and was supported on a base layer of TiO_2 with a 1–2 μm thickness, which had dense and thick structures. The base layer was attached to the FTO coating on the glass. Taking a closer look at the tip of the TiO_2 nanorods, the length of the nanorod spine was measured to be

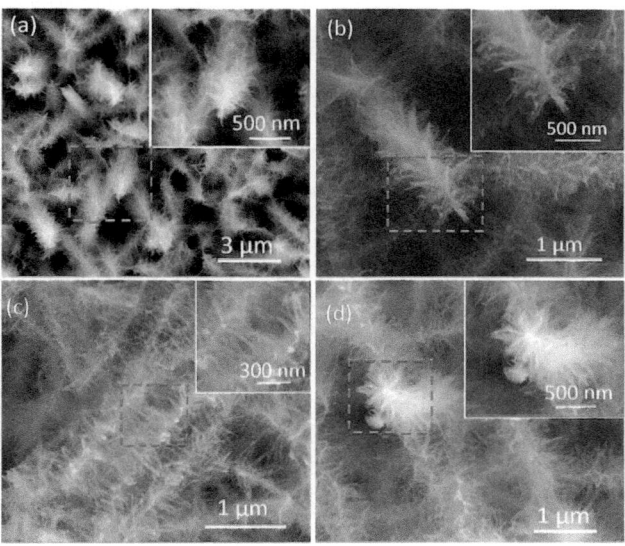

Fig. 3 SEM images, with high magnification inserts indicated by the red border, of (a) as-prepared TiO_2, (b) Ru_{005}–TiO_2, (c) Ru_{01}–TiO_2, and (d) Ru_{05}–TiO_2.

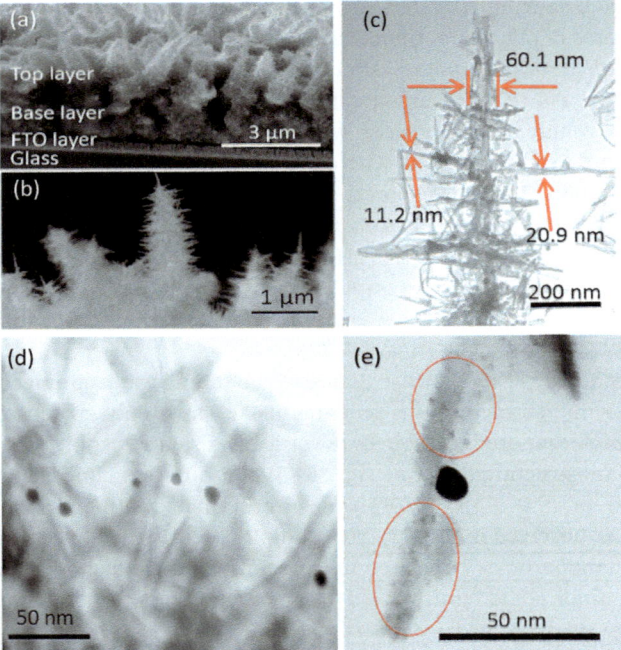

Fig. 4   SEM cross-section views of the as-prepared TiO_2 sample (a and b). A TEM image of a single Ru_{01}–TiO_2 HBN with thickness measurements (c). Low (d) and high (e) magnification HR-TEM images of the Ru_{05}–TiO_2 sample.

between 1–5 µm (Fig. 4b), whereas the nano-branches grown on the spine ranged from 50–300 nm (Fig. 4c). The nano-branch structure was proposed to provide a high surface area exposing more active sites for the photocatalytic reaction and the loading of Ru.[37]

The thickness of the HBN spine and nano-branches and the diameter of the loaded RuO_2 nanoparticles were ~60.1, 20.9, and 11.2 nm, respectively (Fig. 4c). The size of the loaded RuO_2 nanoparticles in the Ru_{05}–TiO_2 sample was ~11.5 nm (Fig. 4d). Furthermore, numerous nanoparticles of a much smaller size compared to the RuO_2 nanoparticles in the vicinity, were observed in the Ru_{05}–TiO_2 sample (Fig. 4e). These highly dispersed nanoparticles present on the TiO_2 HBNs ranged from 1–4 nm in diameter. The EDX mapping on HR-TEM evidenced the presence of RuO_2 nanoparticles distributed along the TiO_2 HBNs. It can also be observed that Ru particles are detected in the lattice of the HBNs (Fig. 5).

X-ray photoelectron spectroscopy (XPS) was performed to investigate the surface properties of the fabricated TiO_2 and Ru_x–TiO_2 thin films and to estimate the amount of Ru loaded on the TiO_2 thin film (Fig. 6). The Ru $3d_{5/2}$ and $3d_{3/2}$ peaks (dark shaded area in Fig. 6), which centred at 281.0 and 285.4 eV, respectively, were present in all of the Ru_x–TiO_2 samples. These peaks were attributed to Ru^{4+}, indicating that the RuO_2 NPs were loaded onto the TiO_2.[46,47] The Ru^{4+} particles seemed to have replaced Ti^{4+} in the HBN lattice as shown from the EDX studies (Fig. 6), as well as from the Raman peak shifts (Fig. 2) and the XRD crystallisation decrease (Fig. 1). The area was analysed and the concentrations of

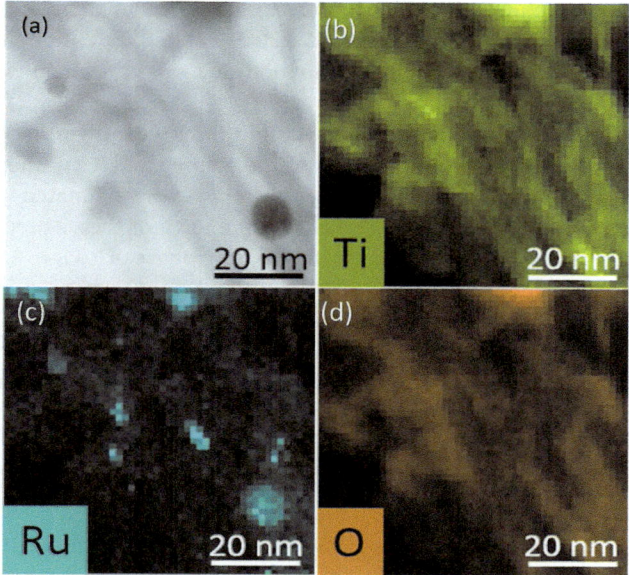

Fig. 5   HR-TEM EDX mapping for the Ru_{05}–TiO_2 sample. (a) TEM image, (b) titania, (c) ruthenium, (d) oxygen element mapping.

Ru were estimated to be 2.82, 1.45, and 1.11 at% for the Ru_{05}–TiO_2, Ru_{01}–TiO_2, and Ru_{005}–TiO_2, respectively. As expected, no Ru peak was observed in the as-prepared TiO_2 sample. Nevertheless, the XPS spectra revealed the peak for K centred at 293.0 eV (data not shown), confirming the presence of potassium

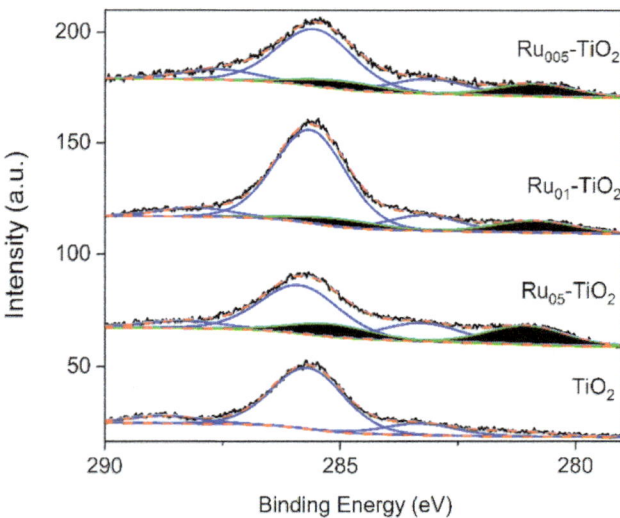

Fig. 6   XPS spectra of the as-prepared TiO_2 and Ru_x–TiO_2 samples. The shaded area represents the Ru area under the curve.

titanate in all of the thin film samples, as observed in the Raman pattern (Fig. 2).[48] The C 1s peaks shown centred at 283 eV, 286 eV, and 288 eV (not shaded) are ascribed to adventitious carbon contamination due to atmosphere exposure of the samples.[49]

The optical properties of the synthesized thin films were investigated via UV-vis spectroscopy. The diffuse reflectance of the as-prepared TiO_2 and Ru_x–TiO_2 samples was measured and the bandgap energy was estimated using the Kubelka–Munk function (Fig. 7a). The bandgap energy of the as-prepared TiO_2 was ~3.6 eV, which is larger than that of the anatase phase of TiO_2 (~3.2 eV). The band gap widening is attributed to the quantum confinement effect of the nanostructures.[50-52] The bandgap energies for the Ru_{005}–TiO_2, Ru_{01}–TiO_2, and Ru_{05}–TiO_2, were ~3.54, 3.53 and 3.49 eV, respectively. Although the shift in the bandgap energy observed in the Ru_x–TiO_2 samples was not significant, the bandgap energy decreased with increasing Ru concentration. This phenomenon is very likely due to the Ru^{4+} present in the crystal lattice of the titania, matching the observations made earlier from EDX and XRD and in the Raman pattern of the

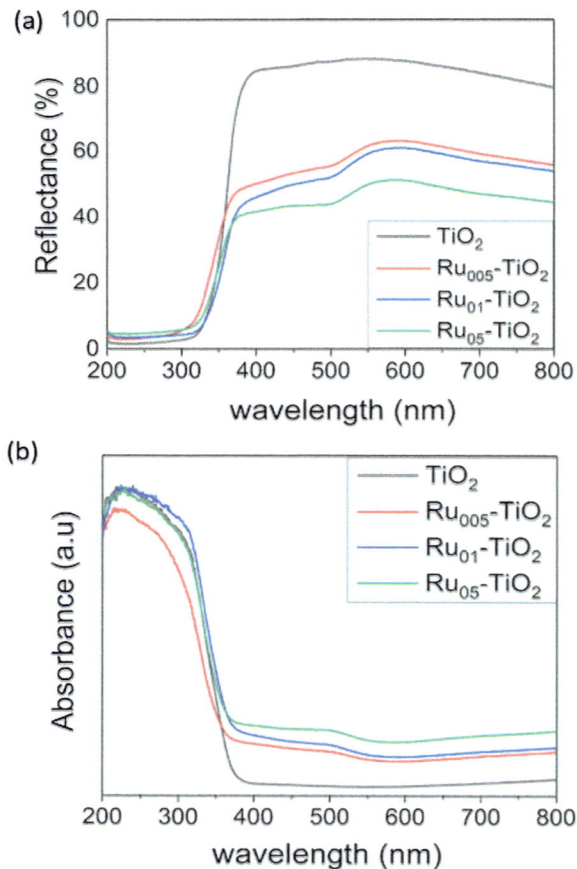

Fig. 7 The diffuse reflectance (a) and absorbance (b) of the as-prepared TiO_2 and the Ru_x–TiO_2 samples.

Ru_{05}–TiO_2 (Fig. 1, 2 and 6, respectively).[53] As confirmed by the XPS, Ru^{4+} was doped into the crystal lattice replacing Ti^{4+}, which explains the increased absorbance in the region of 350–500 nm that is observed in Fig. 7a and b.[53–56] Another observation is that the overall absorbance increased with the Ru loading concentration. The overall increase in absorbance, for wavelengths longer than 500 nm, is attributed to the RuO_2 decorated on the surface of the TiO_2 HBNs.[35] The synergistic effects of the Ru^{4+} doping and RuO_2 decoration on the TiO_2 are shown to greatly enhance the visible light absorption of the Ru_x–TiO_2 samples, making the use of a solar simulator for the photo-generation of cyclic carbonates a viable option.

Photocatalysis studies

The photocatalytic activity of the fabricated samples was evaluated with the photo-conversion of PO into PC under the irradiation of a simulated solar lamp. PC was used as a reference for CCs. A control experiment performed in the absence of photocatalyst and light was conducted. No PC was obtained, as revealed in the FTIR pattern of the control experiment (Fig. 8a). Only the tertiary –CH and anti-symmetry of –CH_3 and –CH_2, positioned at 2847.4, 2921.2 and 2953.1 cm^{-1}, respectively, were observed, and these were attributed to the PO moiety.[57] The as-prepared TiO_2 HBN catalyst, which was placed in the PO solution for 6 h under irradiation of the simulated solar lamp, showed no conversion. In contrast, for the fabricated Ru_x–TiO_2 photocatalyst a trough positioned at 1792 cm^{-1} was observed, which corresponds to one of the characteristic peaks of

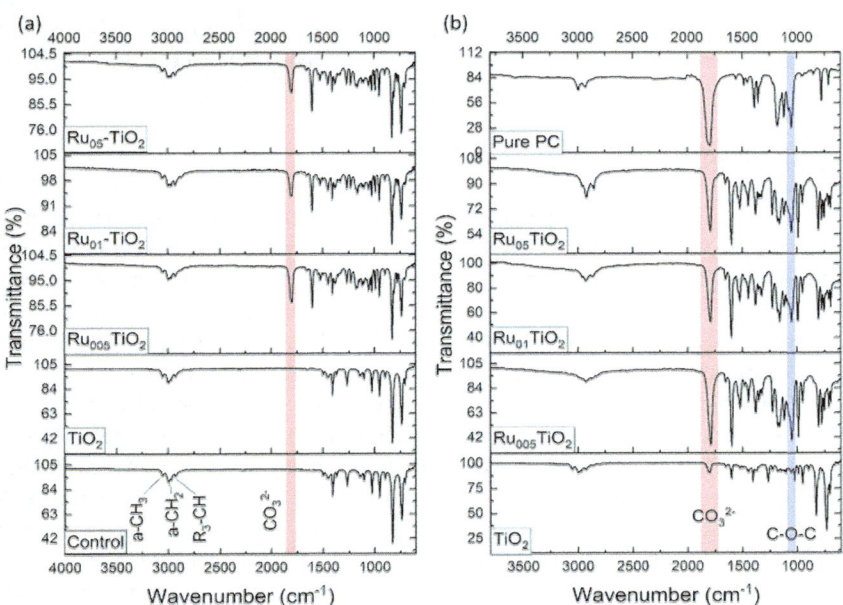

Fig. 8 (a) FTIR patterns of the solution obtained before evaporation from the fabricated Ru_x–TiO_2 and the control experiment without light and photocatalyst; (b) evaporated products with the fabricated Ru_x–TiO_2 photocatalysts and pure cyclic propylene carbonate as reference.

the cyclic carbonate group.[57] The intensity of this trough increased with decreased Ru concentration. This phenomenon indicates that a higher amount of PC was photo-generated with a lower Ru loading (Fig. 8a). When the Ru loading was increased from 0.005 M to 0.01 M, the photo-generation of PC was reduced to ~50% compared to that of $Ru_{005}-TiO_2$ sample. Further increasing the Ru loading, however, did not reduce the photo-generation of PC significantly.

To further investigate the selectivity and purity of the photo-generation of PC, the solution was purified through evaporation to remove the DCM and unreacted PO. Then, the residue solution was analysed using FTIR (Fig. 8b). The intensity of the trough, which was positioned at 1792 cm^{-1}, was enhanced in the purified solutions for all of the Ru_x-TiO_2 samples. The as prepared TiO_2 HBN photo-catalyst presented a weak trough at 1792 cm^{-1}, indicating that a small amount of CCs was produced. This confirms that the reaction is possible without the Ru doping on the TiO_2 HBNs photocatalyst; however, the doping greatly enhances its performance based on the intensity of the trough evident in Fig. 8b. On the other hand, the control experiment solution when evaporated had no residual solution left, as expected. The $Ru_{005}-TiO_2$ sample exhibited the highest intensity among the Ru_x-TiO_2 samples, indicating a high selectivity towards PC. Relative to the intensity of the pure PC trough, the selectivities of each sample towards PC were estimated to be 83, 62 and 57% for the $Ru_{005}-TiO_2$, $Ru_{01}-TiO_2$, and $Ru_{05}-TiO_2$ samples, respectively. Moreover, another trough centred at 1048 cm^{-1}, which was assigned to another characteristic peak of the CC group (C–O–C), was present in the pure PC solution. The same trough was observed in the solution after the photocatalytic reaction with the Ru_x-TiO_2 samples, confirming that PC was photo-generated by the Ru_x-TiO_2 samples. An additional peak centred at 1598 cm^{-1}, which is attributed to the C=N bond, presented in all the solution after the photocatalytic reaction with the Ru_x-TiO_2 samples but was absent in the pure PC solution.[58] The emergence of this peak was very likely due to the dissolution of DMAP, which was used as an additive in the reacting solution.

Conclusions

TiO_2 HBN thin films have been synthesised *via* a 2-step hydrothermal method, doped with Ru^{4+} and decorated with different amounts of RuO_2 nanoparticles. These novel hyper-branch microstructured TiO_2 thin films doped and loaded with Ru^{4+}/RuO_2 exhibited enhanced visible light absorption when compared to the pristine TiO_2 thin film. The size of the loaded RuO_2 nanoparticles ranged from 10 to 15 nm. The best performing amount of Ru loaded onto TiO_2 herein was 0.005 M, which was 1.11 at% Ru relative to TiO_2, as shown in the XPS analysis. The photo-generation of PC using the $Ru_{005}-TiO_2$ sample had an estimated selectivity of 83% towards the CC. Further quantification of the selectivity measurements will be determined using gas chromatography (GC). In the absence of Ru_x-TiO_2 photocatalyst and light, no product was obtained. These novel FTO supported photocatalysts, which eliminate the need for a catalyst separation step, are the first to be reported so far. The photocatalytic approach is a newly proposed concept for the generation of CCs. Compared to the heterogeneous counterparts, which typically require temperatures of up to 80–150 °C and pressures of 10–30 bar, and the homogeneous catalysts that are currently used in industry, at 100–200 °C and 50–100 bar, the photocatalytic approach requires temperatures of up to 55 °C and 2 bar. Therefore, this approach could potentially become a significant advantage for industrial applications.

Conflicts of interest

There are no conflicts to declare.

Acknowledgements

The authors are thankful for the financial support provided by the Engineering and Physical Sciences Research Council EP/K021796/1 and the CRITICAT Centre for Doctoral Training [PhD studentship to SG; Grant code: EP/L016419/1] at Heriot-Watt University. The electron microscopy facility in the School of Chemistry, University of St. Andrews, which is supported by the EPSRC Capital for Great Technologies Grant EP/L017008/1, is acknowledged.

References

1 E. Alper and O. Yuksel Orhan, *Petroleum*, 2017, **3**, 109–126.

2 M. Aresta and A. Dibenedetto, *Dalton Trans.*, 2007, 2975–2992, DOI: 10.1039/b700658f.

3 R. L. Paddock and S. T. Nguyen, *J. Am. Chem. Soc.*, 2001, **123**, 11498–11499.

4 M. North, R. Pasquale and C. Young, *Green Chem.*, 2010, **12**, 1514–1539.

5 J. H. Clements, *Ind. Eng. Chem. Res.*, 2003, **42**, 663–674.

6 S. Fukuoka, I. Fukawa, M. Tojo, K. Oonishi, H. Hachiya, M. Aminaka, K. Hasegawa and K. Komiya, *Catal. Surv. Asia*, 2010, **14**, 146–163.

7 G. Cascio, E. Manghisi, R. Porta and G. Fregnan, *J. Med. Chem.*, 1985, **28**, 815–818.

8 J. W. Comerford, I. D. Ingram, M. North and X. Wu, *Green Chem.*, 2015, **17**, 1966–1987.

9 L. Wang, P. Li, X. Jin, J. Zhang, H. He and S. Zhang, *J. CO2 Util.*, 2015, **10**, 113–119.

10 J. A. Castro-Osma, M. North and X. Wu, *Chem.–Eur. J.*, 2016, **22**, 2100.

11 I. G. Korosteleva, N. A. Markova, N. V. Kolesnichenko, N. N. Ezhova, S. N. Khadzhiev and N. I. Trukhmanova, *Pet. Chem.*, 2013, **53**, 412–417.

12 H. Kawanami and Y. Ikushima, *Chem. Commun.*, 2000, 2089–2090.

13 H. S. Kim, J. J. Kim, B. G. Lee, O. S. Jung, H. G. Jang and S. O. Kang, *Angew. Chem., Int. Ed.*, 2000, **39**, 4096–4098.

14 A. Decortes, A. M. Castilla and A. W. Kleij, *Angew. Chem., Int. Ed.*, 2010, **49**, 9822–9837.

15 M. North and R. Pasquale, *Angew. Chem.*, 2009, **121**, 2990–2992.

16 X.-B. Lu, B. Liang, Y.-J. Zhang, Y.-Z. Tian, Y.-M. Wang, C.-X. Bai, H. Wang and R. Zhang, *J. Am. Chem. Soc.*, 2004, **126**, 3732–3733.

17 F. Jutz, J.-D. Grunwaldt and A. Baiker, *J. Mol. Catal. A: Chem.*, 2008, **279**, 94–103.

18 J. Peng, H.-J. Yang, Y. Geng, Z. Wei, L. Wang and C.-Y. Guo, *J. CO2 Util.*, 2017, **17**, 243–255.

19 V. B. Saptal and B. M. Bhanage, *Curr. Opin. Green Sustain. Chem.*, 2017, **3**, 1–10.

20 A. Zanon, S. Chaemchuen, B. Mousavi and F. Verpoort, *J. CO2 Util.*, 2017, **20**, 282–291.

21 A. Buonerba, A. De Nisi, A. Grassi, S. Milione, C. Capacchione, S. Vagin and B. Rieger, *Catal. Sci. Technol.*, 2015, **5**, 118–123.

22 J. Peng, H.-J. Yang, N. Song and C.-Y. Guo, *J. CO2 Util.*, 2015, **9**, 16–22.

23 V. Preethi and S. Kanmani, *Mater. Sci. Semicond. Process.*, 2013, **16**, 561–575.

24 G. L. Chiarello, M. V. Dozzi and E. Selli, *J. Energy Chem.*, 2017, **26**, 250–258.

25 R. D. Tentu and S. Basu, *Curr. Opin. Electrochem.*, 2017, **5**, 56–62.

26 H. Ahmad, S. K. Kamarudin, L. J. Minggu and M. Kassim, *Renewable Sustainable Energy Rev.*, 2015, **43**, 599–610.

27 D. Kong, J. Z. Y. Tan, F. Yang, J. Zeng and X. Zhang, *Appl. Surf. Sci.*, 2013, **277**, 105–110.

28 J. Z. Y. Tan, Y. Fernández, D. Liu, M. Maroto-Valer, J. Bian and X. Zhang, *Chem. Phys. Lett.*, 2012, **531**, 149–154.

29 J. Zhao, Y. Li, Y. Zhu, Y. Wang and C. Wang, *Appl. Catal., A*, 2016, **510**, 34–41.

30 J. Z. Y. Tan, N. M. Nursam, F. Xia, M.-A. Sani, W. Li, X. Wang and R. A. Caruso, *ACS Appl. Mater. Interfaces*, 2017, **9**, 4540–4547.

31 P. Blanchet and V. Landry, in *Wood Composites*, ed. M. P. Ansell, Woodhead Publishing, 2015, pp. 335–355, DOI: 10.1016/b978-1-78242-454-3.00013-5.

32 H. Koinuma, R. Takahashi, M. Lippmaa, S.-Y. Jeong, Y. Matsumoto, T. Chikyo and S. Suzuki, in *Handbook of Advanced Ceramics*, ed. S. Somiya, Academic Press, Oxford, 2nd edn, 2013, pp. 1103–1124, DOI: 10.1016/b978-0-12-385469-8.00057-5.

33 J. Schneider, M. Matsuoka, M. Takeuchi, J. Zhang, Y. Horiuchi, M. Anpo and D. W. Bahnemann, *Chem. Rev.*, 2014, **114**, 9919–9986.

34 P. K. Prajapati, A. Kumar and S. L. Jain, *ACS Sustainable Chem. Eng.*, 2018, **6**, 7799–7809.

35 J. Tian, X. Hu, N. Wei, Y. Zhou, X. Xu, H. Cui and H. Liu, *Sol. Energy Mater. Sol. Cells*, 2016, **151**, 7–13.

36 F. Yoshitomi, K. Sekizawa, K. Maeda and O. Ishitani, *ACS Appl. Mater. Interfaces*, 2015, **7**, 13092–13097.

37 W.-Q. Wu, H.-S. Rao, H.-L. Feng, X.-D. Guo, C.-Y. Su and D.-B. Kuang, *J. Power Sources*, 2014, **260**, 6–11.

38 Y. Gong, C. Lee and C. Yang, *J. Appl. Phys.*, 1995, **77**, 5422–5425.

39 J. S. Lee, K. H. You and C. B. Park, *Adv. Mater.*, 2012, **24**, 1084–1088.

40 A. León, P. Reuquen, C. Garín, R. Segura, P. Vargas, P. Zapata and P. A. Orihuela, *Appl. Sci.*, 2017, **7**, 49.

41 H. C. Choi, Y. M. Jung and S. B. Kim, *Vib. Spectrosc.*, 2005, **37**, 33–38.

42 T. Ohsaka, *J. Phys. Soc. Jpn.*, 1980, **48**, 1661–1668.

43 Y. Zhang, C. X. Harris, P. Wallenmeyer, J. Murowchick and X. Chen, *J. Phys. Chem. C*, 2013, **117**, 24015–24022.

44 W.-K. Jo, S. Kumar, M. A. Isaacs, A. F. Lee and S. Karthikeyan, *Appl. Catal., B*, 2017, **201**, 159–168.

45 X. Liu and N. Coville, *A Raman Study of Titanate Nanotubes*, 2005.

46 X. Wei, Q. An, Q. Wei, M. Yan, X. Wang, Q. Li, P. Zhang, B. Wang and L. Mai, *Phys. Chem. Chem. Phys.*, 2014, **16**, 18680–18685.

47 S. Kaliaguine, in *Stud. Surf. Sci. Catal.*, Elsevier, 1996, vol. 102, pp. 191–230.

48 Q. Wang, Z. Guo and J. S. Chung, *Mater. Res. Bull.*, 2009, **44**, 1973–1977.

49 M. Descostes, F. Mercier, C. Beaucaire, P. Zuddas and P. Trocellier, *Nucl. Instrum. Methods Phys. Res., Sect. B*, 2001, **181**, 603–609.

50 E. Roduner, *Chem. Soc. Rev.*, 2006, **35**, 583–592.

51 M. A. Siddiqui, V. S. Chandel and A. Azam, *Appl. Surf. Sci.*, 2012, **258**, 7354–7358.

52 T. Takagahara and K. Takeda, *Phys. Rev. B: Condens. Matter Mater. Phys.*, 1992, **46**, 15578–15581.

53 T.-D. Nguyen-Phan, S. Luo, D. Vovchok, J. Llorca, S. Sallis, S. Kattel, W. Xu, L. F. Piper, D. E. Polyansky and S. D. Senanayake, *Phys. Chem. Chem. Phys.*, 2016, **18**, 15972–15979.

54 T. Ohno, F. Tanigawa, K. Fujihara, S. Izumi and M. Matsumura, *J. Photochem. Photobiol., A*, 1999, **127**, 107–110.

55 H. Huang, J. Lin, G. Zhu, Y. Weng, X. Wang, X. Fu and J. Long, *Angew. Chem., Int. Ed.*, 2016, **55**, 8314–8318.

56 M. T. Uddin, Y. Nicolas, C. Olivier, T. Toupance, M. M. Müller, H.-J. Kleebe, K. Rachut, J. Ziegler, A. Klein and W. Jaegermann, *J. Phys. Chem. C*, 2013, **117**, 22098–22110.

57 L. Zhang, X. Luo, Y. Qin and Y. Li, *RSC Adv.*, 2017, **7**, 37–46.

58 V. Kuryanov, M. K. Tokarev, T. A. Chupakhina and V. Y. Chirva, *The Synthesis of N-Acetylglucosamine Heteroaromatic N-β-Glycosides under Phase Transfer Conditions: Part III. l,2,4-Triazolin-3-one Glucosaminides*, 2011.

Faraday Discussions

DISCUSSIONS

Beyond artificial photosynthesis: general discussion

Ryu Abe, 🆔 Mark Bajada, Matthias Beller, 🆔 Andrew B. Bocarsly, 🆔
Julea N. Butt, 🆔 Flavia Cassiola, 🆔 Wolfgang Domcke, 🆔
James R. Durrant, 🆔 Stelios Gavrielides, 🆔 Michael Grätzel,
Leif Hammarström, 🆔 Marta C. Hatzell, 🆔 Burkhard König, 🆔
Akihiko Kudo, 🆔 Moritz F. Kuehnel, 🆔 Ava Lage, Chong-Yong Lee, 🆔
Marcelino Maneiro, 🆔 Shelley D. Minteer, 🆔 Aubrey R. Paris, 🆔
Nicolas Plumeré, 🆔 Joost N. H. Reek, 🆔 Erwin Reisner, 🆔 Souvik Roy,
Christoph Schnedermann, Ravi Shankar, Sergii I. Shylin, 🆔
Wilson A. Smith, 🆔 Han Sen Soo, 🆔 Andreas Wagner
and Dominik Wielend 🆔

DOI: 10.1039/C9FD90022E

Christoph Schnedermann opened a discussion of the paper by Burkhard König: In a broader context, it would be desirable to move towards a more holistic approach by simultaneously optimising the reaction yield and reducing the required photon input. Could you comment on the quantum yields you observe in this and related reactions and how you could potentially optimise them?

Related to the first question, I was wondering if you have considered transitioning from a homogeneous photoredox catalytic system towards a heterogeneous platform and what potential bottlenecks you can envision regarding stability, reaction yield and efficiency?

Burkhard König replied: Photons are used as a reagent in photocatalytic organic synthesis; the quantum yield is therefore of importance. We have not determined the quantum yield of the reported reaction, but for many other organic transformations. Typically for reactions involving radical chain mechanisms good quantum yields, often much larger than 1, are observed. Photocatalytic reactions that require excitation for each catalytic cycle typically show rather small quantum yields, often 0.01 or less. Many photons are wasted and it is certainly a future challenge to design photocatalytic systems that utilize photons more efficiently.

Heterogeneous photocatalysts are more practical than homogeneous photocatalysts, *e.g.* in terms of recycling, separation from the reaction mixture and often stability. So far the number of heterogeneous photocatalysts that have been used in organic synthesis is much smaller than homogeneous photocatalysts, but this may change in the future. The current bottleneck is the limited overlap between materials science (developing potential heterogeneous photocatalysts) and organic synthesis.

Erwin Reisner asked: What is the prospect of using solar irradiation instead of LEDs to drive organic photochemistry?

Burkhard König responded: The intensity of solar irradiation changes when it is overcast. A weather-dependent chemical yield would complicate synthesis. Most chemical plants operate 24/7, but there is no solar irradiation at night to drive the reactions. There may be special applications of simple solar synthesis in some sunny countries, but for photocatalytic chemical production LED light sources are the way to go.

Han Sen Soo remarked: I really appreciate the fact that we are finally bringing together the communities of artificial photosynthesis and photoredox catalysis. I am personally interested in utilizing photoredox catalysis as part of the half-reactions in artificial photosynthesis. Regarding the photoreduction catalysis in the paper, I noticed that Cs_2CO_3 is the optimal catalyst and there seems to be an equilibrium between the phenolate form and the anthrone. Is it not possible to generate and isolate the phenolate form with a strong base and fully characterize the photophysical properties of that photosensitizer? And why do the soluble bases such as the amines perform worse?

Burkhard König replied: The base has two roles in our reaction system: (1) it deprotonates the anthrone to the anthronate, the active photocatalyst, and (2) it neutralizes the HCl, which is a stoichiometric by-product of the reaction. In selecting the best base we had to consider the basicity, the solubility in the reaction mixture and the redox potential of the base. Basic amines are often easy to oxidize and can therefore not be used.
We did not made any attempts to isolate the anthronate anion in pure form.

Han Sen Soo asked: This is a follow-up question. Would the use of a stoichiometric amount of a strong base such as potassium *tert*-butoxide and the addition of a different base also provide the same yields?

Burkhard König answered: Likely yes, but we have not tried such an approach.

Michael Grätzel commented: I wonder about the turnover numbers that can be reached with the molecular catalysts employed in photo-redox catalysis.

Burkhard König responded: The TON that can be achieved with a molecular photocatalyst depends very much on the catalyst itself, the photocatalytic reaction and the reaction conditions. Numbers vary between 10 and 10 000.

Michael Grätzel asked: I wonder how you separate the various products formed during the photo-redox catalytic activation of arylchlorides. Is the sensitizer recuperated from the mixture?

Burkhard König replied: After reaction work-up, flash column chromatography was used to separate remaining or dehalogenated starting material, trapping reagent and sensitizer from the desired products. As most of the employed sensitizers are prone to oxidation in alkaline media and in the presence of air, we could isolate the corresponding anthraquinones in most of the cases. However, GC-MS (EI) analysis of the crude reaction mixtures verified the presence of the anthrone-sensitizers.

Wilson Smith opened a discussion of the paper by Marta C. Hatzell: Really interesting to use RRDE for product detection, that should be an interesting approach moving forward to solve issues with product detection.

My question is about the reaction mechanism you show in your presentation. There, you show that di-nitrogen becomes hydrogenated, where a lot of literature shows that nitrogen needs to adsorb then dissociate, and mono-nitrogen is the active site for hydrogenation. Can you comment on these two pathways, and why you think you have direct hydrogenation of di-nitrogen before nitrogen dissociation?

Marta Hatzell responded: Yes, there are numerous pathways for nitrogen to be fixed, however, most predict that nitrogen reduction occurs through an associative pathway at room temperature and pressure. Currently the proposed reaction pathway in our recent paper[1] suggests a dissociation based pathway may be possible with the carbon radical. We note however that this is a working hypothesis, and we are still exploring if it is practically possible.

1 B. M. Comer, Y.-H. Liu, M. B. Dixit, K. B. Hatzell, Y. Ye, E. J. Crumlin, M. C. Hatzell and A. J. Medford, The Role of Adventitious Carbon in Photo-catalytic Nitrogen Fixation by Titania, *J. Am. Chem. Soc.*, 2018, **140**(45), 15157–15160.

Ava Lage asked: Did you conduct any further characterisation of the recycled catalyst in order to obtain more detailed information on the proposed structural changes, particularly the adsorbed methoxy- or 'other oxidised carbon' species in phase 3?

Marta Hatzell responded: Not at this point in time. We have only indirectly observed the catalyst structure through *ex situ* measurements such as FTIR. We are looking into other approaches to try to explore this reaction *in situ*, but have not started these tests yet.

Ravi Shankar said: You mentioned that you'll be doing some EPR measurements soon. We have done some measurements of our own – what exactly are you hoping to see? Are you looking for a particular paramagnetic signature or will you be doing pulsed measurements to determine lifetimes?

Marta Hatzell replied: We are interested in investigating the role photogenerated CH_3 radical species play in the nitrogen photofixation process. We have designed a series of experiments to evaluate how CH_3 species are photogenerated from various hydrocarbon feed sources in an argon environment. After this, we plan to explore how the carbon radicals interact with a nitrogen gas. Finally, EPR will also enable us to begin to explore the role of oxygen vacancies on the nitrogen photofixation.

Ravi Shankar asked: Going forward, what are the next steps and the key challenges that you envision for this project?

Marta Hatzell responded: The most immediate challenge is in confirming experimentally the theorized catalytic pathway. In addition, we need to begin to explore if there are additional active sites working in concert with carbon radicals to promote nitrogen photofixation. Long-term there is also a need to design catalysts with a high density of these identified actives sites to promote higher rates of ammonia production.

Shelley Minteer remarked: During your presentation, you brought up concerns about contamination/importance of controls/*etc.* This is also a huge issue in the electrocatalytic nitrogen reduction area, as shown from many recent reviews. How are you verifying that the ammonia detected is produced from nitrogen gas?

Marta Hatzell responded: Currently we utilize multiple methods to confirm our results. We use spectrometric approaches, ion chromatography and rotating ring disk electrode experiments. Ideally confirmation through multiple methods provides a degree of certainty that our measurements are correct. We are also looking into exploring the use of isotope labeled gases.

Andreas Wagner remarked: Regarding your RRDE experiments to quantify ammonia, shown in Fig. 2 of the paper (DOI: 10.1039/c8fd00191j): have you done any concentration dependent experiments to perform a linear calibration and be certain that this peak is from NH_4^+ oxidation?

Marta Hatzell responded: We have conducted these experiments and are able to discern differences in the peak height and area as a function of the ammonia concentration.

Moritz F. Kuehnel asked: Could you comment on the reliability of your electrochemical ammonia detection. How do you know that the observed current is actually due to ammonia oxidation and not due to oxidation of some other species in solution? Do you think this approach is more reliable than other methods for quantifying NH_3, such as ion chromatography?

Marta Hatzell responded: We are fairly certain that the peak is ammonia, however, we are investigating this further through a series of tests. We think that the precision and detection limits for ammonia may be more ideal than chromatography or UV/Vis, but are also working to confirm this. We do not envision that this technique will displace other state of the art techniques. Rather we believe that it can complement other techniques. Ultimately we are always looking for multiple techniques that can confirm our findings, and rotating ring disk experiments can aid with this goal.

Flavia Cassiola remarked: The paper is an inspiring revisit to a system reported quite some ago for photocatalytic N_2 fixation. As you pursue simplicity and what you well called frugal engineering, how is the stability of your system coming along? Is your system robust enough (stability and durability) for this concept (simplicity)? What are the developments envisioned for improving stability?

Marta Hatzell answered: Our system is stable, but we have not conducted lifetime tests. In addition, we are currently running in batch mode operation. If we start to investigate longer continuous flow operations we may begin to see catalyst instability and a broad range of products (decrease in selectivity). We know that ammonia can be oxidized on the photocatalyst, and therefore designing the system to prevent back reactions is critical for future system design.

Leif Hammarström opened a discussion of the paper by Joost N. H. Reek: Does your "box" acceptor ever dissociate from the dye after charge separation, or does it rather work as a stationary acceptor group?

Did you record IV-curves? If so, did you see higher open-circuit voltage indicating less recombination (and possibly better fill factors)?

Joost Reek responded: As the current system indeed requires MV and ring-type ("box") compounds as redox mediator, we cannot exclude that the box works as a stationary acceptor only (based on experiments done so far). We are working on systems that can work solely with ring-type mediators, which may lead to more insight into this question.

The IV curve of the system that includes the ring is reported in the ESI of the paper. We didn't include the IV curve of the system that only contains MV as quality of the device (and the IV curve) didn't allow us to draw any conclusions. These type of experiments will be performed with the next generation systems in which photocurrents are much higher, and as a result the IV curve contains more sound information.

Michael Grätzel commented: The sequence of electron transfer reaction induced by light in p-type dye sensitized solar cells you postulated is that the electron is first transferred to the viologene moieties associated with the box and subsequently to freely diffusing methylviologen in solution. Based on redox potentials the second step would encounter a free energy barrier. Could you please comment ?

Joost Reek answered: We have measured the redox properties of the ring and MV, and using the Nernst equilibrium, these potentials and the concentration of the components you can calculate that electron transfer is feasible under the applied conditions. This is described in the ESI of the paper. Still, it would be better to have a system that doesn't need a second redox couple (such as MV in the current system), and we are currently working on that.

Ryu Abe remarked: I was impressed by your study in which a recognition part was introduced to catch and release the box-type redox mediator to control the direction of electron transfer in a p-type dye-sensitized photolectrochemical cell. However, the present system still requires the conventional methyl viologen as the second redox mediator along with the box-one. I understood this unfavorable use of methyl viologen is derived from the fact that the dye molecules were readily desorbed from the surface of NiO when the concentration of box-type shuttle redox mediator was increased. Thus, to achieve your final goal, you must fix the dye molecules rigidly. Do you have any idea for fixing such dye molecules on the surface of p-type NiO?

Joost Reek responded: There are several options to improve the binding of the dye to the NiO surface. One can change the anchoring group from carboxylic acid to a hydroxyamide or a phosphite, and it has been reported that these anchoring groups lead to stronger binding. Another option would be to increase the number of anchoring groups. We will consider these modifications to the dye. Also, initial studies using a neutral version of the ring show that the presence of excess of this ring doesn't lead to leaching. So the first experiments will be focused on this ring.

Matthias Beller commented: You have shown that supramolecular surrounding improved one of your systems by factor of 100. In principle this improvement is due to secondary function or stability. With regards to improving the rate of reaction: are there a lot of works on improving stability, *e.g.* by putting additional shielding around such systems?. Do you know if it is a stability or activity effect that acts?

Joost Reek replied: For the system I showed today the effect is clearly an activity effect. The water oxidation reaction with this catalyst is second order in ruthenium, and as such pre-organization of these catalysts result in rate acceleration. More generally, we have studied many different catalysts in cages, and we have also seen effects on stability. For example, in a recent ChemComm paper[1] we describe a hydrogenase model in a cage, which cannot disproportionate because of the encapsulation, leading to higher stabilities. I think that catalyst encapsulation therefore is a more general strategy that may be beneficial for many different reactions, and further exploration is worthwhile.

1 S. S. Nurttila, R. Zaffaroni, S. Mathew and J. N. H. Reek, *Chem. Commun.*, 2019, **55**, 3081–3084.

Nicolas Plumeré asked: Can you use electrochemical methods to better understand the function of the viologen cage? Is it possible to detect the redox signals of the surface confined viologens? Is the viologen cage really departing the surface upon reduction or is it instead facilitating charge transfer to the methyl viologen in solution without dissociation from the photosensitizers?

Joost Reek answered: Electrochemistry is a powerful tool to get more information on the system. Unfortunately, electrochemistry using the NiO semiconductor material is impossible in the window of interest because of the properties of the semiconductor. When immobilised on gold, you should be able to see the redox signals of the viologen box, in which case you could confirm that the viologen box leaves the ring upon reduction. In related work by Willner (ref. 1) a similar principle on a rotaxane was demonstrated in which viologen box changes station on the rotaxane upon electrochemical reduction.

It is difficult to prove experimentally that the viologen box dissociates upon reduction, rather than that charge transfer occurs to the free viologen without leaving the station (see other comment). We are working on a new system that doesn't need the additional viologen, which should be easier to understand.

1 E. Katz, O. Lioubashevsky and I. Willner, *J. Am. Chem. Soc.*, 2004, **126**(47), 15520–15532.

Michael Grätzel asked: I wonder about the sequential electron transfer postulated for the sensitization process. According to your mechanism the electron is first transferred from the p-type sensitizer to the viologen on the box and from there to the methylviologen which is freely diffused in solution. I would expect that this second transfer has a substantial energy barrier as the redox potential of the viologene on the box is more positive than that of the methylviologen. Hence the eletron could get stuck on the box.

Joost Reek replied: We have measured the redox properties of the ring and MV, and using the Nernst equilibrium, these potentials and the concentration of the components you can calculate that electron transfer is feasible under the applied conditions. This is described in the ESI of the paper. Still, it would be better to have a system that doesn't need a second redox couple (such as MV in the current system), and we are currently working on that.

Sergii Shylin opened a discussion of the paper by Stelios Gavrielides: Your paper is a nice example of work driven by the concept of carbon neutrality in the chemical industry. However, the main drawback of industrial emissions as a source of CO_2 for synthesis is steam, which accompanies CO_2 wastes. Combustion of natural gas produces two H_2O per one CO_2 molecule. This gas mixture can be roughly separated by distillation, but further drying of CO_2 may be challenging on an industrial scale. How sensitive is your reaction (coupling of CO_2 and epoxides) to water traces?

Stelios Gavrielides answered: The reaction is not significantly sensitive to water traces. The Ru-TiO$_2$ HBN photocatalyst is also tolerant to H_2O traces. The addition of water in this reaction has been reported in the literature, with organocatalytic systems, where it was used as a co-catalyst to activate the CO_2/epoxide coupling by enhancing the homogeneous catalyst's (phosphonium halides, ammonium, *etc.*) solubility.

Matthias Beller commented: There are a number of applications at the moment, even on an industrial scale, that use waste stream from factories, *e.g.* polycarbonates. I think that taking water out of the waste stream is no problem, the problem is the stability of the catalysts. Is it really the light, or Lewis function of the TiO$_2$, that does the reaction?

Stelios Gavrielides responded: I agree, taking the water out of the system is not difficult, but also not a necessity because the reaction is not particularly sensitive to small amounts of water. To eliminate doubts related to the origin of the reaction, we conducted control experiments (Fig. 8 in the paper (DOI: 10.1039/c8fd00181b)), under dark conditions, in the presence of photocatalyst, keeping temperature and pressure the same. The photocatalyst was not activated in this case, and we could observe no cyclic carbonate conversion. However, more research is needed in order to gain a better understanding of the reaction pathways.

Matthias Beller commented: Have you ever tried to make polycarbonates?

Stelios Gavrielides replied: We haven't tried to generate polycarbonates yet. Based on the FTIR results alone, we were worried that polycarbonates might be a side product of this reaction, however, after recent NMR spectroscopy of the product, it shows our product is purely cyclic carbonates. However, since poly-carbonates are very valuable chemicals we might in the future try to generate polycarbonates as well.

Michael Grätzel commented: Maybe TiO_2 is not involved in the photo-catalytic process and acts just as a support for the RuO_2 catalyst. RuO_2 is black so it does absorb light but due to its metallic character the photo-excited carriers live only for a short time *i.e.* at most a few picoseconds in this material. Hence the chances are slim that they would induce a chemical reaction.

Stelios Gavrielides answered: The control experiments indicated that the loading of RuO_2 onto TiO_2 worked as the photocatalytic site for the photo-generation of PC although the working mechanism is unclear at this stage and under investigation. Additionally the control experiment using the as-prepared TiO_2 HBN shows a small conversion of CCs which suggests that TiO_2 alone demonstrated better performance than the no-photocatalyst control experiment. Furthermore, the quantum confinement effect of our photocatalyst transforms the band gap from continuous to discrete, which should in principle delay the recombination rate therefore allowing the photo-excited charges enough time to be used in a chemical reaction, at least to some extent. Relevant information can be found in the listed references.[1-3]

1 W.-Q. Wu, H.-S. Rao, H.-L. Feng, X.-D. Guo, C.-Y. Su and D.-B. Kuang, *J. Power Sources*, 2014, **260**, 6–11.
2 I. Musa, F. Massuyeau, L. Cario, J. L. Duvail, S. Jobic, P. Deniard and E. Faulques, *Appl. Phys. Lett.*, 2011, **99**, 243107.
3 M. B. Nardelli, K. Rapcewicz and J. Bernholc, *Appl. Phys. Lett.*, 1997, **71**, 3135.

Ravi Shankar remarked: You mentioned that when you changed the morphology of your material, you observed a change in the band gap. Could you please elaborate a bit more on this? Which morphology change did you observe?

Stelios Gavrielides responded: We are referring to the hyper-branched morphology that was achieved, with the TiO_2 HBNs, which creates a 3D nano-structured nanorod with nano-branches growing on all sides. These nano-branches have a thickness of 5–20 nm. The bandgap of anatase phase TiO_2 in bulk is 3.2 eV. The band gap was found to be widened significantly with this morphology and was calculated to be 3.6 eV. This is attributed to the quantum confinement effect, which is essentially the transition of the continuous energy bands of bulk material to discrete atomic-like energy levels, which has as an effect the band-gap widening and consequently a hypsochromic shift. In more detail, the valence and conduction bands of the semiconductor are no longer continuous and the excited charges are not able to move freely. In other words, the geometrical constraints affect the electron energy. This is observed when the size of a particle is small enough to be in the same magnitude as the de Broglie wavelength of the electron wave function. The quantum confinement effect can change the optical and electronic properties of solids.

Dominik Wielend asked: I have two questions regarding the method and analysis of your FTIR data shown:

Did you record them in a transmission cell configuration or in an ATR configuration?

And regarding the quantification/selectivity values you reported: did you only consider the transmittance values of the band at 1800 cm^{-1} or a combination/ratio of the 1100 and 1800 cm^{-1} bands?

Stelios Gavrielides responded: The reaction solution was measured using an ATR configuration. As mentioned in the paper, the quantification is an estimation since FTIR is not a quantification technique. The values considered were only the values of the 1800 cm^{-1} that correspond to the functional group of the cyclic carbonate. However, ongoing characterisation of the reaction solution using NMR shows that the measurements were fairly accurate; we are currently working on getting more quantifiable information regarding the conversion and yield as well as possible by-products.

Matthias Beller opened the general discussion of the papers by Burkhard König, Marta Hatzell, Joost Reek and Stelios Gavrielides: Considering Dr König's talk: his presentation showed that photocatalysis offers nice possibilities for organic synthesis, which is one of the hottest fields at the moment. These are two very active fields. What can the different fields/communities learn from each other? What can we gain by bringing them together?

Joost Reek replied: That is a good question. I think that the field of photocatalysis for synthesis can learn from the artificial photosynthesis field in the area of device design, materials, and strategies to achieve high efficiencies. In turn, I think that the artificial photosynthesis field is very much focused on water oxidation and proton/CO_2 reduction, whereas there may be many possibilities for energy storage out there that have not been considered (maybe only for smaller scales). So I think that it would be good to keep on bringing these fields together.

Burkhard König replied: I believe both fields can learn and benefit a lot from each other, but there is a gap to bridge. Many materials originally developed for solar fuel applications (hydrogen generation, carbon dioxide utilization *etc.*) may be of high value for organic synthesis. However, the functional group compatibility and the stability of photocatalytic materials in organic reaction mixtures (containing radicals, electrophiles, nucleophiles) may be challenges.

Wolfgang Domcke remarked: In my view, there is an important difference between light-driven photochemical transformations, for example synthesis of fine chemicals, and the oxidation of water and the reduction of carbon dioxide.

The latter two reactions are among the most difficult chemical reactions, since water and carbon dioxide are exceptionally stable and therefore are exceptionally difficult to activate. It is timely to get away from sacrificial reagents and address the real challenges, that is, the oxidation of exactly water and the reduction of exactly carbon dioxide with visible light and without the consumption of other chemicals.

Burkhard König responded: I agree that sacrificial agents should be avoided in any reaction. For photochemical transformations in synthesis it is important to note that there are, in addition to photooxidation and photoreduction, which require a sacrificial oxidant or reductant, many redox-neutral transformations, *e.g.* ATRA reactions, (cyclo)addition or cross-coupling reactions. Such redox-neutral reactions in synthesis do not require sacrificial reagents.

Michael Grätzel commented: Heterogeneous photo-redox catalysis has the advantage over the homogeneous approach in that the products can be readily separated from the catalyst. For example we have published recently together with our colleague Professor Xile Hu and his coworkers a paper in Nature Catalysis where we showed that the CH amination of arenes is photo-electrocatalyzed by a simple mesoscopic Fe_2O_3 electrode.[1]

1 L. Zhang, L. Liardet, J. Luo, D. Ren, M. Grätzel and X. Hu, *Nat. Catal.*, 2019, **2**, 366–373.

Stelios Gavrielides replied: We agree that the heterogeneous photocatalytic approach has significant advantages over the homogeneous counterparts. This was the rationale of using the FTO-grafted photocatalyst for easy separation. Congratulations on a very important and informative paper.

Erwin Reisner addressed Burkhard König: You have mentioned the more complex reaction space in organic photochemistry compared to the simple transformations in artificial photosynthesis. I would like to add another point of view: Organic photochemistry is commonly employing radical chemistry initiated by single-electron transfer events to execute exothermic reactions. In contrast, artificial photosynthesis deals with multi-proton/electron processes, which are also highly endothermic. The latter should therefore be recognised as both thermodynamically and kinetically complex.

Burkhard König responded: I agree that currently the majority of photo-catalytic organic synthesis involves radical reactions and radical or radical ion intermediates. And in most reported reactions the light energy input is used to overcome kinetic barriers of overall exothermic reactions. However, there is no reason not to develop organic reactions involving more than one electron and photon, including endothermic transformations. We do this in our research projects.

Souvik Roy addressed Burkhard König: What's your thought about using the molecular sensitizers as building blocks to build porous materials – then the photocatalysis will occur in cages/confined spaces which will offer scope for tuning the environment.

Burkhard König responded: This is an interesting suggestion combining the selectivity of photocatalytic activation (according to the redox potentials) with confined reaction space as a second element of control. However, the close proximity of the chromophores may alter their photophysical properties and the photochemical stability of some organic dyes may not be good enough for such a material approach to photocatalysis.

Julea Butt addressed Stelios Gavrielides and Joost Reek: An aspect of biological photosynthesis that can provide inspiration for strategies in artificial photosynthesis is the precise spatial arrangement of redox centres having different properties. This facilitates charge separation, and the coupling of single photon events to catalytic chemistry typically requiring multiple electrons. It seems the molecules you describe lend themselves to integration within 3D assemblies that could mimic these aspects of biological photosystems. I welcome your thoughts on the feasibility and utility of such an approach.

Stelios Gavrielides responded: Yes, indeed the spatial arrangement of redox centers would facilitate the charge separation. Further investigation in this regard on the material we have is being undertaken.

Joost Reek replied: I think this is an excellent remark. One of the reasons that the Natural system works so well is that all the molecular components required for photosynthesis are precisely organised in space. Supramolecular chemistry also allows such precise organization of molecules just by programming the building blocks with molecular information that leads to the self assembly. Today's paper is a simple demonstration of that, but you can imagine that more complex systems are also accessible *via* supramolecular assembly strategies. It will take some time and effort, however, to get where we want. I would like to stimulate other scientists to also think about these type of strategies.

Mark Bajada addressed Burkhard König: I agree that using the whole visible (solar) spectrum for the purposes of organic synthesis doesn't make much sense, and from a chemical selectivity perspective, it's better to use LEDs of a specific wavelength. But if we use PVs to generate electricity, which we then use to power an LED of a particular wavelength in order to drive a chemical reaction, why not circumvent the use of LEDs altogether and just look into organic electrochemical synthesis; a field which is also growing rapidly at the moment? Is this research being carried out in your lab?

Burkhard König replied: Electrocatalysis and photocatalysis both use electron transfer for the activation of molecules; therefore similar types of organic transformations can be performed. However, there is one striking difference between the two approaches, at least if we compare electrocatalysis with homogeneous photocatalysis. In electrocatalysis we need solid electrodes and the electron has to cross the solid to liquid phase boundary; in homogeneous photocatalysis the molecular photocatalyst acts as the electron donor or acceptor and no phase boundary has to be crossed. The selection of electrode material, its surface structure and modification is of key importance for stability and performance in electrocatalytic synthesis.

Like many labs around the globe we also look into combinations of photo-catalysis with electrocatalysis.

Leif Hammarström remarked: Another difference between photoredox catal-ysis and electrochemical catalysis is that the latter can often lead to multi-electron redox reactions, as the initial radicals are often easier to oxidize/reduce than the starting material. In photoredox catalysis, this can be controlled as it usually involves single-electron redox initiators.

Burkhard König responded: This is correct. However, electrocatalysis in return has the advantage to offer a wide range of potentials for the electron transfer. Both methods have advantages and limitations.

Matthias Beller addressed Marta Hatzell: With regards to the efficiency, 4.1% should be commercially viable. Ammonia is abundant and so this is very inter-esting for large scale/general applications. What is the timescale for this to be a proper technology? What does the market space look like at low level?

Marta Hatzell responded: We currently do not know what efficiency would aid in promoting commercial viability, as cost would also be a significant issue. We anticipate that low solar-to-ammonia efficiencies may be acceptable due to the rate at which nutrients are utilized. We also intend to look into estimating capital and operating costs to better assess the viability of solar driven fertilizer production. Currently there is little market for dilute fertilizers as the cost to ship dilute fertilizers prohibits their use. However, if produced at the site of use, this could potential enable its use.

Han Sen Soo returned to the discussion of the paper by Joost N. H. Reek: The supramolecular cyclophanes in the paper are very attractive and impressive. In this paper, the application is for p-type DSSCs. However, given that perovskite solar cells are already performing so well with far less research, do you foresee much future in DSSCs? Or are there other applications for the supramolecular systems that you are creating?

Joost Reek responded: It is very difficult to predict what will be applied in the future, and currently both DSSC and perovskite solar cells have advantages and disadvantages. It is true that the efficiencies of perovskite solar cells have increased enormously in past years, but there are no commercial applications yet for several reasons (including stability and lead). On the other hand DSSCs are currently commercially available. Most importantly, as a scientist I would like to create new knowledge and I would like to provide new concepts and new tools, and it is not my primary aim to look at the application. As such, I believe we should not feel hampered or limited in our science by the application.

Han Sen Soo asked: This is a follow-up question. The performance of perov-skite solar cells have already exceeded the best DSSCs, and I can imagine that they will have similar indoor applications with low lighting. Would the two technol-ogies be competitive?

Joost Reek responded: Again, this is very difficult to predict.

Michael Grätzel remarked: This comment addresses the general question of research on dye sensitized solar cells (DSC) and perovskite solar cells (PSC). DSCs are by now a mature technology and cells are sold on the market by a number of companies, such as Exeger, Fujikura, Ricoh and H.Glass. The two main applications of DSCs are in ambient light harvesting to power portable electronic devices such as E-readers, and sensors for the Internet of Things (IoT) market. Here their energy conversion efficiency is substantially higher than that of the best competing technology. The other application is in solar electricity producing translucent glass panels. PSCs emerged from DCSs and are still at a research state. They show very high solar to electric power conversion efficiencies which have reached 23.7%, however more research and long term outdoor field tests are required to address existing notorious stability problems. In addition current formulations of PSCs contain lead and hence their use would be restricted to well protected solar fields.

Leif Hammarström said: This comment is in response to the earlier question on why to continue working on DSSCs when perovskite cells have undergone such great development in recent years. In science, not everyone should go in the same direction. It may sound obvious, but often granting agencies and even colleagues advise or request you to work on what already works best and where lots of scientists already work. This is dangerous for science, and reminds me of the old joke of the drunken man looking for his car keys underneath the lamp post, instead of where he dropped them (in the dark, behind some shrubbery) because it is easier to search where there is already light. With the same argument, people should not have started looking into perovskite cells because silicon (and DSCCs) worked so much better. In particular you young scientists should not fall into this trap, but follow new ideas and research directions.

Matthias Beller replied: This is good advice, go and work on areas that are unknown.

Chong-Yong Lee addressed Marta Hatzell: Regarding your perspective view that 0.1% solar conversion efficiency to ammonia is sufficient for practical application, I am just wondering what do you think about other alternative ways of obtaining this low-level ammonia? For example, by the reduction of nitrous oxide which is present in the atmosphere of some cities such as Beijing and Bangkok? It may achieve advantages of ammonia production as well as treating environmental pollutants.

Marta Hatzell answered: This is definitely another avenue. However, one issue with taking NO_x out of the atmosphere is the energy required to separate NO_x from other atmospheric constituents. Generally, areas which have a high concentration of NO_x also have a high concentration of harmful particulate matter (aerosols). This could contaminate the dilute fertilizers. Ideally, obtaining pure and dilute fertilizers which require minimal separation processes is our current goal.

Aubrey Paris addressed Marta Hatzell: You noted that there seems to be an interest in intentionally producing dilute concentrations of ammonia in solution. Where is this interest coming from? Is it related to the fact that the quantities of ammonia currently used in agriculture are excessively high? If so, do you think that there would be a barrier to entry for dilute ammonia in the agriculture space, since this would result in process and possibly infrastructure changes on farms?

Marta Hatzell responded: Nutrient utilization by a plant is most efficient if fertilizers are supplied when a plant has access to water and sunlight. Therefore, dilute fertilizers solve one issue, as water and nutrients can be supplied at the same time. The challenge is that dilute fertilizers are associated with the infrastructure. A farm would need to have an extensive irrigation system. A second challenge is that shipping dilute fertilizers would most likely be energy intensive and expensive. Therefore dilute fertilizers would most likely only make sense if they were manufactured on site (on farm).

Matthias Beller returned to discussion of the paper by Stelios Gavrielides: In principle it would be more energy efficient to keep the oxygen in the system (reduction costs most of the energy). No change of oxidation state would occur, and the system would be more energy and oxygen rich, which could have intrinsic advantages. Do you have any comments on this?

Stelios Gavrielides answered: I agree energy-wise it would be beneficial to leave oxygen in the system. Since we haven't done any reaction mechanism studies yet, we cannot be sure on how the presence of oxygen could affect the reaction. In the future, we plan to investigate this matter further.

Erwin Reisner returned to discussion of the paper by Marta Hatzell: You have mentioned a desired solar-to-ammonia conversion efficiency of 0.1%. This number appears rather low compared to other efficiency goals in artificial photosynthesis (such as commonly quoted 10% solar-to-H_2 efficiency). Could you please elaborate on this aspect further and explain how this value (0.1%) was determined?

Marta Hatzell answered: This was estimated based on the current nutrient requirements supplied to farms. Currently nitrogen is supplied at a rate of approximately 50–100 kgN ha^{-1} yr^{-1}. With this known nutrient flux, the solar-to-ammonia target was estimated using assumptions regarding solar intensity and capture area.

Andreas Wagner commented: The production of low-concentration ammonia in water *via* solar energy would be fantastic and I am very supportive of this research. I was wondering whether other elements besides nitrogen such as potassium or phosphorus might become the limiting components of plant growth if ammonia is supplied *via* solar chemistry?

Marta Hatzell responded: In developed countries access to phosphorous and potassium are generally easy to acquire. Therefore from a chemical manufacturing perspective nitrogen fixation is the largest bottleneck for fertilizer production. However, if this technology would be envisioned to be a stand-alone fertilizer production system at a farm site, one would need to have the means to ship the additional nutrients (P and K) necessary for plant growth. This is a good point that would need to be addressed on site.

Han Sen Soo remarked: This may be a simplistic question. There are a number of reports about the photocatalytic reduction of N_2 to ammonia recently. However, there has not been much work on photocatalytic or electrocatalytic N_2 oxidation similar to the Ostwald process. The Ostwald process is exothermic, but also consumes a lot of energy and needs ammonia. Are you also looking at photo-catalytic N_2 oxidation reactions?

Marta Hatzell responded: We are looking at these different reactions. There is a large interest in this area as well, however, there are also significant challenges associated with exploring nitrogen oxidation. Many issues regarding contamination exist. We envision seeing similar challenges with nitrogen oxidation, and control experiments will be critical in this area.

Andrew Bocarsly addressed Stelios Gavrielides: Fossil fuels are highly reduced materials and CO_2 is highly oxidized. From a strategic synthetic point of view, aren't we better off making oxidized products from CO_2 and reduced compounds from fossil fuels? Along this line of reasoning, the conversion of CO_2 to cyclic carbonates is a thermodynamically spontaneous reaction, in contrast to fuel formation which has a positive ΔG. Is there a preferred product strategy in this regard?

Stelios Gavrielides responded: The photoreaction of propylene oxide converted into propylene carbonate *via* CO_2 fixation is proposed herein to demonstrate the possible heterogeneous photocatalysis, which is underexplored for this green organic synthesis. We agree that this reaction has a negative ΔG. However, catalysts are widely employed in the industry to improve the yield.

Wilson Smith opened discussion of the concluding remarks by James Durrant: As discussed in the introduction of your talk, there are many approaches to artificial photosynthesis, among them particle-based reactors, photoelectrodes, and PV (or more accurately renewable electricity-based) electrolysis. While it is somewhat common for most people to say they don't want to pick winners, and it is true there is underlying science that translates between different approaches, at what point do we actually need to pick a winner? The timescale for technological large scale solutions is very short, and we need to start making some choices about which approaches are working and should have more attention paid to their upscaling, while other approaches can still benefit from fundamental/basic science.

James Durrant answered: This is an important question for which there is no clear answer. On the one hand, we are addressing an urgent problem, and should be addressing multiple approaches in parallel rather than in series. However you are right that we should be starting to focus our efforts on the most promising front runners. I think to chose these, we need a close engagement between scientists and engineers from both academia and industry, so we can understand both what it appears likely we will be able to achieve scientifically, and what devices and technologies look most promising for scale up and commercial application. In the shorter term, I think that PV plus electrochemistry is the most obvious approach, and indeed is the benchmark we should compare other artificial photosynthetic approaches against. I think we also need to consider carefully what reactions we wish to target. In the long term it is clear we should target water splitting and CO_2 reduction, but in the shorter term, other reactions, including for example selective organic oxidations, may be more commercially attractive.

Erwin Reisner said: I would like to emphasise the importance of bringing industry collaborations on board for academic research and to facilitate tech transfer. Industry partners do not only provide valuable resources, but also offer important questions and much needed expertise and input in the development of real-world processes. I also believe that many energy companies are genuinely very interested in our work, but have only limited contacts to the latest developments in academia. The often perceived paradigm that industry has limited interest in artificial photosynthesis as it's too far from application is not true. I have had very positive interactions with industry and we have just completed a 7-year application oriented basic research programme on solar syngas production half-funded by an industrial partner. The key is pro-active engagement with potential industry partners and reaching out to the right person within the company (and the challenge here is to identify and meet the 'right' person). I would be interested to hear from academic delegates about experience working with industry and the view from industry delegates about working with academics.

Akihiko Kudo replied: In Japan, many companies are interested in artificial photosynthesis. Some companies are carrying out the research by themselves and some are joining a national project.[1] It is crucial to demonstrate a solar water splitting system, not sacrificial H_2 evolution, using a photocatalyst under sunlight with a suitable scale, even if the efficiency (STH) is not satisfying. I am sure that the demonstration will attract much more attention from not only industry people but also politicians.

1 T. Setoyama, T. Takewaki, K. Domen and T. Tatsumi, *Faraday Discuss.*, 2017, **198**, 509–527.

Marcelino Maneiro responded: I completely agree with the importance of promoting collaboration between industry and academia. I think that this type of collaboration is crucial to furthering the transfer of technology to society, in fact I have been trying to do it for many years although with mixed results. My initial experience years ago was not very positive, but contacts in recent years have been much more satisfactory. To be honest, recent state aids to foster this type of collaboration helps without question. Networking events where we can meet people from industry are also extremely useful. Definitively, it is important to connect with the "right" person.

Wilson Smith responded: We have also had very positive experiences working with industry, and have several active and fruitful collaborations with Shell (among others) in the Netherlands. Shell has been a terrific partner by helping us understand issues related to upscaling, while also explicitly funding 'pre-competitive' research that is still far from applications. It is important to have some idea about where we want to take the science we are doing from a practical perspective, so we can learn how to understand and optimize materials and reactions under industrially relevant conditions, which can drastically speed up the timeline or technological development.

Flavia Cassiola asked: When you say leadership and direction, what else do you think we can collaborate more on?

In general, the Energy Industry has become very active in finding solutions towards Net Carbon Footprint. Shell has demonstrated leadership and our commitment to it. Our R&D programs are engaged with academia and we are providing guidance for several key initiatives. We are amazed by the enthusiastic and talented scientists we are collaborating with in our current programs. More and more attention has been paid to techno-economic challenges that each idea must confront and for the engineering aspects. Hopefully, we can continue to collaborate and improve our ways of working.

Flavia Cassiola remarked: Solar fuels include the power-to-X (X being liquids, gas or chemicals). It refers to the synthesis of fuels and chemicals from CO_2 and water with input of renewable electricity. It is very attractive because it uses the existing infrastructure. For fuels, it has high energy density, good stability and it is an excellent option to decarbonize the transport sector (aviation and non-road).

Photosynthetic systems are included in the concept but our efforts in this area revealed critical gaps in the properties of existing materials and the under-standing of synchronized processes occurring in integrated photosystems (light harvesting, atmospheric CO_2 capture and CO_2 conversion *via* (photo)catalysis). We understand the need of considering all options to develop a zero emission mobility and chemical industry. Shell has established key collaborations with academia aiming to work together with researchers to develop solar fuels.

Conflicts of interest

There are no conflicts to declare.

Faraday Discussions

PAPER

Artificial photosynthesis – concluding remarks†

C. Bozal-Ginesta and J. R. Durrant 🆔 *

Received 30th May 2019, Accepted 5th June 2019

DOI: 10.1039/c9fd00076c

This paper follows on from the Concluding Remarks presentation of the 3rd Faraday Discussion Meeting on Artificial Photosynthesis, Cambridge, UK, 25–27th March 2019. It aims to discuss the context for the research discussed at this meeting, starting with an overview of the motivation for research on artificial photosynthesis. It then goes onto analysing the composition and trends in the field of artificial photosynthesis, and its scale relative to other related research areas, primarily using the results of searches of publication databases. As such, we hope it provides helpful insights to researchers in the field.

1. Introduction

Artificial photosynthesis is a term used to describe most widely any man-mediated process which stores sunlight energy in useful, high energy chemicals.[1-4] One of its most studied applications is the production of fuels (so-called solar fuels) to replace fossil fuels and enable the carbon neutral exploitation of the energy subsequently freed to their combustion. Artificial photosynthesis draws inspiration from natural photosynthesis, which selectively and reversibly produces multiple sorts of chemicals under mild conditions, being the main source of energy-rich organic compounds on the planet.[5-7] Artificial photosynthesis is usually composed of three main steps: (1) fast light absorption leading to an excited state, (2) charge generation and charge separation through space, and, finally, (3) slow chemical conversion of substrates and charges into fuels or other chemicals.[8,9]

Since Fukushima and Honda first reported the splitting of water into hydrogen and oxygen with light in 1972,[10] many efforts have been devoted to this goal, but the most promising devices and materials still suffer from fast component degradation and/or large efficiency losses.[11-13] Progress in the field has been limited by scientific factors (i.e. the fundamental understanding of the charge generation and separation processes,[14-16] the available techniques,[17-21] the

Department of Chemistry, Molecular Sciences Research Hub, Imperial College London, 80 Wood Lane, London W12 0BZ, UK. E-mail: j.durrant@imperial.ac.uk

† Electronic supplementary information (ESI) available. See DOI: 10.1039/c9fd00076c

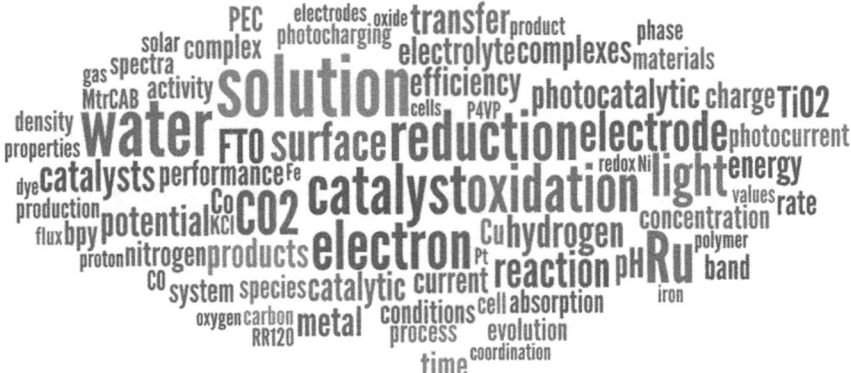

Fig. 1 Word cloud of the text from the scientific papers discussed in the Faraday meeting.[31]

development of better photoabsorbers and catalysts, *etc.*[22-26]) but also by other variables such as the severity of energy and climate crisis,[27,28] differing funding priorities and changing innovation strategies.[29,30] The main motivations to continue working on artificial photosynthesis, however, have remained intact.

In the 3rd Faraday Discussion on Artificial Photosynthesis in Cambridge, UK, different possible approaches to develop a functional and commercially viable solar-to-chemical energy conversion device were covered (Fig. 1). These approaches are based on different classifications of artificial photosynthesis research, which depend on the nature of the materials used (Section 4.2), the degree of system integration (Section 4.1), and the substrates and products involved (Section 4.3). In this article, we will firstly discuss the main motivations for artificial photosynthesis research. Secondly, the different trends in the field will be contextualised based on publication data.

2.   Motivation: the energy problem

Artificial photosynthesis is one strategy to address the growing energy demand worldwide and the dependence of our current energy system on fossil fuels,[32,33] complementary to the development of renewable sources of electrical power, sustainable bio-derived energy and increases in energy efficiency. It offers a potentially low carbon or carbon neutral pathway for the synthesis of sustainable fuels (*e.g.*: 'green' hydrogen from water). It also offers scalable pathways for carbon dioxide reduction and utilisation, as well as the potential for carbon neutral nitrogen reduction for fertiliser manufacture.

A key motivation for artificial photosynthesis concerns the geographical location of energy resources. Fossil fuel reserves are unequally spread geographically.[32,33] The wealthiest countries in the world are the largest energy consumers, and they are typically dependent on the large scale transportation of energy to them, primarily in the form of fossil fuels transported in marine tankers or gas pipes.[32,34,35] Sunlight is the largest available renewable energy resource (Fig. 2).[36] It is geographically relatively dispersed compared to fossil fuels, although also often most intense in regions with less energy demand. Given the

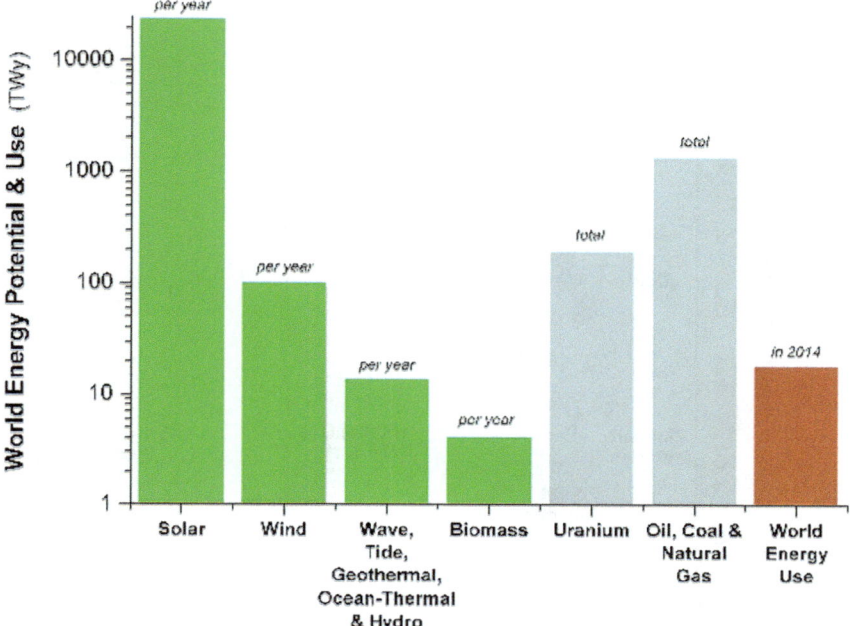

Fig. 2 World energy use and world energy potential of renewable sources yearly (in TW) and non-renewable sources (total recoverable reserves of uranium, oil, coal and natural gas in TWy).[32,35,36]

limitations of long distance electrical power distribution, the storage of solar energy in chemical fuels, and their transportation in tankers or gas pipes presents a scalable pathway to transport renewable energy globally.[37]

A further motivation for artificial photosynthesis results from the intermittency of most renewable electrical power sources. The availability of most renewable energies is variable, depending on the weather conditions, the time of the day, the month and the season. Since peak energy demands occur during the evening and in winter, and because energy-intensive industries and heavy-duty transport cannot rely on fluctuating and unpredictable energy sources, the large scale deployment of renewable energy systems increasingly requires that the energy generated can be stored efficiently and cheaply.[38,39] Electrical power can be stored in batteries and related technologies to address short term (up to days/ weeks) fluctuations in renewable electricity generation, but they lack scalability and their relatively high cost make them unsuitable for long term, and in particular interseasonal, storage.[40,41] Moreover, batteries have a limited volumetric and gravimetric energy density (Fig. 3) compared to chemical fuels.[42,43] This means that whilst batteries are rapidly becoming the energy storage medium of choice for cars, they are likely to remain too heavy to power effectively long-distance marine and aviation sectors. For both of these storage challenges, the efficient storage of renewable energy as chemical fuels (*i.e.* artificial photosynthesis) provides a potentially attractive low carbon solution.

A particular concern with many developing artificial photosynthesis strategies is that the best catalysts converting electrical potential into chemicals, both in

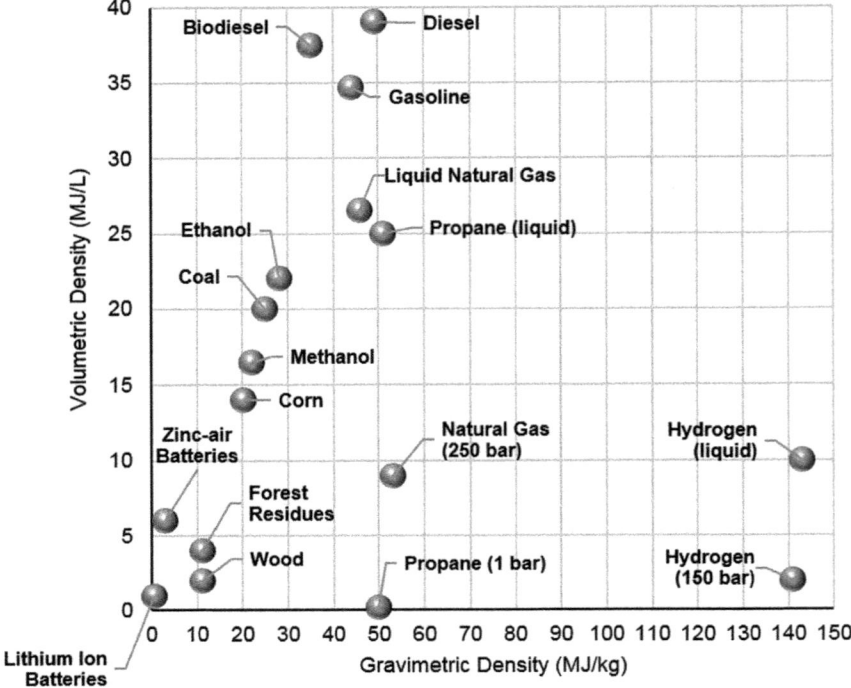

Fig. 3 Volumetric and gravimetric energy density of different fuels and batteries. Adapted from ref. 42 and 43.

electrolyzers and in many direct artificial photosynthesis systems such as photo-electrochemical devices, are based on metals belonging to the platinum group (*i.e.* ruthenium, rhodium, palladium, osmium, iridium and platinum). These metals are classified as critical raw materials by the European Union because of their high supply risk and economic importance, and therefore of limited suitability for large scale mass production.[44-46] This is a potentially limiting constraint for artificial photosynthesis and motivates fundamental research towards developing more sustainable components based on earth-abundant materials.[23,25,26,47,48]

3. Search criteria

The second part of this article is based on an analysis of the artificial photosynthesis literature. The results of any such search depends critically on the choice of search terms, and is therefore only partially objective. Many terms and expressions can be used to refer to processes inherent to artificial photosynthesis, with light-driven catalysis at their heart. Herein, publications are considered to address artificial photosynthesis if the title, abstract or keywords include words with the prefix *photo* followed by *reduction* or *oxidation,* or include the word *electrochemical* or words with the prefix *photo* combined with terms referring to water splitting and CO_2 reduction (see ESI† for the exact search criteria). Different ways to express the same concept have been considered in each one of the searches, including singular and plural terminologies, chemical formulas and the most common synonyms, and broadly yielded similar trends. This search is not perfectly complete, because publications do not always explicitly mention the technical definition of the materials and devices studied, and because we have omitted less common expressions and word combinations. In any case, we hope the selected groups of publications are reasonably representative and indicative of trends in the artificial photosynthesis field.

4. The artificial photosynthesis field

Fig. 4 illustrates the growth in publications in the field of artificial photosynthesis. Whilst the field was initiated in the 1970s with the pioneering work of Fujishima and Honda, it has only grown significantly since the early 1990s, motivated by increased concerns over oil prices and the environmental impact of fossil fuels, as well as scientifically by advances in our understanding of electron transfer in natural photosynthesis and liquid–semiconductor junctions.[34,50–53] Since then, it has grown rapidly, exhibiting an exponential increase in the number of publications (Fig. 4).[49] However, compared with other energy conversion technologies, artificial photosynthesis is still at an early development stage, with relatively few patents and companies compared with related fields such as batteries, electrolysers and solar cells (see Fig. S1†). This perhaps reflects the particularly high level of science challenge associated with the development of artificial photosynthesis technologies.

We turn now to consider the classification by research area of publications to date in the field of artificial photosynthesis.[49] Perhaps unsurprisingly, chemistry is the largest discipline, followed by materials science, engineering, physics, and biochemistry and molecular biology (Fig. 5). On the other hand, electrochemistry stands out as the key area of expertise followed by energy fuels.[54] From the geopolitical perspective, China, United States, the European Union and Japan are at the forefront in artificial photosynthesis research output, with most of these countries having implemented focused programs on this topic in the last 25 years (Fig. 6).[55–57]

4.1. Classification according to device type

The degree of integration of components performing one or more steps (*i.e.* light absorption, charge separation or catalysis) can be a criterion to distinguish

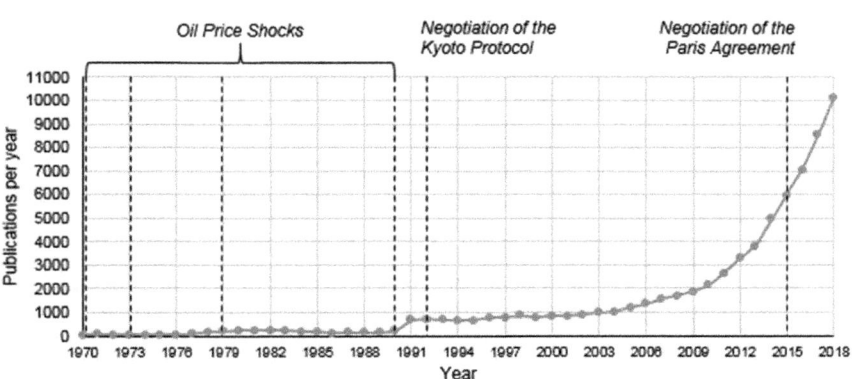

Fig. 4   Historical evolution of publications per year in artificial photosynthesis.[49]

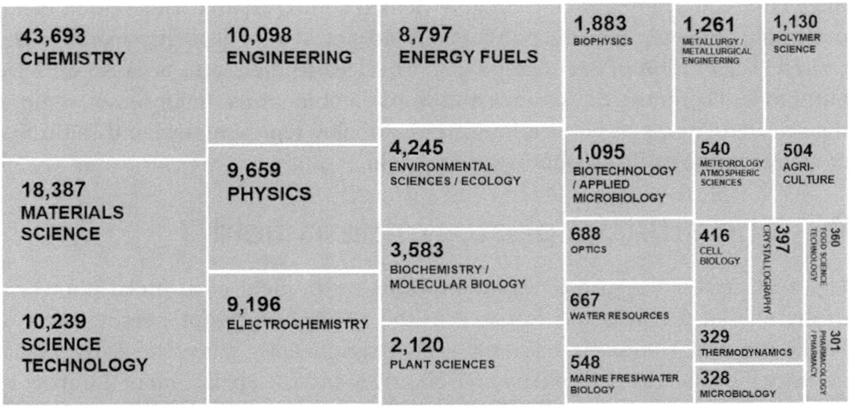

| 43,693 CHEMISTRY | 10,098 ENGINEERING | 8,797 ENERGY FUELS | 1,883 BIOPHYSICS | 1,261 METALLURGY / METALLURGICAL ENGINEERING | 1,130 POLYMER SCIENCE |

Fig. 5 Publications to date in artificial photosynthesis classified by research area (see ESI† for classification criteria).[49]

different research categories. Most simply, three architectures have been proposed for artificial photosynthetic devices.[8,58] The first comprises coupling photovoltaic panels with electrolyzers, which both comprise relatively mature technologies. The second architecture is based on photoelectrochemistry, where light absorbing materials are integrated with catalytically active materials into electrodes that could be used in unbiased light driven electrolyzers. The third architecture is based on photocatalysis, where light absorbing and catalytic materials are suspended in solution as homogeneous suspensions, most typically as particles. Other device architectures, such as photocatalyst sheets or photo-voltaic buried junctions, are attracting increasing interest, but have attracted relatively few publications to date.

Among the total of 74 000 articles published to date related to artificial photosynthesis, 6300 can be readily identified as focusing on devices, junctions and photoelectrodes, and 11 300 on catalysis (including photocatalysis) (see ESI† for search criteria). A much lower proportion can be identified as targeting

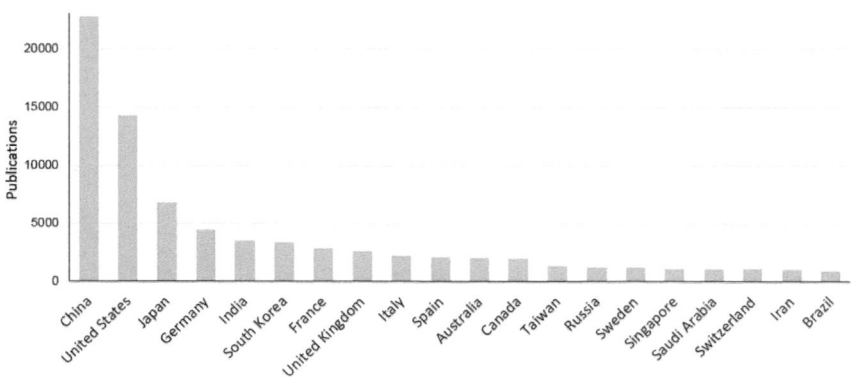

Fig. 6 Publications to date in artificial photosynthesis per country.[49]

photoabsorbers, probably because such publications are rather oriented towards solar cell research and photovoltaics. Fig. 7 represents the historical trends in the proportions of publications related to photocatalytic homogeneous suspensions and photoelectrodes; it is apparent that, since 2010, there has been increasing focus on photoelectrodes rather than photocatalytic suspensions. Relatively few publications can be identified as focusing on combined photovoltaics and electrolyzers, which are most likely to be investigated together primarily at an industrial level, with ~3500 patents on combined photovoltaic and electrolyser systems, significantly more than the number of publications on this approach.

4.2. Classification according to material

Leaving aside genetically modified photosynthetic organisms, the most widely investigated materials studied in the field of artificial photosynthesis as catalysts are earth-abundant metal oxides (e.g. TiO_2, CuO, $BiVO_4$), platinum group metals and metal oxides (e.g. platinum, RuO_2, IrO_2), enzymes, carbon nitrides, metal- and covalent organic frameworks (MOFs and COFs), molecular catalysts, poly-oxometalates, dichalcogenides, perovskites, gallium nitrides and light-absorbing, conjugated polymers (Fig. 8). The first are often regarded as robust and, together with dichalcogenides, MOFs, COFs, semiconducting polymers and carbon nitrides, are used as photocatalysts. POMs have been used both as electrocatalysts and photocatalysts. It is striking from Fig. 8 that carbon nitrides and, to a smaller extent, POMs and dichalcogenides have experienced a particular growth in research activity recently. Carbon nitrides have also been used as dyes, while semiconducting polymers and oxides have been used as charge separation layers. On the other hand, enzymes, which have been investigated in both natural and artificial photosynthesis systems, and molecular catalysts need to be coupled to

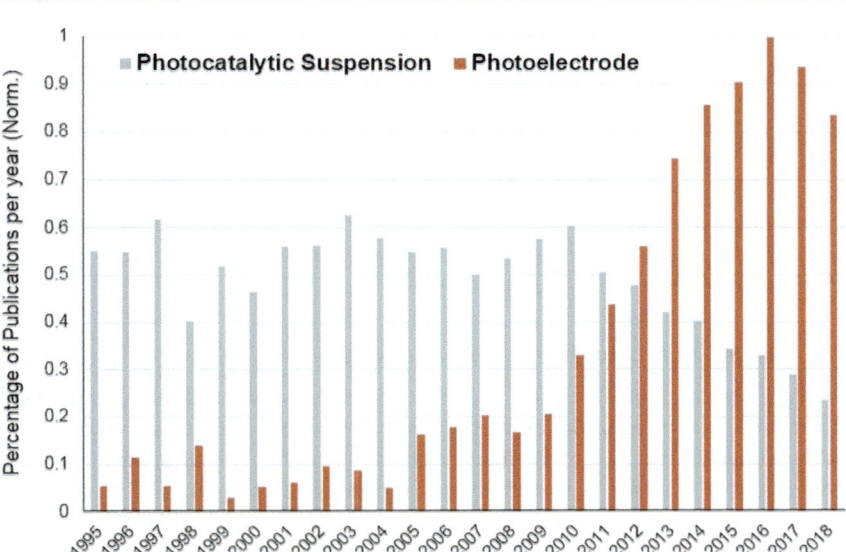

Fig. 7 Normalized percentage of the total amount of publications per year in artificial photosynthesis related to photoelectrodes or photocatalytic suspensions.[49]

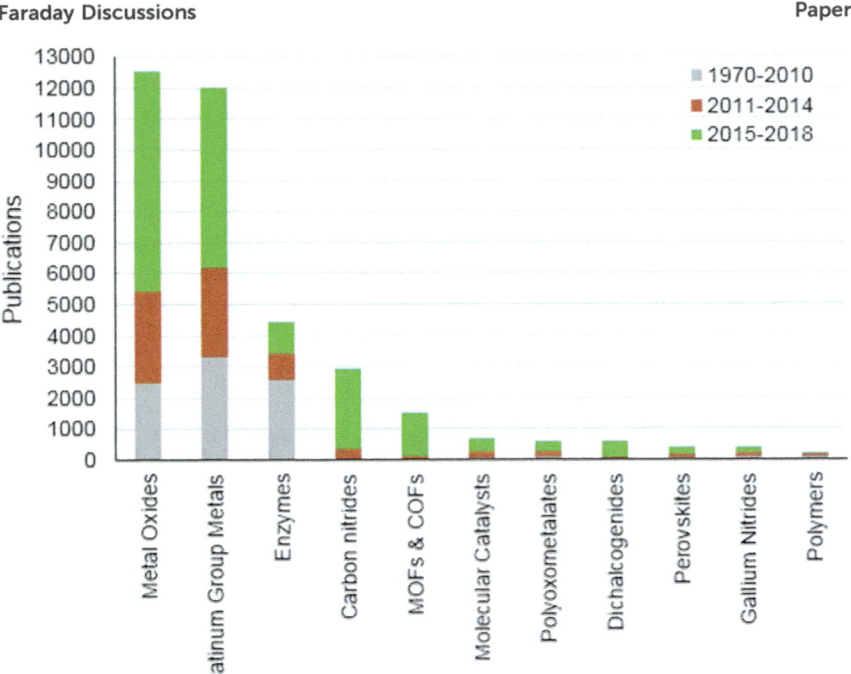

Fig. 8 Publications in artificial photosynthesis related to different materials.[49] We note that publications employing semiconducting silicon were not categorised separately due to overlap with its use as a dopant in metal oxides.

an electron source. Despite being faster, more selective and working at lower overpotentials, they are harder to obtain, typically less stable and more sensitive to the working conditions, which may account for the smaller representation in the field.

Silicon has been excluded because of its wide use as dopant. Furthermore, we have omitted platinum group metals (PGM) and perovskites from the metal oxide search, PGMs and metal oxides from the perovskite search and carbon nitrides from the polymer one to guarantee that catalytic materials are not repeated twice. As a consequence, papers including mutually excluded materials have been filtered out (papers with polymers + carbon nitrides or metal oxides + perovskites). Publications including dyes based on PGM and perovskites or metal oxides, are represented by the PGM category in Fig. 8.

4.3. Classification according to substrates or products

Finally, artificial photosynthesis research can also be sorted according to the type of substrate converted or the final product. Two categories represent ∼90% of the publications in the field. The dominant one is water splitting, involving the reduction of water into hydrogen and its oxidation into oxygen.[23,59,60] The second is CO_2 reduction to form carbon monoxide, formate, formaldehyde, methanol or methane.[61-63] Alternative chemical targets of interest for photo-electrochemical conversion include N_2 fixation into ammonia,[64-66] photo-degradation of organic pollutants[67] or the utilization of low-value industrial side products such as

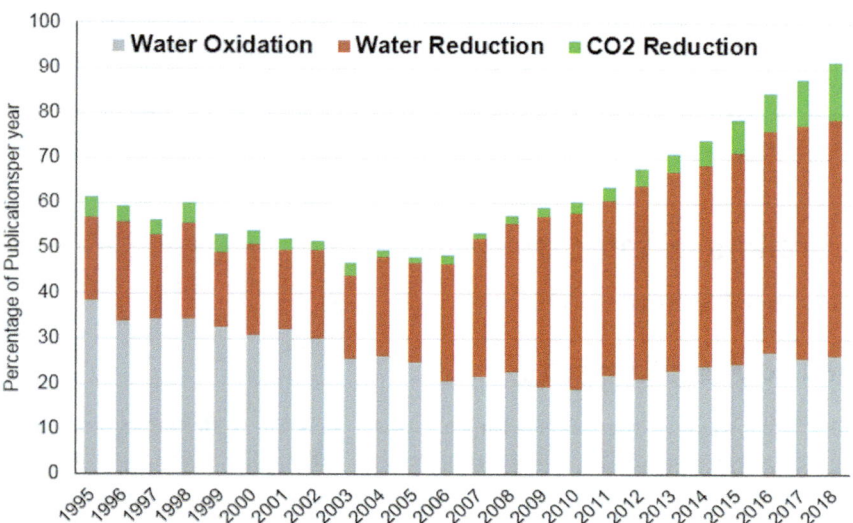

Fig. 9   Percentage of the total amount of publications per year in artificial photosynthesis and photocatalysis exclusively related to CO_2 reduction, water oxidation or water reduction.[49]

glycerol.[68] Fig. 9 shows the percentage of publications focused on water oxidation, water reduction or CO_2 reduction.

While 20 years ago a larger diversity of photocatalysed reactions was studied (including for example a range of organic oxidations), currently most published research in the field of artificial photosynthesis focuses on either CO_2 reduction, water reduction or water oxidation. The proportion of research dedicated to water reduction (or H_2 production) and CO_2 reduction has increased, while water oxidation (or O_2 evolution) has slightly decreased and plateaued at around 25%, although the total amount of publications has increased exponentially in all three sub-areas parallel to the general trend in the field. The evolution of water oxidation research may have been influenced by the progress in understanding Photosystem II in natural photosynthesis,[17,69,70] while research in hydrogen evolution has been favoured by the huge potential of hydrogen as a fuel.[71-74] Last, CO_2 reduction, which can be catalytically more challenging and usually less selective, has gained more attention recently, although still significantly less than that on proton reduction (note thermochemical CO_2 reduction was not included in this analysis).[75-77]

5.   Conclusions

The analyses of the artificial photosynthesis publication data to date reported herein is only limited in scope and the selection of suitable search words. However, even with these limitations, it provides some overview of the field of artificial photosynthesis. Some results of the analysis are particularly striking, such as the rapid growth of interest in carbon nitride photocatalysts, and the continued dominance of publications on proton reduction rather than carbon dioxide reduction, despite major new initiatives on the latter. There is increasing focus on photoelectrodes rather than photocatalyst suspensions, despite some cost projections indicating photocatalyst architectures (either as suspensions or sheets) having the greatest potential to be cost competitive with PV plus electrolysis.[78] The papers presented at the Faraday Discussion meeting addressed many aspects of current research in artificial photosynthesis. Hopefully this article helps to put these studies into context and provides a useful background to this field.

Conflicts of interest

There are no conflicts to declare.

Acknowledgements

CBG acknowledges the EPSRC for the award of DTP studentship, and JRD acknowledges the KAUST funded project OSR-2018-CRG7-3749.2 for financial support.

References

1 N. S. Lewis and D. G. Nocera, *Proc. Natl. Acad. Sci. U. S. A.*, 2006, **103**, 15729–15735.

2 W. Song, Z. Chen, M. K. Brennaman, J. J. Concepcion, A. O. T. Patrocinio, N. Y. Murakami Iha and T. J. Meyer, *Pure Appl. Chem.*, 2011, **83**, 749–768.

3 M. Grätzel, *Nature*, 2001, **414**, 338–344.

4 S. Styring, *Faraday Discuss.*, 2012, **155**, 357–376.

5 J. Barber, *Chem. Soc. Rev.*, 2009, **38**, 185–196.

6 I. McConnell, G. Li and G. W. Brudvig, *Chem. Biol.*, 2010, **17**, 434–447.

7 D. Gust and T. A. Moore, *Science*, 1989, **244**, 35–41.

8 A. J. Cowan and J. R. Durrant, *Chem. Soc. Rev.*, 2013, **42**, 2281–2293.

9 L. Hammarstrom, *Acc. Chem. Res.*, 2015, **48**, 840–850.

10 A. Fukushima and K. Honda, *Nature*, 1972, **238**, 37–38.

11 J. W. Ager, M. R. Shaner, K. A. Walczak, I. D. Sharp and S. Ardo, *Energy Environ. Sci.*, 2015, **8**, 2811–2824.

12 S. Tembhurne, F. Nandjou and S. Haussener, *Nat. Energy*, 2019, **4**, 399–407.

13 O. Khaselev, A. Bansal and J. A. Turner, *Int. J. Hydrogen Energy*, 2001, **26**, 127–132.

14 M. R. Wasielewski, *Acc. Chem. Res.*, 2009, **42**, 1910–1921.

15 T. Hirakawa and P. V. Kamat, *J. Am. Chem. Soc.*, 2005, **127**, 3928–3934.

16 Y. Tachibana, J. E. Moser, M. Graetzel and J. R. Durrant, *J. Phys. Chem.*, 1996, **100**, 20056–20062.

17 M. Suga, F. Akita, K. Hirata, G. Ueno, H. Murakami, Y. Nakajima, T. Shimizu, K. Yamashita, H. Yamamoto, H. Ago and J. R. Shen, *Nature*, 2015, **517**, 99–103.

18 M. H. Zewail, *J. Phys. Chem. A*, 2000, **104**, 5660–5694.

19 A. J. Bard, F. R. F. Fan, D. T. Pierce, P. R. Unwin, D. O. Wipf and F. Zhou, *Science*, 1991, **254**, 68–74.

20 F. F. Abdi, T. J. Savenije, M. M. May, B. Dam and R. van de Krol, *J. Phys. Chem. Lett.*, 2013, **4**, 2752–2757.

21 Z. W. Seh, J. Kibsgaard, C. F. Dickens, I. Chorkendorff, J. K. Norskov and T. F. Jaramillo, *Science*, 2017, **355**, eaad4998.

22 T. Hisatomi, J. Kubota and K. Domen, *Chem. Soc. Rev.*, 2014, **43**, 7520–7535.

23 I. Roger, M. A. Shipman and M. D. Symes, *Nat. Rev. Chem.*, 2017, **1**, 0003.

24 D. L. DuBois, *Inorg. Chem.*, 2014, **53**, 3935–3960.

25 K. Sivula and R. van de Krol, *Nat. Rev. Mater.*, 2016, **1**, 15010.

26 L. Steier and S. Holliday, *J. Mater. Chem. A*, 2018, **6**, 21809.

27 J. Goldemberg, T. B. Johansson, A. K. N. Reddy and R. H. Williams, *Energy for a Sustainable World*, World Resources Institute, 1987.

28 Climate Change 2014, *Synthesis Report, The Intergovernmental Panel on Climate Change*, 2015.

29 Executive Summary, *Commission Staff Working Paper Executive Summary of the Impact Assessment*, European Commission, 2011.

30 Executive Summary, *Quadrennial Technology Review. An Assessment of Energy Technologies and Research Opportunities*, Department of Energy of the United States of America, 2015.

31 wordle.net.

32 http://energyatlas.iea.org, IEA, April 2019.

33 International Energy Agency Publications, *Key World Energy Statistics*, IEA, 2017, https://www.iea.org/publications/freepublications/publication/KeyWorld2017.pdf.

34 J. D. Colgan and J. B. Stockbruegger, *The Oxford Handbook of Energy Politics*, 2018.

35 gapminder.org, April 2019.

36 R. Perez and M. Perez, *Newsletter of the International Energy Agency Solar Heating and Cooling*, 2015, **62**, 4–6.

37 R. Schlögl, De Gruyter, 2013.

38 Committee on Climate Change, *Net Zero Technical report*, Committee on Climate Change, UK, 2019, https://www.theccc.org.uk/publication/net-zero-the-uks-contribution-to-stopping-global-warming/.

39 Federal Ministry for Economic Affairs and Energy, *Net Zero Technical report*, Bundesministerium für Wirtschaft und Energie, Germany, 2017, https://www.bmwi.de/Redaktion/DE/Publikationen/Energie/erneuerbare-energien-in-zahlen-2017.html.

40 T. Napp, H. Hills, M. S. Soltani, J. Bosch and C. Mazur, *18 priority next-generation climate mitigation technologies*, Grantham Institute.

41 B. V. Mathiesen and H. Lund, *IET Renewable Power Generation*, 2009, vol. 3, p. 190.

42 T. Mays and D. Book, *H2FC Supergen Annual Conference*, 2019.

43 Engineering ToolBox, Fossil and Alternative Fuels - Energy Content, 2008, Available at: https://www.engineeringtoolbox.com/fossil-fuels-energy-content-d_1298.html, April 2010.

44 Deloitte Sustainability, British Geological Survey, Bureau de Recherches Géologiques et Minières, Netherlands Organisation for Applied Scientific Research, European Comission Directorate-General for Internal Market, Industry, Entrepreneurship and SMEs, *Study on the review of the list of Critical Raw Materials*, Criticality Assessments, European Comission, 2017, Raw Materials, https://publications.europa.eu/en/publication-detail/-/publication/08fdab5f-9766-11e7-b92d-01aa75ed71a1/language-en.

45 European Comission Directorate-General for Internal Market, Industry, Entrepreneurship and SMEs, *Study on the review of the list of Critical Raw*

Materials, Non-critical Raw Materials Factsheets, European Comission, 2017, Raw Materials, https://publications.europa.eu/en/publication-detail/-/publication/7345e3e8-98fc-11e7-b92d-01aa75ed71a1.

46 APS panel on public affairs & the Materials Research Society, *Energy Critical Elements: Securing Materials for Emerging Technologies*, American Physical Society and Materials Research Society.

47 R. S. Sprick, B. Bonillo, R. Clowes, P. Guiglion, N. J. Brownbill, B. J. Slater, F. Blanc, M. A. Zwijnenburg, D. J. Adams and A. I. Cooper, *Angew. Chem., Int. Ed.*, 2016, **55**, 1792–1796.

48 T. E. Rosser, M. A. Gross, Y.-H. Lai and E. Reisner, *Chem. Sci.*, 2016, **7**, 4024–4035.

49 https://login.webofknowledge.com/, *Web of Science*, April 2019.

50 R. A. Marcus and N. Norman Sutin, *Biochim. Biophys. Acta, Rev. Bioenerg.*, 1985, **811**, 265–322.

51 G. McLendon, *Acc. Chem. Res.*, 1988, **21**, 160–167.

52 A. J. Bard, A. B. Bocarsly, F. R. F. Fan, E. G. Walton and M. S. Wrighton, *J. Am. Chem. Soc.*, 1980, **102**, 3671–3677.

53 H. Gerischer and W. Ekardt, *Appl. Phys. Lett.*, 1983, **43**, 393–395.

54 A. J. Bard and L. R. Faulkner, *Electrochemical Methods: Fundamentals and Applications*, John Wiley & Sons, Inc., 2nd edn, 2001.

55 *Joint Center for Artificial Photosynthesis*, https://solarfuelshub.org/.

56 *UK Solar Fuels Network*, https://www.solarfuelsnetwork.com/.

57 *Swedish Consortium for Artificial photosynthesis*, http://www.solarfuel.se/.

58 J. R. McKone, N. S. Lewis and H. B. Gray, *Chem. Mater.*, 2014, **26**, 407–414.

59 C. C. McCrory, S. Jung, I. M. Ferrer, S. M. Chatman, J. C. Peters and T. F. Jaramillo, *J. Am. Chem. Soc.*, 2015, **137**, 4347–4357.

60 M. G. Walter, E. L. Warren, J. R. McKone, S. W. Boettcher, Q. Mi, E. A. Santori and N. S. Lewis, *Chem. Rev.*, 2010, **110**, 6446–6473.

61 T. Inoue, A. Fujishima, S. Konishi and K. Honda, *Nature*, 1979, **277**, 637–638.

62 F. Studt, I. Sharafutdinov, F. Abild-Pedersen, C. F. Elkjaer, J. S. Hummelshoj, S. Dahl, I. Chorkendorff and J. K. Norskov, *Nat. Chem.*, 2014, **6**, 320–324.

63 G. O. Larrazabal, A. J. Martin and J. Perez-Ramirez, *J. Phys. Chem. Lett.*, 2017, **8**, 3933–3944.

64 J. Yu, Y. Zhang and A. Kudo, *J. Solid State Chem.*, 2009, **182**, 223–228.

65 X. Chen, N. Li, Z. Kong, W.-J. Ong and X. Zhao, *Mater. Horiz.*, 2018, **5**, 9–27.

66 H. Wang, Y. Su, H. Zhao, H. Yu, S. Chen, Y. Zhang and X. Quan, *Environ. Sci. Technol.*, 2014, **48**, 11984–11990.

67 D. Bahnemann, *Sol. Energy*, 2004, **77**, 445–459.

68 V. M. Daskalaki and D. I. Kondarides, *Catal. Today*, 2009, **144**, 75–80.

69 I. D. Young, M. Ibrahim, R. Chatterjee, S. Gul, F. Fuller, S. Koroidov, A. S. Brewster, R. Tran, R. Alonso-Mori, T. Kroll, T. Michels-Clark, H. Laksmono, R. G. Sierra, C. A. Stan, R. Hussein, M. Zhang, L. Douthit, M. Kubin, C. de Lichtenberg, P. Long Vo, H. Nilsson, M. H. Cheah, D. Shevela, C. Saracini, M. A. Bean, I. Seuffert, D. Sokaras, T. C. Weng, E. Pastor, C. Weninger, T. Fransson, L. Lassalle, P. Brauer, P. Aller, P. T. Docker, B. Andi, A. M. Orville, J. M. Glownia, S. Nelson, M. Sikorski, D. Zhu, M. S. Hunter, T. J. Lane, A. Aquila, J. E. Koglin, J. Robinson, M. Liang, S. Boutet, A. Y. Lyubimov, M. Uervirojnangkoorn, N. W. Moriarty, D. Liebschner, P. V. Afonine, D. G. Waterman, G. Evans, P. Wernet,

H. Dobbek, W. I. Weis, A. T. Brunger, P. H. Zwart, P. D. Adams, A. Zouni, J. Messinger, U. Bergmann, N. K. Sauter, J. Kern, V. K. Yachandra and J. Yano, *Nature*, 2016, **540**, 453–457.

70 L. Vogt, D. J. Vinyard, S. Khan and G. W. Brudvig, *Curr. Opin. Chem. Biol.*, 2015, **25**, 152–158.

71 Case Study Report, *Hydrogen Society (Japan), European Comission, Mission-oriented R&I policies: In-depth case studies*, 2018, ISBN 978-92-79-80160-0, DOI: 10.2777/516513.

72 R. M. Pratt, E. W. Freeburg and F. Luzi, DOE, 2017.

73 B. D. James, D. A. DeSantis and G. Saur, *DOE Strategy Analysis*, 2016.

74 I. Staffell, P. Dodds, D. Scamman, A. Velazquez Abad, N. K. Ward, P. Agnolucci, L. Papageorgiou, N. Shah and P. Ekins, *H2FC Supergen*, 2017.

75 Y. Hori, H. Wakebe, T. Tsukamoto and O. Koga, *Electrochim. Acta*, 1994, **39**, 1833–1839.

76 C. D. Windle and R. N. Perutz, *Coord. Chem. Rev.*, 2012, **256**, 2562–2570.

77 W. Tu, Y. Zhou and Z. Zou, *Adv. Mater.*, 2014, **26**, 4607–4626.

78 B. D. James, G. N. Baum, J. Perez and K. N. Baum, *Directed Technologies*, DOE, 2009.

Poster titles

Sonochemical modification of WO_3 and CdS semiconductor nanoparticles for enhanced photocatalytic activity, **D. W. F. Cheung, A. J. Cowan and D. Shchukin,** *The University of Liverpool, UK*

Mechanistic insights for radical processes in artificial photosynthesis and photoredox reactions, **Y. Y. Ng, Z. Hong, L. K.-S. Ng, X. L. Ho, M. Dokic, K. F. Chin and H. S. Soo,** *Nanyang Technological University, Singapore*

Hydrophobicity promotes selective CO_2 reduction on Cu surfaces, **D. Wakerley, S. Lamaison, F. Ozanam, N. Menguy, D. Mercier, P. Marcus, V. Mougel and M. Fontecave,** *Collège de France, France*

Tandem Cu_2O-covered Si microwires photocathodes for solar-to-fuel devices, **P. Kunturu, W. Vijselaar and J. Huskens,** *University of Twente, Enschede, Netherlands*

Proteoliposomes as a modular and controllable platform for efficient bio-hybrid light-harvesting, **A. M. Hancock, S. Meredith, S. D. Connell, L. J. C. Jeuken and P. G. Adams,** *University of Leeds, UK*

Eco-friendly novel complexes of Ni(II) for hydrogen production using solar energy, **F. I. Kamatsos, M. Drosou and C. A. Mitsopoulou,** *National and Kapodistrian University of Athens, Greece*

Theoretical studies on the photocatalytic hydrogen production mechanism by molecular copper complexes, **M. Drosou, G. Ioannidis, D. Tzeli and C. Mitsopoulou,** *University of Athens, Greece*

Frustrated Lewis pairs for CO_2 reduction: an electrochemical and NMR study, **S. Clausing, A. Orthaber, J. Green and J. Wells,** *Uppsala University, Sweden*

Photocatalytic methanol reforming in monolithic titania nanoparticle-based aerogels, **M. Schreck, M. Bertsch, N. Kleger and M. Niederberger,** *ETH Zürich, Switzerland*

Charge accumulation studies in molecular catalysts for solar fuels generation, **C. Bozal-Ginesta, L. Francàs, C. A. Mesa, A. Reynal, D. Antón, A. Eisenschmidt, E. Reisner and J. Durrant,** *Imperial College London, UK*

Optimizing mesostructured silver catalysts for selective carbon dioxide reduction, **S. Suter and S. Haussener,** *EPFL, Switzerland*

Porous boron nitride for combined CO_2 capture and photoreduction, **R. Shankar, M. Sachs, L. Francàs, D. Lubert-Perquel, G. Kerherve, A. Regoutz and C. Petit,** *Imperial College London, UK*

Bifunctional mechanism on multiple active sites and the crucial role of chemical steps in water splitting, **H. C. Nguyën, F. A. Garcés-Pineda, M. de Fez Febré, J. R. Galán-Mascarós and N. López,** *ICIQ, Spain*

Effects of pressure assisting treatment for photoelectrodes consisting of particulate BiXY(X=O,S, Y=Br,I), **N. Nishimura, M. Higashi and R. Abe,** *Asahi Kasei Corporation, UK*

Efficient light-driven hydrogen evolution from water by a quantum dot - triiron molecular catalyst assembly, **C. Li, A. Rahaman, W. Lin, H. Mourad, J. Meng, A. Honarfar, M. Abdellah, K. Zheng and E. Nordlander,** *Lund University, Sweden*

Zn-Cu alloy nanofoams as efficient catalysts for CO_2 reduction to syngas mixtures with potential-independent H_2:CO ratio, **S. Lamaison, D. Wakerley, T. N. Huan, M. Fontecave and V. Mougel,** *Collège de France, France*

Electrocatalytic hydrogen evolution with cobalt–poly(4-vinylpyridine)metallopolymers, **Z. Kap, E. Ülker, S. V. K. Nune and F. Karadas,** *Bilkent University, Turkey*

Solar hydrogen production with trimetallic PNP-complexes, **H. Roithmeyer, J. Pann, W. Viertl, R. Pehn, J. Dutzler, B. Trübenbacher, M. Bendig, B. Krieche and P. Brüggeller,** *Leopold Franzens Universität Innsbruck, Austria*

Patterned Ni-Mo on Si micropillars for unprecedented H_2 production, **W. Vijselaar, P. Westerik, J. Veerbeek, R. Tiggelaar, H. Gardeniers and J. Huskens,** *University of Twente, Netherlands*

Electrochemical capture and release of CO_2 using anthraquinone, **D. Wielend, D. H. Apaydin, D. R. Whang and N. S. Sariciftci,** *Johannes Kepler University Linz, Austria*

Pincer or no pincer, that's the question! – Chemical behaviour and efficiency enhancement of proton relays for hydrogen production in artificial photosynthesis, **W. Viertl, J. Pann, R. Pehn, H. Roithmeyer, M. Bendig, T. Hofer and P. Brüggeller,** *University of Innsbruck, Austria*

Redox mediated water splitting in dye sensitized photo electrochemical cells, **D. F. Bruggeman, and J. N. H. Reek,** *University of Amsterdam, Netherlands*

Developing biohybrid nanoreactors for light-driven reductive chemistry, **S. E. H. Piper, J. van Wonderen, M. J. Soriano-Laguna, A. Stikane, A. J. Gates, L. J. C. Jeuken, E. Reisner and J. Butt,** *University of East Anglia, UK*

The ring launching solar cell: The RING matters, **T. Bouwens, J. Hasenack, S. Mathew and J. N. H. Reek,** *University of Amsterdam, Netherlands*

Towards rational design of Ru-labelled multi-heme cytochromes as photo-transducers of light to electricity, **J. van Wonderen, C. Hall, X. Jiang, K. Adamczyk, D. Li, S. Piper, C. Lau, A. Carof, I. Heisler, L. Jenner, T. Clarke, N. Watmough, I. Sazanovich, M. Towrie, S. Meech, J. Blumberger and J. Butt,** *University of East Anglia, UK*

Accelerated discovery of organic polymer photocatalysts for hydrogen evolution from water through the integration of experiment and theory, **Y. Bai, L. Wilbraham, B. J. Slater, M. A. Zwijnenburg, R. S. Sprick and A. I. Cooper,** *University of Liverpool, UK*

Visible light driven Z-scheme water splitting with transition metal substituted polyoxometalates as shuttle redox mediators, **O. Tomita, H. Naito, Y. Iwase, K. Tsuji, A. Nakada, M. Higashi and R. Abe,** *Kyoto University, Japan*

3D printed microarray electrodes for photoelectrochemical water splitting, **C.-Y. Lee, A. Taylor, S. Beirne and G. G. Wallace,** *University of Wollongong, Australia*

Emulsion polymerisation derived organic photocatalysts for improved light driven hydrogen evolution, **C. M. Aitchison, R. S. Sprick and A. I. Cooper,** *University of Liverpool, UK*

Development of dye-sensitized photoanodes for water oxidation incorporating hetero-trifunctional Prussian blue analogues, **T. Gamze Ulusoy Ghobadi, A. Ghobadi, K. N. Ozvural, E. Ozbay and F. Karadas,** *Bilkent University, Turkey*

Carbon dots derived from biomass as interface power hematite photoanodes for solar water splitting, **Q. Guo, H. Luo, M.-M.Titirici and A. B. J. Sobrido,** *QMUL, UK*

Envisioning photosynthetic systems for commercial production of fuels and valuable chemicals: from H_2 production to CO_2 utilization, **F. Cassiola,** *Shell Technology Center Houston, USA*

Utilising excited state organic anions for photoredox catalysis: Activation of (hetero)aryl chlorides by visible light-absorbing 9-anthrolate anions, **M. Schmalzbauer, I. Ghosh and B. König,** *University of Regensburg, Germany*

Versatile PNP based catalytic systems for artificial photosynthesis, **J. Pann, R. Pehn, W. Viertl, H. Roithmeyer, M. Bendig and P. Brüggeller,** *University of Innsbruck, Austria*

A kinetic barrier to enable semiconductor-free biophotovoltaics, **H. Zhang, D. Buesen, A. Ruff, V. M. Friebe, J. Koc, F. Conzuelo, W. Schuhmann, M. R. Jones, R. N. Frese and N. Plumeré,** *Ruhr-University Bochum, Germany*

Light-induced formation of a surface hetero-junction in photocharged metal-oxide photoanodes, **A. Venugopal and W. Smith,** *Delft University of Technology, Netherlands*

Long-lived excited states of copper photosensitizers containing an extended π-system, **R. Giereth, M. Karnahl and S. Tschierlei,** *Ulm University, Germany*

Leveraging excited-state quantum dynamics to control energy transfer for new photochemistry, **M. A. Allodi, J. P. Otto, S. H. Sohail, L. T. Lloyd, J. Higgins, R. G. Saer, R. E. Wood, S. C. Massey, R. E. Blankenship and G. S. Engel,** *University of Chicago, USA*

Flux synthesis of Bi-based layered oxyhalide photocatalyst for efficient Z-Scheme water splitting under visible light, **K. Ogawa, A. Nakada, O. Tomita, M. Higashi, A. Saeki and R. Abe,** *Kyoto University, Japan*

Enhanced visible-light-driven H_2 evolution over $ZnIn_2S_4$ photocatalyst by surface modification with metal cyanoferrates, **H. Matsuoka, M. Higashi, A. Nakada, O. Tomita and R. Abe,** *Kyoto University, Japan*

Integration of biocatalytic machinery with synthetic electrodes for semi-artificial photosynthesis, **X. Fang, K. Solko, N. Heidary, S. Kalathil and E. Reisner,** *University of Cambridge, UK*

SUNRISE: Solar energy for a circular economy, **L. Hammarström, E. Pastor-Hernandez, J. Durrant and H. de Groot,** *Uppsala University, Sweden*

Photoreduction in optofluidic hollow-core photonic crystal fibre, **P. Koehler, T. Lawson, J. Neises, J. Willkomm, D. Antón García, M. H. Frosz, P. St. J. Russell, E. Reisner and T. Euser,** *University of Cambridge, UK*

Interfacing formate dehydrogenase with metal oxides for reversible electrocatalysis and solar-driven reduction of carbon dioxide, **M. Miller, W. E. Robinson, A. R. Oliveira, N. Heidary, N. Kornienko, J. Warnan, I. A. C. Pereira and E. Reisner,** *University of Cambridge, UK*

Bias-free photoelectrochemical water splitting with photosystem II on a dye-sensitized photoanode wired to hydrogenase, **K. P. Sokol, W. E. Robinson, J. Warnan, N. Kornienko, M. M. Nowaczyk, A. Ruff, J. Z. Zhang and E. Reisner,** *University of Cambridge, UK*

Photoelectrochemical H_2 evolution with a perovskite integrated hydrogenase biocatalyst, **E. Edwardes Moore, V. Andrei, S. Zacarias, I. A. C. Pereira and E. Reisner,** *University of Cambridge, UK*

Photoelectrochemistry of cyanobacterial thylakoid membranes, **J. M. L. T. Wey, X. Chen, C. J. Howe and J. Z. Zhang,** *University of Cambridge, UK*

Iron-doped nickel oxyhydroxide borate modified $BiVO_4$ for selectivity-tunable photoelectrocatalysis of methanol at near-neutral pH, **C.-C. Cheng, Y.-C. Chueh and C.Y. Lin,** *National Cheng Kung University, Chinese Taipei*

Dimer-of-dimers model as a precursor of a highly active catalyst for water photolysis, **M. Maneiro, M. Isabel Fernández-García, L. Rouco, A. M. González-Noya and R. M. Pedrido,** *Universidade de Santiago de Compostela, Spain*

Efficient solar spectrum photosynthetic/catalytic conversion of CO_2 to CH_4 by Cu-sensitized reduced titania, **S. Ali and S.-I. In,** *DGIST, South Korea*

Solar hydrogen generation from ambient humidity: Construction of a scalable device, **G. Zafeiropoulos, H. Johnson, S. Kinge, M. C. M. van de Sanden and M. N. Tsampas,** *Toyota Motor Europe, Netherlands*

Conjugated polymers for photocatalytic hydrogen evolution – Porous or non-porous, **R. S. Sprick, Y. Bai, A. A. Y. Guilbert, M. Zbiri, C. M. Aitchison, L. Wilbraham, Y. Yan, D. J. Woods, M. A. Zwijnenburg and A. I. Cooper,** *University of Liverpool, UK*

Charge transfer in co-crystals of cationic iridium(III) complexes with polyiodides: toward a better understanding of dye-iodine interactions in solar cells, **P. Kalle, S. I. Bezzubov and D. S. Yufit,** *Kurnakov Institute of General and Inorganic Chemistry of the Russian Academy of Sciences, Russian Federation*

Platinum doped carbon dots and its hybridization with TiO_2 for enhanced visible light photocatalytic hydrogen evolution, **H. Luo, A. Jorge Sobrido and M. Titirici,** *Queen Mary University of London, UK*

An artificial photosynthesis approach to the reduction of C-C multiple bonds, **N. Larionova, J. M. Ondozabal and X. C. Cambeiro,** *Queen Mary University of London, UK*

Rational design of polymers for supported fuel catalysis, **J. Warnan, J. J. Leung, J. A. Vigil, B. Reuillard, E. Edwardes Moore and E. Reisner,** *University of Cambridge, UK*

A palladium membrane reactor for organic transformations, **R. Sherbo, A. Kurimoto, R. Delima and C. P. Berlinguette,** *University of British Columbia, Canada*

Enhancement of photocatalytic behaviour of TiO_2 : incorporating and optimizing bimetallic nanostructures, **B. D. Bhuskute, H. Ali-Löytty, C. S. Gopinath and M. Valden,** *Tampere University, Finland*

Photocatalytic water-splitting with carbon nitride materials: The molecular mechanism, **J. Ehrmaier, D. Opalka, A. L. Sobolewski and W. Domcke,** *Technical University of Munich, Germany*

Isolation and characterization of an elusive η2-$[Ru^{IV}\text{-}OO]^{2+}$ intermediate after the O-O bond formation in Ru catalyzed WO: a missing link, **C. Casadevall, V. Martin-Diaconescu, W. R. Browne, F. Franco, N. Cabello, J. Benet-Buchholz and J. Lloret-Fillol,** *ICIQ, Spain*

Flavin conjugated polydopamine: from small molecule photocatalysis towards photoelectrocatalysis applications, **L. Crocker, P. Koehler, A. Kerbs, T. Euser and L. Fruk,** *University of Cambridge, UK*

Identification of the key mechanistic steps involved in the cobalt-catalyzed reduction of CO_2 to CO, **S. Fernández, F. Franco, C. Casadevall, V. Martin-Diaconescu, J. M. Luis and J. Lloret-Fillol,** *Institute of Chemical Research of Catalonia (ICIQ), Spain*

CO_2 photoreduction to CO using Ag-ZrO_2, **H. Zhang, T. Itoi, T. Konishi and Y. Izumi,** *Chiba University, Japan*

Direct observation of polaron formation in Fe_2O_3 PECs, **E. Pastor, J.-S. Park, L. Steier, S. Kim, M. Graetzel, J. R. Durrant, A. Walsh and A. A. Bakulin,** *Imperial College London, UK*

Photocatalytically active ladder polymers, **A. Vogel, M. Foster, L. Wilbraham, C. L. Smith, A. Cowan, M. A. Zwijnenburg, R. S. Sprick, and A. I. Cooper,** *University of Liverpool, UK*

Deuteration of alkynes using a palladium membrane reactor, **A. Kurimoto, R. S. Sherbo, Y. Cao and C. P. Berlinguette,** *University of British Columbia, Canada*

Hybrid semiconductor nanomaterials for photocatalytic organic transformations, **A. Agosti, Y. Nakibli, M. Natali, L. Amirav and G. Bergamini,** *University of Bologna, Italy*

Photocatalytic CO_2-to-CO conversion by a copper(II) quaterpyridine complex, **Z. Guo, F. Yu, Y. Yang, C.-F. Leung, S.-M. Ng, C.-C. Ko, C. Cometto, T.-C. Lau and M. Robert,** *The Education University of Hong Kong, Hong Kong*

Modification of ZnSe quantum dots with a Ni(cyclam) catalyst enables efficient visible-light driven CO_2 reduction in water, **C. D. Sahm, M. F. Kuehnel, G. Neri, J. R. Lee, K. L. Orchard, A. J. Cowan and E. Reisner,** *University of Cambridge, UK*

Tunable syngas production using photoelectrochemical tandem devices, **V. Andrei, B. Reuillard and E. Reisner,** *University of Cambridge, UK*

Next generation $CuCrO_2$ photocathodes for dye-sensitised H_2 production, **C. E. Creissen, J. Warnan, D. Antón-García, F. Odobel and E. Reisner,** *University of Cambridge, UK*

Continuous flow photocatalysis using a nanoengineered carbon nitride material, **M. Bajada, A. Vijeta, A. Savateev, Y. Zhao, M. Antonietti and E. Reisner,** *University of Cambridge, UK*

Biomass photoreforming into H_2 with carbon nitride under benign conditions, **D. S. Achilleos, H. Kasap, A. Huang and E. Reisner,** *University of Cambridge, UK*

Molecular surface engineering for electrocatalytic CO_2 reduction, **A. Wagner, H. K. Ly, N. Heidary, I. Szabó, T. Földes, I. K. Assaf, F. M. Kuehnel, N. Kornienko, J. S. Barrow, K. Sokolowski, M. W. Nau, E. Rosta, I. Zebger, O. Scherman and E. Reisner,** *University of Cambridge, UK*

Hybrid multifunctional materials for solar fuels production by artificial photosynthesis, **A. García-Sánchez, P. Reñones, C. García, E. Alfonso, L. Collado, R. Perez Ruiz, M. Barawi, I. J. Villar-García, M. Liras, F. Fresno and V. A. de la Peña O'Shea,** *IMDEA Energy Institute, Madrid*

Plastic waste as a feedstock for solar-driven H_2 generation, **T. Uekert, M. F. Kuehnel, D. W. Wakerley and E. Reisner,** *University of Cambridge, UK*

Hacking photosynthesis: A high stability pigment–protein biophotocathode, **V. M. Friebe, A. Jóźwiak, M. R. Jones and R. N. Frese,** *Vrije University, Amsterdam*

The Faraday Discussions Poster Prize for the best poster was jointly awarded to Tessel Bouwens of the University of Amsterdam, Netherlands, for her poster on the ring launching solar cell, and Joshua Lawrence of the University of Cambridge, UK, for his poster on photoelectrochemistry of cyanobacterial thylakoid membranes.

List of participants

Professor Ryu Abe, *Kyoto University, Japan*
Dr Demetra Achilleos, *University of Cambridge, United Kingdom*
Mr Amedeo Agosti, *University of Bologna, Italy*
Miss Catherine M. Aitchison, *University of Liverpool, United Kingdom*
Dr Marco A. Allodi, *University of Chicago, USA*
Mr Virgil Andrei, *University of Cambridge, United Kingdom*
Mr Daniel Antón-García, *University of Cambridge, United Kingdom*
Dr Ulf-Peter Apfel, *Ruhr-University Bochum & Fraunhofer UMSICHT, Germany*
Dame Mary Archer, *Member of the Royal Society of Chemistry, United Kingdom*
Mrs Lorna Arens, *Royal Society of Chemistry, United Kingdom*
Dr Gavin Armstrong, *Nature Chemistry, United Kingdom*
Mr Vivek Badiani, *University of Cambridge, United Kingdom*
Mr Yang Bai, *University of Liverpool, United Kingdom*
Mr Mark Bajada, *University of Cambridge, United Kingdom*
Miss Meltiani Belekoukia, *Heriot-Watt University, United Kingdom*
Professor Matthias Beller, *Leibniz Institute for Catalysis, Germany*
Miss Bela Dhananjay Bhuskute, *Tampere University, Finland*
Dr Deirdre Black, *Royal Society of Chemistry, United Kingdom*
Professor Andrew B. Bocarsly, *Princeton University, USA*
Dr Sylvestre Bonnet, *Leiden University, Netherlands*
Miss Tessel Bouwens, *University of Amsterdam, Netherlands*
Miss Carlota Bozal-Ginesta, *Imperial College London, United Kingdom*
Professor Peter Brueggeller, *University of Innsbruck, Austria*
Miss Didjay Bruggeman, *University of Amsterdam, Netherlands*
Dr Darren Buesen, *Ruhr-University Bochum, Germany*
Professor Julea N. Butt, *University of East Anglia, United Kingdom*
Dr Xacobe C. Cambeiro, *Queen Mary University of London, United Kingdom*
Professor Christine A. Caputo, *University of New Hampshire, USA*
Miss Carla Casadevall, *Institute of Chemical Research of Catalonia (ICIQ), Spain*
Dr Flavia Cassiola, *Shell Technology Center Houston, USA*
Mr Nicolas Castel, *University of Cambridge, United Kingdom*
Mr Xiaolong Chen, *University of Cambridge, China*
Mr Daniel Cheung, *University of Liverpool, United Kingdom*
Dr Simon T. Clausing, *Uppsala University, Sweden*
Professor Andrew Cooper, *University of Liverpool, United Kingdom*
Miss Carri Cotton, *Royal Society of Chemistry, United Kingdom*
Mr Charles E. Creissen, *University of Cambridge, United Kingdom*
Mr Leander Crocker, *University of Cambridge, United Kingdom*
Mr Andrew Davidson, *Davidson Analytical Services, United Kingdom*
Professor Víctor A. de la Peña O'Shea, *IMDEA Energy Institute, Spain*
Professor Wolfgang Domcke, *Technische Universität München, Germany*
Mrs Maria Drosou, *University of Athens, Greece*
Professor James R. Durrant, *Imperial College London, United Kingdom*

Ms Esther Edwardes Moore, *University of Cambridge, United Kingdom*
Mr Johannes Ehrmaier, *Technical University of Munich, Germany*
Dr Annika Eisenschmidt, *University of Cambridge, United Kingdom*
Dr Tijmen Euser, *University of Cambridge, United Kingdom*
Dr Yuanxing Fang, *Fuzhou University, China*
Mr Xin Fang, *University of Cambridge, United Kingdom*
Mr Sergio Fernández, *Institute of Chemical Research of Catalonia (ICIQ), Spain*
Dr Vincent Friebe, *VU Amsterdam, Netherlands*
Dr James Gallagher, *Nature Energy, United Kingdom*
Mr Stelios Gavrielides, *Heriot–Watt University, United Kingdom*
Miss Kathryn Gempf, *Royal Society of Chemistry, United Kingdom*
Dr Elizabeth Gibson, *University of Newcastle Upon Tyne, United Kingdom*
Professor Michael Grätzel, *École Polytechnique Fédérale de Lausanne (EPFL), Switzerland*
Mr Domenico Grammatico, *University of Namur / University of Pau and Pays de l'Adour, Italy*
Dr Victor Gray, *University of Cambridge, United Kingdom*
Miss Qian Guo, *Queen Mary University of London, United Kingdom*
Professor Leif Hammarström, *Uppsala University, Sweden*
Mr Ashley Hancock, *University of Leeds, United Kingdom*
Dr Anna Hankin, *Imperial College London, United Kingdom*
Professor Marta C. Hatzell, *Georgia Institute of Technology, USA*
Professor Sophia Haussener, *École Polytechnique Fédérale de Lausanne (EPFL), Switzerland*
Miss Isabelle Heath-Apostolopoulos, *University College London, United Kingdom*
Mr Sam Hillman, *Imperial College London, United Kingdom*
Mr Zhang Hongwei, *Chiba University, Japan*
Professor Jurriaan Huskens, *University of Twente, Netherlands*
Mr Peter Hutton, *Cambridge Management Systems, United Kingdom*
Professor SU-IL In, *Daegu Gyeongbuk Institute of Science and Technology (DGIST), South Korea*
Professor Lars Jeuken, *University of Leeds, United Kingdom*
Miss Hannah Johnson, *Toyota Motor Europe, Belgium*
Mr Evangelos Kalamaras, *Heriot-Watt University, United Kingdom*
Miss Paulina Kalle, *Kurnakov Institute of General and Inorganic Chemistry, Russian Federation*
Mr Fotios Kamatsos, *National and Kapodistrian University of Athens, Greece*
Miss Zeynep Kap, *Bilkent University, Turkey*
Professor Ferdi Karadas, *Bilkent University, Turkey*
Dr Colin King, *Royal Society of Chemistry, United Kingdom*
Mr Philipp Koehler, *University of Cambridge, United Kingdom*
Professor Burkhard König, *University of Regensburg, Germany*
Professor Akihiko Kudo, *Tokyo University of Science, Japan*
Dr Moritz F. Kuehnel, *Swansea University, United Kingdom*
Mr Pramod Patil Kunturu, *MESA+Institute, University of Twente, Netherlands*
Dr Aiko Kurimoto, *University of British Columbia, Canada*
Miss Ava Lage, *University of Cambridge, United Kingdom*
Dr Yi-Hsuan Lai, *National Sun Yat-sen University, Chinese Taipei*
Miss Sarah Lamaison, *Collège de France, France*
Miss Natalia Larionova, *Queen Mary University of London, United Kingdom*
Mr Joshua Lawrence, *University of Cambridge, United Kingdom*
Mr Takashi Lawson, *University of Cambridge, United Kingdom*
Dr Chong-Yong Lee, *University of Wollongong, Australia*

Dr Chi-Fai Leung, *The Education University of Hong Kong, Hong Kong*
Mr Chuanshuai Li, *Lund University, Sweden*
Professor Chia-Yu Lin, *National Cheng Kung University, Chinese Taipei*
Mr Richard Lobo, *Queen Mary University of London, United Kingdom*
Miss Hui Luo, *Queen Mary University of London, United Kingdom*
Professor Marcelino Maneiro, *Universidade de Santiago de Compostela, Spain*
Mr Hikaru Matsuoka, *Kyoto University, Japan*
Ms Melanie Miller, *University of Cambridge, United Kingdom*
Professor Shelley D. Minteer, *University of Utah, USA*
Dr Huu Chuong Nguyen, *Institute of Chemical Research of Catalonia (ICIQ), Spain*
Dr Naoyuki Nishimura, *Asahi Kasei Corporation, United Kingdom*
Dr Wendy Niu, *Royal Society of Chemistry, United Kingdom*
Mr Giovanni Oakes, *University of Cambridge, United Kingdom*
Mr Kanta Ogawa, *Kyoto University, Japan*
Dr Daniel Opalka, *Technical University of Munich, Germany*
Mr Johann Pann, *University of Innsbruck, Austria*
Ms Aubrey R. Paris, *Princeton University, USA*
Dr Ernest Pastor, *Imperial College London, United Kingdom*
Mr Richard Pehn, *University of Innsbruck, Austria*
Mr Samuel Piper, *University of East Anglia, United Kingdom*
Professor Nicolas Plumeré, *Ruhr-University Bochum, Germany*
Mr Chanon Pornrungroj, *University of Cambridge, United Kingdom*
Miss Silvia Pugliese, *eSCALED-Project/ University of Namur / Collège de France, Italy*
Dr Motiar Rahaman, *University of Cambridge, United Kingdom*
Professor Joost H. Reek, *University of Amsterdam, Netherlands*
Professor Erwin Reisner, *University of Cambridge, United Kingdom*
Dr Sebastien Rochat, *University of Bath, United Kingdom*
Mrs Helena Roithmeyer, *Leopold Franzens Universität Innsbruck, Austria*
Dr Souvik Roy, *University of Cambridge, United Kingdom*
Dr Garry Rumbles, *National Renewable Energy Laboratory, USA*
Mr Constantin Sahm, *University of Cambridge, United Kingdom*
Dr Alessandra Sanson, *Consiglio Nazionale delle Ricerche (CNR), Italy*
Mr Matthias Schmalzbauer, *University of Regensburg, Germany*
Dr Christoph Schnedermann, *Cambridge University, United Kingdom*
Ms Murielle Schreck, *ETH Zurich, Switzerland*
Mrs Rosalind Searle, *Royal Society of Chemistry, United Kingdom*
Mr Ravi Shankar, *Imperial College London, United Kingdom*
Dr Wendy Shaw, *Pacific Northwest National Laboratory, USA*
Dr Jeremy Shears, *Shell, United Kingdom*
Miss Rebecca Sherbo, *University of British Columbia, Canada*
Dr Sergii I. Shylin, *Uppsala University, Sweden*
Dr Wilson A. Smith, *TU Delft, Netherlands*
Miss Katarzyna Sokol, *University of Cambridge, United Kingdom*
Dr Han Soo, *Nanyang Technological University, Singapore*
Dr Reiner Sebastian Sprick, *University of Liverpool, United Kingdom*
Miss Anna Stikane, *University of Leeds, United Kingdom*
Dr Jooyoung Sung, *University of Cambridge, United Kingdom*
Mr Brian Tam, *Imperial College London, United Kingdom*
Dr Jeannie Tan, *Heriot-Watt University, United Kingdom*
Mr Warren Thompson, *Heriot-Watt University, United Kingdom*
Dr Osamu Tomita, *Kyoto University, Japan*
Dr Stefanie Tschierlei, *Ulm University, Germany*
Ms Taylor Uekert, *University of Cambridge, United Kingdom*